T0182202

Reaction Kinetics: Exercises, Programs and Theorems

János Tóth • Attila László Nagy • Dávid Papp

Reaction Kinetics: Exercises, Programs and Theorems

Mathematica for Deterministic and Stochastic Kinetics

 Springer

János Tóth
Department of Analysis
Budapest University of Technology
and Economics
Budapest, Hungary

Chemical Kinetics Laboratory
Eötvös Loránd University
Budapest, Hungary

Dávid Papp
Department of Mathematics
North Carolina State University
Raleigh, USA

Attila László Nagy
Department of Stochastics
Budapest University of Technology
and Economics
Budapest, Hungary

Additional material to this book can be downloaded from http://extras.springer.com.

ISBN 978-1-4939-9351-2   ISBN 978-1-4939-8643-9   (eBook)
https://doi.org/10.1007/978-1-4939-8643-9

Printed on acid-free paper

This Springer imprint is published by the registered company Springer Science+Business Media, LLC
part of Springer Nature.
The registered company address is: 233 Spring Street, New York, NY 10013, U.S.A.

Dedicated to our families,
upwards and downwards,
and, last but not least, horizontally.

Preface

Origin More than a quarter of a century ago the senior author (JT) published a book with Péter Érdi on the mathematical models of chemical reactions.[1] Before and after writing that book (and having courses and seminars at the Budapest University of Technology and Economics and at the Eötvös Loránd University, Budapest) we realized that it is almost impossible to do anything in the theory and applications of chemical kinetics without using the computer, even if one is mainly interested in symbolic calculations. Luckily, all three of the present authors learned and taught *Mathematica* from relatively early on, cf. Szili and Tóth;[2] therefore it was quite natural to use this extremely powerful and providently designed language for our purposes. From time to time we wrote codes to solve problems in reaction kinetics, for the simulation of the usual stochastic model (what we have started well before, in the beginning of the seventies of the last century[3]), or for the decomposition of overall reactions[4,5] or for biological modeling.[6] As our programs started to be used by other people we realized that it would be useful to make it available to a wider audience.

However, according to our experience, recent developments in the theory of reaction kinetics, what may also be called **formal reaction kinetics**, are known in the chemists' and chemical engineers' community much less than optimal. Mathematicians are also not familiar enough with this area full of interesting unsolved problems. Therefore, we tried to do the

---

[1]Érdi P, Tóth J (1989) Mathematical models of chemical reactions. Theory and applications of deterministic and stochastic models. Princeton University Press, Princeton, NJ.

[2]Szili L, Tóth J (1996) Mathematics and *Mathematica*. ELTE Eötvös Kiadó, Budapest. http://www.math.bme.hu/~jtoth/Mma/M_M_2008.pdf.

[3]Érdi P, Sipos T, Tóth J (1973) Stochastic simulation of complex chemical reactions by computer. Magy Kém Foly 79(3):97–108.

[4]Papp D, Vizvári B (2006) Effective solution of linear Diophantine equation systems with an application in chemistry. J Math Chem 39(1):15–31.

[5]Kovács K, Vizvári B, Riedel M, Tóth J (2004) Computer assisted study of the mechanism of the permanganate/oxalic acid reaction. Phys Chem Chem Phys 6(6):1236–1242.

[6]Tóth J, Rospars JP (2005) Dynamic modelling of biochemical reactions with applications to signal transduction: principles and tools using *Mathematica*. Biosystems 79:33–52.

impossible: our book is aimed at beginners in the field with some knowledge in introductory calculus, linear algebra, and stochastics who would like to proceed very fast towards reading the original literature and also contribute to the theory and apply it to more and more practical problems. At the same time we are also providing a kind of user's manual to our *Mathematica* program package called **ReactionKinetics**, Version 1.0. The program downloaded freely, and remarks and criticisms leading to Versions 2, 3 etc. are welcome. We also show how the programs work in a CDF (Computable Document Format) files readable by anybody using the freely downloadable program CDFPlayer. One should not forget that the program *Mathematica* is freely available for anyone having a Raspberry Pi computer (even if it does not work as effectively as a paid version on a computer with larger capacity). No deep knowledge of programming is needed; we are convinced that well-written *Mathematica* programs are easy to read, and they are self-explanatory.

Scope and Related Books   Some of the readers might have noted a slight resemblance between the title of our book and that of an early Springer best seller, a classic: Pólya, G. and Szegő, G. *Aufgaben und Lehrsätze aus der Analysis*, (4th edition), Heidelberger Taschenbücher, Springer, (1970) (it would be inappropriate to add this item to the reference list). This is not by pure chance. This may be the only similarity of the authors with Pólya and Szegő: we also wanted to collect as much material as possible from the research literature and present it in unified form to teach the elements and to be available for the researchers as well. This turned out to be a task harder than anticipated in a period of explosive development of this branch of science (after decades of sleeping at the end of last century). It may happen that we do not give an exact reference when posing a problem; however, all the papers we have taken problems from are included in the list of References. We shall almost never give the proof of a theorem but we do try to give the exact references pointing to the first occurrence.

Érdi and Tóth[7] has been used by some readers as a textbook and has also been the starting point of many research papers. Related to that book the present one covers a narrower area, is deeper, is more based on the use of computer, and contains much more figures. That is why we hope it will be suitable for students even more and will give ideas for even more researchers. Let us try to characterize a few related books to help the reader in orientation and further reading. (As to the programs, they will be compared in Chap. 12 at the end of our book because by then the reader will have a certain experience with our program and will also possess the necessary vocabulary.)

---

[7]Érdi P, Tóth J (1989) Mathematical models of chemical reactions. Theory and applications of deterministic and stochastic models. Princeton University Press, Princeton, NJ.

If you need more chemistry—including recipes of exotic reactions—you might consult the books by Pilling and Seakins,[8] Lente,[9] Epstein and Pojman,[10] Scott,[11] Temkin,[12] or Espenson[13] which are almost complementary to our book. Although the main interest of the author of Póta[14] is reaction kinetics, his book contains problems from many other areas of mathematical chemistry as well.

The whole (application oriented) book by Érdi and Lente[15] has been dedicated to stochastic kinetics. The best source for numerical methods needed to deal with problems of kinetics is Turányi and Tomlin.[16] The book by Marin and Yablonsky[17] contains a very large number of concrete examples fully analyzed, but its main goal is to try to rescue the "sinking Atlantis of Soviet Science," as the authors put it: they collect many valuable results which can only be found in hardly accessible (and untranslated) journals in Russian.

Purpose The book is aimed at chemists, engineers, and mathematicians interested in the interdisciplinary field of formal reaction kinetics, an important subfield of mathematical chemistry. Reaction kinetics studies the evolution of reacting species; it can predict time-dependent and stationary behavior, and it can provide a framework for collecting data economically. Hence, it is relevant for everyone working in or studying chemistry, chemical engineering, biochemistry, environmental chemistry, combustion, etc.

Let us say a few words about our package, which is used throughout and taught also to the reader. In all the areas of chemistry where reaction kinetics plays an essential role (biochemistry, combustion, pharmacokinetics, organic and inorganic chemistry) the usual situation is that one has mechanisms, or

[8]Pilling MJ, Seakins PW (1995) Reaction kinetics. Oxford Science Publications, Oxford/New York/Tokyo.

[9]Lente G (2015) Deterministic kinetics in chemistry and systems biology. The dynamics of complex reaction networks. SpringerBriefs in molecular science. Springer, New York.

[10]Epstein I, Pojman J (1998) An introduction to nonlinear chemical dynamics: oscillations, waves, patterns, and chaos. Topics in physical chemistry series. Oxford University Press, New York. http://books.google.com/books?id=ci4MNrwSlo4C.

[11]Scott SK (1991, 1993, 1994) Chemical chaos. International series of monographs on chemistry, vol 24. Oxford University Press, Oxford.

[12]Temkin ON (2012) Homogeneous catalysis with metal complexes: kinetic aspects and mechanisms. Wiley, Chicester.

[13]Espenson JH (2002) Chemical kinetics and reaction mechanisms, 2nd edn. McGraw-Hill, Singapore.

[14]Póta G (2006) Mathematical problems for chemistry students. Elsevier, Amsterdam.

[15]Érdi P, Lente G (2016) Stochastic chemical kinetics. Theory and (mostly) systems biological applications. Springer series in synergetics. Springer, New York.

[16]Turányi T, Tomlin AS (2014) Analysis of kinetic reaction mechanisms. Springer, New York.

[17]Marin GB, Yablonsky GS (2012) Kinetics of chemical reactions. Decoding complexity. Wiley, Weinheim.

complex chemical reactions consisting of a large number of (elementary) reaction steps, and of different chemical species. This situation generates the need to do as much calculations automatically as possible. On the other side, developments in formal reaction kinetics in the last three to four decades made it possible to create tools to give this kind of help to the chemist. Taking all this into consideration it is quite astonishing that so few general packages in reaction kinetics exist. Our package has the following properties:

1. It is based on recent developments in formal reaction kinetics.
2. It has been written in a modern language, *Mathematica* (or Wolfram Language, if you wish), using all the nice features and efficient tools of it.
3. It is of modular structure; therefore the present version can, should, and will be further developed.
4. It includes stochastic simulation methods as well.

Plan After an introduction to time-independent, mainly combinatorial tools, the book covers several mathematical and algorithmic aspects of continuous time (deterministic and stochastic) models of chemical reactions. The book combines the formal presentation of mathematical models and algorithms with the demonstration of the applications using the symbolic and numerical features of *Mathematica*. Each topic is introduced by elementary examples to illustrate the relevant concepts. Then a review of the literature follows, also covering the state of the art. Finally, both simple and real-world examples are analyzed using our *Mathematica* implementation.

Related Disciplines and Intended Audience Within Different Areas

1. Mathematics: linear algebra, linear and integer programming, Diophantine equations, polynomial ordinary differential equations, reaction diffusion equations, graph theory, continuous time discrete state Markov processes.
2. Chemistry: reaction kinetics, physical chemistry, chemical reaction network theory.
3. Industry and technology: combustion, environmental issues, pharmacology.
4. Biology: biochemistry, drug design, enzymology, population biology, systems biology.
5. Government: environmental issues, food or drug safety.

The primary audience will be MSc and PhD students in applied mathematics, chemistry, biochemistry, chemical and environmental engineering. It might also be a textbook for graduate-level courses on mathematical and physical chemistry. We think that newcomers in the field might also learn from it, playing with the programs, and the abundant literature may also help experts. We do not claim that the whole book will be a mandatory text for all the students in the above sciences; however, we are sure that some students from all of the above disciplines will be interested.

Beforethoughts and Afterthoughts We are going to work with clear mathematical concepts, precise definitions, and theorems, nevertheless try to keep in mind the potential applications. Furthermore, we do not want to cover all the areas of reaction kinetics, only those which we know relatively well. We collect and review the corresponding modern literature. Our programs can be used to treat almost all important problems of homogeneous reaction kinetics, both deterministic and stochastic. The nontrivial algorithms we use come partially from our workshop.

The exercises and problems follow very closely the material exposed in the chapters. They can be best used if the reader tries to solve them, and even if (s)he is able to solve a problem, (s)he might learn methods or tricks from the solution presented by us.

Budapest, Hungary   János Tóth
Budapest, Hungary   Attila László Nagy
Raleigh, USA   Dávid Papp
March 25, 2018

Acknowledgments

By reading parts of the manuscript or giving advice, references, examples, etc. during the years or directly while writing the present book M. P. Barnett, R. Csikja, P. Érdi, V. Gáspár, B. Kovács, T. Ladics, M. Lázi, G. Lente, D. Lichtblau, L. Lóczi, M. Moreau, I. Nagy, Z. Noszticzius, Gy. Póta, L. P. Simon, G. Szederkényi, L. Szili, T. Turányi, and J. Zádor offered serious support. Careful reading of our manuscript by Balázs Boros enabled us to make many corrections and improvements. The late colleagues Miklós Farkas, Henrik Farkas, and Gyula Farkas also provided invaluable help.

We have been receiving instant help on LaTeX continuously from F. Wettl and from F. Holzwarth. The use of the mhchem package by M. Hensel helped us write nice chemical formulae. We wouldn't have been able to collect the literature without the inventive and painstaking work by N. Hegedüs and Á. Csizmadia.

The work has partially been supported by the CMST COST Action CM1404 (Chemistry of Smart Energy Carriers and Technologies: SMARTCATS), and in the final time period by the project SNN 125739. The support of our bosses, especially that of M. Horváth, D. Petz, and D. Szász is also acknowledged. Finally, we would like to acknowledge the confidence and patience of our editor, Svetlana Zakharchenko.

Contents

Part III   The Continuous Time Discrete State Stochastic Model

Acronyms and Symbols

The present list of abbreviations and symbols contains those notations which are supposed to be generally known and not defined in the book.

$\mathbf{1}$	A vector with all the components equal to 1, $\mathbf{1}_N \in \mathbb{R}^N$
$\|A\|$	Number of elements (cardinality) of the set A
$]a, b[$	Open interval $\{x \in \mathbb{R} \| a < x < b\}$ under the assumption that $a, b \in \mathbb{R}, a < b$
$\mathscr{B}_\varrho(a)$	Open ball with radius ϱ around the point a
$\mathfrak{B}(A)$	With a statement A it is 1, if the statement is true, it is 0, if the statement is false. Similar to the function **Boole** in *Mathematica*
$\mathbb{C}$	Set of all complex numbers
$\mathrm{diag}(\mathbf{v})$	A diagonal matrix with the components of the vector on the main diagonal
$\mathscr{D}_f$	Domain of the function f
$\partial\Omega$	Boundary of the set Ω
$\partial_n f$	Partial derivative function of the function f with respect to the nth variable
e_n	nth element of the standard basis
$f\|_A$	Restriction of the function f onto the set A: $$\mathscr{D}_{f\|_A} := \mathscr{D}_f \cap A; \forall x \in \mathscr{D}_{f\|_A} : f\|_A(x) = f(x)$$
$(f, g)(x) :=$	$(f(x), g(x))$, for $x \in \mathscr{D}_f \cap \mathscr{D}_g$
$f \circ g$	The composition of the functions f and g defined to be $f(g(x))$ for those elements $x \in \mathscr{D}_g$ for which $g(x) \in \mathscr{D}_f$
$\mathrm{Im}(A)$	Image of the linear map A
$\mathrm{Ker}(A)$	Null space of the linear map A
$\mathscr{M} :=$	$\{1, 2, \ldots, M\}$
$\mathbb{N}$	Set of all natural (here: positive integer) numbers
$\mathbb{N}_0$	Set of all nonnegative integers
$\mathscr{N} =$	$\{1, 2, \ldots, N\}$
$[n]_r$	$:= n(n-1)\ldots(n-r+1)$ with $r \in \mathbb{N}$, **FactorialPower** or falling factorial
$\mathrm{rank}(A)$	Rank of the matrix A
$\mathfrak{R}(z)$	Real part of the complex number $z \in \mathbb{C}$
$\mathbb{R}$	Set of all real numbers

$\mathbb{R}^N$	Set of all N-dimensional vectors with real numbers as components ($N \in \mathbb{N}$)
$\mathbb{R}^{N \times N}$	Set of all $N \times N$ matrices with real numbers as components ($N \in \mathbb{N}$)
$\mathbb{R}^-$	Set of all negative real numbers
$\mathbb{R}^+$	Set of all positive real numbers
$\mathbb{R}_0^+$	Set of all nonnegative real numbers
$\mathscr{R} :=$	$\{1, 2, \ldots, R\}$
$\mathscr{R}_f$	Range of the function f
$\mathbf{X} \succcurlyeq \mathbf{Y}$	The matrix $\mathbf{X} - \mathbf{Y}$ is positive semi-definite.
$\mathbf{X} \succ \mathbf{Y}$	The matrix $\mathbf{X} - \mathbf{Y}$ is positive definite.
$\mathbb{Z}$	Set of all integer numbers
δ_Q	Characteristic or indicator function of the set Q; if $Q = \{m\}$, then one simply writes δ_m. We had to use δ_l also for the deficiency of the lth linkage class.
$\boldsymbol{\chi}_Q$	Characteristic vector of the set Q; $\boldsymbol{\chi}_Q := \begin{bmatrix} \delta_{q_1} & \delta_{q_2} & \ldots & \delta_{q_K} \end{bmatrix}$ if the number of elements of Q is K.

Introduction

1

As usual, introduction is the part of a book written and read last. Here the authors expose what the reader will read (actually, have read). We are not going to be an exception.

The structure of our book is as follows:

Part I treats reactions without models specifying time evolution. Thus, after furnishing the scene, the questions of mass conservation and those of decomposition of overall reactions are treated.

Part II deals with many aspects of the usual deterministic model of reaction kinetics, such as the form, transient and stationary behavior of concentration vs. time curves, and approximations of the induced kinetic differential equation.

Part III is dedicated to the most common stochastic model of reactions. This model is neither microscopic nor macroscopic; it is sometimes referred to as **mesoscopic** referring to the level of description.

Part IV is about the use of numerical methods in reaction kinetics (including estimation problems), about a short history of software packages written for kinetics, and also on the mathematical background to make our book self-contained. Thus, these are certainly not unimportant questions.

There are a few areas really important for the chemical kineticists, such as movements on potential surfaces, molecular dynamics, quantum chemical calculations, statistical methods to evaluate kinetic experiments, etc. Some of the readers might be more interested in exotic phenomena or in more details of Chemical Reaction Network Theory. As to these topics, we must refer the reader to the literature.

Let us mention a few reviews and journals containing relevant material on the mathematical theory of reaction kinetics: Aris (1969) was a starting point for many researchers. Although not as many fields were covered, the review National Research Council et al. (1995) may also be interesting, although the word "kinetics" is only contained once in it. Of the journals we mention first the *Journal of Mathematical Chemistry* and *match (Communications in Mathematical and in Computer Chemistry)*, both of which cover wide areas of mathematical chemistry.

© Springer Science+Business Media, LLC, part of Springer Nature 2018
J. Tóth et al., *Reaction Kinetics: Exercises, Programs and Theorems*,
https://doi.org/10.1007/978-1-4939-8643-9_1

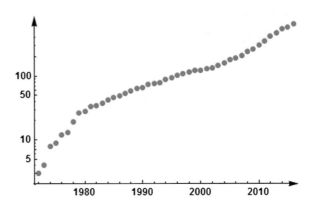

Fig. 1.1 The logarithm of the number of citations of Horn and Jackson (1972) up to a certain year. Note the three phases: up to 1980 and after 2002 and in between

Biomathematical journals or SIAM journals also contain papers relevant in this field. Chemistry has an emphasis in journals such as *International Journal of Chemical Kinetics* or *Reaction Kinetics, Mechanisms, and Catalysis*. The web page
    https://reaction-networks.net/wiki/Mathematics_of_Reaction_Networks
also contains relevant information.

Authors more historically minded than us would write the book as a history of continuous rediscoveries (nonnegativity, adjacency matrix, stochastic model of compartmental systems, simulation of the stochastic model). We shall only make a few remarks in passing these topics.

Let us finish this introduction with an illuminating figure showing the logarithm of the total number of citations up to given years to the fundamental paper by Horn and Jackson (1972) (Fig. 1.1).

1.1   Exercise or Problem

1.1 Find review papers or books to cover the "few areas" mentioned above.

(Solution: page 381.)

References

Aris R (1969) Mathematical aspects of chemical reaction. Ind Eng Chem Fundam 61(6):17–29
Horn F, Jackson R (1972) General mass action kinetics. Arch Ratl Mech Anal 47:81–116
National Research Council, et al (1995) Mathematical challenges from theoretical/computational chemistry. National Academy Press, Washington, DC

Part I

The General Framework

First of all, one has to delineate the physical entity the models of which one is going to deal with: this is a finite set of reaction steps taking place among a finite set of chemical species. Most often we assume that the volume in which the reaction steps take place is constant and so are the pressure and the temperature, as well. Next, we give a very short introduction into the software tool: into *Mathematica* (or Wolfram Language, if you prefer). It would be more appropriate to call this part as a reminder for those who know something about this, because at least a moderate level knowledge of the language is assumed. Part I ends with the definitions and comparisons of different graphs assigned to reactions. These are necessary preliminaries to the treatment of the dynamics of reactions in the chapters to follow.

Preparations

2

Before turning to the formal theory of reaction kinetics, we meditate a bit about the phenomenon we are interested in and also about the assumed circumstances.

2.1   Physical Assumptions

The stage is a vessel of constant volume (one might think of a reactor, a cell, a cell compartment, a biological niche, an engine, a test tube, etc.) at a fixed temperature and pressure, containing a finite number of chemical species. Among these species a finite set of chemical reaction steps take place. We also suppose that mixing is complete; therefore the concentration of the species is the same at all points in the vessel. What we are interested in is the temporal change of the concentrations of the species. It will turn out that several systems not obeying the restrictions can also be put into the framework we use, i.e., in- and outflow, or reactions in a continuously stirred tank reactor (usually abbreviated as CSTR), or some kinds of transport processes, and even several phenomena in population biology including the spread of epidemics, can formally be described by reaction steps. Some of the restrictions will be alleviated below; see in Sect. 6.7.

2.2   The Standard Setting

Let us suppose that in the investigated vessel, a finite number $M \in \mathbb{N}$ of chemical **species** (molecules, radicals, electrons, etc.) denoted by $X(1), X(2), \ldots, X(M)$ is present (indexed by the **species set** $\mathcal{M} := \{1, 2, \ldots, M\}$) and among these species a finite number $R \in \mathbb{N}$ of **reaction step**s

$$\sum_{m \in \mathcal{M}} \alpha(m, r)X(m) \longrightarrow \sum_{m \in \mathcal{M}} \beta(m, r)X(m) \quad (r \in \mathcal{R}), \tag{2.1}$$

© Springer Science+Business Media, LLC, part of Springer Nature 2018
J. Tóth et al., *Reaction Kinetics: Exercises, Programs and Theorems*,
https://doi.org/10.1007/978-1-4939-8643-9_2

(usually identified with the **set of reaction steps** defined as $\mathscr{R} := \{1, 2, \ldots, R\}$)
take place. We call (2.1) for a fixed $r \in \mathscr{R}$ a **reaction step**, and the reaction steps
together form a **reaction**. The nonnegative integer coefficients in (2.1) form the
matrices

$$\boldsymbol{\alpha} = (\alpha(m, r))_{m \in \mathscr{M}; r \in \mathscr{R}} \text{ and } \boldsymbol{\beta} = (\beta(m, r))_{m \in \mathscr{M}; r \in \mathscr{R}}$$

of **stoichiometric coefficients** or **molecularities** expressing the molar proportions
of the species in a reaction step. The **reactant complex vector** of the rth
reaction step is the vector $\boldsymbol{\alpha}(\cdot, r) := [\alpha(1, r)\, \alpha(2, r) \ldots \alpha(M, r)]^{\top}$, its **product
complex vector** is the vector $\boldsymbol{\beta}(\cdot, r) := [\beta(1, r)\, \beta(2, r) \ldots \beta(M, r)]^{\top}$, and the
corresponding **reaction step vector** is the vector $\boldsymbol{\gamma}(\cdot, r) := \boldsymbol{\beta}(\cdot, r) - \boldsymbol{\alpha}(\cdot, r)$. The
rank of the **stoichiometric matrix** $\boldsymbol{\gamma}$ is usually denoted by S.

Remark 2.1 The case when the complex vector is the null vector is not excluded,
and the corresponding complex is the **zero complex**, or **empty complex**, it is
denoted by 0. If it is the reactant complex of a reaction step, then it is usually used to
describe **inflow**; if it is the product complex of a reaction step, then it is usually used
to describe **outflow**, thereby expanding the set of phenomena which can be treated
by the theory.

Definition 2.2 The **stoichiometric subspace** of the reaction (2.1) is the linear space

$$\mathscr{S} := \text{span}\,\{\boldsymbol{\gamma}(\cdot, r) \,|\, r \in \mathscr{R}\}. \tag{2.2}$$

Thus, S is the dimension of $\mathscr{S}$ and also the rank of the stoichiometric matrix. One
can refer to it as to the number of independent reaction steps.

   The formal linear combinations on the left and right sides of the reaction arrows
are the **reactant complex** and the **product complex**, respectively; together they
form the set of **complexes**; $\mathscr{C} := \{\boldsymbol{\alpha}(\cdot, r) \,|\, r \in \mathscr{R}\} \cup \{\boldsymbol{\beta}(\cdot, r) \,|\, r \in \mathscr{R}\}$, the
cardinality of which will be denoted by N below: $N := |\mathscr{C}|$. The matrix with the
complex vectors as columns is the **complex matrix**; it is usually denoted by $\mathbf{Y} \in
\mathbb{N}_0^{M \times N}$.

   It is quite natural to make a few restrictions on the matrices $\boldsymbol{\alpha}, \boldsymbol{\beta}, \boldsymbol{\gamma}$; see, e.g.,
Deák et al. (1992, Sect. 2.1).

Conditions 1

1. All the species participate in at least one of the reaction steps.
2. In every reaction step, at least one of the species changes.
3. All the reaction steps are determined by the pair of its reactant and product
   complexes.

Remark 2.3

1. The last requirement may be too restrictive because if a reaction can proceed through two different transition states (or this reaction step has two **reaction channels**), as in the reaction

$$CH_3CHOH^* + O_2 \rightleftharpoons HO_2^* + CH_3CHO$$

   (Zádor et al. 2009), then one should duplicate this step to exactly represent the mechanistic details of the reaction. However, when using any model (deterministic or stochastic) for the time evolution of the concentrations, the two steps will be contracted (or, lumped) into one.
2. Sometimes we abbreviate reaction steps like $A + X \longrightarrow 2X$ as $X \longrightarrow 2X$ saying that A is an **external species** the concentration change of which will be neglected later. The reason of this might be that these species are present in abundance; therefore their changes are very small during the time period we are describing the process. Another way of wording is if one says that these species form a **pool**. If we want to emphasize that X is not an external species, we may use the term **internal species**. The first form of the reaction step $(A+X \longrightarrow 2X)$ may be called the **genuine reaction step**, while the second form $(X \longrightarrow 2X)$ will simply be called **reaction step**.
3. The presence of a species on both sides of a reaction step is not excluded as the example $X + Y \longrightarrow 2X$ shows. Neither from the dynamical point of view nor from the point of view of stationary points can this reaction "be simplified" into $Y \longrightarrow X$.
4. The assumptions imply that the number of complexes is at least 2.
5. Mathematicians may not like that the complexes are **formal linear combinations**. To make it quite correct (following Feinberg and Horn 1977, pp. 84–85), let us consider the complex vectors as elements of the linear space of functions $\mathbb{R}^{\mathcal{M}}$. The natural **basis** of this space consists of the characteristic functions δ_m of the elements of the species set $\mathcal{M}$. A function $f \in \mathbb{R}^{\mathcal{M}}$ can then be represented as $f = \sum_{m \in \mathcal{M}} f(m)\delta_m$. Now, listen to the jump: when confusion seems unlikely, we shall sometimes write $X(m)$ instead of δ_m, particularly when an element of $\mathbb{R}^{\mathcal{M}}$ is displayed as a linear combination of the natural basis elements of $\mathbb{R}^{\mathcal{M}}$, as, e.g., we write $A_1 + A_2$ instead of $\delta_1 + \delta_2$. This deliberate confusion of species with the corresponding characteristic functions will permit a transition from the rigorous mathematical definition of a reaction to its representation by a graph with complexes as its vertices and reaction steps as its edges: the Feinberg–Horn–Jackson graph (see Definition 3.1).

Definition 2.4 The **molarity** of a complex vector $\mathbf{y} = \begin{bmatrix} y_1 & y_2 & \ldots & y_M \end{bmatrix}^\top \in \mathscr{C}$ is the sum of its stoichiometric coefficients $\sum_{m \in \mathcal{M}} y_m$. A complex is **short**, if its molarity is not more than 2. The **order** of a reaction step is the molarity of its reactant complex, whereas the **order of the reaction** (2.1) is the maximum of the orders of the reaction steps: $\max\{\sum_{m\in\mathcal{M}} \alpha(m,r) \mid r \in \mathcal{R}\}$.

Remark 2.5

1. Reactions of the order not more than two or even reactions only consisting of short complexes are the most important ones, because in this case it is not impossible to have a physical picture about a reaction step: it is the result of a collision, and it is called in this context an **elementary reaction**. The probability of triple collisions is considered to be low. Still, the quite common reaction between nitric oxide and oxygen is of the third order; see Example 6.30. Another case of including seemingly third order reactions is to consider **third bodies** in combustion models. This is a usual trick to consider ambient effects formally without a rigorous physical foundation.
2. Sometimes we use the quadruple $\langle \mathcal{M}, \mathcal{R}, \boldsymbol{\alpha}, \boldsymbol{\beta} \rangle$ to denote the reaction (2.1), and we imply that all the three conditions formulated right now are fulfilled.
3. We might specify a further requirement based on the idea that independent reactions can be treated separately: we might require that the reaction be **connected** in the sense that its Volpert graph is connected; see Sect. 3.3.1).

We shall return more than once to the following examples.

Example 2.6 Consider the reversible **simple bimolecular reaction**

$$A + B \rightleftharpoons C. \tag{2.3}$$

Here we have

$$M = 3, \quad R = 2, \quad X(1) = A, \quad X(2) = B, \quad X(3) = C,$$

$$\boldsymbol{\alpha} = \begin{bmatrix} 1 & 0 \\ 1 & 0 \\ 0 & 1 \end{bmatrix}, \quad \boldsymbol{\beta} = \begin{bmatrix} 0 & 1 \\ 0 & 1 \\ 1 & 0 \end{bmatrix};$$

and Conditions 1 are obviously fulfilled. The stoichiometric subspace is one dimensional ($S = 1$); it is generated by the vector $[\,1\ 1\ -1\,]^{\top}$.

Cases where some or all the reaction steps are irreversible are also often used as models: this is the viewpoint of the mathematician. A kineticists however might say that all the reaction steps are theoretically reversible; in some cases we neglect one from a pair of reactions and that is the case when we speak about irreversible steps: they are nothing else than modeling tools.

Example 2.7 The characteristic quantities of the irreversible **Lotka–Volterra reaction**

$$X \longrightarrow 2X \quad X + Y \longrightarrow 2Y \quad Y \longrightarrow 0 \tag{2.4}$$

are

$$M = 2, \quad R = 3, \quad \alpha = \begin{bmatrix} 1 & 1 & 0 \\ 0 & 1 & 1 \end{bmatrix}, \quad \beta = \begin{bmatrix} 2 & 0 & 0 \\ 0 & 2 & 0 \end{bmatrix};$$

and Conditions 1 are obviously fulfilled again. The stoichiometric subspace is the whole space $\mathbb{R}^2$, thus $S = 2$.

A combustion application of the Lotka–Volterra reaction came from the fact that in the case of a continuous supply of a gasoline–air mixture into the reactor (heated to certain temperature), one can see periodic flashes of the cold flame (1500–2000 K) appearing with a constant frequency. In this case the full combustion does not occur. The oxidation products include aldehydes, organic peroxides, and other compounds. Some regularities have been established for this process. In particular, the flash frequency increases with the increase of oxygen concentration and temperature. In order to explain this effect, Frank-Kamenetskii (1947) (see also Korobov and Ochkov 2011, Chap. 3) suggested the model with two external species

$$A + X \longrightarrow 2X \quad X + Y \longrightarrow 2Y \quad Y \longrightarrow B \tag{2.5}$$

to describe oscillations in cold flames. In this interpretation A is the initial compound, B is the product, X is the superoxide type molecules or radicals, and Y means the aldehyde-type molecules or radicals. A recent analysis of the phenomenon has been given by Mints et al. (1977).

As to oscillation, we shall return to this point in Chap. 8.

Example 2.8 An important model in combustion is the **Robertson reaction**; see Robertson (1966) or Deuflhard and Bornemann (2002, p. 29):

$$A \longrightarrow B \quad 2B \longrightarrow B + C \longrightarrow A + C, \tag{2.6}$$

where $M = 3$, $R = 3$, $X(1) = A$, $X(2) = B$, $X(3) = C$, and

$$\alpha = \begin{bmatrix} 1 & 0 & 0 \\ 0 & 2 & 1 \\ 0 & 0 & 1 \end{bmatrix}, \quad \beta = \begin{bmatrix} 0 & 0 & 1 \\ 1 & 1 & 0 \\ 0 & 1 & 1 \end{bmatrix}, \quad \gamma = \begin{bmatrix} -1 & 0 & 1 \\ 1 & -1 & -1 \\ 0 & 1 & 0 \end{bmatrix}, \quad S = 2;$$

thus the reaction steps are not independent.

Example 2.9 A basic model of enzyme kinetics is the **Michaelis–Menten reaction**; see, e.g., Keleti (1986):

$$E + S \rightleftharpoons C \longrightarrow E + P, \tag{2.7}$$

where E is the enzyme molecule; S is its substrate, the species that the enzyme transforms; C is a temporary complex of the enzyme with the substrate; and finally, P is the product. The last step shows that the enzyme molecule is recovered at the end of the reaction: it is a **catalyst**. The name "enzyme" denotes a catalyst being a protein as to its chemical constitution. Here $M = 4, R = 3, X(1) = E, X(2) = S, X(3) = C, X(4) = P$, and

$$\alpha = \begin{bmatrix} 1 & 0 & 0 \\ 1 & 0 & 0 \\ 0 & 1 & 1 \\ 0 & 0 & 0 \end{bmatrix}, \quad \beta = \begin{bmatrix} 0 & 1 & 1 \\ 0 & 1 & 0 \\ 1 & 0 & 0 \\ 0 & 0 & 1 \end{bmatrix}, \quad \gamma = \begin{bmatrix} -1 & 1 & 1 \\ -1 & 1 & 0 \\ 1 & -1 & -1 \\ 0 & 0 & 1 \end{bmatrix}, \quad S = 2.$$

Sometimes the last step is also assumed to be reversible, but there is an infinite number of further variations on this theme: some with more than one enzyme, some with more than one substrate, one or more inhibitors may also be present, etc.; we refer again to the book by Keleti (1986).

Probably the first reaction to describe **combustion** comes from Mole (1936). His general model is

Example 2.10

$$X + Y \longrightarrow 2X + 2Y \quad X \rightleftharpoons 0 \rightleftharpoons Y, \tag{2.8}$$

where $M = 2, R = 5, X(1) = X, X(2) = Y$, and

$$\alpha = \begin{bmatrix} 1 & 1 & 0 & 0 & 0 \\ 1 & 0 & 0 & 1 & 0 \end{bmatrix}, \quad \beta = \begin{bmatrix} 2 & 0 & 1 & 0 & 0 \\ 2 & 0 & 0 & 0 & 1 \end{bmatrix}.$$

He simplified this model to get one which is easier to treat:

$$2X \longrightarrow 3X \quad X \rightleftharpoons 0, \tag{2.9}$$

where $M = 1, R = 3, X(1) = X$, and $\alpha = \begin{bmatrix} 2 & 1 & 0 \end{bmatrix}, \beta = \begin{bmatrix} 3 & 0 & 1 \end{bmatrix}$. Actually, this reaction is a version with an irreversible step of the Schlögl reaction Schlögl (1972)

$$2X \rightleftharpoons 3X \quad X \rightleftharpoons 0. \tag{2.10}$$

Reactions of nitrogen oxides are equally important in atmospheric chemistry and in combustion. An early model has been constructed by Ogg (1947):

Example 2.11

$$N_2O_5 \rightleftharpoons NO_2 + NO_3 \longrightarrow NO_2 + NO + O_2 \quad NO_3 + NO \longrightarrow 2\,NO_2 \quad (2.11)$$

where

$$M = 5, \, R = 4, \, X(1) = N_2O_5, \, X(2) = NO_2, \, X(3) = NO_3, \, X(4) = NO, \, X(5) = O_2,$$

and

$$\alpha = \begin{bmatrix} 1 & 0 & 0 & 0 \\ 0 & 1 & 1 & 0 \\ 0 & 1 & 1 & 1 \\ 0 & 0 & 0 & 1 \\ 0 & 0 & 0 & 0 \end{bmatrix}, \quad \beta = \begin{bmatrix} 0 & 1 & 0 & 0 \\ 1 & 0 & 1 & 2 \\ 1 & 0 & 0 & 0 \\ 0 & 0 & 1 & 0 \\ 0 & 0 & 1 & 0 \end{bmatrix}, \quad \gamma = \begin{bmatrix} -1 & 1 & 0 & 0 \\ 1 & -1 & 0 & 2 \\ 1 & -1 & -1 & -1 \\ 0 & 0 & 1 & -1 \\ 0 & 0 & 1 & 0 \end{bmatrix}, \quad S = 3.$$

Next, we give a model to describe spontaneous formation of **chirality**; see Frank (1953), Lente (2004), and Barabás et al. (2010).

Molecules containing a carbon atom with four different atoms or radicals attaching to it (an asymmetric carbon atom) have the property that they rotate the plane of polarized light: these are chiral objects, the two forms are **enantiomers** or **optical isomers**. It turns out that almost all the amino acids in nature are left rotating, while sugars are right rotating. It is an unsolved and very important problem of the origin of life where has this asymmetry called chirality come from.

The starting point of all the models is the classical model of Frank:

Example 2.12

$$A \longrightarrow R \quad A \longrightarrow S \quad A + R \longrightarrow 2R \quad A + S \longrightarrow 2S, \quad (2.12)$$

where A is an asymmetric precursor, and R and S are two enantiomers of the same composition, and $M = 3, \, R = 4, \, X(1) = A, \, X(2) = R, \, X(3) = S$, and

$$\alpha = \begin{bmatrix} 1 & 1 & 1 & 1 \\ 0 & 0 & 1 & 0 \\ 0 & 0 & 0 & 1 \end{bmatrix}, \quad \beta = \begin{bmatrix} 0 & 0 & 0 & 0 \\ 1 & 0 & 2 & 0 \\ 0 & 1 & 0 & 2 \end{bmatrix}.$$

Finally, we give an example where there is a negative stoichiometric coefficient and also one which is not an integer. Later—in Chap. 6—we shall speak about the problems which may or may not arise from such coefficients.

Example 2.13 A recent version of the **Oregonator** to describe the **Belousov–Zhabotinsky (BZ) reaction** has been provided by Turányi et al. (1993):

$$X + Y \longrightarrow 2\,P \qquad Y + A \longrightarrow X + P \qquad 2\,X \longrightarrow P + A \qquad (2.13)$$

$$X + A \longrightarrow 2\,X + 2\,Z \qquad X + Z \longrightarrow 0.5\,X + A \qquad Z + M \longrightarrow Y - Z, \qquad (2.14)$$

where

$$X = HBrO_2, \quad Y = Br^-, \quad Z = Ce^{4+},$$

$$A = BrO_3{}^-, \quad P = HOBr, \quad M = CH_2(COOH)_2 \text{ (malonic acid)}.$$

According to our definition (see page 6), the last one is not a reaction step. See, however, Problem 2.1.

> The Belousov–Zhabotinsky reaction is a **chemical oscillator**: a reaction with periodically oscillating concentration of the species. It is also used as a chemical model of nonequilibrium biological phenomena. The mathematical models of the Belousov–Zhabotinsky reaction are of theoretical interest, as well. The short history of the reaction follows.
> In the 1950s B. Belousov noted that in a mixture of potassium bromate, cerium(IV) sulfate, malonic acid, and citric acid in dilute sulfuric acid, the ratio of concentration of the cerium(IV) and cerium(III) ions oscillated, causing the color of the solution to periodically change between a yellow solution and a colorless solution. This is due to the fact that the cerium(IV) ions are reduced by malonic acid to cerium(III) ions, which are then oxidized back to cerium(IV) ions by bromate(V) ions.
> Belousov made two attempts to publish his finding but was rejected on the grounds that he could not satisfactorily explain his results. His work was finally published in a less respectable, not peer-reviewed journal.
> In 1961 A. Zhabotinsky as a graduate student started to investigate the reaction in detail; however, the results were still not widely known in the West until a conference in Prague in 1968.
> Later, researchers realized that the Bray–Liebhafsky reaction discovered in 1921 (Bray 1921) and neglected in the decades to follow was the first oscillatory reaction to be known. The Oregonator is the simplest realistic model of the chemical dynamics of the oscillatory Belousov–Zhabotinsky reaction. It was devised by Field and Noyes working at the University of Oregon (this is where the name "Oregonator" comes from) and is composed of five reaction steps. This reaction is obtained by reduction of the complex chemical mechanism of the Belousov–Zhabotinsky reaction suggested by Field, Kőrös, and Noyes and referred to as the FKN mechanism. See also Turányi et al. (1993).

2.3   An Extremely Rapid Introduction to *Mathematica*

Although we think that the reader has some experience with *Mathematica* and we try to teach her or him on this basis, nevertheless here we are giving a very short introduction to *Mathematica* itself. If needed, this can be completed with the introductory videos offered by the program when *Mathematica* is loaded in, and also with free Internet seminars, or even freely downloadable books, such as Szili and Tóth (1999), etc., or favorites of the present authors, such as Gray (1994,

1996, 1998) and Mangano (2010). For a deeper study, one can also turn to the thick volumes (Trott 2006, 2007, 2013) or can use the recent introductions by the creator of *Mathematica* himself, Wolfram (2015):

> Although the idea of symbolic calculations (especially in the field of logics) goes back to Leibniz, the first working programs only appeared in the sixties of the twentieth century. Nowadays (although a few early systems are surviving) the best known general purpose systems (able also to do numerical calculations) are Maple and *Mathematica*. Beyond the structural advantages of *Mathematica* let us mention that it comes freely when buying the Raspberry Pi computer. Other freely available systems which are worth mentioning are Sage http://www.sagemath.org/ and the less known Spider: http://peterwittek.com/spyder-closer-to-a-mathematica-alternative.html.

Mathematica essentially is a **functional language** that means that it does nothing just evaluates expressions or calculates values of (multiple) function compositions. However, a function might serve really complicated purposes, such as

`JordanDecomposition, NMinimize, NonlinearModelFit`

or `InverseLaplaceTransform` do. All the built-in functions start with capital letters, like

`Sin, ParametricNDSolve, Plot, FindClusters, Manipulate}`.

We followed the same practice when defining our functions like

`Concentrations, WeaklyReversibleQ, SimulationPlot`.

As the above examples suggest, function names in *Mathematica* are usually full English names without abbreviation (certainly with reasonable exceptions as, e.g., `ArcTan`, etc.). The arguments of the functions are put into **[ ]** brackets, like

`Eigenvalues[{{a, b}, {c, d}}]`.

Parentheses like **( )** are only used to break the precedence rule or for grouping terms.

A fundamental data type in *Mathematica* is the **list**; this is how, e.g., vectors and matrices are represented: $\{a,b,a\}$ denotes here the vector $[a\ b\ a]$, whereas

`{{a, b}, {c, d}}`

is the matrix—given as a list of lists (rows)—$\begin{bmatrix} a & b \\ c & d \end{bmatrix}$. Note that in spite of the curly brackets, lists in *Mathematica* are closer to vectors than to sets.

To write more complicated mathematical expressions, one might find the palettes (see the top menu) really helpful.

Recently, the wording is that *Mathematica* is only one of the implementations of **Wolfram Language**. For our purposes it is enough to know that it looks like and works in the same way under any operation system. To evaluate an expression, one has to press ⟨Shift+Enter⟩ or the ⟨Enter⟩ key of the numeric pad.

Actually, it is not, in the sense of the purist, a genuine functional language, rather it is quite an opportunistic language: one can write programs in *Mathematica* in many styles or **programming paradigms**, as in the procedural style; term-rewriting (rule-based) or logic programming also exist here, and one can also use the object-oriented technique, and it is also possible to do **parallel programming**; therefore it is a multiparadigm language.

2.3.1　How to Get Help?

How to get help in case of need? There are many possibilities. One can visit the extremely rich Help item (including the *Virtual Book* and the *Function Navigator*) of the menu at the top of your screen. Having selected a function name and pressing the key $\boxed{\text{F1}}$ leads to the detailed help files containing examples, options, applications, neat examples, etc. Another possibility is to type in either `?SocialMediaData` or `??SocialMediaData` to learn how `SocialMediaData` works. Similarly, `?Parallel*` or `?*ocat*` may also provide interesting answers. Finally, if one is interested in plotting functions, then `?*Plot*` gives you a list of all the functions with names containing the word `Plot` as a substring.

2.3.2　Where to Get a *Mathematica* Function?

1. Thousands of functions are built into *Mathematica*; these are available immediately after the program has been loaded in.
2. Less often used functions, such as `VariationalMethods` or `ANOVA` can be found in separate packages delivered together with the program. If one needs a function of such a package, then she/he should load in the package, using either the `Get` command, `Get["SymbolicC`"]`, or its short form `«SymbolicC``. Another possibility is that the program makes available the function of a package (without actually loading them into the memory). The function `Needs` has this effect: `Needs["ResonanceAbsorptionLines`"]`.
3. Most of the topics (including packages) are covered by tutorials, as, e.g., http://reference.wolfram.com/language/CUDALink/tutorial/Overview.html.
4. A third source of functions is the packages made by users of *Mathematica*. Some of these packages (as, e.g., `SystemModeler`) are sold either by Wolfram Research Inc. or by another party (e.g., `Optica`), some can be found on the Internet (e.g., `ReplaceVariables`), and some are accompanying papers or books. Our package `ReactionKinetics` belongs to this last category.

2.4　First Steps with the Package

The first thing is to download the package `ReactionKinetics` from the website of the Publishing House. Then, let us start using it. As it is a separate package, you need to read it in. Before doing so it is safe to restart *Mathematica* using `Quit[]`.

It is quite common that you would like to have your input (data) and your results in output files, and perhaps you would also put the `ReactionKinetics.m` file in your working directory. Then, it may be useful to add your working directory to the `$Path` by `SetDirectory[NotebookDirectory[]]`. Now we can read in the package: `Needs["ReactionKinetics`"]`. The result of this command is threefold:

1. One can start using the functions of the package.
2. One receives two palettes making the use of it simpler.
3. One receives a message containing some administrative data.

At this point—using the `ReactionKineticsPalette`—we can define, e.g., the Robertson reaction as follows

`robertson = {A -> B, 2 B -> B + C -> A + C}.`

Then, `ReactionsData[robertson]["species", M]` gives the list and the number (or cardinality) of species. Another possibility is to utilize the fact that the Robertson reaction—being popular—is a built-in model; thus `robertson2 = GetReaction["Robertson"]` may be our first step.

Let us see a more complex example. Suppose we are interested in some details of the Lotka–Volterra reaction. The genuine reaction steps are obtained in the following way. `lotkavolterra=GetReaction["Lotka-Volterra"]` gives you

$$A + X \longrightarrow 2X \quad X + Y \longrightarrow 2Y \quad B \longleftarrow Y.$$

If you would like to know some characteristic data of the reaction taking into consideration that A and B are external species, then you might ask this:

`ReactionsData[{Lotka-Volterra}, {A, B}][α, β, γ, M, R].`

Finally, in case you are interested in the capabilities of the program, you may get a quick overview asking for `Information["ReactionKinetics`*"]`. The result will be a nicely formatted table of all the function names of the program acting as links to their description. At the end of the book (in Sect. 12.10), we give a more detailed description of the package.

In the final section of some of the chapters, we propose problems whose solutions help the reader understand functions defined in our package and also to extend them.

2.5 Exercises and Problems

2.1 Construct a reaction with the same induced kinetic differential equation as the model in Example 2.13.

(Solution: page 382.)

2.2 Find *Mathematica* programs or packages on the Internet to investigate reaction kinetics.

(Solution: page 382.)

2.3 Go to demonstrations.wolfram.com and have a look at demonstrations on chemical kinetics.

(Solution: page 382.)

In all the problems below, suppose a reaction is given in the form of a list of irreversible steps, and try to find and algorithm and a *Mathematica* code to solve the problems.

2.4 How would you calculate the set of species and their number in the case of the Robertson reaction?

(Solution: page 383.)

2.5 How would you calculate the set of reaction steps and their number in the case of the reversible Lotka–Volterra reaction (i.e., in the reaction where all the steps are supposed to be reversible)?

(Solution: page 383.)

2.6 Try to sketch a *Mathematica* algorithm for the calculation of the matrices α and β in the case of the simple irreversible bimolecular reaction $A + B \longrightarrow C$.

(Solution: page 383.)

2.7 Use the package to calculate the matrices α and β and the value of S in the case of the **ping-pong reaction** (see Keleti 1986; Azizyan et al. 2013)

$$E + S_1 \rightleftharpoons ES_1 \rightleftharpoons E^*S_1 \longrightarrow E^* + P_1$$
$$E^* + S_2 \rightleftharpoons E^*S_2 \rightleftharpoons ES_2 \longrightarrow E + P_2.$$

An example of the ping-pong reaction is the action of chymotrypsin; if one uses the following definitions, let S_1 be p-nitrophenyl acetate and the first product P_1 be p-nitrophenolate. The second product P_2 is the acetate ion. The action of chymotrypsin is said to be a ping-pong reaction because the binding of the two substrates causes the enzyme to switch back and forth between two states.

(Solution: page 384.)

References

Azizyan RA, Gevorgyan AE, Arakelyan VB, Gevorgyan ES (2013) Mathematical modeling of uncompetitive inhibition of bi-substrate enzymatic reactions. Int J Biol Life Sci Eng 7(10):627–630

Barabás B, Tóth J, Pályi G (2010) Stochastic aspects of asymmetric autocatalysis and absolute asymmetric synthesis. J Math Chem 48(2):457–489

Bray WC (1921) A periodic reaction in homogeneous solution and its relation to catalysis. J Am Chem Soc 43(6):1262–1267

Deák J, Tóth J, Vizvári B (1992) Anyagmegmaradás összetett kémiai mechanizmusokban (Mass conservation in complex chemical mechanisms). Alk Mat Lapok 16(1–2):73–97

Deuflhard P, Bornemann F (2002) Scientific computing with ordinary differential equations. In: Marsden JE, Sirovich L, Golubitsky M, Antmann SS (eds) Texts in applied mathematics, vol 42. Springer, New York

Feinberg M, Horn FJM (1977) Chemical mechanism structure and the coincidence of the stoichiometric and kinetic subspaces. Arch Ratl Mech Anal 66(1):83–97

Frank FC (1953) On spontaneous asymmetric synthesis. Biochim Biophys Acta 11:459–463

Frank-Kamenetskii DA (1947) Diffusion and heat transfer in chemical kinetics. USSR Academy of Science Press, Moscow/Leningrad (in Russian)

Gray JW (1994, 1996, 1998) Mastering Mathematica: programming methods and applications. Academic, London

Keleti T (1986) Basic enzyme kinetics. Akadémiai Kiadó, Budapest

Korobov VI, Ochkov VF (2011) Chemical kinetics: introduction with Mathcad/Maple/MCS. Springer, Moscow

Lente G (2004) Homogeneous chiral autocatalysis: a simple, purely stochastic kinetic model. J Phys Chem A 108:9475–9478

Mangano S (2010) Mathematica cookbook. O'Reilly Media, Beijing

Mints RM, Sal'nikov IE, Sorokina GA, Shishova NA (1977) Autovibrational model of periodic cold flames based on the Lotka–Frank–Kamenetskii scheme. Combust Explosion Shock Waves 13(6):700–707

Mole G (1936) The ignition of explosive gases. Proc Phys Soc 48:857–864

Ogg RAJ (1947) The mechanism of nitrogen pentoxide decomposition. J Chem Phys 15(5):337–338

Robertson HH (1966) The solution of a set of reaction rate equations. In: Walsh JE (ed) Numerical analysis, pp 178–182. Thompson Book Co., Washington, DC

Schlögl F (1972) Chemical reaction models for non-equilibrium phase transitions. Z Phys 253(2):147–161

Szili L, Tóth J (1999) Numerical and symbolic applications of *Mathematica*. Mathematica Pannonica 10(1):83–92

Trott M (2006) The *Mathematica* guidebook for numerics. Springer, New York

Trott M (2007) The *Mathematica* guidebook for symbolics. Springer, New York

Trott M (2013) The *Mathematica* guidebook for programming. Springer, New York

Turányi T, Györgyi L, Field RJ (1993) Analysis and simplification of the GTF model of the Belousov–Zhabotinsky reaction. J Phys Chem 97:1931–1941

Wolfram S (2015) An elementary introduction to the Wolfram language. Wolfram Media, Champaign

Zádor J, Fernandes RX, Georgievskii Y, Meloni G, Taatjes CA, Miller JA (2009) The reaction of hydroxyethyl radicals with O_2: a theoretical analysis of experimental product study. Proc Combust Inst 32:271–277

Graphs of Reactions

3

Since the pioneering work of Pólya (1937), it is obvious that graph theory is a very useful tool for the chemist as well. Recently, the description of molecules with graphs and calculation of their characteristics (they are called **topological indices** in this literature) to predict activities or—more generally—their properties is the topic of a very large number of papers, e.g., in the *Journal of Mathematical Chemistry* or in *match (Communications in Mathematical and in Computer Chemistry)*. It is astonishing that a review with a title promising much more, (Ivanciuc and Balaban 1998) only knows this approach. We are here interested just in the complementary set of graphs: those used in kinetics. It may be worth mentioning that one of the first persons to systematically study the relationships between different graphs and kinetics was Othmer (1981). Finally, we mention the work by Kiss et al. (2017) which, in a certain sense, contains models which are more general than ours: the authors deal with models where the particles (human beings or molecules) are treated individually, and the interactions are described by models similar to the ones used in reaction kinetics.

As to the terminology and fundamentals of graph theory, we propose a few classics, Busacker and Saaty (1965), Harary (1969), Lovász (2007), and Øre (1962), but the definitions and statements needed here have been collected in Sect. 13.5 of the Appendix.

3.1 The Feinberg–Horn–Jackson Graph

Consider the reaction (2.1), and represent it by a directed graph constructed in the following way.

Definition 3.1 Let the set of vertices (consisting of N elements) of the graph be the complexes $\mathscr{C}$, and let us draw an arrow from the complex $\mathbf{y} \in \mathscr{C}$ to $\mathbf{y}' \in \mathscr{C}$, if $\mathbf{y} \longrightarrow \mathbf{y}'$ occurs among the reaction steps. The directed graph composed in this way is the **Feinberg–Horn–Jackson graph** of the reaction. **Weakly connected components** of the Feinberg–Horn–Jackson graph are also called **linkage classes**

© Springer Science+Business Media, LLC, part of Springer Nature 2018
J. Tóth et al., *Reaction Kinetics: Exercises, Programs and Theorems*,
https://doi.org/10.1007/978-1-4939-8643-9_3

here; their number will be denoted by L. Linkage classes are sometimes called **coarse linkage class**es cf. Definition 3.18.

Remark 3.2 The Feinberg–Horn–Jackson graph being a directed graph can also be considered as the graph of the relation "reacts to" within the set of complexes. This relation is never reflexive, and in general it is neither symmetric nor transitive.

If the investigated reaction is given in the form of a list of reaction steps, say, it is the model of the Belousov–Zhabotinsky reaction proposed by Györgyi and Field (1991) (different from the one cited previously in Example 2.13):

$$
\begin{array}{lll}
H + X + Y \longrightarrow 2\,V & A + 2\,H + Y \longrightarrow V + X & 2\,X \longrightarrow V \\
A + H + \frac{1}{2}\,X \longrightarrow X + Z & X + Z \longrightarrow \frac{1}{2}\,X & M + Z \longrightarrow Q \\
V + Z \longrightarrow Y & V \longrightarrow Y & X \longrightarrow 0 \\
Y \longrightarrow 0 & Z \longrightarrow 0 & V \longrightarrow 0
\end{array}
$$

$$\tag{3.1}$$

then the edges of its Feinberg–Horn–Jackson graph can be obtained as

`ReactionsData[{"Belousov-Zhabotinsky"}]["fhjgraphedges"]`

The result might make you disappointed, because it gives no figure. However, if one needs the figure not the graph object, then it can be obtained in the following way:

`ShowFHJGraph[{"Belousov-Zhabotinsky"},`
    `DirectedEdges -> True, VertexLabeling -> True].`

The result can be seen in Fig. 3.1.

Note that this reaction (similarly to Example 2.13) also does not fit into the theory in the strict sense because some of the stoichiometric coefficients are fractions, although here there are no negative stoichiometric coefficients.

Fig. 3.1 The Feinberg–Horn–Jackson graph of a model of the Belousov–Zhabotinsky reaction

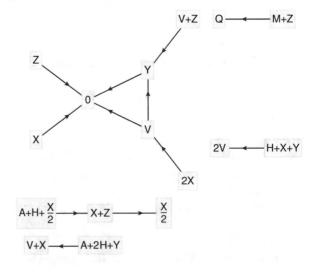

An obvious statement follows.

Remark 3.3 The product of the complex matrix $\mathbf{Y}$ and the incidence matrix $\mathbf{E}$ (see Definition 13.27) of the Feinberg–Horn–Jackson graph is the stoichiometric matrix γ: $\boxed{\mathbf{YE} = \gamma}$.

Let us illustrate the above remark on the example of the Robertson reaction.

Example 3.4 Here the complex matrix is

$$
\begin{array}{c|ccccc}
 & A & B & 2\,B & B+C & A+C \\
\hline
A & 1 & 0 & 0 & 0 & 1 \\
B & 0 & 1 & 2 & 1 & 0 \\
C & 0 & 0 & 0 & 1 & 1
\end{array}
\tag{3.2}
$$

and the incidence matrix is

$$
\begin{array}{c|ccc}
 & A \longrightarrow B & 2\,B \longrightarrow B+C & B+C \longrightarrow A+C \\
\hline
A & -1 & 0 & 0 \\
B & 1 & 0 & 0 \\
2\,B & 0 & -1 & 0 \\
B+C & 0 & 1 & -1 \\
A+C & 0 & 0 & 1
\end{array}
\tag{3.3}
$$

and finally, the stoichiometric matrix is

$$
\begin{array}{c|ccc}
 & A \longrightarrow B & 2\,B \longrightarrow B+C & B+C \longrightarrow A+C \\
\hline
A & -1 & 0 & 1 \\
B & 1 & -1 & -1 \\
C & 0 & 1 & 0
\end{array}
\tag{3.4}
$$

and the verification of the above equality is straightforward (as it is in the general case).

3.1.1 Reversibility, Weak Reversibility

Definition 3.5 If together with the reaction step $\mathbf{y} \longrightarrow \mathbf{y}'$ the reaction step $\mathbf{y}' \longrightarrow \mathbf{y}$ can also be found in the set of reaction steps, then this reaction step is said to be a **reversible reaction step**. If all the reaction steps are reversible, then reaction (2.1) is said to be a **reversible reaction**.

Remark 3.6 Reversibility of a reaction is equivalent to saying that its Feinberg–Horn–Jackson graph—as a relation—is symmetric. A generalization of reversibility will also turn out to be important in many statements below.

Definition 3.7 Suppose that in the reaction (2.1) for all pairs of complexes $\mathbf{y}, \mathbf{y}'$ if there is a series of reaction steps $\mathbf{y} \longrightarrow \mathbf{y}_1 \longrightarrow \mathbf{y}_2 \longrightarrow \cdots \longrightarrow \mathbf{y}_j \longrightarrow \mathbf{y}'$, then there is a series of reaction steps in the backward direction as well: $\mathbf{y}' \longrightarrow \mathbf{z}_1 \longrightarrow \mathbf{z}_2 \longrightarrow \cdots \longrightarrow \mathbf{z}_k \longrightarrow \mathbf{y}$. Then, the reaction is said to be **weakly reversible**.

Remark 3.8

1. Reversibility of a reaction obviously implies weak reversibility, whereas the example of the irreversible triangle reaction in Fig. 3.2 shows that weak reversibility does not imply reversibility.
2. Weak reversibility can also be reformulated in this way: all the directed edges of the Feinberg–Horn–Jackson graph is an edge of at least one directed cycle.
3. Using again the relation theoretical terminology (cf. Appendix, Sect. 13.5), one may say that a reaction is weakly reversible if its transitive closure is symmetric. Transitive closure is obtained by starting from the Feinberg–Horn–Jackson graph so that complexes $\mathbf{y}$ and $\mathbf{y}'$ are connected if there is a series of reaction steps leading from one to the other. If this new relation is symmetric, then there is another series of reaction steps leading to the second complex from the first one, as well. We may call the transitive closure of the Feinberg–Horn–Jackson graph the graph of the "finally reacts to" relation (Fig. 3.3). When speaking of the finally reacts to relation reflexivity is assumed. With all these we have an **equivalence relation** where the equivalence classes are the vertices of the linkage classes.

Fig. 3.2 The irreversible triangle reaction is weakly reversible

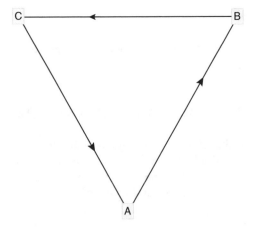

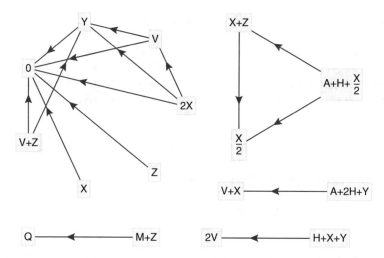

Fig. 3.3 The transitive closure of the Feinberg–Horn–Jackson graph of the model Equation (3.1)

3.1.2 (Generalized) Compartmental Systems

Now let us define special classes of reactions which play important roles in biological applications (transport in the living body, pharmacokinetics, spatial movement of populations) (Jacquez 1999; Brochot et al. 2005), and also in photochemistry (Ohtani 2011).

Definition 3.9 The reaction in which the length of complexes is not more than one is a **compartmental system**. It is a **closed compartmental system**, if it does not contain the empty complex; it is **half-open** if it does contain the empty complex but only as a product complex, i.e., if outflow is allowed, but inflow is excluded.

Remark 3.10 A large part of the literature uses the terms compartmental system and first-order reaction as synonyms; however, according to our definitions (which seem to be more distinguishing), neither the reaction $X \longrightarrow Y + Z$ nor the reaction $X \longrightarrow 2X$ is a compartmental system, although both of them are first-order reactions.

Example 3.11 Measurements were made to find out the effect of additional surfactant materials on the absorption of drugs modeled by salicylic acid (Rácz et al. 1977). To describe the process the simplest possible mathematical model, the compartmental system (or consecutive reaction, if one uses the chemists' term) was used:

$$A \longrightarrow B \longrightarrow C,$$

where A represents the gastric juice, B is a lipoid barrier, and C is the intestinal fluid.

Slight generalizations of the above notions follow.

Definition 3.12 The reaction in which all the complexes contain not more than one species (the length of support of the complexes is not more than one) is a **generalized compartmental system**. It is a **closed generalized compartmental system**, if it does not contain the empty complex; it is **half-open** if it does contain the empty complex but only as a product complex.

Definition 3.13 The reaction in which all the complexes contain not more than one species and no species occurs in more than one complex is a **generalized compartmental system in the narrow sense**.

Closed, half-open, and open generalized compartmental systems in the narrow sense can be defined in the same way as above.

Example 3.14 The reaction $3\,Y \longleftarrow 2\,X \longrightarrow 5\,Y \longrightarrow 4\,X$ is a generalized compartmental system, but not in the narrow sense.

3.1.3 Deficiency

Now we are in the position to define a notion of fundamental importance; see Theorem 8.47.

Definition 3.15 The **deficiency** of the reaction (2.1) is defined to be $\delta := N - L - S$, with N as the number of complexes, L as the number of (weak) components of the Feinberg–Horn–Jackson graph, and S as the rank of γ as given in Definition 2.2 (Fig. 3.4).

Theorem 3.16 *The deficiency is a nonnegative integer for every reaction.*

Proof Let us denote the matrix of complex vectors by $\mathbf{Y}$ and the incidence matrix of the Feinberg–Horn–Jackson graph by $\mathbf{E}$, and then the deficiency is the dimension of $\dim(\mathrm{Ker}(\mathbf{Y}) \cap \mathscr{R}_{\mathbf{E}})$; see Feinberg (1972).   □

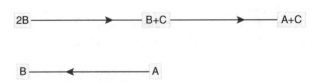

Fig. 3.4 The Feinberg–Horn–Jackson graph of the Robertson reaction having deficiency $\delta = 5 - 2 - 2 = 1$

Theorem 3.17 *If the complex vectors of a reaction (excluding the zero complex) are independent, then the deficiency of the reaction is zero.*

It is an empirical fact that most of the reactions in the classical textbooks on chemistry have zero deficiency. What is more, within a small class of reactions (those only consisting of three short complexes) the fact has been proven using combinatorial methods by Horn (1973b), and it was conjectured to be true for reactions with more complexes. However, reactions with many species and reaction steps in biology (more specifically, in ion channel modeling, see Nagy and Tóth 2012) may have deficiencies different from zero, or even they may have very large deficiencies in areas such as combustion theory (Tóth et al. 2015).

Let us return to Feinberg–Horn–Jackson graphs, and do not forget to visit Sect. 13.5 of the Appendix in case of need.

Definition 3.18 Strongly connected components of the graph of the "finally reacts to" relation are also called **strong linkage class**es of the Feinberg–Horn–Jackson graph. Of these components the ones from which no reaction steps are leading out are the **terminal strong linkage class**es (cf. Definition 3.1); their number will be denoted by T.

Terminal strong linkage class is just another name for **ergodic component** (see Definition 13.30) of the Feinberg–Horn–Jackson graph (Figs. 3.5 and 3.6).

Remark 3.19 Weak reversibility is equivalent to the property that each weak component is strongly connected or that each strong linkage class is terminal. Although for weakly reversible reactions the number T of the ergodic components of the Feinberg–Horn–Jackson graph is the same as the number L of weak components (linkage classes), this property is not sufficient to imply weak reversibility as simple counterexamples can show.

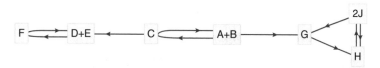

Fig. 3.5 The Feinberg–Horn–Jackson graph of a model from Feinberg and Horn (1977, p. 89)

Fig. 3.6 Strong linkage classes of the Feinberg–Horn–Jackson graph in Fig. 3.5 of which lowermost one A + B $\rightleftharpoons$ C is **not** a terminal strong linkage (ergodic) class. Note also that here $T = 2 > 1 = L$

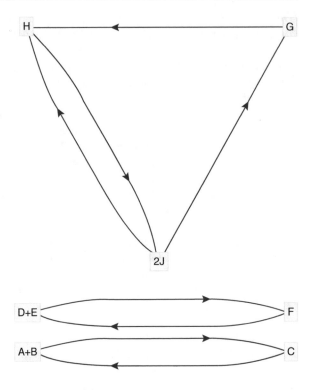

3.2   The Volpert Graph

Another representation with some tradition mainly in the textbooks on biochemistry has formally been given by Volpert and Hudyaev (1985, Chap. XII), and originally by Volpert (1972) and recently by De Leenheer et al. (2006), Banaji and Craciun (2010, 2009), and Donnell and Banaji (2013).

3.2.1   Definition of the Volpert Graph

Definition 3.20 Let the two vertex sets of a **weighted directed bipartite graph** be the species index set $\mathcal{M}$ and the reaction step index set $\mathcal{R}$, respectively; and let us connect vertex $m \in \mathcal{M}$ to vertex $r \in \mathcal{R}$ with an arc of weight $\alpha(m, r)$, if $\alpha(m, r) > 0$, and connect vertex r to vertex m with an arc of weight $\beta(m, r)$ if $\beta(m, r) > 0$. The graph obtained in this way is the **Volpert graph** of the reaction (2.1).

The Volpert graph shows how many molecules of $X(m)$ are needed to the rth reaction step to take place and how many molecules of $X(m)$ are produced in the rth reaction step. In some areas of modeling, e.g., in computer science (but sometimes in reaction kinetics, as well, see Goss and Peccoud 1998; De Leenheer et al. 2006;

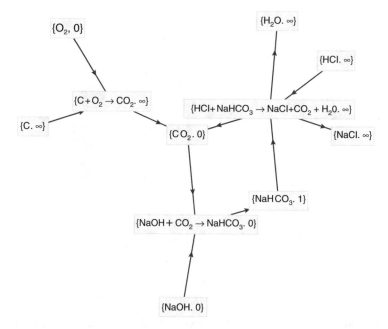

Fig. 3.7 The Volpert graph of a reaction proposed by Petri. All the weights are supposed to be in unity. This graph does contain a directed cycle, while the Feinberg–Horn–Jackson graph of the same reaction does not

Volpert 1972), directed bipartite graphs are called **Petri nets**. The name **DSR graph** is also used (Banaji and Craciun 2009). However, in case of this terminology, it is usual to assign a more active role to one of the sets of vertices than the role of reaction steps (as a kind of transformation) are.

As the legend goes, Petri nets were invented in August 1939 by the German Carl Adam Petri—at the age of 13—for the purpose of describing chemical processes, such as those in Fig. 3.7. However, relationships between the properties of the Petri net of a reaction and its behavior have first been established by Volpert, including the obvious remark that given a weighted, directed bipartite graph, one can immediately write down a reaction having the graph as its Volpert graph.

In the case of larger models, the Volpert graph can be quite large; see, e.g., in Fig. 3.8 or in the figures of the electronic supplement of Tóth et al. (2015).

Remark 3.21 If a reaction is weakly reversible, or if it contains at least a single reversible reaction step, then its Volpert graph contains a cycle: acyclicity of the Volpert graph excludes both weak reversibility and the presence of reversible reaction steps.

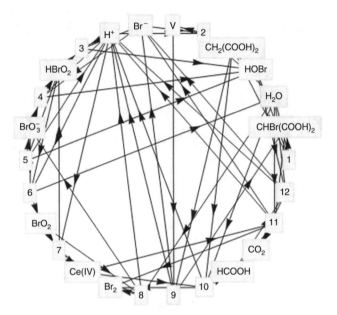

Fig. 3.8 A larger Volpert graph of a model proposed by Clarke for the Belousov–Zhabotinsky reaction, cited by Epstein and Pojman (1998, p. 100)

3.2.2   Indexing

Once we have a Volpert graph, we can assign indices to their vertices which will be really useful later when deciding the acceptability of decompositions of overall reactions in Chap. 5, when studying the behavior of the concentrations vs. time functions around time zero (Theorem 9.1), or infinity (Theorem 8.54).

Definition 3.22 Let $\mathcal{M}_0 \subset \mathcal{M}$ be a subset of the species index set, called the indices of **initial species**.

1. Elements of this set receive zero index: $m \in \mathcal{M}_0$ implies $i(m) = 0$. Zero will also be assigned to those reaction steps r for which all the reactant species are available: $r \in \mathcal{R}$ and $(\forall m \in \mathcal{M}(\alpha(m, r) > 0 \longrightarrow i(m) = 0)$ implies $i(r) = 0)$.
2. By induction, suppose that the indices up to $k \in \mathbb{N}_0$ have been allotted. Then, a species receives the index $k + 1$, if it has no index yet and if there is a reaction step of index k producing it: $\exists r \in \mathcal{R} : (\beta(m, r) > 0 \vee i(r) = k)$. A reaction step receives the index $k + 1$, if it has no index yet and if all the species needed for it to take place are available: $r \in \mathcal{R}$ and $(\forall m \in \mathcal{M} : (\alpha(m, r) > 0 \wedge i(m) \le k)$ implies $i(r) := k)$.
3. As the total number of vertices is finite, the algorithm ends after a finite number of steps. Species and reaction steps without index at termination will receive an index $+\infty$.

Remark 3.23

1. In the applications of the definition, the set of initial species is the set of those species which have a strictly positive initial concentration.
2. The value of an index shows a kind of distance of a vertex from the initial vertices; see Theorem 9.1.
3. If the index of a reaction step is zero, one, or two, then the reaction step may be called **primary**, **secondary**, and **tertiary** (as they take place at level one, two, or three).

3.3 Relationships Between the Feinberg–Horn–Jackson Graph and the Volpert Graph

Different graphs have been defined by the representatives of the two major schools in reaction kinetics. It is quite natural to look for relationships between these.

3.3.1 General Considerations

What does the assumption of connectivity mean? The fact that the Feinberg–Horn–Jackson graph is disconnected only means that one has more than one linkage classes. However, this does not exclude the presence of common species in complexes belonging to different linkage classes.

If the Volpert graph consists of more than one component, then the species and reaction steps belonging to a given connected component can be treated separately, because these do not affect each other in any model. Therefore in most theoretical investigations, one may assume that the Volpert graph is connected. In applications, however, connectedness should be checked.

How is the fact that the Volpert graph is disconnected reflected in the form of the Feinberg–Horn–Jackson graph?

Theorem 3.24 *Suppose that the zero complex is not present among the complexes of a reaction.*

1. *If the Volpert graph has L connected components, then the Feinberg–Horn–Jackson graph has at least L linkage classes.*
2. *If the Feinberg–Horn–Jackson graph has L linkage classes and there are no common species in the linkage classes, then the Volpert graph has exactly L connected components.*

Proof

1. Take two connected components of the Volpert graph. As they have neither species nor reaction steps in common, the reaction steps cannot have common complexes.

2. Consider two linkage classes L_1, L_2 of the Feinberg–Horn–Jackson graph. As each of them is connected from what we have just proved, it follows that they correspond to the connected components V_1, V_2 of the Volpert graph. By the assumption that there are no common species in L_1, L_2, the species vertex sets of V_1, V_2 are disjoint. Hence V_1 and V_2 are not connected by any edge.

□

Example 3.25 Consider the reaction $A \longrightarrow 0 \longrightarrow B$. Then, $L = 1$ but the Volpert graph consists of two components, showing the relevance of exclusion of the empty complex in the above theorem.

Remark 3.26 The Feinberg–Horn–Jackson graph and the Volpert graph of a closed generalized compartmental system in the narrow sense is essentially the same; see Volpert and Hudyaev (1985, p. 609).

Indeed, take the Volpert graph of a compartmental system, and suppose it has the following subgraph: $X(m) \longrightarrow r \longrightarrow X(p)$. Then, this subgraph corresponds to the following edge in the Feinberg–Horn–Jackson graph: $X(m) \longrightarrow X(p)$. On the other hand, if the reaction step r in the reaction is $X(m) \longrightarrow X(p)$, then its Volpert graph should contain the subgraph $m \longrightarrow r \longrightarrow p$. Cases where the empty complex is also involved can be treated similarly. Almost exactly the same proof will do for generalized compartmental systems.

3.3.2   The Species–Complex–Linkage Class Graph

This graph has been introduced by Schlosser and Feinberg (1994) and used to investigate relationships between Feinberg–Horn–Jackson graphs and Volpert graphs by Siegel and Chen (1995).

Definition 3.27 The **species–complex–linkage class graph** or **S–C–L graph** of the mechanism (2.1) is a (undirected) bipartite graph with the following two vertex sets: the species set and the set of linkage classes (weak components of the Feinberg–Horn–Jackson graph). If a complex occurs among the complexes in the linkage class $\mathscr{L}_l$ and contains the species $X(m)$ with a nonzero coefficient, then an edge is drawn which connects the points l and m. Furthermore the edge is labeled with the complex (Fig. 3.9).

Here we only cite structural results by Siegel and Chen (1995)—of which the second one is an explicit statement about the connection of the Feinberg–Horn–Jackson graph and the Volpert graph—later, in Chap. 8 we shall mention results bearing direct relevance on the transient behavior of the concentration time curves.
    The proofs of the following theorems can be found in Siegel and Chen (1995, pp. 487–489).

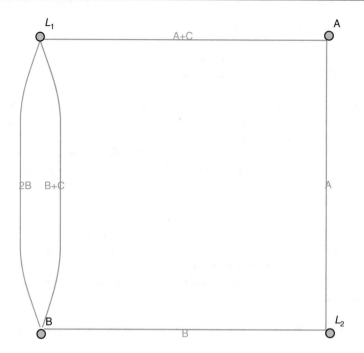

Fig. 3.9 The S–C–L graph of the Robertson reaction

Theorem 3.28 *If the S–C–L graph of a mechanism is acyclic, then the deficiency of the mechanism is zero.*

Theorem 3.29 *If both the S–C–L graph and the Feinberg–Horn–Jackson graph of a mechanism is acyclic, then its Volpert graph is acyclic, as well.*

3.4 The Species Reaction Graph

Crăciun (2002) introduced and several authors (including Crăciun and Feinberg 2006, but see also Definition 3.20) used the concept **species reaction graph**. This is exactly the same as the Volpert graph of the reaction. However, Volpert (1972) in Theorem 8.54 only stated that acyclicity of this graph excludes periodicity and multistability; the mentioned authors made a more detailed analysis of the case when the Volpert graph does have cycles, and they made conclusions on multistationarity and persistence from the types of cycles present in the Volpert graph; see Theorem 8.63. One may say that this way they were able to rigorously investigate the conjectures stated many years earlier by Thomas and King on the role of types of cycles upon exotic behavior in another graph: the influence diagram; see Chap. 8.

3.5   The Complex Graph

As a tool in the investigation of "small" reactions (those with three short complexes), another graph was introduced by Horn (1973a, p. 314).

Definition 3.30 Suppose the reaction (2.1) only consists of short complexes. Let the vertices of the **complex graph** be the elements of the extended species index set $\mathscr{M}_0 := \mathscr{M} \cup \{0\}$, and let us connect the vertices $m, p \in \mathscr{M}_0$ with an edge in case $X(m)+X(p)$ occurs among the complexes. (Here by definition $X(0)$ is the empty complex, $X(m)+X(0) := X(m)$ for all $m \in \mathscr{M}_0$.) The vertices corresponding to $m \in \mathscr{M}$ are called **species vertices**, while the vertex corresponding to 0 is said to be the **root** of the complex graph.

Definition 3.31 Consider two cycles and a path in the complex graph such that in each cycle precisely one vertex is identical with one of the end vertices of the path and that no other vertices are shared with between any two of the aforementioned objects. The two cycles and the path are then said to form a **dumbbell**. The dumbbell is odd, if both of its cycles are odd.

Example 3.32 Let us construct the complex graph of the model by Edelstein (1970)

$$X \rightleftharpoons 2X \quad X + Y \rightleftharpoons Z \rightleftharpoons Y.$$

All the complexes are short, and they—X, $2X$, $X + Y$, Z, Y—can be represented one after another as shown in Fig. 3.10.

Having the complex graphs of the complexes, one is able to put together the complex graph of the whole reaction; see in Fig. 3.11.

Application of the complex graph will be shown in Theorem 8.51 in Chap. 8.

The graphs defined in the present chapter are useful to study the statics and dynamics of the usual deterministic model. Other graphs also useful in this respect can and will only be defined after having at hand the concept of the kinetic differential equation in Chap. 8. Furthermore, other graphs will be defined and used in Chap. 10 on the usual stochastic model of reactions.

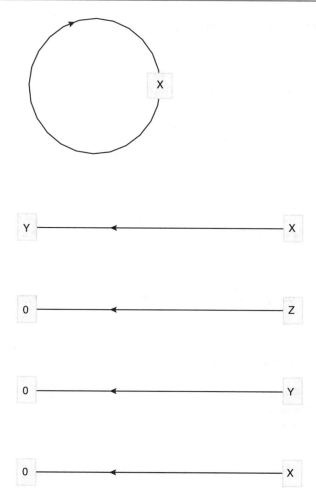

Fig. 3.10 Complex graphs of the individual complexes of the Edelstein model

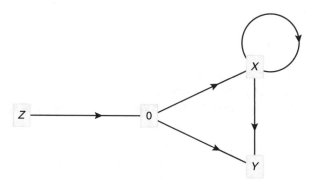

Fig. 3.11 Complex graph of the Edelstein model

3.6   Exercises and Problems

Among the exercises and problems as usual, we propose a few problems whose solution help the reader understand a few functions defined in the package.

   In all the problems below, suppose a reaction is given in the form of a list of irreversible steps.

3.1 Is this property equivalent to weak reversibility: all the vertices of the Feinberg–Horn–Jackson graph are part of at least one directed cycle?

(Solution: page 384.)

3.2 Suppose you are given `fhj`, the Feinberg–Horn–Jackson graph of the reaction. How do you check (by hand and by programs) if it is weakly reversible or not?

(Solution: page 384.)

3.3 How does one produce the tables in Example 3.4 using the program?

(Solution: page 384.)

3.4 Calculate the deficiency of the

1. Michaelis–Menten reaction,
2. Lotka–Volterra reaction,
3. Robertson reaction.

(Solution: page 385.)

3.5 Find cycles in the S–C–L graph of the (irreversible) Lotka–Volterra reaction, thereby confirming that its deficiency cannot be zero.

(Solution: page 385.)

3.6 Prove that the deficiency of a reaction is zero if

1. the number of reaction steps R is 1;
2. the reaction is a compartmental system;
3. the reaction is a generalized compartmental system in the narrow sense.

(Solution: page 385.)

3.7

1. Give one species reaction with positive deficiency
2. Give a first-order reaction with positive deficiency.
3. Give another example showing that a first-order reaction which is not a compart-mental system may also have zero deficiency.

(Solution: page 386.)

Again, suppose that below a reaction is given in the form of a list of irreversible steps, and try to find an algorithm and a *Mathematica* code to solve the problems.

3.8 Try to sketch a *Mathematica* algorithm for the calculation of

1. the set of complexes and their number;
2. the Feinberg–Horn–Jackson graph and
3. the Volpert graph;
4. the number of connected components L of the Feinberg–Horn–Jackson graph;
5. the deficiency.

The best idea might be to write the code for the small special cases we have had up to now, the simple bimolecular reaction, the Robertson reaction, the Lotka–Volterra reaction, etc.

(Solution: page 387.)

3.9 Find a simple version of the glyoxalate cycle on the web and draw its Volpert graph.

(Solution: page 387.)

3.10 Consider the (oversimplified) model for the chlorination of methane:

$$Cl_2 \longrightarrow 2\,Cl^* \quad CH_4 + Cl^* \longrightarrow {}^*CH_3 + HCl \quad {}^*CH_3 + Cl_2 \longrightarrow CH_3Cl + Cl^*$$

and let the set of initial species be $\mathcal{M}_0 := \{Cl_2, CH_4\}$. Calculate the Volpert indices of all the species and reaction steps.

(Solution: page 388.)

3.11 Consider the same model as above, and now let the set of initial species be $\mathcal{M}_0 := \{Cl^*, CH_4\}$. Calculate the Volpert indices of all the species and reaction steps.

(Solution: page 388.)

3.12 Find a cycle in the S–C–L graph of the Petri reaction of Fig. 3.7.

(Solution: page 388.)

3.13 Find an example where the number of linkage classes is larger than the number of the components of its S–C–L graph.

(Solution: page 388.)

3.14 Find an example where the Volpert graph is acyclic and the S–C–L graph is not.

(Solution: page 388.)

References

Banaji M, Craciun G (2009) Graph-theoretic approaches to injectivity and multiple equilibria in systems of interacting elements. Commun Math Sci 7(4):867–900

Banaji M, Craciun G (2010) Graph-theoretic criteria for injectivity and unique equilibria in general chemical reaction systems. Adv Appl Math 44(2):168–184

Brochot C, Tóth J, Bois FY (2005) Lumping in pharmacokinetics. J Pharmacokinet Pharmacodyn 32(5–6):719–736

Busacker RG, Saaty TL (1965) Finite graphs and networks. McGraw-Hill, New York

Crăciun G (2002) Systems of nonlinear equations deriving from complex chemical reaction networks. PhD thesis, Department of Mathematics, The Ohio State University, Columbus, OH

Crăciun G, Feinberg M (2006) Multiple equilibria in complex chemical reaction networks: Ii. The species reaction graph. SIAM J Appl Math 66:1321–1338

De Leenheer P, Angeli D, Sontag ED (2006) Monotone chemical reaction networks. J Math Chem 41(3):295–314

Donnell P, Banaji M (2013) Local and global stability of equilibria for a class of chemical reaction networks. SIAM J Appl Dyn Syst 12(2):899–920

Edelstein BB (1970) Biochemical model with multiple steady states and hysteresis. J Theor Biol 29(1):57–62

Epstein I, Pojman J (1998) An introduction to nonlinear chemical dynamics: oscillations, waves, patterns, and chaos. Topics in physical chemistry series. Oxford University Press, New York. http://books.google.com/books?id=ci4MNrwSlo4C

Feinberg M (1972) Lectures on chemical reaction networks. http://www.che.eng.ohio-state.edu/~feinberg/LecturesOnReactionNetworks/

Feinberg M, Horn FJM (1977) Chemical mechanism structure and the coincidence of the stoichiometric and kinetic subspaces. Arch Ratl Mech Anal 66(1):83–97

Goss PJE, Peccoud J (1998) Quantitative modeling of stochastic systems in molecular biology using stochastic Petri nets. Proc Natl Acad Sci USA 95:6750–6755

Györgyi L, Field RS (1991) Simple models of deterministic chaos in the Belousov–Zhabotinskii reaction. J Phys Chem 95(17):6594–6602

Harary F (1969) Graph theory. Addison–Wesley, Reading

Horn F (1973a) On a connexion between stability and graphs in chemical kinetics. II. Stability and the complex graph. Proc R Soc Lond A 334:313–330

Horn F (1973b) Stability and complex balancing in mass-action systems with three short complexes. Proc R Soc Lond A 334:331–342

Ivanciuc O, Balaban AT (1998) Graph theory in chemistry. In: Schleyer PvR, Allinger NL, Clark T, Gasteiger J, Kollman PA, Schaefer HFI, Schreiner PR (eds) Encyclopedia of computational chemistry. Wiley, Chichester, pp 1169–1190

Jacquez J (1999) Modeling with compartments. BioMedware, Ann Arbor, MI

Kiss IZ, Miller JC, Simon PL (2017) Mathematics of epidemics on networks. From exact to approximate models. Springer, Cham

Lovász L (2007) Combinatorial problems and exercises. AMS Chelsea Publishing, Providence, RI

Nagy I, Tóth J (2012) Microscopic reversibility or detailed balance in ion channel models. J Math Chem 50(5):1179–1199

Ohtani B (2011) Photocatalysis by inorganic solid materials: revisiting its definition, concepts, and experimental procedures. Adv Inorg Chem 63:395–430

Øre O (1962) Theory of graphs. AMS Colloquium Publications, vol 38. AMS, Providence

Othmer HG (1981) A graph-theoretic analysis of chemical reaction networks. I. Invariants, network equivalence and nonexistence of various tpes of steady states. Course notes, 1979

Pólya G (1937) Kombinatorische Anzahlbestimmungen für Gruppen, Graphen und chemische Verbindungen. Acta Math 68(1):145–254

Rácz I, Gyarmati I, Tóth J (1977) Effect of hydrophilic and lipophilic surfactant materials on the salicylic-acid transport in a three compartment model. Acta Pharm Hung 47:201–208 (in Hungarian)

Schlosser PM, Feinberg M (1994) A theory of multiple steady states in isothermal homogeneous CFSTRs with many reactions. Chem Eng Sci 49(11):1749–1767

Siegel D, Chen YF (1995) The S-C-L graph in chemical kinetics. Rocky Mt J Math 25(1):479–489

Tóth J, Nagy AL, Zsély I (2015) Structural analysis of combustion mechanisms. J Math Chem 53(1):86–110

Volpert AI (1972) Differential equations on graphs. Mat Sb 88(130):578–588

Volpert AI, Hudyaev S (1985) Analyses in classes of discontinuous functions and equations of mathematical physics. Martinus Nijhoff Publishers, Dordrecht. Russian original: 1975

Mass Conservation

4

This chapter studies the conservation relations admitted by a reaction, such as the well-known mass-balance equation. After giving a number of characterizations of mass conserving, producing, and consuming reactions, we show how to turn these characterizations into algorithms that identify such reactions and their conservation relations. These relations play an important role in the analysis of dynamic behavior, as we shall see in Chap. 8, where we use them to restrict the possible trajectories. This chapter is primarily based on the survey Deák et al. (1992) and on Schuster and Höfer (1991).

There are two, seemingly rather different, approaches to formalize the intuitive notion of mass conservation. On one hand, if we can assign (abstract) positive "masses" or weights to each species (which may be different from their actual physical mass) in such a way that the total weight of the reactants is equal to the total weight of the products in each elementary step, then the total weight of all constituents is clearly preserved throughout the reaction, regardless of its dynamics. Similarly, if the weights we assign to the species are such that the weight of the products exceeds the weight of the reactants in each step, then the reaction produces mass. This approach is formalized below in Definition 4.2.

The second, somewhat less intuitive, approach to mass conservation considers the effects of individual steps and makes no direct reference to mass. The idea is that if at the end of a sequence of steps we have more of each species than we initially had, the reaction is producing mass. Hence, we shall call a reaction mass producing if such sequences exist. These two dual approaches are closely connected; the connection is formally made in Theorems 4.25 and 4.27.

Finally, let us mention that the concepts in this chapter are connected with stoichiometry. Mass conservation and kinetics may have a slightly different relation as it can be seen in Sect. 8.5.2.1 of Chap. 8.

© Springer Science+Business Media, LLC, part of Springer Nature 2018
J. Tóth et al., *Reaction Kinetics: Exercises, Programs and Theorems*,
https://doi.org/10.1007/978-1-4939-8643-9_4

4.1   Preliminaries: Necessary and Sufficient Conditions

Many model reactions do not adhere to the law of conservation of mass. Example 2.7 raises the question how to understand the first or the third step, because the first one describes creation of mass, whereas the third one—with the **empty complex** on its right-hand side—represents destruction of mass. Actually, such reaction steps are abbreviations, as said above, so the first one is a shorthand for

$$A + X \longrightarrow 2\,X, \tag{4.1}$$

where A is an **external species**, i.e., a species with such a large concentration which may be considered to be unchanged during the time interval we are interested in. Such an external species may be water, a substrate present in large quantity, nutrition, precursor, etc. It will turn out that the **genuine reaction step** (4.1) and the original reaction step $X \longrightarrow 2\,X$ behave in the same way in all of the models.

Allowing steps of the above type makes it possible to describe in- and outflow using formal reaction steps.

Example 4.1 The reaction step $0 \longrightarrow X$ with $M = 1, R = 1, \alpha = [0], \beta = [1]$ describes inflow, while the reaction step $X \longrightarrow 0$ where $M = 1, R = 1, \alpha = [1], \beta = [0]$ describes outflow.

Intuitively, a reaction conserves mass if in each step the total mass on the two sides of the step are equal, i.e., if there exists a componentwise positive vector ρ of "masses" such that for all reaction steps $r \in \mathcal{R}$, we have

$$\sum_{m \in \mathcal{M}} \rho(m)\alpha(m, r) = \sum_{m \in \mathcal{M}} \rho(m)\beta(m, r). \tag{4.2}$$

The notions of mass producing and mass consuming reactions can be defined analogously. We can reformulate these definitions more concisely using the stoichiometric matrix $\gamma = \beta - \alpha$.

Definition 4.2 Reaction (2.1) is **mass conserving** if

$$\exists \rho \in (\mathbb{R}^+)^M \text{ satisfying } \rho^\top \gamma = \mathbf{0}^\top. \tag{4.3}$$

Definition 4.3 Reaction (2.1) is **mass producing** if

$$\exists \rho \in (\mathbb{R}^+)^M \text{ satisfying } \rho^\top \gamma \gneq \mathbf{0}^\top, \tag{4.4}$$

and it is **strongly mass producing** if

$$\exists \rho \in (\mathbb{R}^+)^M \text{ satisfying } \rho^\top \gamma > \mathbf{0}^\top. \tag{4.5}$$

Definition 4.4 Reaction (2.1) is **mass consuming** if

$$\exists \rho \in (\mathbb{R}^+)^M \text{ satisfying } \rho^\top \gamma \lneq \mathbf{0}^\top, \tag{4.6}$$

and it is **strongly mass consuming** if

$$\exists \rho \in (\mathbb{R}^+)^M \text{ satisfying } \rho^\top \gamma < \mathbf{0}^\top. \tag{4.7}$$

Remark 4.5 If we want to be more precise, we shall use the terms **stoichiometrically** mass conserving, mass producing, etc. for the notions introduced here to distinguish them from the notion **kinetically** mass conserving, **kinetically** mass producing, etc. (see Definition 8.33). The first concept is used more often; therefore mass conserving will usually be meant as stoichiometrically mass conserving, and we shall add the adjective kinetically in case there is a possibility of misunderstanding.

Remark 4.6 "Mass conserving" could have also been differentiated into strong and weak mass conserving for the cases when the existence of a positive, respectively, nonnegative vector is assumed.

The strong notions clearly imply the corresponding unqualified properties, but not the way around (Problem 4.1).

With a slight abuse of terms, we will identify the conservation relation with the vector representing the relation; i.e., the vector ρ satisfying (4.3) will also be called a **conservation relation**.

We also remark that, since conservation relations represent linear combinations of concentrations that will turn out to remain constant (see Eq. (8.21)), they are also **linear first integrals** (as opposed to nonlinear or quadratic first integrals) of the kinetic differential equations; see Chap. 8.

It is not trivial to decide if a reaction is mass conserving or not. In fact, early attempts to give simple linear algebraic characterizations of stoichiometric matrices of mass conserving reactions failed. For instance, Ridler et al. (1977) proposed a very simple method to recognize mass conserving reactions, but what they give is in fact only a necessary (but not sufficient) condition, see Problem 4.4.

Using the definitions, characterizing mass conserving, producing, or consuming reactions reduces to checking whether a system of linear equations and inequalities has a solution. It is the presence of inequalities that make linear algebraic characterization difficult. The following simple observation helps in rewriting Definition 4.2 in a form that can be directly handled algorithmically.

Theorem 4.7 *A reaction is mass conserving if and only if*

$$\max_{\lambda \in \mathbb{R}, \rho \in \mathbb{R}^M} \{\lambda \mid \rho^\top \gamma = \mathbf{0}^\top, \rho \geq \lambda \mathbf{1}, \rho^\top \mathbf{1} = 1\} > 0. \tag{4.8}$$

Finding the maximum on the left-hand side of (4.8) is an example of a **linear programming** or **linear optimization** problem (see Sect. 13.4 of the Appendix).

In `ReactionKinetics`, the function `MassConservationRelations` returns a set of conditions which hold among the components of the vector ρ in the definition of mass conservation:

`MassConservationRelations[{"Michaelis-Menten"}]`

returns $\{\rho_E > 0,\ \rho_S > 0,\ \rho_C == \rho_E + \rho_S,\ \rho_P == \rho_S\}$. If the result is `False`, then the reaction is not mass conserving.

The Lotka–Volterra reaction is mass conserving if one considers the genuine reactions steps, and it is not mass conserving if the external species are discarded.

`MassConservationRelations[{"Lotka-Volterra"}]`

returns $\{\rho_A > 0,\ \rho_X == \rho_A,\ \rho_Y == \rho_A,\ \rho_B == \rho_A\}$, while

`MassConservationRelations[{"Lotka-Volterra"},`
`    ExternalSpecies -> {"A", "B"}]`

gives `False`.

Although `GammaLeftNullSpace` gives some of the first integrals, in some cases it may happen to provide a strictly positive first integral, thereby showing that the reaction is mass conserving. For example,

`GammaLeftNullSpace[{"Triangle"},`
`    {a0, b0, c0}, {a, b, c}]`

gives $\{a + b + c == a0 + b0 + c0\}$. Adding reversed reaction steps does not change the results, as it is shown by the fact that

`GammaLeftNullspace[ToReversible[{"Triangle"}],`
`    {a0, b0, c0}, {a, b, c}]`

gives the same result. If one is only interested in the fact whether the reaction is mass conserving or not, then one can use the function `MQ` (see Problem 4.3).

Analogs of Theorem 4.7 can be derived to characterize (strongly) mass producing and consuming reactions as well. We show one example and leave the derivation of the other three to the reader.

Theorem 4.8 *A reaction is mass producing if and only if*

$$\max_{\mu\in(\mathbb{R}_0^+)^M,\lambda\in\mathbb{R}_0^+,\rho\in\mathbb{R}^M}\left\{\lambda\mid\rho^\top\gamma=\mu^\top,\ \mathbf{1}^\top\mu\geq\lambda,\ \rho\geq\mathbf{1}\lambda,\ \mathbf{1}^\top\rho=1\right\}>0.$$
(4.9)

Example 4.9 Consider the Lotka–Volterra reaction from Example 2.4. As the answer to `MassConservationRelations` has shown, this reaction is not mass conserving, the linear program (4.8)—in the present case

$$\max_{\lambda\in\mathbb{R},\begin{bmatrix}\rho_1\\\rho_2\end{bmatrix}\in\mathbb{R}^2}\{\lambda\mid\rho_1=0,\ \rho_2-\rho_1=0,\ -\rho_2=0,\ \rho_1\geq\lambda,\ \rho_2\geq\lambda,\ \rho_1+\rho_2=1\}>0.$$
(4.10)

—is infeasible, and the domain of the function (actually, $\rho \mapsto \lambda(\rho)$) is empty.

On the other hand, observe that—no matter what ρ is—the sum of the three coordinates of $\rho^\top \gamma$ is 0:

$$\begin{bmatrix} \rho_1 & \rho_2 \end{bmatrix} \begin{bmatrix} 1 & -1 & 0 \\ 0 & 1 & -1 \end{bmatrix} \begin{bmatrix} 1 \\ 1 \\ 1 \end{bmatrix} = \rho_1 - \rho_1 + \rho_2 - \rho_2 = 0, \tag{4.11}$$

hence no ρ can satisfy any of the Eqs. (4.4)–(4.7). Consequently, the reaction is neither mass producing nor mass consuming. Equation (4.11) is merely the consequence of the fact that

$$\begin{bmatrix} 1 & -1 & 0 \\ 0 & 1 & -1 \end{bmatrix} \begin{bmatrix} 1 \\ 1 \\ 1 \end{bmatrix} = \begin{bmatrix} 0 \\ 0 \end{bmatrix}. \tag{4.12}$$

Example 4.10 The methanol–formic acid esterification

$$HCOOH + CH_3OH \longrightarrow HCOOCH_3 + H_2O \tag{4.13}$$

is clearly mass conserving, since the numbers of hydrogen, oxygen, and carbon atoms are all preserved. Note, however, that the corresponding ρ vectors in Definition 4.2 are

$$\rho_H^\top = \begin{bmatrix} 2 & 4 & 4 & 2 \end{bmatrix}, \quad \rho_O^\top = \begin{bmatrix} 2 & 1 & 2 & 1 \end{bmatrix}, \quad \text{and} \quad \rho_C^\top = \begin{bmatrix} 1 & 1 & 2 & 0 \end{bmatrix},$$

of which only two satisfy (4.3); ρ_C is not coordinatewise positive. Nevertheless, a third conservation relation, independent from the first two, can be obtained by taking, for example, $\rho_C + \rho_H$ or $\rho_C + \rho_O$ or by observing that the total number of atomic constituents remains unchanged, hence $\rho = \rho_C + \rho_H + \rho_O = \begin{bmatrix} 5 & 6 & 8 & 3 \end{bmatrix}^\top$ is also a conservation relation.

These are not the only (or even the simplest possible) conservation relations of (4.13). Allowing, for the moment, zero coordinates, the vectors

$$\rho_{COH}^\top = \begin{bmatrix} 1 & 0 & 1 & 0 \end{bmatrix}, \quad \rho_{CH_3}^\top = \begin{bmatrix} 0 & 1 & 1 & 0 \end{bmatrix}, \quad \text{and} \quad \rho_{OH}^\top = \begin{bmatrix} 0 & 1 & 0 & 1 \end{bmatrix}, \tag{4.14}$$

which correspond to the conservation of the formyl, methyl, and hydroxyl groups, form another collection of three linearly independent conservation relations, from which we can obtain positive vectors by taking linear combinations of them with positive coefficients, also the one

$$\begin{bmatrix} 46 & 32 & 60 & 18 \end{bmatrix} = 46 \begin{bmatrix} 1 & 0 & 1 & 0 \end{bmatrix} + 14 \begin{bmatrix} 0 & 1 & 1 & 0 \end{bmatrix} + 18 \begin{bmatrix} 0 & 1 & 0 & 1 \end{bmatrix},$$

more familiar to the chemist. (To the mathematician, these are the corresponding molecular weights.)

We can formalize our observation that the reaction (4.13) is "obviously" mass conserving, because the atomic constituents are preserved. First, we need to define what we mean by preservation of atomic constituents in a formal mechanism.

Definition 4.11 We say that the reaction (2.1) **obeys the law of atomic balance** if

1. its species are given by their atomic constituents $A(1), \ldots, A(D)$ through the (formal) linear combinations

$$X(m) = Z(1, m)A(1) + \ldots + Z(D, m)A(D) \quad \forall m \in \mathcal{M},$$

2. $\mathbf{Z}$ has no zero rows or columns, meaning that

   - all of the atoms participate in at least one species, and
   - all the species contain at least one atom, and

3. the matrix $\mathbf{Z}$ satisfies $\mathbf{Z}\boldsymbol{\gamma} = \mathbf{0}$ ($\in (\mathbb{N}_0)^{D \times R}$).

The matrix $\mathbf{Z} \in (\mathbb{N}_0)^{D \times M}$ above is called the **atomic matrix** of the reaction.

Example 4.12 In Example 4.10 $R = 1$, $M = 4$, and each species consists of $D = 3$ constituents; if $A(1) = \mathrm{H}$, $A(2) = \mathrm{O}$, and $A(3) = \mathrm{C}$, then the atomic matrix is

$$\mathbf{Z} = \begin{bmatrix} 2 & 4 & 4 & 2 \\ 2 & 1 & 2 & 1 \\ 1 & 1 & 2 & 0 \end{bmatrix}. \text{ We can see that } \mathbf{Z}\boldsymbol{\gamma} = \begin{bmatrix} 0 \\ 0 \\ 0 \end{bmatrix}, \text{ meaning that the number of each of}$$

three atomic constituents is preserved in the reaction.

Example 4.13 How to do the same with the program? The nice table

	Molecules			
Atoms	HCOOH	CH$_3$OH	HCOOCH$_3$	H$_2$O
C	1	1	2	0
H	2	4	4	2
O	2	1	2	1
Charge	0	0	0	0

is obtained as the output to the input

```
ToAtomMatrix[{"HCOOH", "CH3OH", "HCOOCH3", "H2O"},
    FormattedOutput -> True, Frame -> All].
```

To be sure we also check if the law of atomic balance holds:

```
AtomConservingQ[{"HCOOH"+"CH3OH" -> "HCOOCH3"+"H2O"}].
```

The answer is a reassuring `True`. Let us note that the inverse of `ToAtomMatrix` is `FromAtomMatrix`; try it.

Theorem 4.14 *If a reaction obeys the law of atomic balance, then it is mass conserving.*

Proof The observation that the total number of atomic constituents is preserved translates to the relation $\mathbf{Z}\boldsymbol{\gamma} = \mathbf{0}$. With the choice $\boldsymbol{\rho} := \mathbf{Z}^\top \mathbf{1}$, the law of atomic balance implies that $\boldsymbol{\rho}^\top \boldsymbol{\gamma} = \mathbf{1}^\top \mathbf{Z}\boldsymbol{\gamma} = \mathbf{0}$ and $\boldsymbol{\rho} \geq \mathbf{1} > \mathbf{0}$, since each species has at least one atomic constituent.   □

We remark that if the species have other, non-atomic constituents, charge in particular, the above definition and theorem need to be extended carefully, as negative charge translates to negative entries in the atomic matrix. On the other hand, the total number of electrons will be both positive and preserved in the reaction.

Also note that in some complex reactions, in particular in biochemistry, the precise structure of some of the macromolecules are unknown. This underlines the importance of atom-free models which do not clearly obey the law of atomic balance.

Example 4.10 raises a number of further questions: How many essentially different conservation relations can we find for a mass conserving reaction? What are the simplest ones? Can we characterize all of them? We will return to these questions in Sect. 4.3.

We close this section with a few conditions that guarantee based on the structure of a reaction that it is mass conserving, producing, or consuming. Let us start with a simple observation.

Theorem 4.15 *If a reaction is reversible, then it is neither mass producing nor mass consuming.*

Proof If for some reaction step r and weight vector $\boldsymbol{\rho}$, we have $\boldsymbol{\rho}^\top \boldsymbol{\gamma}(\cdot, r) > \mathbf{0}$, then for the converse reaction r', we have $\boldsymbol{\rho}^\top \boldsymbol{\gamma}(\cdot, r') < \mathbf{0}$ and vice versa. This rules out both $\boldsymbol{\rho}^\top \boldsymbol{\gamma} \gneq \mathbf{0}^\top$ and $\boldsymbol{\rho}^\top \boldsymbol{\gamma} \lneq \mathbf{0}^\top$.   □

As a partial converse, we may consider reactions whose Volpert graph is acyclic:

Theorem 4.16 *Consider a reaction with an acyclic Volpert graph. Then*

- *if none of the reactant complexes is empty, then the reaction is strongly mass consuming;*
- *if none of the product complexes is empty, then the reaction is strongly mass producing.*

We omit the proof, which can be found in Volpert and Hudyaev (1985, pp. 624–626). On the other hand, the property of being mass conserving is preserved under the reversion of reaction steps.

Example 4.17

- The reaction $X + Y \longrightarrow U \quad Y + Z \longrightarrow V$ taken from Volpert and Hudyaev (1985, p. 627), has an acyclic Volpert graph, the empty complex is not present among the reactant complexes, and it is therefore strongly mass consuming. One can choose $\rho = \begin{bmatrix} 1 & 1 & 1 & 1 & 1 \end{bmatrix}^\top$ to show the fulfilment of (4.7). Note that the reaction is also (strongly) mass conserving; choose $\rho = \begin{bmatrix} 1 & 1 & 1 & 2 & 2 \end{bmatrix}^\top$ to show the fulfilment of (4.3). What is more, it is also strongly mass producing: choose $\rho = \begin{bmatrix} 1 & 1 & 1 & 3 & 3 \end{bmatrix}^\top$.
- The modified reaction $0 \longrightarrow U \quad X + Y \longrightarrow U \quad Y + Z \longrightarrow V$ also has an acyclic Volpert graph:

```
AcyclicVolpertGraphQ[{X + Y -> U, Y + Z -> V}]
```

and it is neither strongly mass consuming nor strongly mass conserving. It is however weakly mass conserving; choose the nonnegative vector $\rho = \begin{bmatrix} 0 & 0 & 1 & 0 & 1 \end{bmatrix}$ to show the fulfilment of (4.3).
- The modified reaction $U \longrightarrow 0 \quad X + Y \longrightarrow U \quad Y + Z \longrightarrow V$ also has an acyclic Volpert graph, and it is neither strongly mass producing nor strongly mass conserving.
- The examples suggest that the requirement of acyclicity is quite stringent.

Theorem 4.18 *A reaction $\langle M, R, \alpha, \beta \rangle$ is mass conserving if and only if the reaction $\langle M, R', \alpha, \beta \rangle$ obtained by making every reaction step reversible is mass conserving.*

The following two theorems are from Deák et al. (1992).

Theorem 4.19 *A generalized compartmental system in the narrow sense is mass conserving if and only if it is closed.*

Theorem 4.20 *Consider a first-order reaction (2.1) with $M = R \geq 2$ and $\alpha = I_M$. This reaction is mass conserving if and only if, after an appropriate permutation of the species, the matrix β has the structure*

$$\beta = \begin{bmatrix} P & A \\ 0 & B \end{bmatrix},$$

where β is a matrix with no zero columns, P is a $J \times J$ permutation matrix with some $2 \leq J \leq M$, and B is an upper triangular matrix with zeros in its diagonal and the entries of A are arbitrary.

Remark 4.21 The form of $\boldsymbol{\beta}$ means that the first J species are only transformed into each other, and the species with indices $J + 1, J + 2, \ldots, M$ only react to give species with smaller indices, i.e., we have reaction steps of the following form:

$$X(m) \longrightarrow X(p) \quad (m = 1, 2, \ldots, J; 1 \leq p \leq J) \qquad (4.15)$$

$$X(q) \longrightarrow \sum_{r=1}^{q-1} \beta(r, q)X(r) \quad (q = J + 1, J + 2, \ldots, M). \qquad (4.16)$$

4.2   Mass Conservation and the Stoichiometric Subspace

Mass conserving can be described using less intuitive, more mathematical concepts.

4.2.1   The Stoichiometric Subspace and the Reaction Simplex

Using the notion of stoichiometric subspace (Definition 2.2), one can say, by rephrasing Definition 4.2, that the reaction (2.1) is mass conserving if the orthogonal complement of the stoichiometric subspace contains a positive vector. We will need a further definition.

Definition 4.22 For every $\mathbf{c}_0 \in (\mathbb{R}_0^+)^M \setminus \{0\}$, the set $(\mathbf{c}_0 + \mathcal{S}) \cap (\mathbb{R}_0^+)^M$ is called the **reaction simplex** of $\mathbf{c}_0$ or the **stoichiometric compatibility class** containing the point $\mathbf{c}_0$. Furthermore, $(\mathbf{c}_0 + \mathcal{S}) \cap (\mathbb{R}^+)^M$ is the **positive reaction simplex**.

A fundamental theoretical result by Horn and Jackson (1972) follows.

Theorem 4.23 *A positive reaction simplex is bounded if and only if the reaction is (stoichiometrically) mass conserving.*

Proof

- Assume that the reaction simplex is given by $\mathbf{c}_0 \in (\mathbb{R}^+)^M$. If the system is mass conserving, then there exists $\boldsymbol{\rho} \in (\mathbb{R}^+)^M$ orthogonal to $\mathcal{S}$, meaning that $\boldsymbol{\rho}^\top(\mathbf{x} - \mathbf{c}_0) = 0$ holds for all points $\mathbf{x}$ in the reaction simplex. This, however, implies that $\sum_{m \in \mathcal{M}} \rho_m x_m =: K > 0$. Therefore (with $\rho_{\min} := \min_{m \in \mathcal{M}} \rho_m$) one has $\rho_{\min} \sum_{m \in \mathcal{M}} x_m \leq K$ implying $0 \leq x_m \leq \frac{K}{\rho_{\min}}$, i.e., the reaction simplex is bounded.
- If the simplex is bounded, then it must not contain a nonzero vector with nonnegative coordinates. (If, on the contrary, there existed a nonzero vector $0 \leq \mathbf{x} \in \mathcal{S}$, then $\mathbf{c}_0 + \lambda\mathbf{x}$ would also belong to the simplex for arbitrary large positive λ leading to the contradiction with the fact that the simplex is bounded.) Thus if $\mathcal{S}$ is bounded, then a nonnegative vector in $\mathcal{S}$ can only

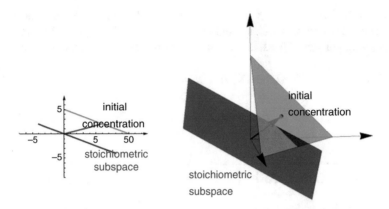

Fig. 4.1 Reaction simplexes of the reaction $2\,X \rightleftharpoons Y$ (left) and that of the Ivanova reaction or of the triangle reaction (right)

be the zero vector. By Theorem 13.22 (Tucker), it is known that there exist nonnegative vectors in $\mathscr{S}$ and $\mathscr{S}^{\top}$, respectively, so that their sum has only positive coordinates. In our case one of the vectors is necessarily zero; therefore the other one only has positive coordinates, and it also belongs to $\mathscr{S}^{\top}$, as it is required in the definition of mass conservation.

- Assume that there is a bounded reaction simplex. Then there exists a positive vector in the orthogonal complement of $\mathscr{S}$; thus the first part of the proof shows that every reaction simplex is bounded.

$\qquad\qquad\qquad\qquad\qquad\qquad\qquad\qquad\qquad\qquad\qquad\qquad\qquad\qquad$ □

Thus, it is enough to check whether one of the simplexes is bounded or not (Fig. 4.1).

Example 4.24 Considering the Lotka–Volterra reaction which is not mass conserving, one can see that all the reaction simplexes are the same unbounded set, the whole first quadrant.

To motivate what follows, recall the last part of Example 4.9. We argued that the reaction cannot be strongly mass producing (or consuming), that is, the vectors ρ of (4.5) and (4.7) **do not exist** by showing that **there exists** a vector z satisfying another linear system, namely, $\gamma z = 0$. This condition was sufficient (although not necessary!) for the reaction to be neither strongly mass producing nor strongly mass consuming. We shall generalize this idea and find similar conditions that are both necessary and sufficient to characterize mass conserving, producing, and consuming reactions. Claims of this type are known in the theory of convex polyhedra as **theorems of the alternative** (connected to such names as Gyula Farkas and Erik Ivar Fredholm), as they assert that one linear system has a solution precisely when another one does not.

There is a greater than simply grammatical distinction between characterizing something by the existence of a solution, rather than by the non-existence of a solution. If
`ReactionKinetics`, or anyone, gives us a solution to a linear system of inequalities,
we can plug that solution back into the inequalities to confirm that it is correct. But what if
we are told that there is no solution? How do we verify that this assertion is correct?

We remind the reader that Gyula Farkas (1847–1930) is known for Farkas' lemma which is
used in linear programming and also for his work on linear inequalities. He had also worked
on the conditions of stability of thermodynamic equilibrium. Our talented young colleague
with the same name, Gyula Farkas (1972–2002), died at the very beginning of his career,
although he has already published almost 30 papers including a few ones on controllability
and observability of reactions, see Farkas (1998a, 1999, 1998b). His main interest was in
bifurcation theory and delay differential equations.

The mass conserving case is the easiest; the remaining cases are discussed in the
next section.

Theorem 4.25   *A reaction is **not** mass conserving if and only if*

$$\exists \mathbf{z} \in \mathbb{R}^R \ \text{satisfying} \ \boldsymbol{\gamma}\mathbf{z} \gneq 0.$$

4.2.2   The Stoichiometric Cone

Evaluating the possible overall effect of a sequence of elementary steps, it is natural
to consider not only the stoichiometric subspace but also its subset that is given by
the **nonnegative** linear combinations of the columns of the stoichiometric matrix.

Definition 4.26   The **stoichiometric cone** of a reaction is the **convex cone** generated by the columns of its stoichiometric matrix, i.e., it is the set

$$\mathscr{K}(\boldsymbol{\gamma}) = \left\{ \sum_{r \in \mathscr{R}} \lambda_r \boldsymbol{\gamma}(\cdot, r) \ \middle| \ \lambda_1, \dots, \lambda_R \geq 0 \right\}. \tag{4.17}$$

The stoichiometric cone of the triangle reaction or that of the Ivanova reaction is
the whole first octant. In Fig. 4.2 two examples are shown.

Clearly, if in a sequence of elementary steps the quantity of each species
increases, the reaction should be considered strongly mass producing. The precise
characterization of strongly mass producing and consuming reactions using the
stoichiometric cone is slightly more convoluted.

Theorem 4.27 (Deák et al. (1992))   *A reaction is **not** mass producing if and only if*

$$\exists \mathbf{y} \lneq 0 \ \text{satisfying} \ \mathbf{y} \in \mathscr{K}(\boldsymbol{\gamma}) \quad \text{or} \quad \exists \mathbf{z} > 0 \ \text{satisfying} \ \boldsymbol{\gamma}\mathbf{z} = 0.$$

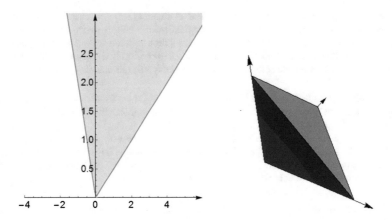

Fig. 4.2   The stoichiometric cone of the reaction $X \longrightarrow 2\,Y \quad X \longrightarrow 3\,X + Y$ and of the triangle reaction

*A reaction is **not** strongly mass producing if and only if*

$$\exists\, \mathbf{y} \lneqq 0 \ satisfying \ \mathbf{y} \in \mathscr{K}(\boldsymbol{\gamma}) \quad or \quad \exists\, \mathbf{z} \gneqq 0 \ satisfying \ \boldsymbol{\gamma}\mathbf{z} = 0.$$

*A reaction is **not** mass consuming if and only if*

$$\exists\, \mathbf{y} \gneqq 0 \ satisfying \ \mathbf{y} \in \mathscr{K}(\boldsymbol{\gamma}) \quad or \quad \exists\, \mathbf{z} > 0 \ satisfying \ \boldsymbol{\gamma}\mathbf{z} = 0.$$

*A reaction is **not** strongly mass consuming if and only if*

$$\exists\, \mathbf{y} \gneqq 0 \ satisfying \ \mathbf{y} \in \mathscr{K}(\boldsymbol{\gamma}) \quad or \quad \exists\, \mathbf{z} \gneqq 0 \ satisfying \ \boldsymbol{\gamma}\mathbf{z} = 0.$$

*The "**or**"s in the above characterizations are **not** exclusive, that is, both the asserted* **y** *and* **z** *may exist simultaneously.*

4.3   Finding Conservation Relations

In this section we answer the questions of how many such relations may exist and how to find the simplest ones.

Equation (4.3) immediately gives an upper bound: since each $\boldsymbol{\rho}$ is from the orthogonal complement of $\mathscr{S}$, the number of linearly independent conservation relations cannot exceed $\dim(\mathscr{S}^{\perp}) = M - \mathrm{rank}(\boldsymbol{\gamma})$ by Remark 13.12. It is not hard to see that this is the exact number for every mass conserving reaction.

Theorem 4.28 *For every reaction the number of linearly independent conservation relations is either* 0 *or* $M - \mathrm{rank}(\boldsymbol{\gamma})$.

Proof Let us use the notation $Q := M - \text{rank}(\boldsymbol{\gamma})$. We can find a basis of $\mathscr{S}^\perp$, that is, vectors $\mathbf{r}_1, \ldots, \mathbf{r}_Q$ which satisfy $\mathbf{r}_q^\top \boldsymbol{\gamma} = 0$ $(q = 1, 2, \ldots, Q)$, by row reducing the stoichiometric matrix. These basis vectors may or may not be coordinatewise positive. If one of the vectors, say $\mathbf{r}_1$, is positive, then we can replace each $\mathbf{r}_q$ by $\boldsymbol{\rho}_q := \mathbf{r}_1 + \varepsilon \mathbf{r}_q$ with a small enough $\varepsilon > 0$ to obtain a new basis of $\mathscr{S}^\perp$ that consists of componentwise positive vectors.   □

The above proof, with some streamlining, can be turned into an algorithm (see Problem 4.3) that finds $M - \text{rank}(\boldsymbol{\gamma})$ independent conservation relations if such a collection exists. The first such relation (which plays the role of $\mathbf{r}_1$ in the proof) can be found using Theorem 4.7. We can then extend $\{\mathbf{r}_1\}$ to a complete basis $\{\mathbf{r}_1, \ldots, \mathbf{r}_Q\}$ of $\mathscr{S}^\perp$ and then construct $\{\boldsymbol{\rho}_1, \ldots, \boldsymbol{\rho}_Q\}$ following the proof.

As Example 4.10 demonstrates, some relations can be obtained by taking nonnegative linear combinations of simpler ones. To be able to talk meaningfully about the "simplest" relations, we need to allow nonnegative coordinates in conservation relations. For the remainder of this chapter, we say that a vector $\boldsymbol{\rho}$ is a **nonnegative mass conservation relation** for a reaction if $\boldsymbol{\rho} \gneq 0$ and $\boldsymbol{\rho}^\top \boldsymbol{\gamma} = 0^\top$.

Observe that if $\boldsymbol{\rho}_1$ and $\boldsymbol{\rho}_2$ are two conservation relations, then by definition every nonzero vector of the form $\boldsymbol{\rho} = \lambda_1 \boldsymbol{\rho}_1 + \lambda_2 \boldsymbol{\rho}_2$ with $\lambda_1 \geq 0$ and $\lambda_1 \geq 0$ is also a conservation relation. (Sets of vectors constructed this way are called **convex cones**.) If both λ_i are nonzero, $\boldsymbol{\rho}$ can be decomposed into two "simpler" relations $\boldsymbol{\rho}_1$ and $\boldsymbol{\rho}_2$. It can be shown that this way the notion of "simplest" conservation relation is well-defined, moreover, that there always exists a unique finite collection of simplest conservation relations.

Theorem 4.29 *For every mass conserving reaction, there exist finitely many nonnegative conservation relations $\boldsymbol{\rho}_1, \ldots, \boldsymbol{\rho}_K$ such that every nonnegative conservation relation $\boldsymbol{\rho}$ can be written as $\boldsymbol{\rho} = \sum_{i=1}^K \lambda_i \boldsymbol{\rho}_i$ for some $\lambda_1, \ldots, \lambda_K \geq 0$. The smallest such collection is unique.*

The unique smallest collection $\boldsymbol{\rho}_1, \ldots, \boldsymbol{\rho}_K$ is called the **generator** of the conservation relations. It is possible that $K > M - \text{rank}(\boldsymbol{\gamma})$, in fact, the difference of the two sides can be very large. Consequently, finding the generator may be a huge undertaking for complex reactions with many (dozens or hundreds of) elementary steps. This is in sharp contrast with the fact that it takes very little time to confirm that the reaction is mass conserving and to find one conservation relation. Its detailed discussion is beyond the scope of this book; we only mention that it is an adaptation of the **Fourier–Motzkin elimination** algorithm from the theory of convex polyhedra (Schrijver 1998, pp. 155–156).

4.4   Exercises and Problems

4.1 Show by example that a mass producing reaction may not be strongly mass producing and a mass consuming reaction may not be strongly mass consuming.

(Solution: page 388.)

4.2 Is it possible that a reaction is simultaneously strongly mass producing and strongly mass consuming? (Note: it is not sufficient to argue that the **same** ρ cannot satisfy both (4.5) and (4.7).)

(Solution: page 389.)

4.3 Use either the function **Maximize** or **LinearProgramming** to implement a one-line *Mathematica* function that checks if a stoichiometric matrix corresponds to a mass conserving reaction. Test it on the reaction

$$A \longrightarrow 2\,M + N_2 \quad A + M \longrightarrow CH_4 + B \quad 2\,M \longrightarrow C_2H_6$$

$$M + B \longrightarrow MED \qquad M + A \longrightarrow C \qquad M + C \longrightarrow TMH$$

(decomposition of azomethane from Smith and Missen 2003) and also on the reaction $A + B \rightleftharpoons C \longrightarrow 2\,D \rightleftharpoons B + E$. Verify that the number of linearly independent conservation relations is $M - \mathrm{rank}(\boldsymbol{\gamma})$.

(Solution: page 389.)

4.4 This problem explores another simple necessary condition for a reaction to be mass conserving.

1. Show that pivoting on the transpose of the stoichiometric matrix does not change whether the reaction corresponding to the matrix is mass conserving or not.
2. Using this observation and the fact that the nonzero columns of the stoichiometric matrix of a mass conserving reaction must have at least one negative and at least one positive entry, Ridler et al. (1977) suggested the following procedure to test if a reaction is mass conserving: find a row echelon form of $\boldsymbol{\gamma}^\top$ by successive pivoting, and verify before each pivot step if the row in which we are to pivot has both a positive and a negative entry. If a row violating this condition is found in the process, the reaction is not mass conserving. Use this test to verify that the reaction

$$A \rightleftharpoons B \rightleftharpoons C \rightleftharpoons A + 2\,D$$

is not mass conserving.

3. If the procedure terminates without finding a violating row, the test is inconclusive, as the following example from Oliver (1980) shows:

$$A \longrightarrow B + D \quad C \longrightarrow A + D \quad C \longrightarrow D.$$

Show that this reaction is not mass conserving, but it passes the above test.

(Solution: page 390.)

4.5 If the Volpert graph of a mechanism is acyclic, then the inequality

$$\rho^\top \gamma < \mathbf{0} \in \mathbb{R}^R \tag{4.18}$$

has a positive solution.

(Solution: page 390.)

References

Deák J, Tóth J, Vizvári B (1992) Anyagmegmaradás összetett kémiai mechanizmusokban (Mass conservation in complex chemical mechanisms). Alk Mat Lapok 16(1–2):73–97

Farkas G (1998a) Local controllability of reactions. J Math Chem 24:1–14

Farkas G (1998b) On local observability of chemical systems. J Math Chem 24:15–22

Farkas G (1999) Kinetic lumping schemes. Chem Eng Sci 54:3909–3915

Horn F, Jackson R (1972) General mass action kinetics. Arch Ratl Mech Anal 47:81–116

Oliver P (1980) Consistency of a set of chemical reactions. Int J Chem Kinet 12(8):509–517

Ridler GM, Ridler PF, Sheppard JG (1977) A systematic method of checking of systems of chemical equations for mass balance. J Phys Chem 81(25):2435–2437

Schrijver A (1998) Theory of linear and integer programming. Wiley, Chichester

Schuster S, Höfer T (1991) Determining all extreme semi-positive conservation relations in chemical reaction systems—a test criterion for conservativity. J Chem Soc Faraday Trans 87(16):2561–2566

Smith WR, Missen RW (2003) Mass conservation implications of a reaction mechanism. J Chem Educ 80(7):833–838

Volpert AI, Hudyaev S (1985) Analyses in classes of discontinuous functions and equations of mathematical physics. Martinus Nijhoff, Dordrecht. Russian original: 1975

Decomposition of Reactions

5

5.1 The Problem: Construction via Deconstruction

Complex reactions occur via pathways of simple reaction steps. It is a fundamental problem of stoichiometry that the overall (or global) reaction is measured, and one would like to reconstruct the underlying network of simple (or elementary) reaction steps. This is a very complex problem with no foolproof recipes for its solution. Epstein and Pojman (1998) devote a chapter of their book to the "art of constructing mechanisms"; in Sect. 5.3 they provide a nine-step procedure that serves as a template for (re)constructing reactions with desired properties:

1. Assemble all relevant experimental data that the mechanism should be able to model. The wider the range of phenomena and conditions, the more believable the mechanism that explains them.
2. Identify the primary stoichiometry of the overall reaction and of any major component processes.
3. Compile a list of chemically plausible species—reactants, products, and intermediates— that are likely to be involved in the reaction.
4. Obtain all available thermodynamic data pertaining to these species.
5. Break down the overall reaction into component processes, consider the likely elementary steps in each of these processes, and examine the literature critically for kinetics data relevant to as many of these processes and steps as can be found.
6. Use known rate constants wherever possible. If none exists, guess a value, or try to isolate from the main reaction a smaller subset of reactions and determine their rate laws.
7. Put all the thermodynamically plausible steps and the corresponding kinetics data together to form a trial mechanism. Use analytic methods, for example, the steady-state approximation, as well as chemical intuition, to derive a qualitative picture of how the system works.
8. Use numerical methods, such as numerical integration and bifurcation analysis, to simulate the experimental results. If serious discrepancies arise, consider the possibility of missing steps, erroneous data, or false assumptions, and return to the previous steps to reformulate the mechanism.

© Springer Science+Business Media, LLC, part of Springer Nature 2018
J. Tóth et al., *Reaction Kinetics: Exercises, Programs and Theorems*,
https://doi.org/10.1007/978-1-4939-8643-9_5

9. Continue to refine and improve the mechanism, testing it against all new experimental results, particularly those carried out under conditions very different from those of the data used to construct the mechanism. The greatest success of a mechanism is not to give agreement with data already in hand, but to predict successfully the results of experiments that have not yet been carried out.

The authors also show examples demonstrating how these principles can be applied in practice.

Several steps of this process are purely mathematical or can be accelerated by combinatorial considerations, utilizing the graphs of Chap. 3 and some additional tools that will be developed in this chapter. A simplified template for the decomposition of a reaction to elementary steps may consist of the following steps:

1. determine the combinatorially possible species and select the chemically acceptable ones,
2. determine the combinatorially possible reaction steps, and select the chemically acceptable ones,
3. find those decompositions of the given overall reaction which are combinatorially feasible, and then select the chemically acceptable ones.

Besides these steps a number of related questions will be raised and answered below, including methods to find the minimal cycles of a reaction.

The greatest difficulty in coming up with meaningful mechanisms is to incorporate in the procedure all domain-specific expertise, all available thermodynamic data about the participating species, and experimental data about the mechanism itself. A more reasonable goal is to provide a generous list of (perhaps implausible or even impossible) solutions, all of which adhere to the basic conservation laws (e.g. all reaction steps must conserve the atomic constituents and charge) and all other combinatorial constraints imposed by the problem (e.g. all steps must be of order at most two); and leave all further processing to the chemist. At this abstract level, the last two steps of the above template are almost equivalent, as we shall see immediately.

5.2   Reaction Steps and Decompositions

Suppose that a reaction step called the **overall reaction** is given, which we wish to decompose to simpler (in some sense elementary) steps. We are also given a list of species that are reactants, products, or possible intermediate species (in the chemical sense) in the reaction. We assume that the atomic constituents of these species are known or that at least the species are represented by constituents that are conserved in every reaction step. There is no conceptual difference between the two forms of this assumption, so we will assume that the species are given by an atomic matrix as in Definition 4.11. Note that in this spirit charge is also considered to be an "atomic" constituent.

5.2.1   Constructing Elementary Reaction Steps

The definition of "elementary" steps varies, for our purposes we fix it the following way.

Definition 5.1 A reaction step is said to be an **elementary reaction step** if it is of order at most two.

Our first goal is to construct all elementary steps that might take place between a set of species that are given by their atomic matrix. Thermodynamic considerations are ignored in this search; instead, we are interested in finding all hypothetical steps that obey the law of atomic (and charge) balance. (Recall Definition 4.11.)

Proposition 5.2 *The number of possible reactant complexes of elementary steps is limited to* $2M + \binom{M}{2}$.

Proof The total number of all possible nonempty short complexes is the number of monomolecular complexes $X(1), X(2), \ldots, X(M)$; plus the number of homobimolecular complexes $2X(1), 2X(2), \ldots, 2X(M)$; plus the number of the heterobimolecular complexes $X(1) + X(2), \ldots, X(M - 1) + X(M)$. The empty complex is excluded if mass conservation is to be kept.   □

To generate all elementary steps, it suffices to find all possible product complexes for each of these reactant complexes. A D-dimensional integer vector, the **atomic vector**, describing the atomic structure of the complex is associated with each reactant complex the same way as the column $\mathbf{Z}(\cdot, m)$ of the atomic matrix (Definition 4.11) corresponds to the mth species. If $\mathbf{d}$ is the atomic vector of a reactant complex, then finding all product complexes amounts to finding all (nonnegative integer) solutions $\mathbf{x} \in \mathbb{N}_0^M$ of the system of linear equations

$$\mathbf{Z}\mathbf{x} = \mathbf{d}. \tag{5.1}$$

To rule out solutions in which the same species appears on both sides (which may be called a **direct catalyst**), we can modify the above equation by deleting those columns of $\mathbf{Z}$ that correspond to the species of the reactant complex.

> Equations like (5.1) are sometimes called **linear Diophantine equations**, where Diophantine refers to the property that only nonnegative integer solutions are sought, after the Greek mathematician **Diophantus of Alexandria** who was one of the first mathematicians to study such equations.

Example 5.3 To determine the set of combinatorially feasible elementary reaction steps between the six species $\{H, O_2, OH, O, H_2, H_2O\}$, we need to solve $6 + 6 + \binom{6}{2} = 27$ systems of small linear Diophantine equations corresponding to the 27 nonempty reactant complexes. Even this tiny problem has altogether 53 solutions, although not without seeming redundancy: By our definition the reaction step $2\,O_2 \longrightarrow 4\,O$ is an elementary reaction step, and will be among the solutions, along with $O_2 \longrightarrow 2\,O$. Note that these two steps are stoichiometrically equivalent, although they are different from the point of view of kinetics.

Let us write down the equations and the solutions, e.g., for the reactant complex $OH + H_2$. If the first component of the atomic vectors corresponds to H and the second one to O, then $\mathbf{d} = \begin{bmatrix} 3 \\ 1 \end{bmatrix}$, and $\mathbf{Z} = \begin{bmatrix} 1 & 0 & 1 & 0 & 2 & 2 \\ 0 & 2 & 1 & 1 & 0 & 1 \end{bmatrix}$; therefore Eq. (5.1) reads as

$$x_1 + x_3 + 2x_5 + 2x_6 = 3, \quad 2x_2 + x_3 + x_4 + x_6 = 1.$$

The general solution to this system is

$$x_1 = 3 - x_3 - 2x_5 - 2x_6, \quad x_2 = \frac{1}{2}(1 - x_3 - x_4 - x_6).$$

Considering that each x_i must be a nonnegative integer, the solutions can be easily enumerated. We obtain five solutions,

$$\begin{bmatrix} 1 & 0 & 0 & 0 & 0 & 1 \end{bmatrix}^\mathsf{T}, \begin{bmatrix} 3 & 0 & 0 & 1 & 0 & 0 \end{bmatrix}^\mathsf{T},$$

$$\begin{bmatrix} 1 & 0 & 0 & 1 & 1 & 0 \end{bmatrix}^\mathsf{T} \begin{bmatrix} 2 & 0 & 1 & 0 & 0 & 0 \end{bmatrix}^\mathsf{T},$$

$$\begin{bmatrix} 0 & 0 & 1 & 0 & 1 & 0 \end{bmatrix}^\mathsf{T},$$

corresponding to the reaction steps

$$OH + H_2 \longrightarrow H + H_2O \quad OH + H_2 \longrightarrow 3\,H + O \quad OH + H_2 \longrightarrow H + O + H_2$$

$$OH + H_2 \longrightarrow 2\,H + OH \quad OH + H_2 \longrightarrow OH + H_2$$

of which we may discard the last three on the basis that they are direct catalytic steps (meaning here that the same species occurs on both sides). The fifth reaction step also violates the second of Conditions 1 in Chap. 2 according to which at least one species should change in each reaction step.

Ruling out direct catalytic steps means that we delete the columns of $\mathbf{Z}$ corresponding to the species in the reactant complex; therefore the system modifies to

$$x_1 + 2x_6 = 3, \quad 2x_2 + x_4 + x_6 = 1,$$

leading to the general solution $x_1 = 3 - 2x_6$, $x_2 = \frac{1}{2}(1 - x_4 - x_6)$, and to the nonnegative integer solutions $\begin{bmatrix} 1 & 0 & 0 & 1 \end{bmatrix}^\mathsf{T}$ and $\begin{bmatrix} 3 & 0 & 1 & 0 \end{bmatrix}^\mathsf{T}$, corresponding to the first and second reaction steps above.

Before discussing algorithms to systematically and efficiently enumerate the solutions of these equations, we show that the decomposition of reactions is another instance of the same general problem.

5.2.2   Decomposition of Overall Reactions to Elementary Steps

Take a nonempty short complex $\mathbf{y}$ with atomic vector $\mathbf{d}$. Then each solution $\mathbf{x}$ of (5.1) determines an elementary reaction with reactant complex $\mathbf{y}$; the corresponding reaction vector is $\mathbf{x} - \mathbf{y} \in \mathbb{Z}^M$. We can construct a reaction whose steps are the elementary steps obtained in the previous section, and its stoichiometric matrix $\boldsymbol{\gamma}$ has $\mathbf{x} - \mathbf{y}$ as one of its columns. Suppose that altogether R elementary steps were found for the given reactant complexes, and they are put as columns into the matrix $\boldsymbol{\gamma}$. A linear combination of these elementary steps, with coefficients $\mathbf{z}$, is a decomposition of the overall reaction obeying the law of atomic (and charge) balance if and only if

$$\boldsymbol{\gamma}\mathbf{z} = \mathbf{b}, \quad \mathbf{z} \in \mathbb{N}_0^R, \tag{5.2}$$

where $\mathbf{b} \in \mathbb{Z}^M$ is the reaction vector of the overall reaction. Again, we are primarily interested in nonnegative integer solutions of this equations, although nonnegative rational solutions also can be interpreted as decompositions. Hence, this is again a linear Diophantine equation.

Example 5.4 Consider the overall reaction $H + O_2 + 3\,H_2 \longrightarrow 3\,H + 2\,H_2O$ and the set of elementary steps

$$H + O_2 \longrightarrow O + OH \quad O + H_2 \longrightarrow H + OH \quad OH + H_2 \longrightarrow H + H_2O$$

$$2\,H \longrightarrow H_2 \quad H + OH \rightleftharpoons H_2O$$

The Eq. (5.2) takes the form

$$\begin{bmatrix} -1 & 1 & 1 & -2 & -1 & 1 \\ -1 & 0 & 0 & 0 & 0 & 0 \\ 1 & -1 & 0 & 0 & 0 & 0 \\ 1 & 1 & -1 & 0 & -1 & 1 \\ 0 & -1 & -1 & 1 & 0 & 0 \\ 0 & 0 & 1 & 0 & 1 & -1 \end{bmatrix} \mathbf{z} = \begin{bmatrix} 2 \\ -1 \\ 0 \\ 0 \\ -3 \\ 2 \end{bmatrix}.$$

The rows correspond to the species in the following order: H, O_2, O, OH, H_2, H_2O. Since $\boldsymbol{\gamma}(\cdot, 5) + \boldsymbol{\gamma}(\cdot, 6) = \mathbf{0}$, the equation has infinitely many nonnegative integer solutions. However, the only **minimal solution** is $\mathbf{z} = \begin{bmatrix} 1 & 1 & 2 & 0 & 0 & 0 \end{bmatrix}^\top$, corresponding to the decomposition

$$H + O_2 + 3\,H_2 \longrightarrow 3\,H + 2\,H_2O \quad = \quad \begin{array}{l} 1 \times H + O_2 \longrightarrow O + OH \\ 1 \times O + H_2 \longrightarrow H + OH \\ 2 \times OH + H_2 \longrightarrow H + H_2O \end{array} \quad .$$

Clearly, Eqs. (5.1) and (5.2) are of very similar form, yet, they possess very different qualities. Note the following differences:

- In elementary step generation, multiple relatively small systems need to be solved, all of which share the coefficient matrix, the atomic matrix, unless we remove the columns corresponding to the reactant complex to avoid the reactant species appearing as products. However, in decomposition a single large-scale system is solved.
- In elementary step generation, all rows but the one corresponding to the electric charge consist of nonnegative numbers; in decomposition every row and column may contain entries of different sign.
- In elementary step generation, the number of solutions is always finite; in decomposition it is often infinite.

Only the last claim is not immediate. To see that there are finitely many elementary reaction steps, note that there are only finitely many reactant complexes, and for each reactant complex, there are only finitely many product complexes with the same atomic constituents. In more mathematical terms, if in Eq. (5.1) the mth species satisfies

$$\mathbf{Z}(\cdot, m) \geq 0 \quad \text{and} \quad Z(d, m) > 0 \tag{5.3}$$

for some $d \in \{1, \ldots, D\}$, then $0 \leq x(m) \leq d(r)/Z(d, m)$, so the mth component of every solution can be shown to be bounded. By the definition of the atomic matrix $\mathbf{Z}$, (5.3) holds for every species except possibly for the free electron. Obviously, if all other components of every solution can be bounded, the number of free electrons can also be bounded. Thus, the elementary reaction steps correspond to integer vectors from a bounded set, proving that there are only finitely many of them.

Elaborating on the potentially infinite number of decompositions, if a sequence of steps forms a cycle in which no species are generated or consumed, then that sequence can be added to any decomposition an arbitrary number of times, leading to an infinite number of different decompositions. A straightforward application of Dickson's lemma 13.15 yields that the converse is also true. For the purposes of this section, we can formally define a **cycle** as a vector $\mathbb{N}_0^R \ni \mathbf{v} \neq \mathbf{0}$ satisfying $\boldsymbol{\gamma}\mathbf{v} = \mathbf{0}$, and with this notion we have the following lemma.

Lemma 5.5 *The number of decompositions is finite if and only if the elementary reactions cannot form a cycle. Furthermore, the number of decompositions that do not contain a cycle is always finite.*

The lemma motivates two further problems related to decompositions: first, one needs to be able to decide whether the elementary reactions can form cycles or not, which amounts to the solution of a linear programming problem (See Sect. 5.3.1). Second, if cycles do exist, the goal changes from generating all decompositions to generating all **minimal** (i.e., cycle-free) decompositions and all **minimal cycles** (i.e., cycles that cannot be expressed as a sum of two cycles), which brings us back to linear Diophantine equations.

Another simple approach to limit the number of solutions is to impose an upper bound on the number of elementary steps.

In the next section, we discuss algorithms for the solution of each of these problems.

5.3 Solving Linear Diophantine Equations

The solution of linear Diophantine equations is very difficult (even finding a *single* solution to such an equation is **NP-hard**), and no generally efficient polynomial algorithm is known for it.

For some questions an algorithm can be found to provide an answer in time which is a polynomial function of the size of the problem; these questions form the complexity class **P**. For some questions an algorithm can be find to check in **polynomial time** in the size of the input if an answer is good or bad; these questions form **NP**. One of the Millenium Problems (the prize for the solution of which would be one million US dollars offered by the Clay Mathematical Institute) is whether **NP = P** or **NP ≠ P**. A problem is **NP-hard** means that it is at least as hard as the hardest problems in **NP**. More precisely, a problem is NP-hard when every problem in **NP** can be reduced in polynomial time to it. As a consequence, finding a polynomial algorithm to solve any NP-hard problem would give polynomial time algorithms for all the problems in **NP**.

Rather than trying to advocate an all-purpose method, we present a few different methods and indicate which one is expected to be the most effective when solving different problems. (The methods we are to present were found to be sufficiently fast for problems up to a few dozen species and few hundred elementary steps.)

5.3.1 Homogeneous Equations: Finding Minimal Cycles of a Mechanism

There is a trivial infinite procedure to find all minimal solutions of a homogeneous linear Diophantine equation or equivalently to find all minimal cycles of a reaction: It consists of enumerating all sequences of steps of length one, two, etc., adding every possible step in every iteration to those sequences which do not yet contain a cycle. In other words, if the equation is $\mathbf{Ax} = 0$, we enumerate $\mathbf{A}(\cdot, 1), \mathbf{A}(\cdot, 2), \ldots,$ then $2\mathbf{A}(\cdot, 1), \mathbf{A}(\cdot, 1) + \mathbf{A}(\cdot, 2), \ldots,$ and so on, each time adding another column to each sum enumerated so far. Whenever a sum totalling the zero vector, i.e., a sequence containing a new cycle is found, the cycle is saved, and the corresponding search direction is discarded, as it may not contain further minimal cycles. The main problem with this approach is that it may never terminate. The algorithm of Contejean and Devie (1994) is a clever modification of this simple procedure which is not only faster but also terminates after a finite number of iterations. The geometric idea behind it is illustrated in Fig. 5.1. For algorithmically versatile people, we show a simple (although not very efficient) *Mathematica* code:

```
MyContejeanDevie[A_] :=
Block[{P, Q, n = Length[A], B = {}},
    P = IdentityMatrix[n];
    While[P =!= {},
        B = Union[B, Select[P, ZeroVectorQ[#.A] &]];
        Q = Select[Complement[P, B], Function[p, And
            @@(!ComponentwiseLessEqualQ[#, p]&/@B)]];
        P = DeleteCases[Union@@Table[
            If[(q.A).(A[[i]]) < 0,
                q + UnitVector[n,i], Null], {q, Q}, {i,n}], Null];
    ]; B
]
```

Mathematica has essentially three structures to collect series of commands into a single function, (called **scoping constructs**) these are **Module**, **Block** and **With**. The reason we have used **Block** here is that it is usually faster than **Module**, and **With** has only restricted capability (with some advantages in other cases.)

We omit the straightforward code of the function **ZeroVectorQ** (which tests whether its argument is the zero vector). **ComponentwiseLessEqualQ** tests whether its first argument is componentwise less than or equal to its second argument:

```
ComponentwiseLessEqualQ[a_, b_]:=And @@ Thread[a <= b],
```

and **UnitVector[n,i]**, which is the n-dimensional ith unit vector.

MyContejeanDevie[A] returns all minimal solutions of the homogeneous system $\mathbf{Ax} = \mathbf{0}$, $\mathbf{x} \in \mathbb{N}_0^R$. Candidate sequences are kept in the list **P**; the minimal solutions are collected in **B**. The key difference from the infinite procedure outlined above is the condition **(q.A).(A[[i]]) < 0**: It means that during the search we only take steps that turn back toward the origin (see Fig. 5.1). Obviously, if a sequence can be extended so that it returns to the origin, then this extension must contain at least one step in this direction, and without loss of generality, such a step can be the next one. Consequently this method will certainly enumerate all solutions.

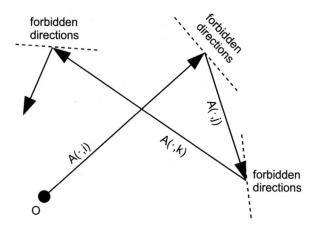

Fig. 5.1 Illustration of the Contejean–Devie algorithm. In each iteration every sequence of steps is extended by steps that turn back toward the origin

Theorem 5.6 (Contejean and Devie (1994)) *The Contejean–Devie algorithm terminates and returns the list of all minimal solutions of* $\mathbf{Ax} = \mathbf{0}$, $\mathbf{x} \in \mathbb{N}_0^R$.

We omit the proof, as it is rather tedious. The difficult part is to show that the algorithm terminates.

This algorithm can also be adapted to solve the inhomogeneous system $\mathbf{Ax} = \mathbf{b}$, $\mathbf{x} \in \mathbb{N}_0^R$. We simply replace $\mathbf{A}$ with $\mathbf{A}' = [\mathbf{b} \ \mathbf{A}]$ and solve the system $\mathbf{A}'\mathbf{x} = \mathbf{0}$ with the above algorithm, except that we never increase the first component of the candidate solutions above 1. Papp and Vizvári (2006) suggested a further improvement for the case when the length of the solution is bounded: If we are only interested in solutions whose components sum to at most n, the set of "forbidden directions" on Fig. 5.1 can be extended.

5.3.2 Enumerating Minimally Dependent Subsets

A problem related to finding minimal cycles is the following: given a matrix $\mathbf{A}$, find all (inclusionwise) **minimal linearly dependent subsets** (or **simplexes**) of its columns. The most obvious application of this problem is the generation of reaction steps: the minimal linearly dependent subsets of the columns of the atomic matrix $\mathbf{Z}$ correspond to reaction steps with the smallest sets of species. Many of these steps are expected to be elementary. Of course, this assumes that steps are reversible and that they do not include direct catalysts.

In Szalkai (1997) a simple elementary method is proposed for the solution of this problem. It relies on two observations: first, recognizing linearly dependent and minimal linearly dependent subsets is easy using linear algebra (Problem 5.2). Second, as we enumerate (some of) the subsets searching for solutions, linearly independent sets of vectors shall be extended by adding another vector, while dependent ones, if they are not minimally dependent, should be shrunk by the removal of a vector. Szalkai's additional observation is that this can be done in a systematic manner that does not even require that we maintain a set of candidate solutions. It enumerates a subset of the vectors, taking obvious shortcuts in such a way that no minimally dependent subset is avoided, yielding a very memory-efficient algorithm.

Example 5.7 Consider the set of species $\{H, O_2, OH, O, H_2, H_2O\}$, with the corresponding atomic matrix $\mathbf{Z} = \begin{bmatrix} 1 & 0 & 1 & 0 & 2 & 2 \\ 0 & 2 & 1 & 1 & 0 & 1 \end{bmatrix}$. The set of columns of the matrix $\mathbf{Z}$ has 63 nonempty subsets, 14 of which are minimal linearly dependent. To get some intuition about how the algorithm works, let us follow the first couple of its steps on this example.

1. We start by taking the set of species $\{H\}$, which is singleton, hence independent. We add the next species, O_2.
2. We have now $\{H, O_2\}$. As its type is still independent, we add the next species, OH.
3. We have now $\{H, O_2, OH\}$. We determine its type, and we find that it is a simplex (minimal linearly dependent); the corresponding reaction step is $2H + O_2 \rightleftharpoons 2OH$.

    Since this was a simplex subset, we proceed by replacing the last species with the next one, O.
4. We have now $\{H, O_2, O\}$. It is dependent, but not minimally. A somewhat counterintuitive step follows: We do not drop a species from the dependent set, but again we proceed by replacing the last species with the next one, H_2. (We do this until there are no more species left; see the next step.)
5. After a few steps identical to the previous one, the working set $\{H, O_2, H_2O\}$ is reached. We have found our second minimally linearly dependent subset; the corresponding reaction step is $4H + O_2 \rightleftharpoons 2H_2O$.
6. Recall that H_2O is the last species, hence we cannot proceed as before, by replacing H_2O with the next species. Hence, we drop H_2O and replace the previous species, O_2, by its successor, OH.
7. We have now $\{H, OH\}$. It is independent, we add the next species, O.
8. We have now $\{H, OH, O\}$. We found our next minimal linearly dependent subset; the corresponding reaction step is $H + O \rightleftharpoons OH$.

Carrying on this way, we find a total of 14 minimal linearly dependent subsets of the six species. They correspond to the reactions

OH $\rightleftharpoons$ H + O	$O_2 \rightleftharpoons 2\,O$	$H_2O \rightleftharpoons O + H_2$
$2\,OH \rightleftharpoons H_2 + O_2$	$H_2 \rightleftharpoons 2\,H$	$H_2O \rightleftharpoons H + OH$
$2\,OH \rightleftharpoons 2\,H + O_2$		$H_2O \rightleftharpoons 2\,H + O$
$2\,OH \rightleftharpoons O + H_2O$		$2\,H_2O \rightleftharpoons 4\,H + O_2$
$2\,OH \rightleftharpoons 2\,O + H_2$		$2\,H_2O \rightleftharpoons 2\,H_2 + O_2$
$4\,OH \rightleftharpoons 2\,H_2O + O_2$		$2\,H_2O \rightleftharpoons 2\,OH + H_2$

5.3.3 Solving Linear Diophantine Equations Using Linear Programming

We return to the problem of enumerating all solutions of a (nonhomogeneous) linear Diophantine equation

$$\mathbf{Ax} = \mathbf{b}, \quad \mathbf{x} \geq 0, \quad \mathbf{x} \in \mathbb{Z}^R.$$

We assume that the number of solutions is finite, which is the case when we are generating elementary steps from a given set of species or when we are enumerating all decompositions of an overall reaction that consists of a bounded number of steps.

The algorithm presented here was proposed by Papp and Vizvári (2006). It first determines the general solution of $\mathbf{Ax} = \mathbf{b}$, over the reals, in a parametric form. Then taking into account the nonnegativity constraints and the integrality condition, it enumerates the solutions. The method relies upon the following simple lemma.

Lemma 5.8 *Let* $r = \text{rank}(\mathbf{A})$. *Then the vector space of the solutions of the homogeneous system* $\mathbf{Ax} = \mathbf{0}$ *has a basis* $\{\mathbf{b}_1, \mathbf{b}_2, \ldots, \mathbf{b}_r\}$ *such that the matrix* $[\mathbf{b}_1\, \mathbf{b}_2\, \ldots\, \mathbf{b}_r]$ *has an* $r \times r$ *diagonal submatrix containing positive integers in the diagonal.*

If one particular solution of the inhomogeneous equation $\mathbf{Ax} = \mathbf{b}$ is

$$\mathbf{p} = [p_1\ p_2\ \ldots\ p_R]^\top, \quad (p_i \in \mathbb{Q}),$$

then the general solution of this inhomogeneous equation can be written in the form

$$\mathbf{x} = \mathbf{p} + \mathbf{b}_1 a_1 + \ldots + \mathbf{b}_r a_r \quad (a_i \in \mathbb{R}).$$

Now our task reduces to determining those vectors $\mathbf{a} = [a_1\ a_2\ \ldots\ a_r]^\top$ for which each component of $\mathbf{x}$ is a nonnegative integer.

Assume that $\beta_i = (\mathbf{b}_i)_i \in \mathbb{N}$ is the positive integer component of $\mathbf{b}_i$ mentioned in the lemma (the diagonal element of the matrix consisting of the basis). Then the ith component of $\mathbf{x}$ is $x_i = p_i + \beta_i a_i$, which can only be an integer if a_i is a rational number of the form

$$(t - p_i)/\beta_i, \quad t \in \mathbb{Z}.$$

If bounds on each of the x_i are known, then only a finite number of possible solutions remain to check. The bounds can be determined by linear programming: The variables x_i need to be minimized and maximized subject to all the linear constraints we have on the solutions, including of course the equations $\mathbf{Ax} = \mathbf{b}$, the nonnegativity constraints $\mathbf{x} \geq \mathbf{0}$, and the bound on the size of the solutions, if we have one.

We shall call this algorithm the **LP-based enumerative** method.

A word on implementation: the number of candidate solutions in the rectangular box determined by the lower and upper bounds on the x_i is likely to be too high to enumerate and test explicitly all of them, unless the problem is very small. A more efficient approach is to first determine the bounds for one variable, say x_1, only. Then for every possible value of x_1 determine separate bounds for x_2, etc. This procedure can be implemented in many different ways. The simplest possible implementation uses linear programming twice in every iteration. The implementation can be made more efficient in several ways, the details of most of them are beyond the scope of this book. In particular, advanced linear programming software are capable of **warmstarting**, that is, accelerating the solution of a linear program that is obtained by some small modification of an already solved linear program. The LP-based enumerative method consists almost entirely of solving such linear programs: problems involving the same constraints and only different objective functions, and problems that are identical to earlier ones except that one of the previous variables has its value fixed.

If the set of solutions is unbounded, then we can still use the LP-based enumerative algorithm to obtain every solution not longer than some given constant. Although there is no way to obtain only minimal solutions directly with this method, it is possible to first bound the size of minimal solutions using Theorem 5.9 below, then find all solutions up to that size, and then throw away the solutions that are not minimal. We close our discussion on the LP-based enumerative method by showing how the sizes of the minimal solutions can be bounded.

Theorem 5.9 *Consider the homogeneous linear Diophantine equation* $\mathbf{Ax} = \mathbf{0}$ *with* $r = \mathrm{rank}(\mathbf{A})$. *Let* D_r *be the minor of* $\mathbf{A}$ *of order r that has the largest absolute value. Similarly, let* D_r' *be the minor of order $r + 1$ of* $\begin{bmatrix} 1 \dots 1 \\ \mathbf{A} \end{bmatrix}$ *with the largest absolute value. Then the following inequalities hold for every minimal solution* $\mathbf{m}$ *of the homogeneous equation:*

$$\|\mathbf{m}\|_1 \leq (1 + \max_i \|A(\cdot, i)\|_1)^r, \tag{5.4}$$

$$\|\mathbf{m}\|_1 \leq (n - r)|D_r'|, \tag{5.5}$$

$$\|\mathbf{m}\|_\infty \leq (n - r)\left(\frac{\|\mathbf{A}\|_1}{r}\right)^r, \tag{5.6}$$

$$\|\mathbf{m}\|_\infty \leq (n - r)|D_r|. \tag{5.7}$$

The bound (5.7) is always sharper than (5.6), but the right-hand sides of (5.5) and (5.7) are not computable in practice. For larger systems only the first and the third bounds are easily computable. Note that this theorem yields a second proof of the fact that the number of minimal solutions is finite.

5.3.4 Reducing the Search Space

The high computational cost of the enumeration algorithms means that efforts to reduce the number of species and elementary steps both using thermodynamic information and combinatorial considerations are worthwhile. In the remaining of the section, we discuss a few approaches that help reducing the size of the problem.

5.3.4.1 Finding Unavoidable Steps

In the decomposition of overall reactions, there might be some elementary steps which must take part in every decomposition. These steps can be subtracted from the overall reaction, decreasing the size of the solutions. This is particularly useful for solution methods that perform well on problems with small solutions, such as the Contejean–Devie method. These elementary reactions and the number of times they must take place in the decompositions can be found using linear programming.

For the rth reaction step ($1 \le r \le R$), consider the linear program

$$\min_{\mathbf{z} \in \mathbb{R}^R} \{z_r \mid \boldsymbol{\gamma}\mathbf{z} = \mathbf{b},\ \mathbf{z} \ge 0\}. \tag{5.8}$$

Here $\mathbf{b}$ is the atomic vector of the overall reaction; the equality constraints are the balance equations from (5.2). The optimal value of this linear program (rounded up to the nearest integer) is a lower bound on the number of times the rth step must take place in every decomposition of the overall reaction. This is a rather fast method, as it only requires the solution of R linear programs. More advanced linear programming techniques can further accelerate this approach, but that is again beyond the scope of this book.

5.3.4.2 Eliminating Combinatorially Impossible Steps and Species Using Linear Programming

Every decomposition method becomes faster if elementary steps that cannot take part in any decomposition are eliminated. As a first filter, we should rule out as many steps as possible based on chemical evidence, e.g., if they violate some thermodynamical constraint. More mathematical approaches include the methods presented in this paragraph and the next.

The approach used to bound the coefficients of the steps from below may be used again. In order to get upper bounds on these coefficients, it suffices to simply change the objective in (5.8) from the minimization of z_r to the maximization of z_r. If the maximum is less than 1, we can eliminate the rth step. More sophisticated variants of this idea were proposed by Kovács et al. (2004) and Papp and Vizvári (2006), which we outline next. The same ideas can be used to obtain decompositions as well (see Sect. 5.3.5).

Given a subset of reaction steps $\mathscr{R}' \subseteq \mathscr{R}$, we may ask the question if there exists a decomposition that has at least one step with index included in $\mathscr{R}'$. A sufficient condition for an affirmative answer is given by the following lemma.

Lemma 5.10 *Let* **b** *be the atomic vector of an overall reaction and* $\mathscr{R}' \subseteq \mathscr{R}$ *be the index set of a collection of elementary steps. Denote by* $\chi_{\mathscr{R}'} \in \mathbb{N}^R$ *the characteristic vector of* $\mathscr{R}'$, *and consider the linear program*

$$\min_{\mathbf{z} \in \mathbb{R}^R} \{ \chi_{\mathscr{R}'}^\top \mathbf{z} \mid \gamma \mathbf{z} = \mathbf{b}, \ \mathbf{z} \geq 0, \ \chi_{\mathscr{R}'}^\top \mathbf{z} \geq 1 \}. \tag{5.9}$$

If this linear program has no feasible solution, then every decomposition of the overall reaction uses exclusively steps whose indices are not *in* $\mathscr{R}'$.

Proof Suppose by contradiction that a decomposition with at least one step in $\mathscr{R}'$ exists. If $\mathbf{z} \geq 0$ is the corresponding vector, then $\gamma \mathbf{z} = \mathbf{b}$ holds by definition, furthermore $y_r \geq 1$ for at least one $r \in \mathscr{R}'$, ensuring $\chi_{\mathscr{R}'}^\top \mathbf{z} \geq 1$. Hence, this vector $\mathbf{z}$ satisfies all the constraints of the linear program and makes the objective function value at least 1. □

This lemma can be used iteratively as follows. Starting with $\mathscr{R}' = \mathscr{R}$, use Lemma 5.10 to verify if there may exist any decompositions at all. If the answer is negative, stop, and return $\mathscr{R}'$ as the index set of steps we can eliminate, as they cannot take part in any decomposition. If the lemma gives a positive answer, a decomposition may exist involving those steps whose corresponding components in the optimal solution $\mathbf{z}$ of (5.9) are positive. We cannot eliminate these steps, so we remove them from $\mathscr{R}'$ and recourse, trying to find a decomposition involving at least one step still included in $\mathscr{R}'$. We continue this until the set $\mathscr{R}'$ becomes empty (and we conclude that no steps can be eliminated), or we find a nonempty $\mathscr{R}'$ whose reactions cannot be involved in any decomposition. It is clear that this procedure terminates after at most R iterations. Two *Mathematica* implementations of this method are set as Problems 5.6 and 5.7 below. The function **Omittable** takes **b** and γ and applies Lemma 5.10.

5.3.4.3 Eliminating Combinatorially Impossible Steps and Species Using Volpert Indices

The previous algorithm does not take all combinatorial information into consideration: It uses the balance equations, but it does not take into account what the initial species are. Recall that a decomposition is chemically infeasible if any of its species has zero concentration during the whole reaction and that by Theorems 8.6 and 8.14, a species has a constant zero concentration if and only if it gets infinite index in the Volpert indexing procedure defined in Definition 3.22.

Although the cited theorem talks about the indexing of the steps of a decomposition that has already been found, with a slight modification, it can also be used to identify species and reactions that cannot take part in any decomposition. All one has to do is to consider all species and all elementary reactions as if they all took part in a decomposition. There is no overall reaction, and in the indexing process, the exact indices do not matter, only whether they are finite. After indexing the species and the reaction steps, those with infinite index can be eliminated, as they cannot take part in any decomposition.

Example 5.11 Recall Example 5.4. Regardless of the overall reaction, if OH and H_2O are the only species initially available (which correspond to the third and sixth rows in the atomic matrix), then a cascade of deductions follows, corresponding to the steps of the Volpert indexing process:

1. with only the initial species (who get index 0) being available, only one reaction step may take place: $H_2O \longrightarrow H + OH$, producing one additional species, H. Hence, this step gets index 0, and H gets index 1.
2. With H also available, two further reaction steps may take place: $H + OH \longrightarrow H_2O$, which does not yield any new species, and $2H \longrightarrow H_2$, produces a new species: H_2. Both of these steps get index 1; the species H_2 gets index 2.
3. With H_2 also available, yet another step may take place: $OH + H_2 \longrightarrow H + H_2O$; this step gets index 2. It does not produce any species that had not been indexed already, hence all other reaction steps and species get an infinite index, and the indexing process is over.

With these initial species, O and O_2 cannot be produced by any sequence of reaction steps, and hence they can be eliminated. Similarly, the first two reaction steps may never occur, and they can be eliminated. This is how we get this result.

```
Row[VolpertIndexing[{"H" + "O2" -> "O" + "OH",
    "O" + "H2" -> "H" + "OH",
    "OH" + "H2" -> "H" + "H2O", 2 "H" -> "H2",
    "H" + "OH" <=> "H2O"}, {"OH", "H2O"},
  Verbose -> True]]
```

And the result is as follows:

Species	Indices	Reaction steps	Indices
H_2O	0	$H_2O \longrightarrow H + OH$	0
OH	0	$H + OH \longrightarrow H_2O$	1
H	1	$2H \longrightarrow H_2$	1
H_2	2	$H_2 + OH \longrightarrow H + H_2O$	2
O	∞	$H_2 + O \longrightarrow H + OH$	∞
O_2	∞	$H + O_2 \longrightarrow O + OH$	∞

We invite the reader to try another set of initial species (Problem 5.5).

5.3.5   Partial Enumeration

The algorithms of Sects. 5.3.1 and 5.3.3 have aimed at enumerating all solutions or all minimal solutions of a linear Diophantine equation. Their running time may be high, and it may also take time until they return even the first solution. We revisit the idea of Sect. 5.3.4.2 and turn it into a heuristic to quickly generating some decompositions. We may not find all (or a huge enough number of) decompositions with this method, but whatever we find, we find it much faster than with the previous algorithms.

Recall that a decomposition corresponds to a vector $\mathbf{z}$ satisfying the constraints $\boldsymbol{\gamma}\mathbf{z} = \mathbf{b}, \mathbf{z} \geq 0, \mathbf{z} \in \mathbb{Z}^R$. The first two constraints are linear equations and inequalities; we can find vectors that satisfy them by solving a linear program:

$$\min_{\mathbf{z} \in \mathbb{R}^R} \{\mathbf{c}^\top \mathbf{z} \mid \boldsymbol{\gamma}\mathbf{z} = \mathbf{b}, \mathbf{z} \geq 0\}, \tag{5.10}$$

with an arbitrary $\mathbf{c} \in \mathbb{R}^R$. If $\mathbf{c} \geq 0$, the linear program cannot be unbounded, i.e., either there are no vectors $\mathbf{z}$ satisfying the constraints or the linear program has an optimal solution. It can be shown that if the optimal solution exists, then there exists a rational optimal solution with at most M nonzero components. (Linear programming softwares typically return such a solution.) If the nonzero components happen to be integers, then we have found a decomposition. Even if the solution is noninteger, we can interpret it as a decomposition, as for an appropriate $n \in \mathbb{N}_0$, $n\mathbf{z}$ is an integer, and it corresponds to a decomposition of n times the original reaction.

> If the solution is not an integer, there exist techniques that help us find an integer solution via the solution of a sequence of further linear programs; these algorithms belong to the realm of **integer programming**, and are beyond the scope of this book.

The linear program (5.10) can find at most one decomposition. But since the vector $\mathbf{c}$ is arbitrary, we can solve it for a variety of vectors, and some of them may give a new decomposition.

We can also use the idea of Sect. 5.3.4.2 and introduce a further parameter: a subset $\mathscr{R}' \subseteq \mathscr{R}$ of reaction steps from which at least one step must take part in the next decomposition. We can find such a decomposition by solving

$$\min_{\mathbf{z} \in \mathbb{R}^R} \{\mathbf{c}^\top \mathbf{z} \mid \boldsymbol{\gamma}\mathbf{z} = \mathbf{b}, \mathbf{z} \geq 0, \boldsymbol{\chi}_S^\top \mathbf{z} \geq 1\}, \tag{5.11}$$

and recourse after removing from S the index of every nonzero component of the last optimal $\mathbf{z}$, until the linear program becomes infeasible. Each iteration yields a new decomposition (or a rational vector that we may or may not wish to interpret as a decomposition). These (fractional) decompositions have the property that every reaction step that may occur in a decomposition has a nonzero coefficient in at least one of the decompositions found. We call such a collection of vectors a **covering decomposition set**. Every cost vector $\mathbf{c}$ has an associated covering decomposition set, but different **c**'s may not yield different sets. By generating several of them, we can rapidly obtain a number of decompositions. Unfortunately there does not seem to be a way to systematically generate every decomposition this way.

To generate decompositions the function `CoveringDecompositionSet` (or `Decompositions`) can also be used.

Example 5.12 The command

```
CoveringDecompositionSet[
    {"O_2" + 2 "H_2" -> "H" + "H_20"},
    {"H" + "O_2" -> "O" + "OH",
    "O" + "H_2" -> "H" + "OH",
    2 "OH" + "H_2" -> "H" + "H_20"},
    ObjectiveFunction -> GreedySelection,
    Verbose->True]
```

produces a decomposition.

5.3.6 Summary and Comparison of Algorithms

In elementary step generation, a number of small linear Diophantine equations need to be solved, all of which have a finite number of solutions, which are all minimal. To solve these equations, in principle we may use either the Contejean–Devie algorithm or the LP-based enumerative method. As the solutions are typically very small, the Contejean–Devie algorithm is suitable for most problems. It will run faster than the LP-based enumerative method, whose bottleneck in these problems is likely to be the solution of a large number of linear programs. There is hardly any point in using the partial enumeration algorithm, as all solutions can be enumerated even for large problems.

In decomposition problems the sizes of the solutions can be much larger. The search space reduction methods of Sect. 5.3.4 are strongly recommended, although keep in mind that the linear programming-based identification of unavoidable and omittable reaction steps and species only accelerates the Contejean–Devie algorithm; it does not help the LP-based enumerative method which incorporates the same constraints in its LP formulation as the elimination algorithm. On the other hand, both the Contejean–Devie and the LP-based enumerative method benefit from the elimination of reaction steps and species using Volpert indexing.

The number of reaction steps in the decompositions can be bounded from below using linear programming (Problem 5.8). This bound may also guide us in choosing a suitable algorithm. If the smallest decomposition consists of K steps, then the Contejean–Devie algorithm will enumerate all sequences of length $K - 1$ without steps in forbidden directions, and the number of such sequences can be very large unless K or R is small. In this case the LP-based enumerative method is the best choice, followed by the variant of the Contejean–Devie method that enumerates only solutions up to a given size.

A real-life application of our methods was to obtain chemically acceptable decompositions of the overall autocatalytic reaction of permanganate/oxalic acid into elementary steps (see Kovács et al. 2004). As a chemical result, we obtained mathematical justification of the well-known autocatalytic nature and of the less-known crucial role of radical CO^{2-}. By inspecting the obtained decompositions, one can find reaction steps inaccessible to chemical intuition, and these point out the direction of further experimental investigations.

5.4   Exercises and Problems

5.1 Using the species of Example 5.3, find the elementary steps starting from the reactant complex $H + H_2O$.

(Solution: page 391.)

5.2 Give a simple linear algebraic method to test whether a set of vectors is minimal linearly dependent. Give a *Mathematica* implementation. It should not be more than a single line.

(Solution: page 391.)

5.3 Write a *Mathematica* code for the algorithm by Szalkai described above. Use *Mathematica*'s pattern matching mechanism and fixed-point mechanisms.

(Solution: page 392.)

5.4 (Szalkai (1997)) How many minimal linearly dependent subsets can we select from the atomic vectors of the species CO, CO_2, O_2, H_2, CH_2O, CH_3OH, C_2H_5OH, $(CH_3)_2CO$, CH_4, CH_3CHO, H_2O?

(Solution: page 392.)

5.5 Verify that in the mechanism of Example 5.11, no reaction steps or species can be eliminated if the species initially available are $\{H, H_2, O_2\}$.

(Solution: page 392.)

5.6 Write a *Mathematica* code for the function **Omittable** based on Lemma 5.10.

(Solution: page 393.)

5.7 The **ReactionKinetics** function **Omittable** (see p. 68) can be more easily implemented using the **CoveringDecompositionSet** function introduced in Sect. 5.3.5. How?

(Solution: page 393.)

5.8 Set up a linear programming problem whose solution bounds from below the number of reaction steps in the decompositions of an overall reaction.

(Solution: page 393.)

References

Contejean E, Devie H (1994) An efficient incremental algorithm for solving systems of linear Diophantine equations. Inf Comput 113(1):143–172

Epstein I, Pojman J (1998) An introduction to nonlinear chemical dynamics: oscillations, waves, patterns, and chaos. Topics in physical chemistry series. Oxford University Press, New York. http://books.google.com/books?id=ci4MNrwSlo4C

Kovács K, Vizvári B, Riedel M, Tóth J (2004) Computer assisted study of the mechanism of the permanganate/oxalic acid reaction. Phys Chem Chem Phys 6(6):1236–1242

Papp D, Vizvári B (2006) Effective solution of linear Diophantine equation systems with an application in chemistry. J Math Chem 39(1):15–31

Szalkai I (1997) Lineáris algebra, sztöchiometria és kombinatorika (Linear algebra, stoichiometry and combinatorics). Polygon 7(2):35–51

Part II

The Continuous Time Continuous State Deterministic Model

Different forms of the usual deterministic model: a (special kind of) polynomial differential equation is introduced. The behavior of its stationary and transient solutions are studied symbolically and numerically. Beyond the exact results symbolic and numerical approximations are presented.

The Induced Kinetic Differential Equation

6

6.1 Introduction

Our main interest here and in the next two chapters is in the (transient and longtime) behavior of the trajectories of the induced kinetic differential equations of reactions. Before dealing with this problem, we review the different forms of the induced kinetic differential equation, because each of the individual forms has a special advantage.

6.2 Heuristic Derivation of the Induced Kinetic Differential Equation

Let us consider as an example the Robertson reaction (Robertson 1966):

$$A \xrightarrow{k_1} B \quad 2B \xrightarrow{k_2} B + C \xrightarrow{k_3} A + C. \tag{6.1}$$

Let us denote the molar concentrations of the individual species by the corresponding lowercase letters: $a(t) := [A](t)$, $b(t) := [B](t)$, $c(t) := [C](t)$. As to the effect of the first reaction step on the concentration $a(t)$ in the "small" time interval $]t, t + h[$, one may assume that it is as follows: $a(t+h) = a(t) - \Phi(a(t), h) + \varepsilon(h)h$, where we try to choose the function Φ to be the simplest possible one, i.e. we assume it is proportional to both the concentration $a(t)$ and to the length h of the time interval, and we denote the proportionality factor by k_1, i.e. $\Phi(a(t), h) := k_1 a(t)h$. The fact that the length of the interval is small is reflected in two assumptions:

1. First, the concentration is "essentially the same" at the beginning and at the end of the interval: in mathematical terms a is continuous; there is no jump in the concentration.

© Springer Science+Business Media, LLC, part of Springer Nature 2018
J. Tóth et al., *Reaction Kinetics: Exercises, Programs and Theorems*,
https://doi.org/10.1007/978-1-4939-8643-9_6

2. Second, the error $\varepsilon(h)h$ of the approximation is so small that it tends to zero even if divided by the length h of the interval when h tends to zero: $\lim_0 \varepsilon = 0$.

Proceeding in the same way, we get $b(t+h) = b(t) + k_1 a(t)h + \varepsilon(h)h$ as the effect of the first reaction step on concentration $b(t)$, because the gain in $b(t)$ is the same as the loss in $a(t)$.

How to calculate the effect of the second step? Had we the step $B+D \longrightarrow B+C$ then an argument similar to that above would give the following relations:

$$b(t+h) = b(t) + (1-1)k_2 b(t)d(t)h + \varepsilon(h)h$$

$$c(t+h) = c(t) + (1-0)k_2 b(t)d(t)h + \varepsilon(h)h$$

$$d(t+h) = d(t) + (0-1)k_2 b(t)d(t)h + \varepsilon(h)h.$$

Why? Because again we assumed that the changes of concentrations are proportional to everything on what it may depend (taken a relatively simplistic point of view). Therefore, it is a straightforward assumption that the step $B+B \longrightarrow B+C$ causes the following changes:

$$b(t+h) = b(t) - k_2 b(t)b(t)h + \varepsilon(h)h$$

$$c(t+h) = c(t) + k_2 b(t)b(t)h + \varepsilon(h)h.$$

Returning to the original full Robertson reaction, we get

$$a(t+h) = a(t) - k_1 a(t)h + k_3 b(t)c(t)h + \varepsilon(h)h$$

$$b(t+h) = b(t) + k_1 a(t)h - k_2 b(t)^2 h - k_3 b(t)c(t)h + \varepsilon(h)h$$

$$c(t+h) = c(t) + k_2 b(t)^2 h + \varepsilon(h)h.$$

Subtracting the first terms of the right-hand sides and dividing through by h, we get

$$\frac{a(t+h) - a(t)}{h} = -k_1 a(t) + k_3 b(t)c(t) + \varepsilon(h)$$

$$\frac{b(t+h) - b(t)}{h} = +k_1 a(t) - k_2 b(t)^2 - k_3 b(t)c(t) + \varepsilon(h)$$

$$\frac{c(t+h) - c(t)}{h} = +k_2 b(t)^2 + \varepsilon(h).$$

Tending to zero with h, the right-hand sides do have a limit (because of the assumption $\lim_0 \varepsilon = 0$); therefore the left-hand sides should also have one, which can only be the derivative of the corresponding functions:

$$\dot{a}(t) = -k_1 a(t) + k_3 b(t)c(t)$$
$$\dot{b}(t) = +k_1 a(t) - k_2 b(t)^2 - k_3 b(t)c(t)$$
$$\dot{c}(t) = +k_2 b(t)^2,$$

or—with functions instead of function values—

$$\dot{a} = -k_1 a + k_3 bc \quad \dot{b} = k_1 a - k_2 b^2 - k_3 bc \quad \dot{c} = k_2 b^2.$$

This is the **induced kinetic differential equation** of the Robertson reaction. It can be obtained in the program as follows:

```
DeterministicModel[{"Robertson"}, {k1, k2, k3},
   {a, b, c}]
```

To be more precise one should use different ε functions in the different equations—without any additional gain.

Let us add a few words on the units of the rate coefficients. In order to have the same units on both sides of the above equations k_1 should have the unit time^{-1}, whereas k_2 and k_3 should have the unit time^{-1}mass^{-1}volume.

This is the way how a chemist heuristically constructs a differential equation to describe the time evolution of chemical species in reactions. Now we are going to give formal definitions, and the reader should check on many examples to convince herself/himself if the formal definition is leading to the result expected by the chemist's intuition. Finally, we remark that because the volume has been supposed to be constant, we could have argued using mass instead of concentrations.

6.3   Equivalent Forms of the Induced Kinetic Differential Equations

Different authors use different forms of the same general form of the induced kinetic differential equation, but this fact will turn out to have more advantages than disadvantages.

6.3.1   Reaction Steps Emphasized

Let us recapitulate the general form (2.1) of reactions here:

$$\sum_{m \in \mathcal{M}} \alpha(m, r)X(m) \longrightarrow \sum_{m \in \mathcal{M}} \beta(m, r)X(m) \quad (r \in \mathcal{R}). \tag{6.2}$$

We want to formulate a system of ordinary differential equations as a model, but this formulation if specialized to the Robertson example above will be a bit more general.

Suppose that we are also given functions called **kinetics** (in the narrow sense, as opposed to the name of the branch of science) $w_r \in \mathscr{C}^1(\mathbb{R}^M, \mathbb{R}_0^+)$, $(r \in \mathscr{R})$ corresponding to the rth reaction step with the following properties.

Conditions 2

1. If all the species needed to step r are available, then step r proceeds, i.e. if for all $m \in \mathscr{M} : \alpha(m, r) > 0$ implies $c_m > 0$, then $w_r(\mathbf{c}) > 0$;
2. If at least one species needed to step r is missing, then step r does not take place, i.e. if there exists $m \in \mathscr{M} : \alpha(m, r) > 0$ and $c_m = 0$, then $w_r(\mathbf{c}) = 0$.

Definition 6.1 With all the above definitions and assumptions, the usual continuous time, continuous state deterministic model, the **induced kinetic differential equation** of the reaction (2.1) is the autonomous ordinary differential equation

$$\dot{c}_m(t) = \sum_{r \in \mathscr{R}} (\beta(m, r) - \alpha(m, r)) w_r(c_1(t), c_2(t), \ldots, c_M(t)) \quad (m \in \mathscr{M}).$$

$$(6.3)$$

Let us introduce $\mathbf{w} := \begin{bmatrix} w_1 & w_2 & \ldots & w_R \end{bmatrix}^\top$ and $\mathbf{c}(t) := \begin{bmatrix} c_1(t) & c_2(t) & \ldots & c_M(t) \end{bmatrix}^\top$; then we can repeat the equations in vectorial form as

$$\dot{\mathbf{c}}(t) = (\boldsymbol{\beta} - \boldsymbol{\alpha}) \cdot \mathbf{w}(\mathbf{c}(t)) = \boldsymbol{\gamma} \cdot \mathbf{w}(\mathbf{c}(t)). \tag{6.4}$$

This equation is valid for all $t \in \mathbb{R}$ for which the solution can be defined. The equation can also be written down for functions or in **global form** (as opposed to the **local form** used in Eq. (6.4)) as

$$\dot{\mathbf{c}} = (\boldsymbol{\beta} - \boldsymbol{\alpha}) \cdot \mathbf{w} \circ \mathbf{c} = \boldsymbol{\gamma} \cdot \mathbf{w} \circ \mathbf{c}. \tag{6.5}$$

(Here we use the $\circ$ sign to denote composition of functions; see the list of notations on page xxiii.)

Remark 6.2

1. Equation (6.3) expresses a **fundamental assumption of homogeneous reaction kinetics**: The effects of the different reaction steps on the concentration changes of the species are independent of each other; the effects sum up (more precisely, they are calculated as linear combinations with the differences of the stoichiometric coefficients as coefficients).

2. One might say that (6.2) shows the reaction steps, some of which might come from **genuine reaction step**s, i.e. from steps where **external species** of constant concentration are also present. These are species which are present in such quantities that they practically do not change during the reaction, e.g. water, solvent or some precursors or substrates, etc. If one wants to emphasize the presence of external species, then the expression **internal species** is used to name the entities which are usually called species. The genuine reaction step may express a real physical process, like collision. Consider the trypsin-catalyzed transformation of trypsinogen described by Trypsin + Trypsinogen $\longrightarrow$ 2 Trypsin. Here trypsinogen is in excess, and in any usual model, it might be considered as an external species; therefore one can study instead of this genuine reaction step the reaction step Trypsin $\longrightarrow$ 2 Trypsin. Let us note by passing that Trypsin here is a direct catalyst.
3. In- and outflow can also be described by genuine reaction steps containing external species, like A $\longrightarrow$ X and X $\longrightarrow$ A, which can be replaced in the next step of modeling by 0 $\longrightarrow$ X and X $\longrightarrow$ 0, respectively.

Most of the models are formulated with a special form of the kinetics, with the mass action form, in which the stoichiometric coefficients of the reactant complexes play a special role.

Definition 6.3 Suppose that we are given positive numbers called **reaction rate coefficient**s $k_r \in \mathbb{R}^+$ where $r \in \mathscr{R}$, corresponding to each reaction step, and suppose the kinetics is defined in the following way:

$$w_r(\mathbf{c}) := k_r \mathbf{c}^{\alpha(\cdot, r)} \quad (r \in \mathscr{R}). \tag{6.6}$$

(The definitions and properties of less frequent vectorial operations can be found in Sect. 13.2 of the Appendix.) Then the kinetics is said to be of the **mass action** type.

Remark 6.4

1. Mass action kinetics possesses both properties of Conditions 2 (see Problem 6.3).
2. The induced kinetic differential equation of a reaction with mass action type kinetics takes the following form:

$$\dot{c}_m(t) = \sum_{r \in \mathscr{R}} (\beta(m, r) - \alpha(m, r)) k_r \prod_{p=1}^{M} c_p(t)^{\alpha(p, r)} \quad (m \in \mathscr{M}), \tag{6.7}$$

or in the global form using vectorial notations including the notation $\odot$ for **Hadamard–Schur product** (see again Sect. 13.2 of the Appendix)

$$\dot{\mathbf{c}} = (\beta - \alpha) \cdot \mathbf{k} \odot \mathbf{c}^\alpha = \gamma \cdot \mathbf{k} \odot \mathbf{c}^\alpha, \tag{6.8}$$

with $\mathbf{k} := \begin{bmatrix} k_1 \ k_2 \ \ldots \ k_R \end{bmatrix}^\top$ (see Problem 6.1). One can see more clearly the role of **pure monomials** from the form

$$\dot{\mathbf{c}} = (\boldsymbol{\gamma} \cdot \mathrm{diag}(\mathbf{k})) \cdot \mathbf{c}^\alpha. \tag{6.9}$$

There are several ways to deduce the mass action form of the kinetics from **functional equation**s.

The aim of Garay (2004) is to characterize monomials of the form $k_r \prod_{p=1}^M c_p^{\alpha(p,r)}$ by a functional equation. He shows that under appropriate conditions (similar to but not identical with those used here) the solution to the functional equation

$$\frac{w(\mathbf{c} \odot \mathbf{F})}{w(\mathbf{c})} = \mathbf{F}^{\mathbf{r(c)}} \tag{6.10}$$

is $w(\mathbf{c}) = k\mathbf{c}^\mathbf{r}$ with positive real number $k \in \mathbb{R}^+$ and with a vector of nonnegative real numbers $\mathbf{r} \in (\mathbb{R}_0^+)^M$. Equation (6.10) is the result of heuristic application of Euler discretisation of an obvious identity.

Tóth and Érdi (1978) on page 243 give a verbal description of the properties which should naturally be obeyed by the reaction rates. Let us consider the reaction X $\longrightarrow$ C, where C is an arbitrary complex. In this case it is quite natural to assume that the reaction rate w of this step is nonnegative, monotone, and additive in the concentration of X:

$$w(c_1 + c_2) = w(c_1) + w(c_2). \tag{6.11}$$

The nontrivial solutions of this functional equation—**the Cauchy equation**—is $w(c) = kc$. Problem 6.2 shows the solution of a more complicated case.

If one is interested in the factors in Eq. (6.8) separately, then the following construct may be used.

```
gamma = ReactionsData[{"Lotka-Volterra"}, {"A","B"}][{γ}]
alpha = ReactionsData[{"Lotka-Volterra"}, {"A","B"}][{α}]
gamma.{k1, k2, k3} Times @@ ({x, y}^alpha).
```

At this point one cannot help adding a few more words about units. The concentration of the species is usually measured in the units $\mathrm{mol\ dm^{-3}}$, and time is measured in seconds; therefore the unit one has on the left-hand side of (6.7) is $\mathrm{mol\ dm^{-3}\ s^{-1}}$ (in general terms, mass volume^{-1} time^{-1}.) As the stoichiometric coefficients are pure numbers, terms $k_r \prod_{p=1}^M c_p(t)^{\alpha(p,r)}$ should also have the same unit $\mathrm{mol\ dm^{-3}\ s^{-1}}$, but as the unit of $\prod_{p=1}^M c_p(t)^{\alpha(p,r)}$ is $\left(\mathrm{mol\ dm^{-3}}\right)^{\sum_{p \in \mathcal{M}} \alpha(p,r)}$, therefore to get a dimensionally correct equation, k_r should have the unit

$$\left(\mathrm{mol\ dm^{-3}}\right)^{1-\sum_{p \in \mathcal{M}} \alpha(p,r)} \mathrm{s^{-1}},$$

or introducing the notation

$$o(r) := \sum_{p \in \mathcal{M}} \alpha(p, r) \tag{6.12}$$

for the order of the rth reaction step (see Definition 2.4): $(\text{mol dm}^{-3})^{1-o(r)} \text{s}^{-1}$. Specially, one has the following units.

Order of the reaction step	Unit of the reaction rate coefficient
0	mol dm^{-3}s^{-1}
1	s^{-1}
2	mol^{-1}dm^3 s^{-1}

If a kinetics different from the mass action type is used, it is the responsibility of the modeler (or the user of our programs) to choose the appropriate unit for the reaction rate coefficients.

We shall return to the units when discussing the change of units in connection with the stochastic model in Chap. 10. In very special cases, a change of unit might also help decrease the stiffness of a model (see Problem 9.25).

6.3.2 Complexes Emphasized

In the form of the induced kinetic differential equation (6.3), one can see that a reaction step vector $\gamma(\cdot, r) := \beta(\cdot, r) - \alpha(\cdot, r)$, where $r \in \mathcal{R}$, may correspond to many different reaction steps. Therefore, an alternative description of the same equation arises with the concepts defined here.

Definition 6.5

- The different complex vectors put into the columns of a matrix form the **complex matrix**: $\mathbf{Y} := [\mathbf{y}_1 \ \mathbf{y}_2 \ \cdots \ \mathbf{y}_N]$.
- The **complex index set** is $\mathcal{N} := \{1, 2, \ldots, N\}$.
- The complex matrix defines a linear map from $\mathbb{R}^N$, the **species space**, into $\mathbb{R}^M$, the **complex space**: $\mathbf{Y} : \mathbb{R}^N \to \mathbb{R}^M$.

These names come from the fact that elements of the standard basis $\mathbf{e}_1, \mathbf{e}_2, \ldots, \mathbf{e}_M$ in $\mathbb{R}^M$ correspond to the species, while elements of the standard basis $\mathbf{f}_1, \mathbf{f}_2, \ldots, \mathbf{f}_N$ in $\mathbb{R}^N$ correspond to the complexes. The role of the linear map described by $\mathbf{Y}$ is to provide the complex vectors by $\mathbf{Y}\mathbf{f}_n = \mathbf{y}_n \in \mathbb{R}^M$.

As before, the state of the reaction at time t is characterized by the vector $\mathbf{c}(t) \in \mathbb{R}^M$ of concentrations of the (internal) species. The complex corresponding to the complex vector $\mathbf{y}_n$ for all $n \in \mathcal{N}$ is denoted (as usual, see Chap. 2) by C_n. The rate of the reaction step

$$C_q \xrightarrow{r_{nq}} C_n \tag{6.13}$$

only depends on the concentration vector. As pressure, volume, and temperature are assumed to be constant, this rate is given by the function $r_{nq} \in \mathscr{C}(\mathbb{R}^M, \mathbb{R}_0^+)$. (In this construction this function—the same as above, it is only the indexing what differs—is called **kinetics**.) If the reaction (6.13) is not present, then for all $\mathbf{c} \in \mathbb{R}^M : r_{nq}(\mathbf{c}) = 0$ holds; furthermore, for all $\mathbf{c} \in \mathbb{R}^M : r_{nn}(\mathbf{c}) = 0$.

Further definitions follow.

Definition 6.6

- The **reaction rate matrix** at the concentration $\mathbf{c} \in \mathbb{R}^M$ is the matrix with reaction rates as entries: $\mathbf{R}(\mathbf{c}) \in (\mathbb{R}_0^+)^{N \times N}$.
- The **creation rate** of the complex C_n is $\displaystyle\sum_{q \in \mathcal{N}} r_{nq}(\mathbf{c}) \in \mathbb{R}_0^+$.
- The **annihilation rate** of the complex C_n is $\displaystyle\sum_{q \in \mathcal{N}} r_{qn}(\mathbf{c}) \in \mathbb{R}_0^+$.
- The **formation rate** of the complex C_n is $\displaystyle\sum_{q \in \mathcal{N}} r_{nq}(\mathbf{c}) - \sum_{q \in \mathcal{N}} r_{qn}(\mathbf{c}) \in \mathbb{R}$,
- The **complex formation vector** is $\mathbf{g}(\mathbf{c}) := (\mathbf{R}(\mathbf{c}) - \mathbf{R}^\top(\mathbf{c}))\mathbf{1}_N \in \mathbb{R}^N$.
- The **species formation vector** is

$$\mathbf{f}(\mathbf{c}) := \mathbf{Y}\mathbf{g}(\mathbf{c}) = \sum_{n \in \mathcal{N}} \sum_{q \in \mathcal{N}} r_{nq}(\mathbf{c})(\mathbf{y}_n - \mathbf{y}_q)$$

$$= \sum_{r \in \mathcal{R}} (\boldsymbol{\beta}(\cdot, r) - \boldsymbol{\alpha}(\cdot, r))w_r(\mathbf{c}) \in \mathbb{R}^M, \tag{6.14}$$

cf. Problem 6.8. The mth component of this vector gives the formation rate of species $X(m)$.

- In this formulation the kinetics is said to be of the **mass action type**, if

$$r_{nq}(\mathbf{c}) = k_{nq}\mathbf{c}^{\mathbf{y}_q} \tag{6.15}$$

for some $k_{nq} \in \mathbb{R}_0^+$ where $n, q \in \mathcal{N}$. The number k_{nq} is said to be the **reaction rate coefficient** of the reaction step (6.13); they are only positive for the reaction steps (6.13) occurring among the reaction steps, and they comprise the corresponding concentration products of external species, if needed.

Example 6.7 Consider the genuine reaction steps of the reversible Lotka–Volterra reaction

$$A + X \underset{\kappa_{-1}}{\overset{\kappa_1}{\rightleftharpoons}} 2X \quad X + Y \underset{\kappa_{-2}}{\overset{\kappa_2}{\rightleftharpoons}} 2Y \quad Y \underset{\kappa_{-3}}{\overset{\kappa_3}{\rightleftharpoons}} B.$$

The part of the induced kinetic differential equation of this reaction for the concentrations of X and Y reads as

$$\dot{x} = \kappa_1 a x - \kappa_{-1} x^2 - \kappa_2 x y + \kappa_{-2} y^2 \quad \dot{y} = \kappa_2 x y - \kappa_{-2} y^2 - \kappa_3 y + \kappa_{-3} b.$$

Assuming that the concentration of A and B is constant, having the values a_0 and b_0, respectively, one can introduce the reaction rate coefficients of the reaction steps (as opposed to the genuine reaction steps) as follows:

$$k_1 := \kappa_1 a_0 \quad k_{-1} := \kappa_{-1} \quad k_2 := \kappa_2 \quad k_{-2} := \kappa_{-2} \quad k_3 := \kappa_3 \quad k_{-3} := \kappa_{-3} b_0,$$

and with these the induced kinetic differential equation of the reaction

$$X \underset{k_{-1}}{\overset{k_1}{\rightleftharpoons}} 2X \quad X + Y \underset{k_{-2}}{\overset{k_2}{\rightleftharpoons}} 2Y \quad Y \underset{k_{-3}}{\overset{k_3}{\rightleftharpoons}} 0$$

will be the same as the induced kinetic differential equation of the above reaction consisting of genuine reaction steps.

Let us consider the reaction rate coefficients as entries in the matrix $\mathbf{K} \in (\mathbb{R}_0^+)^{N \times N}$. For $k_{nq} > 0$, let us define the nonzero **reaction step vector** $\mathbf{x}_{nq} := \mathbf{y}_n - \mathbf{y}_q$ corresponding to the reaction step (6.13), i.e., $\mathbf{x}_{nq}$ is one of the columns of the stoichiometric matrix $\mathbf{y}$. The notions stoichiometric space, mass conservation, reaction simplex, and positive reaction simplex have all been defined before in Chap. 4. Using the previous definitions and notations, we have the following relations:

$$\mathbf{R}(\mathbf{c}) = \mathbf{K} \mathrm{diag}(\mathbf{c}^{\mathbf{Y}}) \in \mathbb{R}^{N \times N}, \tag{6.16}$$

$$\mathbf{g}(\mathbf{c}) = (\mathbf{K} - \mathrm{diag}(\mathbf{K}^\top \mathbf{1}_N))\mathbf{c}^{\mathbf{Y}} \in \mathbb{R}^N, \tag{6.17}$$

$$\mathbf{f}(\mathbf{c}) = \sum_{n \in \mathcal{N}} \sum_{q \in \mathcal{N}} k_{nq} \mathbf{c}^{\mathbf{y}(q)} \mathbf{x}_{nq} = \mathbf{Y}(\mathbf{K} - \mathrm{diag}(\mathbf{K}^\top \mathbf{1}_N))\mathbf{c}^{\mathbf{Y}} \in \mathscr{S}, \tag{6.18}$$

(where again Sect. 13.2 of the Appendix may help), and finally, the induced kinetic differential equation of the reaction (6.13) in coordinates in the general case is

$$\boxed{\dot{c}_m(t) = \sum_{n \in \mathcal{N}} \sum_{q \in \mathcal{N}} r_{nq}(\mathbf{c}(t))(y_n^m - y_q^m),} \tag{6.19}$$

(where y_n^m is the mth component of the vector $\mathbf{y}_n$), or—using vectorial notations—

$$\dot{\mathbf{c}}(t) = \mathbf{f}(\mathbf{c}(t)) = \sum_{n \in \mathcal{N}} \sum_{q \in \mathcal{N}} r_{nq}(\mathbf{c}(t)) \mathbf{x}_{nq}. \tag{6.20}$$

The equation can again be formulated in global form using the notation $\circ$ for the composition of functions as

$$\dot{\mathbf{c}} = \mathbf{f} \circ \mathbf{c} = \sum_{n \in \mathcal{N}} \sum_{q \in \mathcal{N}} (r_{nq} \circ \mathbf{c}) \mathbf{x}_{nq}. \tag{6.21}$$

In case the kinetics is of the mass action type, these equations specialize into

$$\boxed{\dot{c}_m(t) = \sum_{n \in \mathcal{N}} \sum_{q \in \mathcal{N}} k_{nq} \mathbf{c}(t)^{\mathbf{y}_q} (y_n^m - y_q^m),} \tag{6.22}$$

or in the global form using vectorial notations

$$\dot{\mathbf{c}} = \sum_{n \in \mathcal{N}} \sum_{q \in \mathcal{N}} k_{nq} \mathbf{c}^{\mathbf{y}_q} (\mathbf{y}_n - \mathbf{y}_q), \tag{6.23}$$

or even more tersely

$$\dot{\mathbf{c}} = \mathbf{Y} \left(\mathbf{K} - \mathrm{diag}(\mathbf{K}^\top \mathbf{1}_N) \right) \mathbf{c}^{\mathbf{Y}}. \tag{6.24}$$

Remark 6.8

- Equation (6.23) can be reformulated in a seemingly symmetric way

$$\dot{\mathbf{c}} = \sum_{n \in \mathcal{N}} \sum_{q \in \mathcal{N}} \mathbf{y}_q (k_{nq} \mathbf{c}^{\mathbf{y}_q} - k_{qn} \mathbf{c}^{\mathbf{y}_n}),$$

  which however contains many zero reaction rate coefficients.
- In general, to calculate the vector $\mathbf{c}^{\mathbf{Y}}$, calculations of N products of powers are needed, whereas to calculate the vector $\mathbf{c}^{\alpha}$, one needs to calculate only as many products of powers as the number of different reactant complexes.
- The $\left(\mathbf{K} - \mathrm{diag}(\mathbf{K}^\top \mathbf{1}_N) \right) \in \mathbb{R}^{N \times N}$ expression or sometime its negative has many names in graph theory; it may be called **Laplacian matrix**, **admittance matrix**, **Kirchhoff matrix**, or **discrete Laplacian** of the Feinberg–Horn–Jackson graph with weights k_{nq}.

6.3.3 Network Structure Emphasized

The form of the induced kinetic differential equation proposed by Othmer (1985) is similar to (6.5) or (6.8) but is capable to express the network structure of the reaction more explicitly. However, it was only Boros (2008, 2013) who was able to fully parlay this form. Othmer introduces the incidence matrix $\mathbf{E}$ in the following way.

Definition 6.9 The **incidence matrix** (or, more precisely, the **vertex-edge incidence matrix**) $\mathbf{E}$ of the reaction (2.1) is an $N \times R$ matrix with entries E_{nr} defined as follows:

$$E_{nr} := \begin{cases} +1, & \text{if } C_n \text{ is the product complex of reaction step } r; \\ -1, & \text{if } C_n \text{ is the reactant complex of reaction step } r; \\ 0, & \text{otherwise.} \end{cases} \tag{6.25}$$

With this notation the induced kinetic differential equation of (2.1) in the general case is

$$\dot{\mathbf{c}} = \mathbf{Y}\mathbf{E}\,\mathbf{w} \circ \mathbf{c}, \tag{6.26}$$

and if specialized to the mass action case, one gets

$$\dot{\mathbf{c}} = \mathbf{Y}\mathbf{E}(\mathbf{k} \odot \mathbf{c}^{\alpha}). \tag{6.27}$$

Example 6.10 Consider the Lotka–Volterra reaction

$$X \xrightarrow{k_1} 2X \quad X + Y \xrightarrow{k_2} 2Y \quad Y \xrightarrow{k_3} 0. \tag{6.28}$$

In this case one has

$$\mathbf{Y} = \begin{bmatrix} 1 & 2 & 1 & 0 & 0 & 0 \\ 0 & 0 & 1 & 2 & 1 & 0 \end{bmatrix}, \quad \mathbf{E} = \begin{bmatrix} -1 & 0 & 0 \\ 1 & 0 & 0 \\ 0 & -1 & 0 \\ 0 & 1 & 0 \\ 0 & 0 & -1 \\ 0 & 0 & 1 \end{bmatrix}, \quad \mathbf{k} = \begin{bmatrix} k_1 \\ k_2 \\ k_3 \end{bmatrix},$$

therefore the right-hand side of the induced kinetic differential equation is

$$
\mathbf{YE(k \odot c^{\alpha})} = \begin{bmatrix} 1 & 2 & 1 & 0 & 0 & 0 \\ 0 & 0 & 1 & 2 & 1 & 0 \end{bmatrix} \begin{bmatrix} -1 & 0 & 0 \\ 1 & 0 & 0 \\ 0 & -1 & 0 \\ 0 & 1 & 0 \\ 0 & 0 & -1 \\ 0 & 0 & 1 \end{bmatrix} \left(\begin{bmatrix} k_1 \\ k_2 \\ k_3 \end{bmatrix} \odot \mathbf{c}^{\begin{bmatrix} 1 & 1 & 0 \\ 0 & 1 & 1 \end{bmatrix}} \right)
$$

$$
= \begin{bmatrix} 1 & -1 & 0 \\ 0 & 1 & -1 \end{bmatrix} \begin{bmatrix} k_1 x \\ k_2 x y \\ k_3 y \end{bmatrix} = \begin{bmatrix} k_1 x - k_2 x y \\ k_2 x y - k_3 y, \end{bmatrix}
$$

as expected. Here, $\mathbf{c} = \begin{bmatrix} x \\ y \end{bmatrix}$.

Once we have $\mathbf{Y}$ and the reaction steps at hand, it is not difficult to construct the incidence matrix $\mathbf{E}$ (see Problem 6.4).

6.3.4   Reaction Rate Coefficients Emphasized

Any form of the induced kinetic differential equation in the mass action case suggests that the right-hand side is a linear function of the reaction rate coefficients. First, we demonstrate this fact in the case of the Mole reaction (2.8). Its induced kinetic differential equation (of the mass action type; if not said otherwise, we shall always assume this) is as follows:

Example 6.11

$$
\dot{x} = k_1 x y - k_2 x + k_{-2} \quad \dot{y} = k_1 x y + k_3 - k_{-3} y. \tag{6.29}
$$

By heuristic inspection one can rewrite the right-hand side of the (6.29) in the following way:

$$
\begin{bmatrix} \dot{x} \\ \dot{y} \end{bmatrix} = \begin{bmatrix} xy & -x & 1 & 0 & 0 \\ xy & 0 & 0 & 1 & -y \end{bmatrix} \cdot \begin{bmatrix} k_1 \\ k_2 \\ k_{-2} \\ k_3 \\ k_{-3} \end{bmatrix} \tag{6.30}
$$

How to formalize this seemingly obvious factorization?

Theorem 6.12 *The species formation vector of the mechanism* $\langle \mathcal{M}, \mathcal{R}, \alpha, \beta, \mathbf{k} \rangle$ *(endowed with mass action type kinetics) is the same as*

$$
\begin{bmatrix}
\gamma(1,1)\mathbf{c}^{\alpha(\cdot,1)} & \gamma(1,2)\mathbf{c}^{\alpha(\cdot,2)} & \cdots & \gamma(1,R)\mathbf{c}^{\alpha(\cdot,R)} \\
\gamma(2,1)\mathbf{c}^{\alpha(\cdot,1)} & \gamma(2,2)\mathbf{c}^{\alpha(\cdot,2)} & \cdots & \gamma(2,R)\mathbf{c}^{\alpha(\cdot,R)} \\
\vdots & \vdots & & \vdots \\
\gamma(M,1)\mathbf{c}^{\alpha(\cdot,1)} & \gamma(M,2)\mathbf{c}^{\alpha(\cdot,2)} & \cdots & \gamma(M,R)\mathbf{c}^{\alpha(\cdot,R)}
\end{bmatrix}
\cdot
\begin{bmatrix}
k_1 \\
k_2 \\
\vdots \\
k_R
\end{bmatrix},
\tag{6.31}
$$

or in vectorial form

$$
\gamma \, \mathrm{diag}(\mathbf{c}^{\alpha})\mathbf{k}.
\tag{6.32}
$$

Proof First, let us denote the right-hand side of the induced kinetic differential equation (6.8) by $\mathbf{f}(\mathbf{c}, \mathbf{k})$ and show that it is linear in its second argument. Indeed,

$$
\mathbf{f}(\mathbf{c}, \mathbf{k}_1 + \mathbf{k}_2) = \mathbf{f}(\mathbf{c}, \mathbf{k}_1) + \mathbf{f}(\mathbf{c}, \mathbf{k}_2) \text{ and } \forall \lambda \in \mathbb{R} : \mathbf{f}(\mathbf{c}, \lambda\mathbf{k}) = \lambda\mathbf{f}(\mathbf{c}, \mathbf{k}),
$$

i.e. $\mathbf{f}(\mathbf{c}, \cdot) : (\mathbb{R}_0^+)^R \longrightarrow \mathbb{R}^M$ is linear; therefore it can be written in the following way:

$$
\mathbf{f}(\mathbf{c}, \mathbf{k}) = \mathbf{G}(\mathbf{c})\mathbf{k}.
\tag{6.33}
$$

Second, the matrix of a linear operator in the standard basis can be determined in such a way that one determines the images of the basis vectors:

$$
\mathbf{G}(\mathbf{c}) = \begin{bmatrix} \mathbf{f}(\mathbf{c}, \mathbf{e}_1) & \mathbf{f}(\mathbf{c}, \mathbf{e}_2) & \cdots & \mathbf{f}(\mathbf{c}, \mathbf{e}_R) \end{bmatrix},
$$

where $\mathbf{e}_1, \mathbf{e}_2, \ldots, \mathbf{e}_R \in (\mathbb{R}_0^+)^R$ are the elements of the standard basis of $\mathbb{R}^R$. Hence one obtains $\mathbf{f}(\mathbf{c}, \mathbf{e}_r) = \gamma \cdot \mathbf{e}_r \odot \mathbf{c}^{\alpha}$ $(r \in \mathcal{R})$, which is the same as the rth component of the coefficient matrix in (6.32) before $\mathbf{k}$.

Although the above derivation is not without moral, an immediate proof comes from the fact that $\mathbf{k} \odot \mathbf{c}^{\alpha} = \mathrm{diag}(\mathbf{c}^{\alpha}) \cdot \mathbf{k}(= \mathrm{diag}(\mathbf{k}) \cdot \mathbf{c}^{\alpha})$. $\qquad \square$

How to calculate the coefficient matrix in (6.32) using the program? Here is an example ending in the result `True`.

```
rhs[{k1_, k2_, k3_}] :=
    RightHandSide[{"Lotka-Volterra"},
        {k1, k2, k3}, {x, y},
        ExternalSpecies -> {"A", "B"}]
G = Transpose[rhs /@ IdentityMatrix[3]];
rhs[{u, v, w}] == G.{u, v, w}
```

The knowledge obtained here will turn out to be really useful when solving one of the most important inverse problems: that of estimating the reaction rate coefficients (see Chap. 11).

6.3.5   Evolution Equation for the Reaction Extent

The concept of the **extent of reaction steps** is often used in classical textbooks on reaction kinetics, although without a general definition (Atkins and Paula 2013, p. 201). The extent of the rth reaction step expresses how often this reaction step has taken place. (It will turn out that its stochastic counterpart is exactly the number of reaction steps having occurred up to a given time.) The following form of the induced kinetic differential equation $\mathbf{c}(t) = \mathbf{c}(0) + \boldsymbol{\gamma} \int_0^t \mathbf{w}(\mathbf{c}(s)) \, \mathrm{d}s$ shows that this is exactly $\int_0^t \mathbf{w}(\mathbf{c}(s)) \, \mathrm{d}s$; therefore we introduce the formal definition.

Definition 6.13   The reaction extent is defined to be $\mathbf{U}(t) := \int_0^t \mathbf{w}(\mathbf{c}(s)) \, \mathrm{d}s$ for all t in the domain of the solution of the induced kinetic differential equation.

To have an evolution equation for the reaction extent, let us take the derivative of its defining equality to get $\dot{\mathbf{U}}(t) = \mathbf{w}(\mathbf{c}(t)) = \mathbf{w}(\mathbf{c}(0) + \boldsymbol{\gamma}\mathbf{U}(t))$ to which one can add if needed the initial condition $\mathbf{U}(0) = \mathbf{0} \in \mathbb{R}^R$.

Let us see an example for the differential equation of the reaction extent.

```
{alpha, gamma} =
    ReactionsData[{"Michaelis-Menten"}]["α", "γ"];
Thread[{u1'[t], u2'[t], u3'[t]} ==
    {k1, k2, k3} Times @@ (({e0, s0, 0, 0}
        + gamma.{u1[t], u2[t], u3[t]})^alpha)]
```

The concept of reaction extent naturally appears in stochastic models as well (see later in Sect. 10.2.2).

Finally, we mention that a code to deal with reactions should start with **parsing**, with the translation of reaction steps into a (deterministic or stochastic) model, as it was realized very early in the history of computer programs aimed at studying reactions (see Chap. 12).

6.4   Usual Categorial Beliefs

Large classes of nonlinear differential equations have been defined by different authors, and sometimes they are believed to incorporate the class of induced kinetic differential equations. Here we show that this is not the case in general, even if we restrict ourselves to the case of mass action type kinetics.

6.4.1 Generalized Lotka–Volterra Systems

The classical authors of population dynamics, Lotka and Volterra, elaborated models to describe oscillatory behavior of biological populations (see, e.g. Lotka 1925). When oscillatory chemical reactions started to arouse interest, the Lotka–Volterra model became a starting point that is why some mathematicians think that at least the generalized Lotka–Volterra model is capable of describing induced kinetic differential equations of—at least—second-order reactions. Let us investigate this statement in detail.

Definition 6.14 The differential equation

$$\dot{x}_m = x_m \left(\sum_{p \in \mathcal{M}} a_{mp} x_p + b_m \right) \quad (m \in \mathcal{M}) \tag{6.34}$$

or, shortly, $\dot{\mathbf{x}} = \mathbf{x} \odot (\mathbf{Ax} + \mathbf{b})$ is a **generalized Lotka–Volterra system**, if $M \in \mathbb{N}; a_{mp}, b_m \in \mathbb{R}$ for $m, p \in \mathcal{M}$. The a_{mp} entry of the **community matrix** $\mathbf{A}$ expresses the effect of the pth species on the mth species, while the component b_m of the vector $\mathbf{b}$ expresses birth or death depending on whether it is positive or negative.

The technical terms in the above definition reveal that generalized Lotka–Volterra systems are mainly, although not exclusively, used in population dynamics.

Example 6.15 Obviously, neither the induced kinetic differential equation $\dot{c} = c^3$ of the reaction $3\,\mathrm{X} \xrightarrow{1} 4\,\mathrm{X}$, nor the induced kinetic differential equation

$$\dot{a} = -k_1 a + k_3 bc \quad \dot{b} = k_1 a - k_2 b^2 - k_3 bc \quad \dot{c} = -k_2 b^2$$

of the Robertson reaction, nor even the induced kinetic differential equation $\dot{c} = 1$ of the reaction $0 \xrightarrow{1} \mathrm{X}$ is of the generalized Lotka–Volterra form. Still—no wonder—that the induced kinetic differential equation of the Lotka–Volterra reaction is a generalized Lotka–Volterra system.

It is not only that the class of kinetic differential equations is not a proper subset of Lotka–Volterra systems, but the case is just the opposite: All Lotka–Volterra systems are kinetic differential equations, because they contain no negative cross effect (see Definition 6.24 below). One can even explicitly define a reaction having Eq. (6.34) as its induced kinetic differential equation as follows:

$$\mathrm{X}_m + \mathrm{X}_p \xrightarrow{a_{mp}} ((1 + \mathrm{sign}(a_{mp}))\mathrm{X}_m + \mathrm{X}_p \quad \mathrm{X}_m \xrightarrow{b_m} (1 + \mathrm{sign}(b_m))\mathrm{X}_m.$$

Another direction of generalization of the Lotka-Volterra model has been introduced and investigated by Farkas and Noszticzius (1985), Dancsó et al. (1991), and Boros et al. (2017a,b).

6.4.2 Kolmogorov Systems

A possible generalization of Lotka–Volterra systems can be obtained when the second factor on the right-hand side of (6.34) is not necessarily linear, just an arbitrary continuously differentiable function; see, e.g., Hirsch et al. (2004), pp. 246–253 (where one cannot find the name of Kolmogorov) or Sigmund (2007).

Definition 6.16 The differential equation

$$\dot{x}_m = x_m \cdot g_m \circ \mathbf{x} \quad (m \in \mathcal{M}) \tag{6.35}$$

or, shortly, $\dot{\mathbf{x}} = \mathbf{x} \odot (\mathbf{g} \circ \mathbf{x})$ is a **Kolmogorov system**, where $M \in \mathbb{N}$; $g_m \in \mathscr{C}^1(\mathbb{R}^M)$ for $m \in \mathcal{M}$ and $\mathbf{g} = \begin{bmatrix} g_1 & g_2 & \cdots & g_M \end{bmatrix}^{\top}$.

Example 6.17 Obviously, neither the induced kinetic differential equations of the Example 6.15 nor the induced kinetic differential equation

$$\dot{x} = a + k_1 x^2 y - bx - x \quad \dot{y} = -k_1 x^2 y + bx \tag{6.36}$$

of the Brusselator by Prigogine and Lefever (1968)

$$0 \underset{1}{\overset{a}{\rightleftharpoons}} X \quad 2X + Y \xrightarrow{k_1} 3X \quad X \xrightarrow{b} Y$$

is a Kolmogorov system. The induced kinetic differential equation

$$\dot{x} = k_1 x^2 y \quad \dot{y} = k_2 x y^2$$

of the reaction $2X + Y \xrightarrow{k_1} 3X + Y \quad X + 2Y \xrightarrow{k_2} X + 3Y$ is a Kolmogorov equation which is not a Lotka–Volterra system.

Let us note that a Kolmogorov system with any polynomial $\mathbf{f}$ is the induced kinetic differential equation of some reaction, because they cannot contain negative cross effect (see Definition 6.24 below).

6.4.3 Monotone Systems

Definition 6.18 The differential equation

$$\dot{\mathbf{x}} = \mathbf{f} \circ \mathbf{x} \tag{6.37}$$

is a **monotone system**, Hirsch and Smith (2005, 2006) and Smith (1995, 2008), if $\mathbf{y}_0 \leq \mathbf{z}_0$ implies that

$$\mathbf{y}(t) \leq \mathbf{z}(t) \quad (t \in \mathscr{D}_\mathbf{y} \cap \mathscr{D}_\mathbf{z}), \tag{6.38}$$

where $\mathbf{y}$ and $\mathbf{z}$ are solutions to (6.37) with initial conditions $\mathbf{y}(0) = \mathbf{y}_0$ and $\mathbf{z}(0) = \mathbf{z}_0$, respectively.

Theorem 6.19 *Equation* (6.37) *is monotone if and only if for all* $\overline{\mathbf{x}} \in \mathscr{D}_f$, *one has that the off-diagonal elements of* $\mathbf{f}'(\overline{\mathbf{x}})$ *are nonnegative (in other words* $\mathbf{f}'(\overline{\mathbf{x}})$ *is a Metzler-matrix).*

It is easy to find induced kinetic differential equations having and also those not having this property. The induced kinetic differential equations of first-order reactions always are monotonous systems, because the coefficient matrix is a Metzler-matrix as a consequence of the lack of negative cross effect.

Remark 6.20 Monotonicity above can be generalized as follows. In the Definition 6.18, $\mathbf{y}_0 - \mathbf{x}_0 \in (\mathbb{R}_0^+)^M$ implies $\mathbf{y}(t) - \mathbf{x}(t) \in (\mathbb{R}_0^+)^M$. Instead of $(\mathbb{R}_0^+)^M$ one can take another **cone** $\mathscr{K} \subset \mathbb{R}^M$ to arrive at the definition of monotonicity with respect to an arbitrary cone.

Banaji (2009) investigates the problem in this more general setting and provides sufficient conditions to ensure monotonicity of the induced kinetic differential equation of a reaction. The goal of the author is rather finding cones which a given reaction preserves than to find reactions which preserve the most important cone, the first orthant.

Remark 6.21 An important property of monotone systems is that if $\mathbf{f}(\mathbf{0}) = \mathbf{0}$, then solutions starting from nonnegative initial vectors remain nonnegative throughout their domain, cf. Theorem 8.6.

An induced kinetic differential equation is almost never a monotone system with respect to the first orthant; consider, e.g., the induced kinetic differential equation of the Lotka–Volterra reaction (see Problem 6.13). See however Problem 6.12.

Example 6.22 Suppose that in the reaction (De Leenheer et al. 2006)

$$C_1 \rightleftharpoons C_2 \rightleftharpoons \cdots \rightleftharpoons C_n \rightleftharpoons C_{n+1} \rightleftharpoons \cdots \rightleftharpoons C_{N-1} \rightleftharpoons C_N$$

at least one of the complexes C_n is nontrivial, and each species is part of precisely one complex. Then the induced kinetic differential equation of the reaction (with mass action type kinetics or even with kinetics which is slightly more general) can be transformed into a cooperative system, i.e., into a system belonging to a subclass of monotone systems—with all the benefits of knowledge on monotone systems relating longtime behavior (Smith 1995, 2008).

6.5   Polynomial and Kinetic Differential Equations

We learned the definition of an induced kinetic differential equation and we also learned what it is not. Definitions are needed to exactly tell what it is.

6.5.1   Polynomial and Quasipolynomial Differential Equations

Definition 6.23 Let $M \in \mathbb{N}$, and let $\Omega \subset \mathbb{R}^M$ be a connected open set, $\mathbf{f} : \Omega \longrightarrow \mathbb{R}^M$ be a function with the property, that all its coordinate functions f_m are polynomials of all its variables. (Cf. Problem 6.14.) Then,

$$\dot{\mathbf{x}} = \mathbf{f} \circ \mathbf{x} \tag{6.39}$$

is said to be a **polynomial differential equation**, or a **polynomial system** for short.

Polynomial differential equations have the advantageous property that the Taylor series of their solutions can explicitly be written down. Should one finish studying (generalized) polynomial equations at this point, let alone kinetic differential equations? Not at all, fortunately. First of all, the explicit solution gives one of the solutions and says nothing about the domain of the solution. Truncating the Taylor series, one is able to get quantitative information or even elaborate a numerical method (Brenig et al. 1996), but it gives no information about the qualitative properties (such as stability of equilibria, existence of periodic solutions, etc.). Think of the fact that the Taylor series of the sine function does not immediately show the periodicity of the function. Still, we think that this formula should deserve more interest than it does up to the present.

Another area of investigation of the above mentioned authors (including A. Goriely, as well; see e.g. Brenig and Goriely 1989) is to find appropriate forms of polynomial differential equations. The earlier author Beklemisheva (1978) investigated the same class of models with very similar tools. Generalized or quasi-polynomial systems have also been used for control purposes by Magyar et al. (2008) and Szederkényi et al. (2005).

6.5.2   The Absence of Negative Cross Effect

Definition 6.24 Consider the (6.39) polynomial system, and suppose that for all $m \in \mathcal{M}$, the coordinate function f_m has the property that its value at any argument $\mathbf{c} \in \mathbb{R}^M$ (after simplification) contains only terms with negative sign which do depend on c_m. Then the given polynomial system is said to have **no negative cross effect**.

Example 6.25 The **Lorenz equation**

$$\dot{x} = \sigma y - \sigma x \quad \dot{y} = \varrho x \boxed{-xz} \quad \dot{z} = xy - \beta z \qquad (6.40)$$

or the equation of the **harmonic oscillator**

$$\dot{x} = y \quad \dot{y} = \boxed{-x} \qquad (6.41)$$

does have negative cross effect; the terms showing this are put in a box.

Example 6.26 The induced kinetic differential equation of the Lotka–Volterra reaction (6.28)

$$\dot{x} = k_1 x - k_2 xy \quad \dot{y} = k_2 xy - k_3 y \qquad (6.42)$$

has no negative cross effect, as all the terms with negative sign $-k_2 xy, -k_3 y$ do depend on the corresponding variable x and y, respectively.

How to check this property? We give two solutions.

```
CrossEffectQ[polyval_,vars_] :=
    Module[{M = Length[vars]},
        And @@ (Map[Not[Negative[#]]&,
        Flatten[MapThread[ReplaceAll,
        {MonomialList[polyval, vars],
        Thread[vars -> #]& /@
        (1 - IdentityMatrix[M])}]]])].
```

And now let us use the newly defined function.

```
CrossEffectQ[{d, c - 4y x^2 + 5x y + 6z + 7w,
    a x + 2y, -b x y}, {x, y, z, w}]
```

The answer is—as expected—depending on the signs of the parameters.

```
!Negative[d] && !Negative[c] && !Negative[a] &&
    !Negative[-b] .
```

A more easily readable version leads to the same result.

```
CrossEffectQ2 [polyval_, vars_] :=
Module [{M = Length [vars], L},
    L [i_] := If [Head [polyval [[i]]] === Plus,
    Apply [List, polyval [[i]]], {polyval [[i]]}]
    /. MapThread [Rule, {vars,
    ReplacePart [ConstantArray [1, M], {i} -> 0]}];
And @@ Map [# >=0&, Flatten [Table [L [i], {i, 1, M}]]]
]
```

Using this function for the same example

```
CrossEffectQ2 [{d, c - 4y x^2 + 5x y + 6z + 7w,
    a x + 2y, -b x y}, {x, y, z, w}]
```

we obtain the result

```
d >= 0 && c >= 0 && a >= 0 && -b >= 0.
```

Now we are in the position to characterize the induced kinetic differential equations' reactions endowed with mass action type kinetics.

Theorem 6.27 (Hárs and Tóth (1979)) *A polynomial system is the induced kinetic differential equation of a reaction endowed with mass action type kinetics if and only if it has no negative cross effects.*

Proof First, we show that the induced kinetic differential equation of a reaction endowed with mass action type kinetics cannot contain negative cross effect. Let us consider the form (6.7), i.e.,

$$\dot{c}_m(t) = \sum_{r \in \mathscr{R}} (\beta(m, r) - \alpha(m, r)) k_r \prod_{p=1}^{M} c_p(t)^{\alpha(p,r)} \quad (m \in \mathscr{M}). \tag{6.43}$$

If the term $(\beta(m, r) - \alpha(m, r)) k_r \prod_{p=1}^{M} c_p(t)^{\alpha(p,r)}$ has a negative sign, then $\beta(m, r) < \alpha(m, r)$, and this inequality together with the fact that $0 \leq \beta(m, r)$ implies that $0 < \alpha(m, r)$, i.e., the product $\prod_{p=1}^{M} c_p(t)^{\alpha(p,r)}$ contains the factor $c_m(t)^{\alpha(m,r)}$, where the exponent is strictly positive.

Second, suppose the right-hand side of the polynomial system

$$\dot{\mathbf{x}} = \mathbf{Z}\mathbf{x}^{\mathbf{a}}, \tag{6.44}$$

where $\mathbf{Z} \in \mathbb{R}^{M \times R}, \mathbf{a} \in \mathbb{N}_0^{M \times R}, M, R \in \mathbb{N}$ does not contain negative cross effect, which means that $Z(m, r) < 0$ implies that $a(m, r) \geq 1$, and let us show that there is a reaction having the given polynomial system as its induced kinetic differential equation. We construct the reaction steps and also give the reaction rate coefficients of the reaction which one may call the **canonical realization** of the given polynomial system.

A term on the right-hand side of the mth equation which has a positive sign is of the form

$$Z(m, r) \prod_{p \in \mathcal{M}} x_p^{a(p,r)} \tag{6.45}$$

with $Z(m, r) \in \mathbb{R}^+$. The reaction step $\mathbf{a}(\cdot, r) \xrightarrow{Z(m,r)} \mathbf{a}(\cdot, r) + \mathbf{e}_m$ contributes to the induced kinetic differential equation with the single term (6.45) and nothing else.

A term on the right-hand side of the mth equation which has a negative sign is of the form

$$Z(m, r) \prod_{p \in \mathcal{M}} x_p^{a(p,r)} \tag{6.46}$$

with $-Z(m, r) \in \mathbb{R}^+$ and $a(m, r) \geq 1$. The reaction step $\mathbf{a}(\cdot, r) \xrightarrow{-Z(m,r)} \mathbf{a}(\cdot, r) - \mathbf{e}_m$ contributes to the induced kinetic differential equation with the single term (6.46) and nothing else. Thus we have constructed in an algorithmic way an inducing reaction to the given polynomial system without negative cross effect. □

A good review of our earlier results has been given by Chellaboina et al. (2009). Applications of the above characterization have been given from and outside our group, as well (see the later chapters).

As to the meaning of the theorem, one may put it also this way: A large subset of polynomial equations are kinetic. In connection with this, we can mention the paper by Kowalski (1993) asserting that practically all nonlinear systems in the open first orthant can be formulated as kinetic differential equations. If one restricts the investigations onto the open first orthant, one can also transform polynomial differential equations into kinetic ones generalizing the idea of Samardzija et al. (1989) given by Crăciun: One can multiply the right-hand sides with the product of all the variables.

Let us also mention that Dickenstein and Millán (2011), Remark 2.1, is a good paragraph on the different formulations of the induced kinetic differential equation used by different authors.

6.5.3 Quadratic Right-Hand Sides

It is known that polynomial differential equations can be transformed via introduction of new variables into quadratic polynomial equations (Myung and Sagle 1992). The advantage of this transformation is that polynomial differential equations with quadratic polynomials on their right-hand side can be investigated using methods of **nonassociative algebras** (Markus 1960; Kaplan and Yorke 1979; Myung and Sagle 1992). The question obviously arises if starting from a kinetic differential equation, is it possible to remain within this set when one applies such a transformation?

Theorem 6.28 (Halmschlager and Tóth (2004)) *Given a kinetic differential equation of a mass action type mechanism, there exists a smooth transformation of the variables in which the equation has a homogeneous quadratic polynomial without negative cross effect as a right-hand side.*

Proof Instead of giving a formal proof here (cf. Problems 6.16 and 6.17), we show an example the steps of which can obviously be carried out in the general case, as well.

Let us consider the induced kinetic differential equation

$$\dot{x} = a + x^2 y - (b+1)x \quad \dot{y} = bx - x^2 y, \tag{6.47}$$

of the Brusselator $0 \xrightarrow{a} X \quad 2X + Y \xrightarrow{1} 3X \quad X \xrightarrow{b} Y \quad X \xrightarrow{1} 0$, where $a, b \in \mathbb{R}^+$. Let z be such that $\dot{z} = 0$, $z(0) = 1$; then introducing the variables

$$\xi_1 := x^2 \quad \xi_2 := y^2 \quad \xi_3 := z^2$$
$$\xi_4 := xy \quad \xi_5 := xz \quad \xi_6 := yz \tag{6.48}$$
$$\xi_7 := x \quad \xi_8 := y \quad \xi_9 := z$$

one gets the following system:

$$
\begin{aligned}
\dot{\xi}_1 &= 2x(a + x^2 y - (b+1)x) = 2a\xi_7\xi_9 + 2\xi_1\xi_4 - 2(b+1)\xi_1\xi_9 \\
\dot{\xi}_2 &= 2y(bx - x^2 y) & = 2b\xi_7\xi_8 - 2\xi_1\xi_2 \\
\dot{\xi}_3 &= 0 & = 0 \\
\dot{\xi}_4 &= (a + x^2 y - (b+1)x)y \\
&\quad + x(bx - x^2 y) & = a\xi_8\xi_9 + \xi_1\xi_2 - (b+1)\xi_4\xi_9 + b\xi_7^2 - \xi_1\xi_4 \\
\dot{\xi}_5 &= (a + x^2 y - (b+1)x)z & = a\xi_9^2 + \xi_1\xi_6 - (b+1)\xi_5\xi_9 \\
\dot{\xi}_6 &= (bx - x^2 y)z & = b\xi_7\xi_9 - \xi_1\xi_6 \\
\dot{\xi}_7 &= a + x^2 y - (b+1)x & = a\xi_9^2 + \xi_1\xi_8 - (b+1)\xi_7\xi_9 \\
\dot{\xi}_8 &= bx - x^2 y & = b\xi_7\xi_9 - \xi_1\xi_8 \\
\dot{\xi}_9 &= 0 & = 0.
\end{aligned}
$$

□

Remark 6.29 Allowing fractional or even negative stoichiometric coefficients might be useful sometimes as in the model

$$\dot{x} = -k_1 xy + k_2 ay - 2k_3 x^2 + k_4 ax - 0.5 k_5 xz \tag{6.49}$$

$$\dot{y} = -k_1 xy - k_2 ay + k_6 mz \tag{6.50}$$

$$\dot{z} = +2k_4 ax - k_5 xz - 2k_6 mz \tag{6.51}$$

of a version of the Oregonator (Turányi et al. 1993)

$$X + Y \xrightarrow{k_1} P \qquad Y + A \xrightarrow{k_2} X + P \qquad 2X \xrightarrow{k_3} P + A \tag{6.52}$$

$$X + A \xrightarrow{k_4} 2X + 2Z \qquad X + Z \xrightarrow{k_5} 1/2X + A \qquad Z + M \xrightarrow{k_6} Y - Z \tag{6.53}$$

(where X = $HBrO_2$, Y = Br^-, Z = Ce_4^+, A = BrO_3^-, P = HOBr, M = malonic acid) of the Belousov–Zhabotinsky reaction. Still, the induced kinetic differential equation can be considered as the induced kinetic differential equation of a reaction because the right-hand side contains no negative cross effect (see Problem 2.1).

In some cases, however, this may lead to the appearance of negative cross effects, as the simplest possible example $0 \xrightarrow{1} -X$ with the induced kinetic differential equation $\dot{c} = -1$ shows.

6.5.4 Examples

Let us see a few examples of reactions with different kinetics.

Example 6.30 (Mass Action Kinetics)

1. Decomposition of nitrogen penta-oxide follows a first-order reaction step.

$$N_2O_5 \xrightarrow{k} 2NO_2 + \frac{1}{2}O_2 \qquad w([N_2O_5], [NO_2], [O_2]) := k[N_2O_5].$$

2. Decomposition of ammonium nitrite in aqueous solution also follows a first-order reaction step.

$$NH_4NO_2 \xrightarrow{k} N_2 + 2H_2O \qquad w([NH_4NO_2], [N_2], [H_2O]) := k[NH_4NO_2].$$

3. Decomposition of nitrogen peroxide follows a second-order reaction step.

$$2NO_2 \xrightarrow{k} 2NO + O_2 \qquad w([NO_2], [NO], [O_2]) := k[NO_2]^2.$$

4. The reaction between nitric oxide and oxygen is a third-order reaction.

$$2\,NO + O_2 \xrightarrow{k} 2\,NO_2 \quad w([NO], [O_2], [NO_2]) := k[NO]^2[O_2].$$

Example 6.31 (Non-mass Action Kinetics)

1. Although we have a second-order reaction, the rate being not of the mass action type is a zeroth-order function of the relevant concentration. Ammonia (NH_3) gas decomposes over platinum catalyst to nitrogen gas (N_2) and hydrogen gas (H_2).

$$2\,NH_3 \xrightarrow{Pt} N_2 + 3\,H_2 \quad w([NH_3], [N_2], [H_2]) := k.$$

2. Although we have a third-order reaction, the rate being not of the mass action type is a second-order function of the relevant concentrations. Nitrogen dioxide (NO_2) gas reacts with fluorine gas (F_2) to give nitrosyl fluoride.

$$2\,NO_2 + F_2 \longrightarrow 2\,NO_2F \quad w([NO_2], [F_2], [NO_2F]) := k[NO_2][F_2].$$

3. Although we have a second-order reaction, the rate being not of the mass action type is a fractional order function of the relevant concentration. Hydrogen (H_2) gas reacts with bromine (Br_2) gas to give hydrogen bromide vapor.

$$H_2 + Br_2 \longrightarrow 2\,HBr \quad w([H_2], [Br_2], [HBr]) := k[H_2][Br_2]^{1/2}.$$

4. Let us see another example of a reaction with a reaction rate of fractional order.

$$CH_3CHO \longrightarrow CH_4 + CO \quad w([CH_3CHO], [CH_4], [CO]) := k[CH_3CHO]^{3/2}.$$

5. In enzyme kinetics very often rational functions are used as rate functions, e.g.,

$$w(s) := \frac{k_1 e_0 s}{K_M + s}$$

   can be used as the rate of the reaction S $\longrightarrow$ P with positive constants k_1, e_0, K_M. The rate is positive if s is positive and zero if $s = 0$; thus Conditions 2 are fulfilled.

6. The above special type of non-mass action kinetics, the Michaelis–Menten-type (or Holling-type) kinetics, is often used in population biology (Hsu et al. 2001; Kiss and Kovács 2008; Kiss and Tóth 2009; May 1974).

7. The reaction $H_2 + Br_2 \longrightarrow 2\,HBr$ is usually described in such a way that the rate of production of HBr is given by the expression

$$\frac{k[H_2][Br_2]^{3/2}}{K[HBr] + [Br_2]}.$$

(Actually, this rate comes as a result of an approximation; see Problem 9.26.) Again, this rate is zero, if either $[H_2]$ or $[Br_2]$ is zero, and is positive if both are positive.

6.6   The Graphs of Element Fluxes

In this section we add a graph to the ones defined in Chap. 3, namely, graphs useful in following the fluxes of the individual elements. Here we follow the clear description by Turányi and Tomlin (2014), pp. 54–55; see also Turns (2000), pp. 165–168, and the original source by Revel et al. (1994) or the verbal description by Orth et al. (2010). If the number of atoms is $D \in \mathbb{N}$, then we shall altogether have D graphs, one for each atom.

Let us consider the reaction (2.1) with arbitrary kinetics, $X(m) = Z(1, m)A(1) + \cdots + Z(D, m)A(D) \quad \forall m \in \mathcal{M}$, with the atoms $A(1), A(2), \ldots, A(D)$ and suppose also that the number of each atom is the same on the two sides of the reaction steps, i.e., the reaction (2.1) obeys the law of atomic balance, i.e., $\mathbf{Z}\boldsymbol{\gamma} = \mathbf{0} \in \mathbb{N}_0^{D \times R}$ (Definition 4.11). Let us construct a weighted directed graph for the atom $A(d)$, where $d \in \{1, 2, \ldots, D\}$ in the following way. The vertices of the graphs will be the species, and two species $X(p)$ and $X(q)$ are connected with an arrow if there is a directed path from $X(p)$ to $X(q)$ of length two in the Volpert graph of the reaction. (More explicitly, if $X(q)$ is directly produced from $X(p)$, i.e., there is a reaction step r with the reactant complex containing the species $X(p)$ and the product complex containing $X(q)$.) This arrow receives the weight

$$\mathrm{Flux}_{p,q}^{d,r}(t) := \frac{Z(d,p)\alpha(p,r)Z(d,q)\beta(q,r)}{\sum_{m \in \mathcal{M}} Z(d,m)\alpha(m,r) + \sum_{m \in \mathcal{M}} Z(d,m)\beta(m,r))} w_r(\mathbf{c}(t)) \qquad (6.54)$$

$$(p, q \in \mathcal{M}; d \in \{1, 2, \ldots, D\}; r \in \mathcal{R}\})$$

for times $t \in \mathscr{D}_{\mathbf{c}}$, i.e., in the domain of $\mathbf{c}$.

Remark 6.32 Let us analyze this expression.

1. Following Turányi and Tomlin (2014) and differently from Revel et al. (1994), we dropped the dot above the flux (which might have directly emphasized that the flux is a velocity-type quantity).

2. The numerator shows that the larger the flux is, the more atoms we have on both sides of the reaction step. (The flux is proportional to these numbers.)
3. The denominator contains the total number of $A(d)$ atoms on the two sides of the rth reaction step.
4. The formula also shows that in case of reversible reaction steps, the weights will be the same in both directions.
5. Fluxes of radicals can also be studied by the corresponding flux graph; however, in this case it is the user (the chemist) who is to provide the coefficients of the radicals in the individual species (or to form the **generalized atomic matrix**): It cannot be calculated automatically from the numbers of the individual atoms even if the constitution of the radical is given.
6. Note that the weights depend on the initial concentrations and also on the reaction rates or—in case of mass action kinetics—on reaction rate coefficients.
7. The weights are time dependent as well; to calculate these the solution of the induced kinetic differential equation is needed. However, if one is only interested in the ratios of the atomic fluxes, then in some very special cases, it may happen that $w_r(\mathbf{c}(t))$ is not needed (see Example 6.33 below).
8. The weights might also be time independent, if one calculates the reaction rates at a **stationary point**, i.e., at a solution $\mathbf{c}_*$ of $\mathbf{f}(\mathbf{c}) = \mathbf{0}$, cf. Definition 7.3.
9. If there are more than one reaction steps in which $X(p)$ is transformed into $X(q)$, then the fluxes are to be summed up:

$$\text{Flux}_{p,q}^d(t) := \sum_{r \in \mathscr{R}} \text{Flux}_{p,q}^{d,r}(t).$$

10. Suppose some of the reaction steps are reversible. In this case we have two choices. We can proceed as usual to get the flux graphs. The second alternative is that for all pairs of species connected with a directed path of length two in the Volpert graph, we calculate the **net flux**, i.e., the differences of the fluxes in the two directions, and we draw an arrow from $X(p)$ to $X(q)$ with the difference as the weight in case this difference is positive and in the opposite direction if it is negative. (In this case the weight will be the negative of the differences between reaction rates.) As it may change sign during the reaction, the structure of the graph may depend on time.

Example 6.33 Let us consider the methanol–formic acid esterification Example (4.13):

$$\text{HCOOH} + \text{CH}_3\text{OH} \longrightarrow \text{HCOOCH}_3 + \text{H}_2\text{O}, \qquad (6.55)$$

and suppose the reaction rate is given by the function $w \in \mathscr{C}^1(\mathbb{R}^4, \mathbb{R}_0^+)$. Here we have the $R = 1$ reaction steps among $M = 4$ species composed of $D = 3$ atoms.

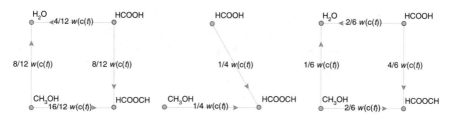

Fig. 6.1 Fluxes of H, C, and O in the reaction (4.13)

The matrices of stoichiometric coefficients and the atomic matrix are

$$
\alpha = \begin{bmatrix} 1 \\ 1 \\ 0 \\ 0 \end{bmatrix}, \quad
\beta = \begin{bmatrix} 0 \\ 0 \\ 1 \\ 1 \end{bmatrix}, \quad
\mathbf{Z} = \begin{bmatrix} 2 & 4 & 4 & 2 \\ 1 & 1 & 2 & 0 \\ 2 & 1 & 2 & 1 \end{bmatrix},
$$

if the species and atoms are numbered in the order of their appearance. Fulfilment of obeying the law of atomic balance is shown by

$$
\mathbf{Z} \cdot \gamma = \mathbf{Z} \cdot (\beta - \alpha) = \begin{bmatrix} 0 \\ 0 \\ 0 \end{bmatrix}.
$$

As we have three atoms, we have also three graphs for element fluxes. (The total number of H, C, and O atoms, respectively, is 6, 2, and 3.) All are shown in Fig. 6.1. If one is only interested in the ratios of the fluxes, then the common factor $w(\mathbf{c}(t))$ can be deleted from all the edges. This is not the case in more complicated examples with more than one reaction step (see Sect. 6.8 of Exercises and Problems).

6.7 Temperature and Diffusion

In Chap. 2 we described restrictions as to the generality of the investigated models. Here we add a few words on disregarded physical effects.

6.7.1 Temperature

Temperature dependence does not seem to be important practically when modeling metabolic processes, but it is strongly needed in atmospheric chemistry and combustion. (Theoretically it is important everywhere, because all the reaction steps are either **exothermic** or **endothermic**: either requiring or producing heat.)

Let us consider the simple first-order decomposition of a species, the concentration of which at time t is denoted by $c(t)$, and assume that the reaction rate coefficient k does depend on the temperature T, and this dependence can be described by the **Arrhenius form** (Nagy and Turányi (2011); Tomlin et al. (1992), Volpert and Hudyaev (1985), p. 611):

$$k(T) = k_0 e^{-\frac{A}{RT}}, \tag{6.56}$$

where the constants k_0, A, R are as follows:

- k_0 is the **preexponential factor** having a positive numerical value,
- A is the **activation energy** (in J/mol) of the reaction step having a positive numerical value in general (although in exceptional cases it can also be negative),
- $R = 8.314 \frac{J}{mol \cdot K}$ is the **universal gas constant**.

Then the usual model to describe this process is

$$\dot{c}(t) = -k_0 e^{-\frac{A}{RT(t)}} c(t) \tag{6.57}$$

$$\dot{T}(t) = k_0 e^{-\frac{A}{RT}} c(t) Q - a(T(t) - T_a), \tag{6.58}$$

where the further constants Q, a, T_a are as follows:

- Q is the thermal effect: the ratio of the standard molar reaction step enthalpy ΔH and the constant pressure heat capacity C_p, Turányi and Tomlin (2014), p. 11; having the unit K/mol and being positive, if the step is **endothermic**, and negative, if the step is **exothermic**,
- a is the **coefficient of heat transfer**, reciprocal of the **characteristic time** of the system, having the unit $\frac{1}{s}$ and being positive,
- T_a is the **ambient temperature** with the unit K.

Remark 6.34

- The first term in the second equation expresses the effect of reaction step on the temperature of the ambience. If $Q < 0$, then the reaction step is **exothermic**: Heat is absorbed from the surroundings. If $Q > 0$, then the reaction step is **endothermic**: Heat is released into the surroundings.
- The second term describes **Newton's law of cooling**: If the temperature of the reaction vessel is higher than the ambient temperature, then it will cool down; if it is lower, then it will warm up.
- Including temperature dependence in this form (or in a similar way in more complicated cases) causes no problem in the numerical treatment, but only a part of the theory we are going to explicate in the present book can directly be applied for this situation.

The left side of Fig. 6.2 shows that the decay is much faster in the case of an endothermic reaction. The right-hand side shows that the temperature tends to a constant value, but this is higher than the initial one in the case of an endothermic reaction and lower than the initial one in the case of an exothermic reaction.

The difference of concentrations with changing and constant (300 K) temperature can be seen on Fig. 6.3. If the reaction is endothermic, then the concentration is always smaller with changing temperature than with a constant temperature. However, if the reaction is exothermic, then the concentration is always larger with changing temperature than with a constant temperature.

Now, let us formulate the evolution equation for reactions involving temperature effects. The kinetics is assumed to be of the mass action type; the temperature dependence is allowed to be of the **generalized Arrhenius type** (see, e.g., Nagy

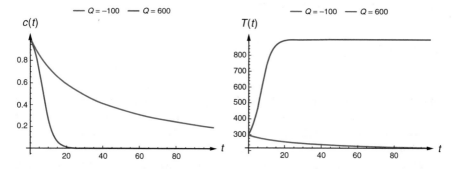

Fig. 6.2 The change of concentration and temperature in a reaction described by (6.57) and (6.58) with $k_0 = R = 1$, $A = 1000$, $a = 0$; $T(0) = 300$, $c(0) = 1$ and different values of Q

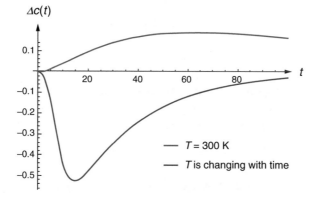

Fig. 6.3 The difference of concentrations with changing and constant (300 K) temperature with $k_0 = R = 1$, $A = 1000$, $a = 0$; $T(0) = 300$, $c(0) = 1$, and different values of Q

and Turányi 2011).

$$\dot{c}_m(t) = \sum_{r \in \mathscr{R}} \gamma(m, r) k_r^0 T(t)^{n_r} e^{-\frac{A_r}{RT(t)}} \mathbf{c}(t)^{\alpha(\cdot, r)} \quad (m \in \mathscr{M})$$

$$\dot{T}(t) = -\left(\sum_{r \in \mathscr{R}} \sum_{m \in \mathscr{M}} Q_r k_r^0 T(t)^{n_r} e^{-\frac{A_r}{RT(t)}}\right) - a(T(t) - T_a)$$

Here n_r is a parameter usually obtained by fitting the generalized Arrhenius form to time-dependent data on the reaction rate coefficient; all the other parameters have similar meaning as above; if they have the index r, then they may be different for different reaction steps.

The first term of the second equation expresses the fact that the reaction step changes the temperature, whereas the second term describes Newton's law of cooling as a result of contact with the ambience.

6.7.2   Diffusion

Diffusion is simpler or more complicated, as you like. Most of our models are **concentrated variable models**, to use the engineering term, expressing the fact that **spatial inhomogeneities** are disregarded. From time to time, we shall also consider diffusion, at least a special form of it: the one without cross-diffusion (see page 107). These models are also known by the name **distributed variable models** as opposed to concentrated variable models as, e.g., ordinary differential equations. The simple first-order decomposition accompanied with diffusion is usually described with the following distributed variable model which is the partial differential equation:

$$\frac{\partial c(t, x)}{\partial t} = D \frac{\partial^2 c(t, x)}{\partial x^2} - kc(t, x), \tag{6.59}$$

where

- the positive real constant D is the **diffusion coefficient** with the unit $\frac{m^2}{s}$,
- x is the (here, one-dimensional) spatial variable,
- the positive real constant k is the reaction rate coefficient.

Remark 6.35

- To make the model well defined, some additional initial and/or boundary conditions are also to be specified.
- Equations of the form similar to Eq. (6.59) are also used to describe spatiotemporal changes of temperature, as well, and they are also used in probability

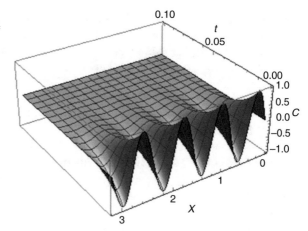

Fig. 6.4 Solution of (6.59) with $D = 1, k = 1, c(0, x) = e^{-0.01x} \sin(8x), c(t, 0) = 0, c(t, \pi) = 0$

theory to describe the change of the probability density of some stochastic processes.

- As to the general treatment of reaction-diffusion processes, mathematicians prefer the book by Smoller (1983). More application oriented are the books by Britton (1986) and Fife (1979).

Figure 6.4 shows the **smoothing effect of diffusion**: The initial waves die out very soon, at around $t = 0.04$.

6.7.3 Temperature and Diffusion Together

An equation (of which the applicability should carefully be investigated in all cases) unifying the effects of temperature and diffusion is

$$\frac{\partial c_m(t, \mathbf{x})}{\partial t} = D_m \Delta c_m(t, \mathbf{x}) + \sum_{r \in \mathcal{R}} \gamma(m, r) k_0^r T(t, \mathbf{x})^{n_r} e^{-\frac{A_r}{RT(t, \mathbf{x})}} \mathbf{c}(t, \mathbf{x})^{\alpha(\cdot, r)},$$

$$(m \in \mathcal{M})$$

$$\frac{\partial T(t, \mathbf{x})}{\partial t} = \lambda \Delta T(t, \mathbf{x}) - \sum_{r \in \mathcal{R}} Q_r k_0^r T(t, \mathbf{x})^{n_r} e^{-\frac{A_r}{RT(t, \mathbf{x})}} \mathbf{c}(t, \mathbf{x})^{\alpha(\cdot, r)},$$

where λ is the **heat diffusion coefficient**.

Note that mixing is not taken into consideration, and all the concentrations only affect their own diffusion: **Cross-diffusion** is excluded.

Actually, the best solution would be to start from **thermodynamic** principles and present models as special cases, systems with special constitutive relations. However, it is not possible to find a generally accepted thermodynamic (as opposed to thermo**static**) theory including reactions and physical processes as well (cf. Érdi 1978; Keszei 2011; Kirkwood and Oppenheim 1961; Matolcsi 2005; Rock 2013; Schubert 1976; Vándor 1952). This fact is the more/is even more interesting if we realize that a reacting system should be by definition the object of thermodynamics, because it is a kind of general (macroscopic) physical system theory.

6.8   Exercises and Problems

6.1 Show the equivalence of Eqs. (6.7) and (6.8).

(Solution: page 393.)

6.2 Determine the form of the reaction rate of the reaction step $X + Y + Z \longrightarrow C$ (where C is an arbitrary complex) assuming that it is a nonnegative, monotone, and additive function of all the species concentrations present in the reactant complex.

(Solution: page 393.)

6.3 Prove that mass action kinetics possess both the properties of Condition 2.

(Solution: page 394.)

6.4 How do you construct the vertex-edge incidence matrix of a reaction?

(Solution: page 394.)

6.5 Show that with the notations of the present chapter $\mathbf{YE} = \boldsymbol{\gamma}$ holds.

(Solution: page 394.)

6.6 Construct the adjacency matrix (also called vertex-incidence matrix) of the Michaelis–Menten reaction by hand, and design a general algorithm or a *Mathematica* code to calculate it.

(Solution: page 395.)

6.7 Find the third factor in Eq. (6.24) in the special case of (6.29).

(Solution: page 395.)

6.8 Using the definition $\mathbf{g(c)} := (\mathbf{R(c)} - \mathbf{R}^\top(\mathbf{c}))\mathbf{1}_N$ of the complex formation vector, show that the two forms of the species formation vector $\mathbf{f(c)}$—i.e., $\mathbf{Yg(c)}$ and

$$\sum_{n \in \mathcal{N}} \sum_{q \in \mathcal{N}} r_{nq}(\mathbf{c})(\mathbf{y}_n - \mathbf{y}_q)$$

—are the same.

(Solution: page 395.)

6.9 Calculate explicitly the coefficient matrix $\mathbf{G}$ in (6.33) standing before the vector of rate coefficients.

(Solution: page 395.)

6.10 Show that the induced kinetic differential equation of the reaction in Fig. 6.5 is a generalized Lotka–Volterra system.

(Solution: page 396.)

6.11 Find a reaction with a Kolmogorov system as its induced kinetic differential equation which is not a generalized Lotka–Volterra system.

(Solution: page 396.)

6.12 Show that the induced kinetic differential equation of the consecutive reaction $A \xrightarrow{1} B \xrightarrow{1} C$ is a monotone system.

(Solution: page 396.)

6.13 Show that the induced kinetic differential equation of the Lotka–Volterra reaction is not a monotone system (with respect to the first orthant). Hint: Use the fact of the existence of a nonlinear first integral.

(Solution: page 396.)

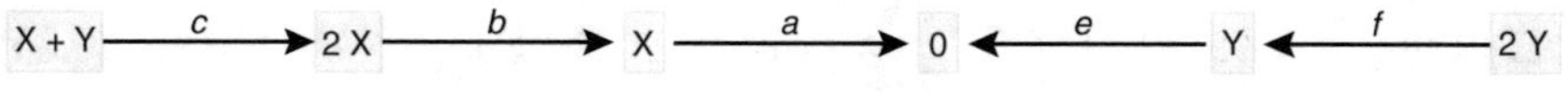

Fig. 6.5 A reaction with induced kinetic differential equation that is a generalized Lotka–Volterra system

6.14 Show that a function $\mathbf{f} = (f_1, f_2) : \mathbb{R}^2 \longrightarrow \mathbb{R}^2$ which is a polynomial in all of its variables is a **multivariate polynomial**, i.e., it is of the form

$$f_m(\mathbf{x}) = \sum_{|\mathbf{k}|=0}^{G_m} a_{\mathbf{k}} \mathbf{x}^{\mathbf{k}} \quad (m = 1, 2)$$

with appropriate integers $G_m \in \mathbb{N}_0$.

(Solution: page 398.)

6.15 Find the solution of the induced kinetic differential equation of the autocatalytic reaction $2\,X \xrightarrow{\;1\;} 3\,X$ in the form of a Taylor series.

(Solution: page 399.)

6.16 Try to find much less number of variables in which the induced kinetic differential equation (6.47) of the Brusselator can be transformed into a homogeneous quadratic differential equation which is also kinetic.

(Solution: page 399.)

6.17 Provide a constructive proof of Theorem 6.28.

(Solution: page 399.)

6.18 Using the program **ReactionKinetics**, show that the law of atomic balance holds in the case of the methanol–formic acid reaction Eq. (4.13).

(Solution: page 400.)

6.19 Construct the flux graph of the radical OH in the reaction (4.13) assuming the reaction rate is given by the function w.

(Solution: page 400.)

6.20 Construct the graphs of atomic fluxes for the reaction

$$H+O_2 \underset{w_{-1}}{\overset{w_1}{\rightleftarrows}} OH+O \quad O+H_2 \underset{w_{-2}}{\overset{w_2}{\rightleftarrows}} OH+H \quad OH+H_2 \underset{w_{-3}}{\overset{w_3}{\rightleftarrows}} H+H_2O \quad (6.60)$$

and assuming the reaction rates as shown.

(Solution: page 400.)

6.21 Show that no reaction containing only three complexes induces the differential equation (8.28) with $a, b, c > 0$.

(Solution: page 400.)

6.22 Reproduce Fig. 6.2 by modifying the reaction from first-order decay to first-order autocatalysis in Eqs. (6.57) and (6.58).

(Solution: page 402.)

6.23 Suppose the dth-order decay $(d \in \mathbb{N}_0)$ $d\,X \longrightarrow 0$ and the temperature change of a reaction step is described by the following set of differential equations:

$$\dot{c} = -e^{-\frac{a}{T}} b c^d \quad \dot{T} = e^{-\frac{a}{T}} c^d$$

(in accordance with (14.36), but the interaction with the neighborhood is neglected and $n = 0$ is assumed), where the constants a, b are positive and their meaning can be found out by comparison with the equations in Sect. 6.7.1. Find c 'as a function of' T, i.e., being interested in solutions for which $\mathscr{R}_T \subset \mathbb{R}^+$. Show that there exists a function C such that $C \circ T = c$. Find also the time behavior of c and T separately.

(Solution: page 403.)

6.9 Open Problems

1. Suppose an induced kinetic differential equation is of the Kolmogorov form. Can we say something about the relevant characteristics (deficiency, number of linkage classes, reversibility and weak reversibility, acyclicity of the Volpert graph etc.) of the underlying reaction?
2. About the transformation leading to homogeneous quadratic kinetic polynomial one also has a few questions.
   a. Can the transform be specialized to be unique with some kinds of chemically relevant restrictions?
   b. What is the minimal number of new variables to arrive at a quadratic kinetic equation? What freedom do we have in choosing the new variables?
   c. How are the relevant characteristics (deficiency, number of linkage classes, reversibility and weak reversibility, acyclicity of the Volpert graph etc.) transformed?
   d. Is it possible to get such a transform which can be realized by a reaction with a prescribed property?
   e. Can the above questions be answered numerically (e.g., using the methods by Szederkényi and coworkers; see, e.g., Johnston et al. 2012a,b, 2013; Szederkényi 2010; Szederkényi and Hangos 2011), or can they be answered

symbolically for any values of the parameters in the original induced kinetic differential equation?

f. Suppose we have a statement about the qualitative or quantitative properties of the transform. What consequences can be drawn about the original solutions from this knowledge?

3. Which are the exact necessary and sufficient conditions to be inflicted on the stoichiometric coefficients leading to a kinetic differential equation? (We have seen that negative sign is not excluded.)

4. Delineate large classes of reactions having an induced kinetic differential equation which is a generalized Lotka–Volterra system, a Kolmogorov system, a monotone system.

5. Relations between the graph of element fluxes and the other graphs defined earlier are to be investigated.

References

Atkins P, Paula JD (2013) Elements of physical chemistry. Oxford University Press, Oxford

Banaji M (2009) Monotonicity in chemical reaction systems. Dyn Syst 24(1):1–30

Beklemisheva LA (1978) Classification of polynomial systems with respect to birational transformations I. Differ Uravn 14(5):807–816

Boros B (2008) Dynamic system properties of biochemical reaction systems. Master's thesis, Eötvös Loránd University, Budapest and Vrije Universiteit, Amsterdam

Boros B (2013) On the positive steady states of deficiency-one mass action systems. PhD thesis, School of Mathematics, Director: Miklós Laczkovich, Doctoral Program of Applied Mathematics, Director: György Michaletzky, Department of Probability Theory and Statistics, Institute of Mathematics, Faculty of Science, Eötvös Loránd University, Budapest

Boros B, Hofbauer J, Müller S (2017a) On global stability of the Lotka reactions with generalized mass-action kinetics. Acta Appl Math 151(1):53–80

Boros B, Hofbauer J, Müller S, Regensburger G (2017b) The center problem for the Lotka reactions with generalized mass-action kinetics. In: Qualitative theory of dynamical systems, pp 1–8

Brenig L, Goriely A (1989) Universal canonical forms for time-continuous dynamical systems. Phys Rev A 40(7):4119–4122

Brenig L, Codutti M, Figueiredo A (1996) Numerical integration of dynamical systems using Taylor expansion. In: Posters of ISSAC'96 (International symposium on symbolic and algebraic computation). http://cso.ulb.ac.be/~mcodutti/Publications/issac96/poster.html

Britton NF (1986) Reaction-diffusion equations and their applications to biology. Academic Press, London

Chellaboina V, Bhat SP, Haddad WM, Bernstein DS (2009) Modeling and analysis of mass-action kinetics. IEEE Control Syst Mag 29:60–78

Dancsó A, Farkas H, Farkas M, Szabó G (1991) Investigations into a class of generalized two-dimensional Lotka–Volterra schemes. Acta Appl Math 23:107–127

De Leenheer P, Angeli D, Sontag ED (2006) Monotone chemical reaction networks. J Math Chem 41(3):295–314

Dickenstein A, Millán MP (2011) How far is complex balancing from detailed balancing? Bull Math Biol 73:811–828

Érdi P (1978) Racionális kémiai termodinamika (Rational chemical thermodynamics). ELTE TTK Kémiai Kibernetikai Laboratórium, Budapest

Farkas H, Noszticzius Z (1985) Generalized Lotka-Volterra schemes and the construction of two-dimensional explodator cores and their liapunov functions via critical hopf bifurcations. J Chem Soc, Faraday Trans 2 81:1487–1505

Fife PC (1979) Mathematical aspects of reacting and diffusing systems. Springer, Berlin

Garay BM (2004) A functional equation characterizing monomial functions used in permanence theory for ecological differential equations. Univ Iagell Acta Math 42:69–76

Halmschlager A, Tóth (2004) Über Theorie und Anwendung von polynomialen Differential-gleichungen. In: Wissenschaftliche Mitteilungen der 16. Frühlingsakademie, Mai 19–23, 2004, München-Wildbad Kreuth, Deutschland, Technische und Wirtschaftswissenschaftliche Universität Budapest, Institut für Ingenieurweiterbildung, Budapest, pp 35–40

Hárs V, Tóth J (1979) On the inverse problem of reaction kinetics. In: Farkas M (ed) Colloquia mathematica societatis János Bolyai. Qualitative theory of differential equations, vol 30, pp 363–379

Hirsch MW, Smith H (2005) Monotone maps: a review. J Differ Equ Appl 11(4):379–398

Hirsch MW, Smith H (2006) Chapter 4 monotone dynamical systems. In: A Cañada PD, Fonda A (eds) Handbook of differential equations: ordinary differential equations, vol 2, North-Holland, pp 239–357. https://doi.org/10.1016/S1874-5725(05)80006-9. http://www.sciencedirect.com/science/article/pii/S1874572505800069

Hirsch MW, Smale S, Devaney RL (2004) Differential equations, dynamical systems, and an introduction to chaos. Pure and applied mathematics, vol 60. Elsevier/Academic, Amsterdam/Cambridge

Hsu SB, Hwang TW, Kuang Y (2001) Global analysis of the Michaelis–Menten-type ratio-dependent predator-prey system. J Math Biol 42(6):489–506

Johnston MD, Siegel D, Szederkényi G (2012a) Computing linearly conjugate chemical reaction networks with minimal deficiency. In: SIAM conference on the life sciences–SIAM LS 2012, 7–10 August, San Diego, USA, pp MS1–2–45

Johnston MD, Siegel D, Szederkényi G (2012b) A linear programming approach to weak reversibility and linear conjugacy of chemical reaction networks. J Math Chem 50:274–288

Johnston MD, Siegel D, Szederkényi G (2013) Computing weakly reversible linearly conjugate chemical reaction networks with minimal deficiency. Math Biosci 241:88–98

Kaplan JL, Yorke JA (1979) Nonassociative, real algebras and quadratic differential equations. Nonlinear Anal Theory Methods Appl 3(1):49–51

Keszei E (2011) Chemical thermodynamics: an introduction. Springer, New York

Kirkwood JG, Oppenheim I (1961) Chemical thermodynamics. McGraw-Hill, New York

Kiss K, Kovács S (2008) Qualitative behaviour of n-dimensional ratio-dependent predator prey systems. Appl Math Comput 199(2):535–546

Kiss K, Tóth J (2009) n-dimensional ratio-dependent predator-prey systems with memory. Differ Equ Dyn Syst 17(1–2):17–35

Kowalski K (1993) Universal formats for nonlinear dynamical systems. Chem Phys Lett 209(1–2):167–170

Lotka AJ (1925) Elements of physical biology. Williams and Wilkins, Baltimore

Magyar A, Szederkényi G, Hangos KM (2008) Globally stabilizing feedback control of process systems in generalized Lotka–Volterra form. J Process Control 18(1):80–91

Markus L (1960) Quadratic differential equations and non-associative algebras. Ann Math Studies 45:185–213

Matolcsi T (2005) Ordinary thermodynamics. Akadémiai Kiadó, Budapest

May RM (1974) Stability and complexity in model ecosystems. Princeton University Press, Princeton

Myung HC, Sagle AA (1992) Quadratic differential equations and algebras. In: Bokut LA, Ershov YL, Kostrikin AI (eds) Proceedings of the international conference on algebra dedicated to the Memory of A. I. Mal'cev, 1989, Akademgorodok, Novosibirsk, RSFSR, Contemporary Mathematics, vol 131 (2), AMS, Providence, pp 659–672

Nagy T, Turányi T (2011) Uncertainty of Arrhenius parameters. Int J Chem Kinet 43(7):359–378

Orth JD, Thiele I, Palsson BØ (2010) What is flux balance analysis? Nat Biotechnol 28(3):245–248

Othmer H (1985) The mathematical aspects of temporal oscillations in reacting systems. In: Burger M, Field RJ (eds) Oscillations and traveling waves in chemical systems. Wiley, New York, pp 7–54

Prigogine I, Lefever R (1968) Symmetry breaking instabilities in dissipative systems, II. J Chem Phys 48:1695–1700

Revel J, Boettner JC, Cathonnet M, Bachman JS (1994) Derivation of a global chemical kinetic mechanism for methane ignition and combustion. J Chim Phys 91(4):365–382

Robertson HH (1966) Walsh JE (ed) The solution of a set of reaction rate equations. Thompson Book, Toronto, pp 178–182

Rock PA (2013) Chemical thermodynamics. University Science Books, Herndon

Samardzija N, Greller LD, Wasserman E (1989) Nonlinear chemical kinetic schemes derived from mechanical and electrical dynamical systems. J Chem Phys 90(4):2296–2304

Schubert A (1976, in Hungarian) Kinetics of homogeneous reactions. Műszaki Könyvkiadó, Budapest

Sigmund K (2007) Kolmogorov and population dynamics. In: Charpentier E, Lesne A, Nikolski NK (eds) Kolmogorov's heritage in mathematics. Springer, Berlin, pp 177–186

Smith HL (1995, 2008) Monotone dynamical systems: an introduction to the theory of competitive and cooperative systems. Mathematical surveys and monographs. AMS, Providence

Smoller J (1983) Shock waves and reaction-diffusion equations. Springer, New York

Szederkényi G (2010) Computing sparse and dense realizations of reaction kinetic systems. J Math Chem 47:551–568

Szederkényi G, Hangos KM (2011) Finding complex balanced and detailed balanced realizations of chemical reaction networks. J Math Chem 49:1163–1179

Szederkényi G, Hangos KM, Magyar A (2005) On the time-reparametrization of quasi-polynomial systems. Phys Lett A 334(4):288–294

Tomlin AS, Pilling MJ, Turányi T, Merkin J, Brindley J (1992) Mechanism reduction for the oscillatory oxidation of hydrogen: sensitivity and quasi-steady-state analyses. Combust Flame 91:107–130

Tóth J, Érdi P (1978) Models, problems and applications of formal reaction kinetics. A kémia újabb eredményei 41:227–350

Turányi T, Tomlin AS (2014) Analysis of kinetic reaction mechanisms. Springer, Berlin

Turányi T, Györgyi L, Field RJ (1993) Analysis and simplification of the GTF model of the Belousov–Zhabotinsky reaction. J Phys Chem 97:1931–1941

Turns SR (2000) An introduction to combustion. Concepts and applications. MacGraw-Hill, Boston

Vándor J (1952, in Hungarian) Chemical thermodynamics, vol I./1. Publishing House of the Hungarian Academy of Sciences, Budapest

Volpert AI, Hudyaev S (1985, Russian) Analyses in classes of discontinuous functions and equations of mathematical physics. Martinus Nijhoff, Dordrecht. Original: 1975

Stationary Points

7

7.1 Introduction

Constant solutions of the induced kinetic differential equation of a mechanism may have no physical relevance at all. A minimal requirement is that they should be nonnegative (see Example 7.4). In some cases one may be interested in strictly positive stationary points. The next problem is if such a stationary point is unique in some sense or not. One should like to know something about the stability properties of a stationary point. These will be the topics of the next chapter. As a preparation of these topics, however, we study structural properties of the underlying reaction which are closely connected to stability, as well.

Let us start with formal definitions to fix the notation to be used below. Note that we are dealing with mass action type kinetics if not said otherwise.

7.2 Stationary Points

Let us start with a general definition. Let $M \in \mathbb{N}, \mathbf{f} \in \mathscr{C}^1(T, \mathbb{R}^M)$, where $T \subset \mathbb{R}^M$ is a connected open set: a domain.

Definition 7.1 The point $\mathbf{x}_* \in T$ is a **stationary point** of the differential equation $\dot{\mathbf{x}}(t) = \mathbf{f}(\mathbf{x}(t))$, if $\mathbf{f}(\mathbf{x}_*) = \mathbf{0}$ holds.

Remark 7.2

1. Although in the theory of differential equations, **equilibrium** is the word used for this concept, we prefer using the expression "stationary point" in connection with applications and also avoid using **steady state** because the expressions "equilibrium" and "steady state" are heavily loaded with undefined and unwanted connotations. Sometimes we use **fixed point**; this wording expresses the fact that

© Springer Science+Business Media, LLC, part of Springer Nature 2018
J. Tóth et al., *Reaction Kinetics: Exercises, Programs and Theorems*,
https://doi.org/10.1007/978-1-4939-8643-9_7

a stationary point is the fixed point of the map $\mathbf{x}_0 \mapsto \Phi(t, \mathbf{x}_0)$ for all $t \in \mathbb{R}$, where $\Phi(t, \mathbf{x}_0)$ is the solution of a differential equation at time t which starts from the initial point $\mathbf{x}_0 \in T$. The map Φ is also called the **solution operator** of the differential equation.

2. A stationary point may be identified with a constant solution of the differential equation.

Next, using the definitions and notations introduced in Chaps. 4 and 6, we specialize the above definition for the case of induced kinetic differential equations of reactions endowed with mass action type kinetics and introduce a few further definitions. Before that, let us have another look at the right-hand side of the induced kinetic differential equation (the species formation vector) of (2.1) assuming mass action type kinetics. It can be written as follows:

$$\mathbf{f}(\mathbf{c}) = \mathbf{Y}\left(\mathbf{R}(\mathbf{c}) - \mathbf{R}(\mathbf{c})^\top\right)\mathbf{1}_N \tag{7.1}$$

with the matrix of reaction rates $\mathbf{R}(\mathbf{c}) := \mathbf{K} \cdot \operatorname{diag}(\mathbf{c}^\mathbf{Y})$ calculated from the matrix $\mathbf{K}$ of the reaction rate coefficients in the usual way. Using the complex formation vector

$$\mathbf{g}(\mathbf{c}) := (\mathbf{K} - \operatorname{diag}(\mathbf{K}^\top \mathbf{1}_N))\mathbf{c}^\mathbf{Y} = \left(\mathbf{R}(\mathbf{c}) - \mathbf{R}(\mathbf{c})^\top\right)\mathbf{1}_N, \tag{7.2}$$

as well, we get for the species formation vector the following expression:

$$\mathbf{f}(\mathbf{c}) = \mathbf{Y}\mathbf{g}(\mathbf{c}). \tag{7.3}$$

Now we are in the position to introduce—step by step—a series of definitions related to the mass action type mechanism defined by $\langle \mathcal{M}, \mathbf{Y}, \mathbf{K} \rangle$ (or, equivalently, by $\langle \mathcal{M}, \mathcal{R}, \boldsymbol{\alpha}, \boldsymbol{\beta}, \mathbf{k} \rangle$) following Horn and Jackson (1972).

Definition 7.3 The concentration $\mathbf{c}_* \in \mathbb{R}^M$ is a **stationary point** of the induced kinetic differential equation $\dot{\mathbf{c}} = \mathbf{f} \circ \mathbf{c}$ if $\mathbf{f}(\mathbf{c}_*) = \mathbf{0}$ holds. The set of positive stationary points will be denoted as

$$E := \{\mathbf{c}_* \in (\mathbb{R}^+)^M | \mathbf{f}(\mathbf{c}_*) = \mathbf{0}\} = \{\mathbf{c}_* \in (\mathbb{R}^+)^M | \mathbf{g}(\mathbf{c}_*) \in \operatorname{Ker}(\mathbf{Y})\}.$$

7.3   Existence of Nonnegative and Positive Stationary Points

To illustrate the problem, let us mention first that if one has two, two-variable quadratic polynomials and tries to find their roots by expressing one of them as the function of the other and substituting the result into the other, then in general one may arrive at a degree **sixth** polynomial and not at a quartic (degree fourth) one as one might expect.

The reason to be still optimistic is that a new tool to solve systems of polynomial equations has been invented in the sixties of the last century by Buchberger (2001); Buchberger and Winkler (1998): the **Gröbner basis**. It is known that a system of linear equations $Ax = b$ can be solved in such a way that the coefficient matrix A is transformed into, say, upper triangular form, and then one solves the last equation containing a single variable, then substitutes the result into the penultimate equation which turns by this substitution into a one-variable equation which can easily be solved, etc. The theory of Gröbner basis makes the same transformation possible for systems of polynomial equations, as well. This surely cannot mean that higher than fourth order equations can symbolically be solved, this procedure only provides a one-variable equation which can either symbolically or numerically be solved, then the results can be substituted into the next equation reducing it to a one variable equation again etc. Even if the numerical methods cannot be avoided in complicated cases, the problems become much easier to handle. Problem 7.1 will show how this procedure works.

No general theory on the existence and number of positive (respectively, nonnegative) roots of multivariate polynomials seems to exist. There are some methods, like the one by Pedersen et al. (1993), based on algebraic geometry, but usually they lead to calculations with high costs. The method of semi-definite programming (see page 160) is also used in this area (Lasserre et al. 2008). No wonder that the first relevant results in this direction come from people interested in reaction kinetics (Millán 2011; Millán et al. 2012).

Let us see a few special examples and general theorems on the existence of stationary points.

Example 7.4

1. The reaction $0 \longrightarrow X$ has no stationary point.
2. The mechanism $3\,X \underset{1/2}{\overset{1/2}{\rightleftharpoons}} X$ having the induced kinetic differential equation $\dot{x} = -x^3 + x$ has three stationary points: $-1, 0, 1$.
3. The differential equation $\dot{x} = xy+x+y+1 \quad \dot{y} = 2(xy+x+y+1)$ is a kinetic differential equation, and it has a unique stationary point: $(-1, -1)$.

These examples show that neither the existence, nor the uniqueness, nor the nonnegativity (positivity) of the stationary point(s) follow immediately from the form of the induced kinetic differential equation of a reaction. One needs a few general statements.

Theorem 7.5 (Wei (1962)) *Stoichiometrically mass conserving reactions do have nonnegative stationary points.*

Proof The nonnegative reaction simplexes are closed and convex sets. If the reaction is mass conserving, then they are also bounded by Theorem 4.23; thus they are compact and convex sets. Choose the reaction simplex containing c_0. The solution operator $\Phi(t, \cdot)$ of the reaction in question maps this set into itself and is continuous; therefore according to Brouwer's fixed-point theorem (Theorem 13.35 in the Appendix), it has a fixed point.   □

Remark 7.6

- Subconservativity is also enough to ensure the existence of a fixed point, although in this case the positive reaction simplexes may be unbounded. The proof is left to the reader as Problem 7.2.
- Mass conserving reactions do not necessarily obey a positive stationary point as the example $X \longrightarrow Y$ shows. Note that while the stationary point of the induced kinetic differential equation is $\begin{bmatrix} 0 & x_* \end{bmatrix}^\top$ with arbitrary $x_* \in \mathbb{R}$, the stationary point of the corresponding initial value problem is $\begin{bmatrix} 0 & x_0 + y_0 \end{bmatrix}^\top$ with arbitrary $x_0, y_0 \in \mathbb{R}$.
- A sufficient condition of positivity of the stationary point is given by Kaykobad (1985) for the case of generalized Lotka–Volterra systems (see Sect. 6.4.1).

The good news is that the special form of the right-hand sides of induced kinetic differential equations allows to have a few general statements on the existence of stationary points.

Theorem 7.7 (Orlov and Rozonoer (1984b)) *Reversible reactions have at least one positive stationary point in each positive stoichiometric compatibility class.*

The proof of the theorem is based on a series of lemmas and also on the first one of two papers by Orlov and Rozonoer (1984a).

Simon (1995) investigated the $M = 2$ case in more details, beyond having as a corollary the above statement he also proved that the trajectories of the induced kinetic differential equation of a reversible reaction remain in a closed bounded set bounded away from zero.

The question obviously arises whether it is possible to substitute reversibility with weak reversibility. There exists a paper and a manuscript on the topics.

Theorem 7.8 (Boros (2013)) *Weakly reversible deficiency one mechanisms have at least one positive stationary point in each positive reaction simplex.*

This theorem is probably true without the restriction to the deficiency one, as the manuscript Deng et al. (2011) (which is has been rigorously proved by Boros (2017) at the moment of closing the manuscript of this book) asserts: Weakly reversible mechanisms have a positive stationary point in each positive stoichiometric compatibility class.

7.4 Uniqueness of Stationary Points

Uniqueness of the stationary point has to be defined carefully. The mechanism $2X \underset{k_{-1}}{\overset{k_1}{\rightleftharpoons}} Y$ has stationary points of the form $\begin{bmatrix} x_* & 2k_1/k_{-1}x_*^2 \end{bmatrix}^\top$ of which those with $x_* \geq 0$ are physically meaningful. However, it has a single stationary point in all of the positive reaction simplexes $\left\{ \begin{bmatrix} x_0 & y_0 \end{bmatrix}^\top + \lambda \begin{bmatrix} 2 & -1 \end{bmatrix}^\top \mid \lambda \in \mathbb{R} \right\} \cap (\mathbb{R}^+)^2$, as Fig. 7.1 shows. It means that typically it is not the induced kinetic differential equation but the **initial value problem** describing the evolution of the reaction which may determine the stationary point uniquely. This is the sense of uniqueness we are interested in.

There are a lot of sufficient and necessary conditions of uniqueness; most of them are closely related to the description of the transient behavior of the solutions of the induced kinetic differential equation of a mechanism; therefore part of these topics will be picked up again in the next chapter. Here we mention a few criteria and counterexamples which are easy to understand and apply the concepts introduced up to this point.

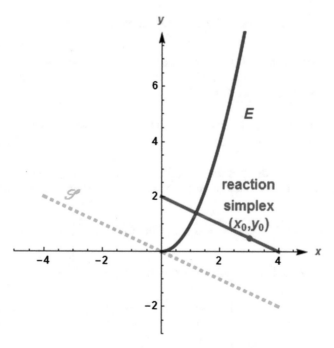

Fig. 7.1 Stationary points and positive reaction simplexes in the reaction $2X \underset{k_{-1}}{\overset{k_1}{\rightleftharpoons}} Y$

The example by Horn and Jackson (1972), p. 110, shows that even a mass conserving reaction may have more than one stationary point in each positive reaction simplex (see Problem 7.3).

Remark 7.9 Let us consider the form $\dot{\mathbf{c}} = \boldsymbol{\gamma}\,(\mathbf{k} \odot \mathbf{c}^\alpha)$ with the usual notations. Then, an intermediate step toward calculating the stationary points might be to find positive vectors in the right null space of $\boldsymbol{\gamma}$ which might be interpreted as **stationary rates of the reaction steps (steady-state flux rates)**. Obviously, $\mathbf{c}_* \in \mathbb{R}^M$ is a stationary point if and only if $\mathbf{w}_* := \mathbf{k} \odot \mathbf{c}_*^\alpha$ is a stationary rate. As the next step, one may try to solve the equation $\mathbf{w}_* := \mathbf{k} \odot \mathbf{c}_*^\alpha$—or, equivalently, $\ln(\frac{\mathbf{w}_*}{\mathbf{k}}) = \boldsymbol{\alpha}^\top \ln(\mathbf{c}_*)$— for $\mathbf{c}_*$.

Example 7.10 Consider the induced kinetic differential equation of the Lotka–Volterra reaction (2.4) written in the form:

$$\begin{bmatrix} \dot{x} \\ \dot{y} \end{bmatrix} = \begin{bmatrix} 1 & -1 & 0 \\ 0 & 1 & -1 \end{bmatrix} \begin{bmatrix} k_1 x \\ k_2 x y \\ k_3 y \end{bmatrix}$$

Then, as the right null space of $\boldsymbol{\gamma}$ is generated by the vector $\begin{bmatrix} 1 & 1 & 1 \end{bmatrix}^\top$, one has to solve $a\begin{bmatrix} 1 & 1 & 1 \end{bmatrix}^\top = \begin{bmatrix} k_1 x_* & k_2 x_* y_* & k_3 y_* \end{bmatrix}^\top$ to get $x_* = a/k_1$, $x_* y_* = a/k_2$, $y_* = a/k_3$ which can only hold if $a = k_1 k_3 / k_2$ leading to the known result: $x_* = k_3/k_2$, $y_* = k_1/k_2$. Note that in this case the stationary point is independent from the initial concentrations, cf. Sect. 7.10.

The method proposed in the previous example is delusory, as the next example shows.

Example 7.11 Upon determining a basis of the kernel of $\boldsymbol{\gamma}$, one sees that the positive stationary flux rates of the induced kinetic differential equation of the reaction $3X \underset{k_2}{\overset{k_3}{\rightleftarrows}} 2X \quad X \underset{k_0}{\overset{k_1}{\rightleftarrows}} 0$ are of the form $p\begin{bmatrix} 1 \\ 0 \\ 0 \\ 1 \end{bmatrix} + q\begin{bmatrix} 0 \\ 0 \\ 1 \\ 1 \end{bmatrix} + r\begin{bmatrix} 0 \\ 1 \\ 0 \\ -1 \end{bmatrix}$

with $p, q, r > 0$, $p + q > r$. As the next step, one has to solve

$$\begin{bmatrix} p \\ q \\ r \\ p+q-r \end{bmatrix} = \begin{bmatrix} k_3 c_*^3 \\ k_2 c_*^2 \\ k_1 c_* \\ k_0 \end{bmatrix}.$$

Eliminating the variables p, q, r from this system of equations, we arrive at the original equation for the stationary points $0 = k_3 c_*^3 - k_2 c_*^2 + k_1 c_* - k_0$.

Schuster and Schuster (1991) gave an algorithm and a C program to find the stationary reaction rates (in a slightly more general setting) together with a biochemical example based on tools of convex analysis. The algorithm and the code are modifications of those given for deciding mass conservation (Schuster and Höfer 1991). This can easily be understood if one considers that to prove stoichiometric mass conservativity, one has to find positive vectors in the **left** null space of the matrix $\boldsymbol{\gamma}$ (see Chap. 4).

Póta and Stedman (1995) shows multistationarity in a realistic system: The stationary behavior of a reaction describing the nitric acid-hydroxylamine reaction is studied under conditions of a continuously fed stirred tank reactor and with kinetics of not the mass action type. The authors show that the reaction has three positive stationary states in an unbounded region of the feed concentration. As opposed to this, in the example which can be seen in Fig. 7.2 (and in other cases in the literature, as well), multistationarity could only be observed for a bounded region (interval) of the parameters (Fig. 7.3).

It is also worth mentioning the paper by Shiu (2008) in which she has found the smallest reaction showing multistationarity with respect to the number of complexes, the number of connected components of the Feinberg–Horn–Jackson graph the number of species, and the dimension of an invariant polyhedron. See also the related papers by Joshi and Shiu (2015), Wilhelm (2009), and Mincheva and Roussel (2007) and page 184.

Fig. 7.2 The Feinberg–Horn–Jackson graph of the Horn–Jackson reaction admitting multiple stationary states for $0 < \varepsilon < 1/6$

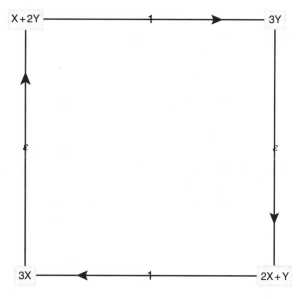

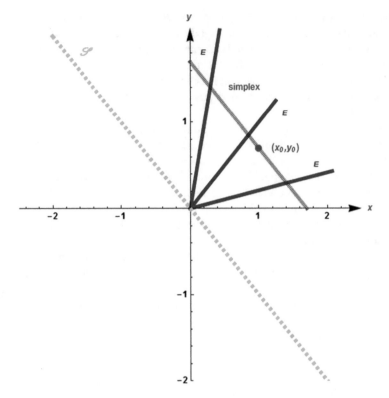

Fig. 7.3 Multiple stationary points in all positive reaction simplexes for $\varepsilon = 1/12$ in the Horn–Jackson reaction of Fig. 7.2

7.5   Stationary Points and First Integrals

Calculating the physically realistic stationary points of an induced kinetic differential equation is a multistep process. Once we have a candidate to be a stationary point (solutions to $\mathbf{f}(\mathbf{c}_*) = \mathbf{0}$) which is also nonnegative, we should also check if it is consistent with the first integrals of the induced kinetic differential equation. Although we are aware of Wei's theorem 7.5, in the general case, the situation is more complicated.

Let us consider a few examples.

Example 7.12

1. The mere existence of a linear first integral does not imply the existence of a positive stationary point as the example by Othmer (1985), p. 11, shows. The reaction $X \longrightarrow 2X + Y$ has a linear first integral: $(x, y) \mapsto x - y$, and its stationary points are $(0, y_*)$ with arbitrary $y_* \in \mathbb{R}$. (Let us mention that

Othmer (1985) calls the linear first integrals **kinematic invariants**.) Note that the first component of the solution of the induced kinetic differential equation being $t \mapsto x_0 e^t$ tends to infinity as $t \to +\infty$. This counterintuitive behavior can also be found in the case of the very simple reaction $X \longrightarrow 2X$, as well: The only stationary point is zero, but the solutions starting from a positive initial concentration tend to infinity as $t \to +\infty$.

2. An even uglier example follows. The reaction $0 \longrightarrow X + Y$ has a linear first integral: $(x, y) \mapsto x - y$, and it has no stationary points at all.

3. It may happen (see Definition 8.33) that an induced kinetic differential equation has a linear first integral which is not defined by an element of the left kernel space of y. If this is the case, then one has an (some) additional linear first integral(s) to check.

4. The existence of nonlinear first integrals may help further narrow the set of stationary points to physically acceptable cases. Consider the reaction given by Volpert and Hudyaev (1985) in pp. 627–629:

$$X + Y \longrightarrow U \quad Y + Z \longrightarrow V. \tag{7.4}$$

A basis of the left null space of y is $\begin{bmatrix} 1\ 0\ 0\ 1\ 0 \end{bmatrix}$, $\begin{bmatrix} 0\ 1\ 0\ 1\ 1 \end{bmatrix}$, $\begin{bmatrix} 0\ 0\ 1\ 0\ 1 \end{bmatrix}$, if the variables are taken in the order X, Y, Z, U, V. (Their sum is a positive vector corresponding to the definition of stoichiometric mass conservation.) Suppose that the initial concentrations are $\begin{bmatrix} x_0\ y_0\ z_0\ 0\ 0 \end{bmatrix}$ with $x_0, y_0, z_0 > 0$; and let us denote the coordinates of the stationary point by x_*, y_*, z_*, u_*, v_*. The nonnegative stationary points corresponding to these first integrals are:

- if $\boxed{y_* \neq 0}$ it is $\begin{bmatrix} 0\ y_0 - x_0 - z_0\ 0\ x_0\ z_0 \end{bmatrix}$ assuming that $x_0 + z_0 < y_0$,
- if $\boxed{y_* = 0}$ they are $\begin{bmatrix} \xi\ 0\ x_0 - y_0 + z_0 - \xi\ x_0 - \xi\ \xi - (x_0 - y_0) \end{bmatrix}$ with all values of ξ satisfying $\max\{0, x_0 - y_0\} \leq \xi \leq \min\{x_0, x_0 - y_0 + z_0\}$. This can only happen if $x_0 + z_0 \geq y_0$.

This is not the end of the story, because the induced kinetic differential equation

$$\dot{x} = -k_1 xy \quad \dot{y} = -k_1 xy - k_2 yz \quad \dot{z} = -k_2 yz \quad \dot{u} = k_1 xy \quad \dot{v} = k_2 yz$$

has a **nonlinear first integral** too in the open first orthant: $\begin{bmatrix} x\ y\ z\ u\ v \end{bmatrix} \mapsto x^{k_2}/z^{k_1}$. (It is Problem 7.8 to verify this.) Therefore the value of ξ can uniquely be determined from the requirement

$$\frac{(x_*)^{k_2}}{(z_*)^{k_1}} = \frac{(\xi)^{k_2}}{(x_0 - y_0 + z_0 - \xi)^{k_1}} = \frac{(x_0)^{k_2}}{(z_0)^{k_1}}.$$

No other first integral (independent from those given) can exist. (Independence of nonlinear functions is given in Definition 13.36.)

The function **StationaryPoints** only finds those nonnegative stationary points which are consistent with the linear first integrals but not necessarily with the nonlinear ones; therefore all other restrictions should be explicitly taken into consideration. When symbolic solutions are looked for, one may meet problems originated in the complexity of the task.

7.6   Complex Balance

Nonnegative stationary points may have further advantageous properties from the point of view of dynamic behavior. We go on with studying these properties, while in the next chapter, we treat theorems utilizing these properties.

Definition 7.13 The mechanism is **complex balanced at the positive stationary point** c_* if

$$g(c_*) = \left(K - \text{diag}(K^\top 1_N) \right) c_*^Y = 0 \in \mathbb{R}^N \tag{7.5}$$

holds, as well. The set of positive stationary points where the mechanism is complex balanced will be denoted as $C := \{c \in (\mathbb{R}^+)^M \mid g(c_*) = 0\}$. If the mechanism is complex balanced at all positive stationary points, i.e., if $E = C$, then it is complex balanced.

The name comes from the property that at a complex balanced stationary point c_*, the creation rate and the annihilation rate of each complex are the same, i.e., for all $n \in \mathcal{N}$ $\sum_{q \in \mathcal{N}} r_{nq}(c_*) = \sum_{q \in \mathcal{N}} r_{qn}(c_*)$ holds, cf. (7.2).

Theorem 7.14 (Horn and Jackson (1972)) *If a mechanism with mass action type kinetics is complex balanced at a positive stationary state, then the mechanism is complex balanced.*

Theorem 7.15 (Horn (1972)) *A reaction is complex balanced for any choice of reaction rate coefficients if and only if it is weakly reversible and its deficiency is zero.*

Dickenstein and Millán (2011), Proposition 4, also give a necessary and sufficient condition of complex balancing without needing to have a stationary state.

The paper by van der Schaft et al. (2015) provides a further necessary and sufficient condition for complex balancing using algebraic graph theory which can be checked constructively. (The reader may be interested in the series of papers written by the group which can be found in the reference list of this paper.)

Example 7.16

- The example $X \xrightarrow{k_1} Y \quad 2Y \xrightarrow{k_2} 2X$ similar to the mechanisms treated by Wegscheider (1901/1902) (and neither weakly reversible, nor of deficiency zero) shows that a mechanism may have a positive stationary point without being complex balanced.

All the complexes are either only created or only annihilated; thus the mechanism cannot be complex balanced. However, the points of the form $\left[2k_2c\ \sqrt{k_1c}\right]^\top$ with arbitrary $c \in R_0^+$ are stationary points; they are positive for $c \in \mathbb{R}^+$. It is also true that all positive reaction simplexes contain exactly one positive stationary point. See also Problem 7.5.

- The mechanism $X \underset{k_{-1}}{\overset{k_1}{\rightleftharpoons}} Y$ $2Y \underset{k_{-2}}{\overset{k_2}{\rightleftharpoons}} 2X$ is reversible, therefore it is also weakly reversible, but its deficiency being $4 - 2 - 1 = 1$, it cannot be complex balanced for all choices of reaction rate coefficients. The rate of production and that of annihilation of the complexes at stationary points is only equal if $(k_1/k_{-1})^2 = k_{-2}/k_2$. (Check it.)
- The irreversible triangle reaction is complex balanced. This can be shown by using Theorem 7.15, since its deficiency is zero, and clearly it is weakly reversible.

The most important consequence of complex balancing relating the stationary points follows.

Theorem 7.17 *If a mechanism is complex balanced, then there is exactly one stationary point in each positive reaction simplex (stoichiometric compatibility class).*

Remark 7.18 A sufficient condition of complex balance has been given by Orlov (1980), whereas a simple necessary condition is weak reversibility. Furthermore, a sufficient and necessary condition in the case of mass action type kinetics is given in Theorem 7.15.

7.7   The Complex Matrix and the Complex Formation Function

In order to enlighten the existence of stationary points, it is useful to study the relationship between $\mathrm{Ker}(\mathbf{Y})$ and $\mathscr{R}_\mathbf{g}$ (both subsets of $\mathbb{R}^N$) following Horn and Jackson (1972). The logically possible relationships are as follows.

1. $\mathrm{Ker}(\mathbf{Y}) \cap \mathbf{g}((\mathbb{R}^+)^M) = \emptyset$
2. $\mathrm{Ker}(\mathbf{Y}) \cap \mathbf{g}((\mathbb{R}^+)^M) = \{\mathbf{0}\}$
3. $\mathbf{0} \notin \mathrm{Ker}(\mathbf{Y}) \cap \mathbf{g}((\mathbb{R}^+)^M) \neq \emptyset$
4. $\{\mathbf{0}\} \subsetneqq \mathrm{Ker}(\mathbf{Y}) \cap \mathbf{g}((\mathbb{R}^+)^M)$

Examples follow.

1. All three reactions: $X \longrightarrow 2X$ and $0 \longrightarrow X$ and $X \longrightarrow Y$ with whatever reaction rate coefficients.

   - First, $\mathbf{Y} = \begin{bmatrix} 1 & 2 \end{bmatrix}$; therefore $\text{Ker}(\mathbf{Y}) = \{a \begin{bmatrix} 2 & -1 \end{bmatrix}^\top \mid a \in \mathbb{R}\}$, the vectors of which include the zero vector and also vectors with components of different sign and with a ratio $2 : -1$. However, $\mathbf{g}(x) = kx \begin{bmatrix} -1 & 1 \end{bmatrix}^\top$, thus $\mathbf{g}(\mathbb{R}^+) = \{kx \begin{bmatrix} -1 & 1 \end{bmatrix}^\top \mid x \in \mathbb{R}^+\}$, and therefore the intersection is empty.
   - Second, $\mathbf{Y} = \begin{bmatrix} 0 & 1 \end{bmatrix}$; therefore $\text{Ker}(\mathbf{Y}) = \{a \begin{bmatrix} 1 & 0 \end{bmatrix}^\top \mid a \in \mathbb{R}\}$, the vectors of which are the vectors with zero as their second component. However, $\mathbf{g}(x) = k \begin{bmatrix} -1 & 1 \end{bmatrix}^\top$; thus $\mathbf{g}(\mathbb{R}^+) = \{k \begin{bmatrix} -1 & 1 \end{bmatrix}^\top\}$ is a single vector with nonzero second components; therefore the intersection is empty.
   - Last, $\mathbf{Y} = \begin{bmatrix} 1 & 0 \\ 0 & 1 \end{bmatrix}$; therefore $\text{Ker}(\mathbf{Y}) = \{\mathbf{0}\}$. However, $\mathbf{g}(x, y) = k \begin{bmatrix} -x & y \end{bmatrix}^\top$; thus $\mathbf{g}(\mathbb{R}^{+2})$ only has vectors with components of different sign.

2. The irreversible triangle reaction, see in Fig. 3.2. Here $\mathbf{Y} = \begin{bmatrix} 1 & 0 & 0 \\ 0 & 1 & 0 \\ 0 & 0 & 1 \end{bmatrix}$; therefore $\text{Ker}(\mathbf{Y}) = \{\mathbf{0}\}$. However, $\mathbf{g}(x, y, z) = \begin{bmatrix} k_3z - k_1x & k_1x - k_2y & k_2y - k_3z \end{bmatrix}^\top$; thus $\mathbf{0} \in \mathscr{R}_{\mathbf{g}|_{\mathbb{R}^{+3}}}$. (Remember the notation of restriction; see the notations on page xxiii.)

3. The irreversible version of the Wegscheider reaction, (Wegscheider 1901/1902): $X \xrightarrow{k_1} Y \quad 2Y \xrightarrow{k_2} 2X$. Here

$$\mathbf{Y} = \begin{bmatrix} 1 & 0 & 0 & 2 \\ 0 & 1 & 2 & 0 \end{bmatrix}, \text{ therefore } \text{Ker}(\mathbf{Y}) = \{\begin{bmatrix} -2d & -2c & c & d \end{bmatrix}^\top \mid c, d \in \mathbb{R}\},$$

Furthermore, $\mathbf{g}(x, y) = \begin{bmatrix} -k_1x & k_1x & -k_2y^2 & k_2y^2 \end{bmatrix}^\top$. Now let us find real numbers c, d and positive numbers x, y so that

$$\begin{bmatrix} -2d \\ -2c \\ c \\ d \end{bmatrix} = \begin{bmatrix} -k_1x \\ k_1x \\ -k_2y^2 \\ k_2y^2 \end{bmatrix}.$$

Obviously, with fixed $c < 0$, the following cast will do: $d := -c, x := -2c/k_1, y := \sqrt{-c/k_2}$.

4. This cannot occur with mass action kinetics (see Horn and Jackson (1972)). An example with no mass action kinetics has been taken from the cited paper as Problem 7.7

Remark 7.19 In the case of the triangle reaction and also of the reaction X $\longrightarrow$ Y one has:

$$\mathrm{Ker}(\mathbf{Y}) = \{\mathbf{0}\}, \tag{7.6}$$

thus we can only have the first two cases here. If there is a positive stationary point, then it must be complex balanced. Equation (7.6) is fulfilled if and only if the complex vectors are linearly independent, what can only occur when $M \geq N$. Two important special cases are the closed compartmental system and the closed generalized compartmental system in the narrow sense (see Chap. 3.). Let us also mention that instead of complex balanced system, the expression **toric dynamical system** is used in algebra (see, e.g., Craciun et al. 2009 and the references therein).

7.8 Detailed Balance

The next property is even more restrictive; therefore logically it should come here. Historically however, this is the older concept.

7.8.1 A Short History

After such men as Maxwell and Boltzmann, and before Einstein, at the beginning of the twentieth century, it was Wegscheider (1901/1902) who gave the formal kinetic example in the left part of Fig. 7.4 to show that in some cases the existence of a positive stationary state alone does not imply the equality of all the individual forward and backward reaction rates in equilibrium: A relation (in this case $k_1/k_2 = k_3/k_4$) should hold between the reaction rate coefficients to ensure this (see Problem 7.12). Equalities of this kind will be called (and later exactly defined) as

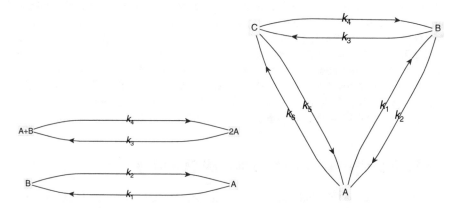

Fig. 7.4 The Wegscheider reaction and the reversible triangle reaction

spanning forest conditions below. Let us emphasize that violation of this equality does not exclude the existence of a positive stationary state; it may exist and be unique for all values of the reaction rate coefficients (see also Problem 7.12). A similar statement holds for the reversible triangle reaction in Fig. 7.4. The necessary and sufficient condition for the existence of such a positive stationary state for which all the reaction steps have the same rate in the forward and backward direction is now $k_1 k_3 k_5 = k_2 k_4 k_6$. Equalities of this kind will be called (and later exactly defined) as **circuit conditions** below. Again, violation of this equality does not exclude the existence of a positive stationary state; it may exist and be unique for all values of the reaction rate coefficients (see the details in Problem 7.11).

These examples are qualitatively different from, e.g., the simple bimolecular reaction which has the same stationary reaction rate in both directions no matter what the values of the reaction rate coefficients are (see Problem 7.10).

A quarter of a century after Wegscheider, the authors Fowler and Milne (1925) formulated in a very vague form a general principle called the **principle of detailed balance** stating that in real thermodynamic equilibrium, all the subprocesses (whatever they mean) should be in dynamic equilibrium (whatever this means) separately in such a way that they do not stop but proceed with the same velocity in both directions. Obviously, this also means that time is reversible at equilibrium; that is why this property may also be called **microscopic reversibility**, although it may be appropriate to reserve this expression for a similar property of the stochastic model (see Chap. 10). A relatively complete summary of the early developments was given by Tolman (1925).

The modern formulation of the principle accepted by IUPAC (Gold et al. 1997) essentially means the same (given that the **principle of charity** is applied):

> "The principle of microscopic reversibility at equilibrium states that, in a system at equilibrium, any molecular process and the reverse of that process occur, on the average, at the same rate."

In addition, we note that in the present chapter, we only have in mind deterministic models (surely not speaking of the general but vague formulation of Fowler and Milne cited above). Turning to stochastic models, one possible approach is to check the fulfilment of microscopic reversibility in the following way. Let us suppose we have some measurements on a process, and present the data with reversed time; finally use a statistical test to see if there is any difference. This is an absolutely correct approach and has also been used in the field of channel modeling (Rothberg and Magleby 2001). Another approach will be treated in Chap. 10, Sect. 10.5.2.

7.8.2 Rigorous Treatment of Detailed Balancing

A restriction on the positive stationary points even stricter than complex balancing follows. We would like to formulate the concept in such a way that at a detailed balanced stationary point (which can only exist in a reversible reaction), all pairs of reaction–antireaction step pairs proceed with the same rate in both directions. In order to introduce the definitions, the general description of reactions will be slightly modified to fit the case when all the reaction steps are reversible.

Definition 7.20 (Horn and Jackson (1972)) The mass action type induced kinetic differential equation of the reversible mechanism

$$\sum_{m \in \mathcal{M}} \alpha(m, p) X(m) \rightleftharpoons \sum_{m \in \mathcal{M}} \beta(m, p) X(m) \quad (p \in \mathcal{P} := \{1, 2, \dots, P\}) \quad (7.7)$$

with $P \in \mathbb{N}$ **pairs of reaction steps** is

$$\dot{\mathbf{c}}_m(t) = \sum_{p=1}^{P} (\beta(m, p) - \alpha(m, p))(k_p \mathbf{c}(t)^{\boldsymbol{\alpha}(\cdot, p)} - k_{-p} \mathbf{c}(t)^{\boldsymbol{\beta}(\cdot, p)}) \quad (7.8)$$

The mechanism is **detailed balanced at the positive stationary point $\mathbf{c}_*$** if for all $p \in \mathcal{P}$

$$k_p \mathbf{c}_*^{\boldsymbol{\alpha}(\cdot, p)} = k_{-p} \mathbf{c}_*^{\boldsymbol{\beta}(\cdot, p)} \quad (7.9)$$

or

$$\mathbf{R}(\mathbf{c}_*) - \mathbf{R}^\top(\mathbf{c}_*) = \mathbf{K} \operatorname{diag}(\mathbf{c}_*^{\mathbf{Y}}) - \operatorname{diag}(\mathbf{c}_*^{\mathbf{Y}}) \mathbf{K}^\top = \mathbf{0} \quad (7.10)$$

holds. The set of positive stationary points where the mechanism is detailed balanced will be denoted as

$$D := \{\mathbf{c}_* \in (\mathbb{R}^+)^M \mid \mathbf{R}^\top(\mathbf{c}_*) = \mathbf{R}(\mathbf{c}_*)\}.$$

If the mechanism is detailed balanced at all positive stationary points (or, equivalently, if $D = C = E \neq \emptyset$), then it is **detailed balanced**.

Note that $\boldsymbol{y}$ is an $M \times P$ matrix here, i.e., each reaction–antireaction pair is represented by a single column.

Theorem 7.21 *If a mechanism is detailed balanced at a positive stationary state $\mathbf{c}_*$, then there exists a single stationary point in each positive stoichiometric compatibility class which is detailed balanced; thus, in special, the mechanism is detailed balanced.*

Proof We follow here Volpert and Hudyaev (1985), pp. 635–639. If $S = M$, then there is a unique positive reaction simplex, the first orthant of the species space. According to the assumption, $\boldsymbol{y}^\top \mathbf{u} = \ln(\boldsymbol{\kappa})$ with

$$\boldsymbol{\kappa} = \left[\frac{k_1}{k_{-1}} \frac{k_2}{k_{-2}} \cdots \frac{k_P}{k_{-P}} \right] = \left[\kappa_1 \ \kappa_2 \ \dots \ \kappa_P \right]$$

has a solution $\mathbf{u}_*$, and because $\boldsymbol{\gamma}$ is of full rank (consequently $M \leq R$), this solution is unique; thus $\mathbf{c}_* = e^{\mathbf{u}_*}$ should hold.

If $S < M$, then without restricting generality, one can assume that the first S rows of $\boldsymbol{\gamma}^\top$ are linearly independent. To have a (positive) detailed balanced stationary point means that it is possible to solve the following system of equations:

$$\sum_{m \in \mathcal{M}} \gamma(m, r) u_m = \ln(\kappa_r) \quad (r = 1, 2, \ldots, S) \tag{7.11}$$

$$\sum_{m \in \mathcal{M}} \varrho_m^r e^{u_m} = \sum_{m \in \mathcal{M}} \varrho_m^r c_{0,m} \quad (r = 1, 2, \ldots, M - S), \tag{7.12}$$

where $\mathbf{c}_0 = \begin{bmatrix} c_{0,1} & c_{0,2} & \cdots & c_{0,M} \end{bmatrix}$ is the fixed initial concentration and the vectors $\boldsymbol{\varrho}^r = [\varrho_1^r \; \varrho_2^r \; \cdots \; \varrho_M^r]^\top$ are linearly independent solutions of

$$\boldsymbol{\varrho}^\top \boldsymbol{\gamma} = \mathbf{0}, \tag{7.13}$$

i.e., (not necessarily positive) linear first integrals of the induced kinetic differential equation. Equations (7.11) and (7.12) are individually solvable, because solvability of (7.11) immediately follows from our assumption and because for all $\bar{\mathbf{c}}$ such that $\bar{\mathbf{c}} \in (\mathbf{c}_0 + \mathscr{S}) \cap (\mathbb{R}^+)^M$ $\ln(\bar{\mathbf{c}})$ is a solution to (7.12). We are looking for a common solution.

If $\mathbf{u}_0$ is a particular solution to the inhomogeneous equation (7.11), then its general solution can be written in the form

$$u_m(\boldsymbol{\xi}) = u_{0m} + \sum_{r=1}^{M-S} \xi_r \varrho_m^r \quad (m \in \mathcal{M}), \tag{7.14}$$

or

$$\mathbf{u}(\boldsymbol{\xi}) - \mathbf{u}_0 = \boldsymbol{P}^\top \boldsymbol{\xi}, \tag{7.15}$$

with arbitrary components $\xi_1, \xi_2, \ldots, \xi_{M-S}$ of $\boldsymbol{\xi} \in \mathbb{R}^{M-S}$. Here $\boldsymbol{P}_{(R-S) \times M} := (\varrho_m^r)$. Of this set of solutions, we shall select a solution to (7.12) and will show that it is unique. To this purpose let us introduce the function

$$G(\mathbf{u}) := \sum_{m \in \mathcal{M}} (e^{u_m} - \bar{c}_m u_m), \tag{7.16}$$

where $\bar{\mathbf{c}}$ is as above. The function G is bounded from below and $\lim_{|\mathbf{u}| \to +\infty} G(\mathbf{u}) = +\infty$. As $\boldsymbol{\xi} = (\boldsymbol{PP}^\top)^{-1} \boldsymbol{P}(\mathbf{u}(\boldsymbol{\xi}) - \mathbf{u}_0)$, one has that $\lim_{|\boldsymbol{\xi}| \to +\infty} |\mathbf{u}(\boldsymbol{\xi})| \to +\infty$. Therefore it is also true that $\lim_{|\boldsymbol{\xi}| \to +\infty} G(\mathbf{u}(\boldsymbol{\xi})) = +\infty$. Thus the composite function $\boldsymbol{\xi} \mapsto \overline{G}(\boldsymbol{\xi}) := G(\mathbf{u}(\boldsymbol{\xi}))$ restricted onto the solution set of (7.11) has a minimum at some point $\boldsymbol{\xi}_0$, where one has

$$0 = \frac{\partial \overline{G}(\boldsymbol{\xi})}{\partial \xi_r} = \sum_{m \in \mathcal{M}} \frac{\partial G}{\partial u_m} \frac{\partial u_m}{\partial \xi_r} \tag{7.17}$$

$$= \sum_{m \in \mathcal{M}} (e^{u_m(\boldsymbol{\xi})} - \overline{c}_m) \varrho_m^r = \sum_{m \in \mathcal{M}} (\varrho_m^r e^{u_m(\boldsymbol{\xi})} - \varrho_m^r \overline{c}_m)), \tag{7.18}$$

which means that (7.12) is fulfilled. Summing up, $\mathbf{u}(\boldsymbol{\xi}_0)$ defines a common solution of (7.12) and (7.11).

To prove uniqueness, let us start from (7.17). This equation shows that any common solution of (7.12) and (7.11) is determined by (7.14) at some stationary point of the composition function $\overline{G}$. The matrix whose elements are the second derivatives $\frac{\partial^2 \overline{G}}{\partial \xi_r \partial \xi_s} = \sum_{m \in \mathcal{M}} e^{u_m(\boldsymbol{\xi})} \varrho_m^r \varrho_m^s$ $(r, s = 1, 2, \ldots, S - M)$ of this function is positive definite, as for arbitrary $\eta_1, \eta_2, \ldots, \eta_{M-S}$, we have

$$\sum_{r,s=1}^{M-S} \frac{\partial^2 \overline{G}}{\partial \xi_r \partial \xi_s} \eta_r \eta_s = \sum_{m \in \mathcal{M}} e^{u_m(\boldsymbol{\xi})} \left(\sum_{r=1}^{M-S} \varrho_m^r \eta_r \right)^2 \geq 0, \tag{7.19}$$

and since the rows of the matrix $\mathbf{P}^\top$ are linearly independent, equality can only hold for $\eta_1 = 0, \eta_2 = 0, \ldots, \eta_{M-S} = 0$. Therefore the function $\overline{G}$ is strictly convex; thus it only has a single stationary point, a minimum, providing the unique common solution of (7.12) and (7.11). $\qquad \square$

We are especially interested in reactions which are detailed balanced for some choices of the reaction rate constants, and also in the restrictions upon the rate constants which ensure detailed balancing. Problems 7.10–7.12 will show examples for both cases. However, checking the condition by hand is not always easy to carry out. General equivalent conditions are needed. Two approaches will be shown.

7.8.2.1 Conditions of Detailed Balancing: Circuits and Spanning Forests

A set of necessary and sufficient conditions have been formulated in the following way by Feinberg (1989). Consider the reaction (7.7), and suppose first that we have chosen an arbitrary spanning forest (see Definition 13.29) of the Feinberg–Horn–Jackson graph. It is possible to find a set of $P - (N - L)$ independent circuits induced by the choice of the spanning forest. For each of these circuits, we write an equation which asserts that the product of the rate constants in the clockwise direction and the counterclockwise direction is equal. Thus we have $P - (N - L)$ equations: the **circuit conditions**.

Next, these equations are supplemented with the δ **spanning forest conditions** as follows (where δ is the deficiency of the reaction). Suppose that the edges of the spanning forest have been given an orientation. Then there are δ independent nontrivial solutions to the vector equation $\sum_{i,j} a_{ij}(\mathbf{y}_j - \mathbf{y}_i) = \mathbf{0}$, where the sum is taken for all reaction steps in the oriented spanning forest and $\mathbf{y}_j - \mathbf{y}_i$ is the corresponding reaction step vector. (Note the difference between the numbering of the reaction step vectors when the complex vectors $\mathbf{y}_n$ ($n \in \mathcal{N}$) are used or when one uses the vectors $\boldsymbol{\gamma}(\cdot, r)$.) With the a_{ij} coefficients obtained as solutions, the spanning forest conditions are $\prod k_{ij}^{a_{ij}} = \prod k_{ji}^{a_{ij}}$, where k_{ij} are the corresponding rate coefficients.

If all the widely accepted **necessary** conditions, the circuit conditions, are complemented with the spanning forest conditions, then they form a set of necessary and sufficient conditions for detailed balancing in mass action systems of arbitrary complexity.

Theorem 7.22 (Feinberg) *The reaction (7.7) is detailed balanced for all those choices of the reaction rate constants which satisfy the* $P-(N-L)$ *circuit conditions and the* δ *spanning forest conditions.*

Example 7.23 In the case of the reversible triangle reaction (Fig. 7.4), $P = 3$, $N = 3$, $L = 1$, therefore one has one circuit condition (expressing the equality of the product of the reaction rate coefficients taken in two different directions), and as $\delta = 0$, no spanning forest conditions exist. In the case of the reversible Wegscheider reaction (Fig. 7.4), $P = 2$, $N = 4$, $L = 2$, therefore one has no circuit condition, and as $\delta = 1$, one has a single spanning forest condition. To determine this, let us take as a directed spanning forest the following reaction: $A \longrightarrow B \quad A + B \longrightarrow 2A$. Then, one has to determine the solutions to the following system of equations:

$$a_{12} \left(\begin{bmatrix} 0 \\ 1 \end{bmatrix} - \begin{bmatrix} 1 \\ 0 \end{bmatrix} \right) + a_{34} \left(\begin{bmatrix} 2 \\ 0 \end{bmatrix} - \begin{bmatrix} 1 \\ 1 \end{bmatrix} \right) = \mathbf{0} \in \mathbb{R}^2. \tag{7.20}$$

The solution is (actually, the solutions are, as we have a homogeneous linear system) $a_{12} = a_{34}$; thus the spanning forest condition in this case is $k_1^1 k_4^1 = k_2^1 k_3^1$.

Remark 7.24 Unlucky as it may be, the circuit conditions are called *spanning tree method* in Colquhoun et al. (2004).

Remark 7.25 There are three interesting special cases generalizing the examples.

1. For a reversible mass action reaction which has a deficiency zero, the circuit conditions alone become necessary and sufficient for detailed balancing. The reason why the circuit conditions were generally accepted as sufficient as well is that a large majority of models are of zero deficiency. This case is exemplified by the triangle reaction.
2. For networks with no nontrivial circuits, that is, in which there are just $N - L$ reaction pairs and so $P - (N - L) = 0$, the circuit conditions are vacuous. Therefore, the spanning forest conditions **alone** are necessary and sufficient for detailed balancing. The example by Wegscheider belongs to this category.

3. Finally, if a reversible network is circuitless and has a deficiency of zero, both the circuit conditions and the spanning forest conditions are vacuous. The system is detailed balanced, regardless of the values of the rate constants. Such is a compartmental system with no circuits in the FHJ-graph, the simple bimolecular reaction or the ion channel model in Érdi and Ropolyi (1979).

Now we have learned that in the case of chemical reactions, the general principle of detailed balancing can only hold if and only if both the spanning tree conditions and the circuit conditions are fulfilled. However, it became a general belief among people dealing with reaction kinetics (especially since Shear 1967) that the circuit conditions alone are not only necessary but also sufficient for all kinds of reactions: Wegscheider's example proving the contrary was not known well enough. It was Feinberg (1989) who gave the definitive solution of the problem in the area of formal kinetics: He clearly formulated, proved, and applied the two easy-to-deal-with sets of conditions which together make up a necessary and sufficient condition of detailed balance (for the case of mass action kinetics). In other words, he completed the known necessary condition (the circuit conditions)—which might however be empty in some cases—with another condition (the spanning forest conditions) making this sufficient, as well.

We have also seen above that the reason why the false belief is widespread is that in case of reactions with deficiency zero, the circuit conditions alone are also **sufficient** not only necessary, and most textbook examples have zero deficiency.

Neither the present authors, nor the above cited IUPAC document assert that the principle should hold without any further assumptions; for us it is an important hypothesis the fulfilment of which should be checked individually in each reaction considered.

7.8.2.2 Application of the Fredholm's Alternative Theorem
The advantage of Feinberg's approach is that it identifies two separate conditions— the circuit conditions and the spanning forest conditions—using graph theory. However, there is another, simpler way to find necessary and sufficient conditions. This approach has been formulated by Vlad and Ross (2009) and also by Joshi (2015). Dickenstein and Millán (2011) gave the clearest formulation and presented an unnecessarily complicated proof.

Theorem 7.26 *The reaction (7.7) is detailed balanced if and only if with all the elements* $\mathbf{a}$ *of the (right) kernel of* $\boldsymbol{\gamma}$ *one has* $\boldsymbol{\kappa}^{\mathbf{a}} = 1$ *with* $\boldsymbol{\kappa} := \left[\frac{k_1}{k_{-1}}, \frac{k_2}{k_{-2}}, \dots, \frac{k_P}{k_{-P}} \right]^\top$.

Proof According to Theorem 13.19, the existence of a positive solution $\mathbf{c}_*$ to $k_p \mathbf{c}_*^{\alpha(\cdot, p)} = k_{-p} \mathbf{c}_*^{\beta(\cdot, p)}$ $(p \in \mathscr{P})$ or (what is the same) the existence of a solution to $\boldsymbol{\gamma}^\top \ln(\mathbf{c}_*) = \ln(\boldsymbol{\kappa})$ is equivalent to the requirement that for all such vector $\mathbf{a} \in \mathbb{R}^P$ for which $\boldsymbol{\gamma}\, \mathbf{a} = \mathbf{0}$ holds, $\mathbf{a}^\top \ln(\boldsymbol{\kappa}) = 0$ should also hold. This, however, means that $\boldsymbol{\kappa}^{\mathbf{a}} = 1$. $\qquad\square$

Fig. 7.5 Circuit and spanning forest conditions both play a role

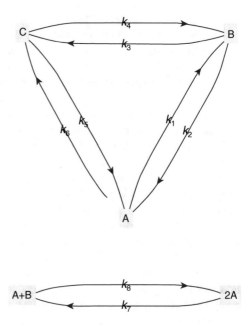

Remark 7.27 The two approaches are compared by Dickenstein and Millán (2011), Proposition 3, who show that Feinberg grouped these conditions more structurally into two sets of conditions.

Example 7.28 Let us consider the (somewhat artificial, still useful) mechanism `compli` Fig. 7.5.

The application of the following then gives one of the spanning trees of the Feinberg–Horn–Jackson graph.

```
FindSpanningTree[ReactionsData[{compli}]["fhjgraphedges"]]
```

A spanning tree is $C \longleftarrow A \longrightarrow B \quad 2A \longrightarrow A + B$. Making the spanning tree boils down to two simple rules which can be formulated as follows. First, keep only one of the reaction steps of the reversible step pairs; and second, discard one edge of each circle in the resulting graph. These clearly lead to a forest, and since all the complexes were kept, it spans the Feinberg–Horn–Jackson graph.

Let us use the first approach. As here $P - (N - L) = 4 - (5 - 2) = 1$, one has a single circuit condition: $k_1 k_3 k_5 = k_2 k_4 k_6$. The deficiency is $\delta = N - L - S = 5 - 2 - 2 = 1$; therefore one has a single spanning forest condition: $k_2 k_7 = k_1 k_8$. The second approach requires to find two independent vectors in the right null space of

$$
y = \begin{bmatrix} -1 & -1 & 0 & 1 \\ 1 & 1 & -1 & 0 \\ 0 & 0 & 1 & -1 \end{bmatrix} \text{ which are } \begin{bmatrix} 1 \\ 0 \\ 1 \\ 1 \end{bmatrix} \text{ and } \begin{bmatrix} -1 \\ 1 \\ 0 \\ 0 \end{bmatrix}.
$$

The first one gives $k_1^1 k_7^0 k_3^1 k_5^1 = k_2^1 k_8^0 k_4^1 k_6^1$, i.e., the circuit conditions, while the second one gives $k_1^{-1} k_7^1 k_3^0 k_5^0 = k_2^{-1} k_8^1 k_4^0 k_6^0$, i.e., the spanning forest conditions.

The one line code

```
DetailedBalanced[complicated,
    {k1, k2, k7, k8, k3, k4, k5, k6}]
```

provides the same result.

An interesting relationship has been found between detailed balance and complex balance by Dickenstein and Millán (2011) as Theorem 1.1.

Theorem 7.29 (Dickenstein and Millán (2011)) *If the circuit conditions are satisfied in a reversible mechanism, then detailed balancing and complex balancing are equivalent.*

Remark 7.30

- The statement is a generalization of two other statements: If there are no cycles in the undirected graph of a reversible reaction, then detailed balancing and complex balancing coincide. If the deficiency of the reaction is zero, then the circuit conditions alone imply detailed balancing.
- A mechanism may be complex balanced and not detailed balanced, as the irreversible triangle reaction shows. Even if one takes the reversible triangle reaction, it is unconditionally complex balanced, but to be detailed balanced, it should also fulfil the circuit condition which in this case is that the product of the reaction rate coefficients should be the same in both directions. Cf. also Example 7.16 where another reversible reaction is shown which is conditionally complex balanced and also Problem 8.13.

7.8.3 Applications

Examples taken from applications show that in realistic models the conditions may be quite complicated, that is why the function **DetailedBalance** may be useful.

7.8.3.1 Finding a Detailed Balanced Realization

The induced kinetic differential equation does not uniquely determine the inducing mechanism; therefore it is a useful idea to look for a mechanism of simple structure if a kinetic differential equation is given. To give a more specific example, let us consider the following problem. Given a kinetic differential equation, find an inducing mechanism which is reversible and detailed balanced and has a deficiency zero. This can be done in some cases using the method by Szederkényi and Hangos (2011) (see Problem 7.13). The topic is naturally connected to uniqueness questions as well (see Sect. 11.5).

7.8.3.2 Models of Ion Channels in Nerve Membranes

There is a difference in electric potential between the interior of nerve cells and the interstitial liquid. An essential part of the system controlling the size of this potential difference is the system of ion channels: pores made up from proteins in the membranes through which different ions may be transported via active and passive transport, thereby changing the potential difference in an appropriate way.

Recent papers on formal kinetic models of ion channel gating show that people in this field think that the principle of detailed balance or microscopic reversibility should hold. (However, some authors do not consider the principle of microscopic reversibility indispensable, e.g., Naundorf et al. (2006), Supplementary Notes 2, Figure 3SI(a), page 4, provides a channel model which is not even reversible, let alone detailed balanced.) This may be supported either by an out of the blue theoretical argument (they should obey the laws of thermodynamics) or by a practical one (if the principle holds, one should measure less reaction rate coefficients because one also has the constraints implied by the principle). The second argument seems to be the more important one in the papers by Colquhoun et al. (2004) and Burzomato et al. (2004). However, the principle is applied in an imprecise way: First, only the necessary part consisting of the circuit conditions is applied, and second, the models are formulated in a way that they do not obey the principle of mass conservation. In the papers Nagy et al. (2009) and Nagy and Tóth (2012) (note that A. L. Nagy and I. Nagy are different authors), the models are transformed into mass conserving ones, and the full set of necessary and sufficient conditions are applied. The main result is that in classes of models including all the known ion channel examples are compartmental models; therefore they have zero deficiency at the beginning, and being transformed into a mass conserving model, they have no circuits; therefore one has only to test the spanning forest conditions. It is not less interesting that the spanning forest conditions obtained for the transformed models are literally the same as the circuit conditions for the original models.

7.8.3.3 A Model for Hydrogen Combustion

Let us see how our program **DetailedBalanced** works in the case of a more complicated example: in the field of combustion modeling, famous for working with "large" models.

The reaction proposed by Kéromnès et al. (2013) to describe H_2 combustion contains 9 species and 21 pairs of reaction steps, together forming a reversible reaction. In this reaction the necessary and sufficient conditions of detailed balance are as follows:

$$k_{26}k_{27}k_{42} = k_{25}k_{28}k_{41}, \quad k_4k_{13}k_{39} = k_3k_{14}k_{40}, \quad k_2k_{14}k_{17} = k_1k_{13}k_{18},$$

$$k_3k_8k_{11} = k_4k_7k_{12}, \quad k_2k_4k_5 = k_1k_3k_6, \quad k_2k_4k_9k_{21} = k_1k_3k_{10}k_{22},$$

$$k_2k_8k_9k_{19} = k_1k_7k_{10}k_{20}, \quad k_2k_9k_{14}k_{15} = k_1k_{10}k_{13}k_{16},$$

$$k_2k_8k_9k_{13}k_{23}k_{37}^2 = k_1k_7k_{10}k_{14}k_{24}k_{38}^2, \quad k_2k_8k_9k_{14}k_{24}k_{35}^2 = k_1k_7k_{10}k_{13}k_{23}k_{36}^2,$$

$$k_1 k_8 k_{10} k_{13} k_{24} k_{33}^2 = k_2 k_7 k_9 k_{14} k_{23} k_{34}^2, \quad k_1 k_8 k_9 k_{13} k_{24} k_{29}^2 = k_2 k_7 k_{10} k_{14} k_{23} k_{30}^2,$$

$$k_2 k_8 k_9 k_{13} k_{24} k_{25}^2 = k_1 k_7 k_{10} k_{14} k_{23} k_{26}^2, \quad k_2 k_4^2 k_8 k_9 k_{13} k_{24} k_{31}^2 = k_1 k_3^2 k_7 k_{10} k_{14} k_{23} k_{32}^2.$$

7.9   Stationary Points and Symmetry

There is a nice uniform formulation of the introduced concepts (Horn and Jackson 1972, p. 93).

Let, as usual, M be the number of species and N the number of complexes in a reaction, and let $\mathbf{R}(\mathbf{c})$ be the matrix of reaction rates at the concentration $\mathbf{c}$ so that $r_{nq}(\mathbf{c})$ is the reaction rate of the reaction $C(q) \longrightarrow C(n)$. Let Λ be a function which maps $N \times N$ matrices into some finite dimensional vector space, and let us call the mechanism **symmetric** at $\mathbf{c}_* \in (\mathbb{R}^+)^M$ with respect to Λ, if $\Lambda(\mathbf{R}(\mathbf{c}_*)) = \Lambda(\mathbf{R}^\top(\mathbf{c}_*))$ holds. The transposition of $\mathbf{R}$ physically corresponds to reversal of time, since under this operation the rates of each reaction–antireaction pair are exchanged. Thus if the mechanism is symmetric at $\mathbf{c}_*$, then the value of Λ at this concentration remains invariant under reversal of time.

1. If $\Lambda(\mathbf{R}(\mathbf{c}_*)) := \mathbf{Y}\mathbf{R}(\mathbf{c}_*)\mathbf{1}_N$, then the mechanism is symmetric at $\mathbf{c}_*$ if and only if $\mathbf{c}_*$ is a positive stationary point.
2. If $\Lambda(\mathbf{R}(\mathbf{c}_*)) := \mathbf{R}(\mathbf{c}_*)\mathbf{1}_N$, then the mechanism is symmetric at $\mathbf{c}_*$ if and only if it is complex balanced at $\mathbf{c}_*$.
3. If $\Lambda(\mathbf{R}(\mathbf{c}_*)) := \mathbf{R}(\mathbf{c}_*)$, then the mechanism is symmetric at $\mathbf{c}_*$ if and only if it is detailed balanced at $\mathbf{c}_*$.

Note that $\mathbf{X} \mapsto \Lambda(\mathbf{X})$ is linear function in all of the three cases and also that the right-hand side of the induced kinetic differential equation is an antisymmetric function of $\mathbf{R}(\mathbf{c})$; let this function be $\mathbf{L}$, i.e., $\mathbf{f}(\mathbf{c}) = \mathbf{L}(\mathbf{R}(\mathbf{c}))$. Then one has $\mathbf{L}(\mathbf{R}(\mathbf{c}) + \mathbf{R}^\top(\mathbf{c})) = \mathbf{0} \in \mathbb{R}^M$ for all concentrations, and $\Lambda(\mathbf{R}(\mathbf{c}) + \mathbf{R}^\top(\mathbf{c})) = \mathbf{0}$ for those concentrations at which the mechanism is symmetric. We are not aware of any continuation of this concept for the deterministic model. Note also that the above formulation (which has some resemblance to dynamic symmetries, see Sect. 8.9 below) has not used the fact that the concentration vector in question is a stationary point. However, detailed balance and the kinship with the corresponding notion in the case of the stochastic model (which one might like to call **microscopic reversibility** to emphasize the difference) are straightforward. (See, e.g., Joshi 2015.)

Finally, let us mention that detailed balance has been extended by Gorban and Yablonsky (2011) as a limiting case for cases when some of the reaction steps are irreversible in such a way that irreversible reactions are represented as limits of reversible steps.

7.10   Absolute Concentration Robustness

Although here we treat a very special property of the stationary points, we dedicate a separate section to absolute concentration robustness because of its importance.

The present section can be considered as another one on the delicate properties of stationary points from formal viewpoint. If however one thinks of the possible applications, it relates a kind of stability of some systems in the everyday sense of the word. In this section we follow the Shinar and Feinberg (2010).

Let us start with two simple examples.

Example 7.31 The initial value problem to describe the simple reversible reaction $X \underset{k_{-1}}{\overset{k_1}{\rightleftharpoons}} Y$ of deficiency zero being

$$\dot{x} = -k_1 x + k_{-1} y, \quad \dot{y} = k_1 x - k_{-1} y, \quad x(0) = x_0, \quad y(0) = y_0$$

has the stationary points of the form $\xi \begin{bmatrix} k_{-1} & k_1 \end{bmatrix}^\top$ with any $\xi \in \mathbb{R}$. Of these stationary points, the one reached from the initial point (x_0, y_0) (fitting to the first integral $\begin{bmatrix} x & y \end{bmatrix}^\top \mapsto x + y$) is $(x_0 + y_0) \begin{bmatrix} \frac{k_{-1}}{k_{-1}+k_1} & \frac{k_1}{k_{-1}+k_1} \end{bmatrix}^\top$. This vector is always positive, if $x_0 + y_0$ is positive (i.e., something has been present initially: a not too restrictive assumption), and all its coordinates do depend on the value $x_0 + y_0$.

Example 7.32 The initial value problem of the reaction (an irreversible skeleton of the Wegscheider reaction which can also be interpreted as the transformation of the inactive form of a protein Y into the active form X)

$$X + Y \xrightarrow{k_1} 2Y \quad Y \xrightarrow{k_2} X \tag{7.21}$$

of deficiency one being

$$\dot{x} = -k_1 xy + k_2 y, \quad \dot{y} = k_1 xy - k_2 y, \quad x(0) = x_0, \quad y(0) = y_0$$

has positive stationary points only if $x_0 + y_0 > \frac{k_2}{k_1}$, and then the stationary point is $\begin{bmatrix} \frac{k_2}{k_1} & x_0 + y_0 - \frac{k_2}{k_1} \end{bmatrix}^\top$.

In contrast to the previous example, the first coordinate of the stationary point is independent of the total initial concentration of the species. This is the kind of robustness or stability we are interested in here (Fig. 7.6).

Thus, here we are interested in mechanisms in which some concentrations are protected against (even large) changes of the initial concentrations of the species.

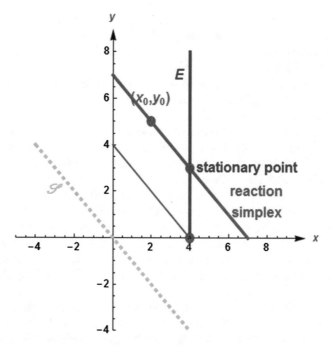

Fig. 7.6 The first component of the stationary point does not depend on the initial concentration in case of the reaction (7.21)

Definition 7.33 A mechanism shows **absolute concentration robustness** for a species if the concentration of that species is identical in every positive stationary concentration.

A set of easy to apply sufficient conditions to ensure absolute concentration robustness has been collected by Shinar and Feinberg (2010).

Theorem 7.34 *Suppose a mechanism with deficiency one has a positive stationary state. If the underlying reaction has two nonterminal complex vectors that only differ in species Z, then the system shows absolute concentration robustness for the species Z.*

Remember, that the notions used here have been defined in Chap. 3, except **nonterminal complex** which is a complex not in a terminal strong linkage class (ergodic component). One may also need the Definition 13.30.

Remark 7.35

1. In Example 7.32 the nonterminal complexes are $X + Y$ and Y; as they only differ in X, the concentration of this species does not depend on the initial concentrations.

2. A mass conserving reaction with deficiency zero can never show absolute concentration robustness (Shinar et al. 2009). In the case $\delta = 2$, mechanisms with and without absolute concentration robustness exist, as the supporting material to Shinar and Feinberg (2010) shows.
3. With the theorem by Boros (2013) at hand, one would easily fall into a trap: If weak reversibility together with deficiency one implies the existence of a positive stationary point, then perhaps it would be enough to assume weak reversibility instead of the existence of a positive stationary point. Alas, this is leading to nowhere! Weak reversibility is equivalent to not having nonterminal nodes in the Feinberg–Horn–Jackson graph; therefore a weakly reversible reaction can never fulfil the last condition of Theorem 7.34.
4. The function **AbsoluteConcentrationRobustness** investigates if these sufficient conditions hold in a mechanism or not. How does it work? To check the conditions of the theorem, we need a program what we have not needed up to now: We have to find the strong components of the Feinberg–Horn–Jackson graph, select the nonterminal strong components, calculate the differences of all the nonterminal complexes, and see if there is any consisting of a single species.

Shinar and Feinberg apply the theorem to biological systems which as experimentally shown do show absolute concentration robustness. The first one is the *Escherichia coli* EnvZ-Ompr system, where the phosphorylated form of the response-regulator is the species with absolute concentration robustness. The second one is the *Escherichia coli* IDHKP-IDH glyoxylate bypass regulation system. Here the species showing absolute concentration robustness is IDH: the active, unphosphorylated TCA cycle enzyme isocitrate dehydrogenase. (Let us note by passing that the whole metabolic graph of the *Escherichia coli* can be found this way:

```
ExampleData[{"NetworkGraph",
    "MetabolicNetworkEscherichiaColi"}].)
```

The authors give a more detailed biological analysis and provide a useful reference list of the problem in Shinar and Feinberg (2011). Another application of the theory for a realistic biochemical problem and a *Mathematica* program to carry out the calculations can be found in Dexter and Gunawardena (2013). More details on the biological background can be found here: Kitano (2004), Barkai and Leibler (1997), Blüthgen and Herzel (2003), and Ferrell (2002).

Another approach to robustness has been given by Li and Rabitz (2014) for models used to describe gene networks. Let us also mention that robustness is closely related to sensitivity: Shinar et al. (2009), Turányi (1990), and Turányi and Tomlin (2014). Loosely speaking a model is robust if it is insensitive to changes of such parameters as the initial concentration and reaction rate coefficients.

A case when the stationary point strongly depends on the parameters is shown as Problem 7.19.

Finally, we mention that the connection between stationary points and oscillation is much less simple as it is usually thought (see Tóth 1999).

7.11 Exercises and Problems

7.1 Find the solutions of the system of polynomial equations

$$x(1 - 2x + 3y + z) = 0, \ y(1 + 3x - 2y + z) = 0, \ z(2 + y - z) = 0$$

in such a way that transform the system into triangular form using the *Mathematica* function **GroebnerBasis** and then find the roots of the system.

(Solution: page 403.)

7.2 Show that mass consuming reactions do have a nonnegative stationary point.

(Solution: page 404.)

7.3 Show that the Horn–Jackson reaction in Fig. 7.2 with $0 < \varepsilon < \frac{1}{6}$ has three positive stationary points in each positive reaction simplexes.

(Solution: page 404.)

7.4 Redo the calculations on the reaction (7.4) using the programs.

(Solution: page 404.)

7.5 Find conditions on the reaction rate coefficients under which the reversible mechanism

$$X \underset{k_{-1}}{\overset{k_1}{\rightleftharpoons}} Y \quad 2X \underset{k_{-2}}{\overset{k_2}{\rightleftharpoons}} 2Y. \tag{7.22}$$

with the usual mass action type kinetics is complex balanced.

(Solution: page 405.)

7.6 If in a reversible mechanism with mass action kinetics, the number of species and the number of reaction steps are the same ($M = R$) and the stoichiometric matrix γ is not singular (invertible), then the mechanism has a unique positive stationary point.

(Solution: page 405.)

7.7 Consider the reaction (7.22) with the reaction rates defined as follows:

$$w_1(x, y) := x^2, \ w_{-1}(x, y) := y, \ w_2(x, y) := x, \ w_{-2}(x, y) := y^2,$$

and show that it has a positive stationary point at which it is complex balanced and another one at which it is not.

(Solution: page 405.)

7.8 Verify that the function $\begin{bmatrix} x & y & z & u & v \end{bmatrix}^\top \mapsto x^{k_2}/z^{k_1}$ is the first integral of the induced kinetic differential equation

$$\dot{x} = -k_1 xy \quad \dot{y} = -k_1 xy - k_2 yz \quad \dot{z} = -k_2 yz \quad \dot{u} = k_1 xy \quad \dot{v} = k_2 yz$$

in the open first orthant. Show also that beyond this and the linear ones given above, the equation has no other first integral, and the existing ones are independent.

(Solution: page 406.)

7.9 Can a reaction with the zero complex present have a positive stationary point?

(Solution: page 406.)

7.10 Show that the simple bimolecular reaction

$$A + B \underset{k_{-1}}{\overset{k_1}{\rightleftharpoons}} C \tag{7.23}$$

is detailed balanced for any choice of the reaction rate coefficients.

(Solution: page 407.)

7.11 Prove that the reversible triangle reaction in Fig. 7.4 has a single positive stationary point for all values of the reaction rate coefficients. Prove also that the mechanism is detailed balanced at this stationary point if and only if $k_1 k_3 k_5 = k_2 k_4 k_6$ holds.

(Solution: page 408.)

7.12 Prove that the reversible Wegscheider reaction in Fig. 7.4 has a single positive stationary point for all values of the reaction rate coefficients. Prove also that the mechanism is detailed balanced at this stationary point if and only if $k_1 k_4 = k_2 k_3$ holds.

(Solution: page 409.)

7.13 Show that the induced kinetic differential equation of the irreversible reaction with deficiency one $3\,Y \xrightarrow{1} 3\,X \xrightarrow{0.5} 2\,X + Y$ has the same induced kinetic differential equation as the reversible detailed balanced reaction $3\,Y \underset{0.5}{\overset{1}{\rightleftharpoons}} 3\,X$ with deficiency zero.

(Solution: page 409.)

7.14 Find the nonnegative stationary points of the (irreversible) Lotka–Volterra reaction (6.28). Find a nonlinear first integral in the open first quadrant and show that the first integral takes its minimum at the positive stationary point.

(Solution: page 409.)

7.15 Do the same as in the Problem 7.14 above for the **Ivanova reaction**:

$$X+Y \xrightarrow{k_1} 2\,Y \quad Y+Z \xrightarrow{k_2} 2\,Z \quad Z+X \xrightarrow{k_3} 2\,X. \qquad (7.24)$$

(Solution: page 410.)

7.16 Show that the Lotka–Volterra reaction has absolute concentration robustness with respect to all the species.

(Solution: page 410.)

7.17 Show that the special case of the Oregonator in Fig. 7.7 has deficiency two; still it has absolute concentration robustness with respect to all the species.

(Solution: page 410.)

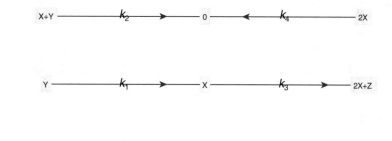

Fig. 7.7 A special case of the Oregonator

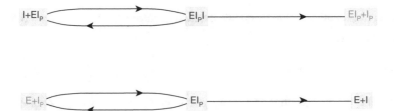

Fig. 7.8 The Feinberg–Horn–Jackson graph of the IDHK-IDP reaction. Terminal complexes are pink, nonterminal complexes are blue

7.18 Assuming that the IDHKP-IDH mechanism in Fig. 7.8 has a positive stationary state proves that it shows absolute concentration robustness with respect to the species I (iodine).

(Solution: page 410.)

7.19 The dimensionless form of the induced kinetic differential equation of a reaction is

$$\dot{x} = -xy + f(1-x) \quad \dot{y} = xy + y\left(\left(1 - \frac{f_2}{f}\right)y_0 - y\right) + (f - f_2)y_0 - fy.$$
$$(7.25)$$

Study the dependence of the x_* component of the stationary state(s) on the parameter f in the interval $[e^{-3}, e^{-1}]$ when the other parameters are fixed as follows.

1. | $f_2 = 1/135$ $y_0 = 10/27$ |
2. | $f_2 = 0.001$ $y_0 = 0.25$ |
3. | $f_2 = 0.001$ $y_0 = 0.29$ |

(See Li and Li 1989 and also Póta 2006, p. 4, and Ganapathisubramanian and Showalter 1984.)

(Solution: page 411.)

7.12 Open Problems

1. Kinetically mass conserving (see Definition 8.33) reactions are not necessarily stoichiometrically mass conserving, as the example $X \xrightarrow{k} 0$ $X \xrightarrow{k} 2X$ shows. Do they always have a nonnegative stationary point?

2. Continuing the solution of Problem 7.6 shows the existence in all the remaining cases reproducing the result by Orlov and Rozonoer (1984b) based on the form of the induced kinetic differential equation.
3. Formulate a necessary and sufficient condition ensuring that the set of concentrations defining stationary flux rates are the same as the set of stationary states (assuming nonnegativity or positivity in both cases).
4. Under what conditions is it possible to have absolute concentration robustness with respect to all the species for reactions with $\delta \leq 2$?

References

Barkai N, Leibler S (1997) Robustness in simple biochemical networks. Nature 387:913–917

Blüthgen N, Herzel H (2003) How robust are switches in intracellular signaling cascades? J Theor Biol 225:293–300

Boros B (2013) On the existence of the positive steady states of weakly reversible deficiency-one mass action systems. Math Biosci 245(2):157–170

Boros B (2017) On the existence of positive steady states for weakly reversible mass-action systems. arxivorg

Buchberger B (2001) Gröbner bases: a short introduction for systems theorists. In: EUROCAST, pp 1–19. http://dx.doi.org/10.1007/3-540-45654-6_1

Buchberger B, Winkler F (1998) Gröbner bases and applications, vol 251. Cambridge University Press, Cambridge

Burzomato V, Beato M, Groot-Kormelink PJ, Colquhoun D, Sivilotti LG (2004) Single-channel behavior of heteromeric $\alpha 1\beta$ glycine receptors: an attempt to detect a conformational change before the channel opens. J Neurosci 24(48):10924–10940

Colquhoun D, Dowsland KA, Beato M, Plested AJR (2004) How to impose microscopic reversibility in complex reaction mechanisms. Biophys J 86(6):3510–3518

Craciun G, Dickenstein A, Shiu A, Sturmfels B (2009) Toric dynamical systems. J Symb Comput 44(11):1551–1565

Deng J, Jones C, Feinberg M, Nachman A (2011) On the steady states of weakly reversible chemical reaction networks. arXiv preprint arXiv:11112386

Dexter JP, Gunawardena J (2013) Dimerization and bifunctionality confer robustness to the isocitrate dehydrogenase regulatory system in Escherichia coli. J Biol Chem 288(8):5770–5778

Dickenstein A, Millán MP (2011) How far is complex balancing from detailed balancing? Bull Math Biol 73:811–828

Érdi P, Ropolyi L (1979) Investigation of transmitter-receptor interactions by analyzing postsynaptic membrane noise using stochastic kinetics. Biol Cybern 32(1):41–45

Feinberg M (1989) Necessary and sufficient conditions for detailed balancing in mass action systems of arbitrary complexity. Chem Eng Sci 44(9):1819–1827

Ferrell JEJ (2002) Self-perpetuating states in signal transduction: positive feedback, double-negative feedback and bistability. Curr Opin Cell Biol 14(2):140–148

Fowler RH, Milne EA (1925) A note on the principle of detailed balancing. Proc Natl Acad Sci USA 11:400–402

Ganapathisubramanian N, Showalter K (1984) Bistability, mushrooms, and isolas. J Chem Phys 80(9):4177–4184

Gold V, Loening KL, McNaught AD, Shemi P (1997) IUPAC compendium of chemical terminology, 2nd edn. Blackwell Science, Oxford

Gorban AN, Yablonsky GS (2011) Extended detailed balance for systems with irreversible reactions. Chem Eng Sci 66(21):5388–5399

Horn F (1972) Necessary and sufficient conditions for complex balancing in chemical kinetics. Arch Ratl Mech Anal 49:172–186

Horn F, Jackson R (1972) General mass action kinetics. Arch Ratl Mech Anal 47:81–116

Joshi B (2015) A detailed balanced reaction network is sufficient but not necessary for its Markov chain to be detailed balanced. Discrete Contin Dyn Syst Ser B 20(4):1077–1105

Joshi B, Shiu A (2015) A survey of methods for deciding whether a reaction network is multistationary. Math Model Nat Phenom 10(5):47–67

Kaykobad M (1985) Positive solutions of positive linear systems. Linear Algebra Appl 64:133–140

Kéromnès A, Metcalfe WK, Heufer K, Donohoe N, Das A, Sung CJ, Herzler J, Naumann C, Griebel P, Mathieu O, Krejci MC, Petersen EL, Pitz J, Curran HJ (2013) An experimental and detailed chemical kinetic modeling study of hydrogen and syngas mixture oxidation at elevated pressures. Comb Flame 160(6):995–1011

Kitano H (2004) Biological robustness. Nat Rev Genet 5:826–837

Lasserre JB, Laurent M, Rostalski P (2008) Semidefinite characterization and computation of zero-dimensional real radical ideals. Found Comput Math 8(5):607–647

Li R, Li H (1989) Isolas, mushrooms and other forms of multistability in isothermal bimolecular reacting systems. Chem Eng Sci 44(12):2995–3000

Li G, Rabitz H (2014) Analysis of gene network robustness based on saturated fixed point attractors. EURASIP J Bioinform Syst Biol 2014(1):4

Millán MSP (2011) Métodos algebraicos para el estudio de redes bioquímicas. PhD thesis, Universidad de Buenos Aires en el area Ciencias Matematicas, Buenos Aires, 167 pp.

Millán MP, Dickenstein A, Shiu A, Conradi C (2012) Chemical reaction systems with toric steady states. Bull Math Biol 74(5):1027–1065

Mincheva M, Roussel MR (2007) Graph-theoretic methods for the analysis of chemical and biochemical networks. I. Multistability and oscillations in ordinary differential equation models. J Math Biol 55(1):61–86

Nagy I, Tóth J (2012) Microscopic reversibility or detailed balance in ion channel models. J Math Chem 50(5):1179–1199

Nagy I, Kovács B, Tóth J (2009) Detailed balance in ion channels:applications of Feinberg's theorem. React Kinet Catal Lett 96(2):263–267

Naundorf B, Wolf F, Volgushev M (2006) Unique features of action potential initiation in cortical neurons. Nature 440:1060–1063

Orlov VN (1980) Kinetic equations with a complex balanced stationary point. React Kinet Catal Lett 14(2):149–154

Orlov VN, Rozonoer LI (1984a) The macrodynamics of open systems and the variational principle of the local potential I. J Frankl Inst 318(5):283–314

Orlov VN, Rozonoer LI (1984b) The macrodynamics of open systems and the variational principle of the local potential II. Applications. J Frankl Inst 318(5):315–347

Othmer H (1985) The mathematical aspects of temporal oscillations in reacting systems. In: Burger M, Field RJ (eds) Oscillations and traveling waves in chemical systems. Wiley, New York, pp 7–54

Pedersen P, Roy MF, Szpirglas A (1993) Counting real zeros in the multivariate case. In: Computational algebraic geometry. Birkhäuser, Boston, pp 203–224

Póta G (2006) Mathematical problems for chemistry students. Elsevier, Amsterdam

Póta G, Stedman G (1995) An uncommon form of multistationarity in a realistic kinetic model. J Math Chem 17(2–3):285—289

Rothberg BS, Magleby KL (2001) Testing for detailed balance (microscopic reversibility) in ion channel gating. Biophys J 80(6):3025–3026

Schuster S, Höfer T (1991) Determining all extreme semi-positive conservation relations in chemical reaction systems—a test criterion for conservativity. J Chem Soc Faraday Trans 87(16):2561–2566

Schuster S, Schuster R (1991) Detecting strictly detailed balanced subnetworks in open chemical reaction networks. J Math Chem 6(1):17–40

Shear D (1967) An analog of the Boltzmann H-theorem (a Liapunov function) for systems of coupled chemical reactions. J Theor Biol 16:212–225

Shinar G, Feinberg M (2010) Structural sources of robustness in biochemical reaction networks. Science 327(5971):1389–1391

Shinar G, Feinberg M (2011) Design principles for robust biochemical reaction networks: what works, what cannot work, and what might almost work. Math Biosci 231(1):39–48

Shinar G, Alon U, Feinberg M (2009) Sensitivity and robustness in chemical reaction networks. SIAM J Appl Math 69(4):977–998

Shiu A (2008) The smallest multistationary mass-preserving chemical reaction network. Lect Notes Comput Sci 5147:172–184. Algebraic Biology

Simon LP (1995) Globally attracting domains in two-dimensional reversible chemical dynamical systems. Ann Univ Sci Budapest Sect Comput 15:179–200

Szederkényi G, Hangos KM (2011) Finding complex balanced and detailed balanced realizations of chemical reaction networks. J Math Chem 49:1163–1179

Tolman RC (1925) The principle of microscopic reversibility. Proc Natl Acad Sci USA 11:436–439

Tóth J (1999) Multistationarity is neither necessary nor sufficient to oscillations. J Math Chem 25:393–397

Turányi T (1990) KINAL: a program package for kinetic analysis of complex reaction mechanisms. Comput Chem 14:253–254

Turányi T, Tomlin AS (2014) Analysis of kinetic reaction mechanisms. Springer, Berlin

van der Schaft A, Rao S, Jayawardhana B (2015) Complex and detailed balancing of chemical reaction networks revisited. J Math Chem 53(6):1445–1458

Vlad MO, Ross J (2009) Thermodynamically based constraints for rate coefficients of large biochemical networks. Syst Biol Med 1(3):348–358

Volpert AI, Hudyaev S (1985) Analyses in classes of discontinuous functions and equations of mathematical physics. Martinus Nijhoff Publishers, Dordrecht (Russian original: 1975)

Wegscheider R (1901/1902) Über simultane Gleichgewichte und die Beziehungen zwischen Thermodynamik und Reaktionskinetik homogener Systeme. Zsch phys Chemie 39:257–303

Wei J (1962) Axiomatic treatment of chemical reaction systems. J Chem Phys 36(6):1578–1584

Wilhelm T (2009) The smallest chemical reaction system with bistability. BMC Syst Biol 3(90):9

Time-Dependent Behavior of the Concentrations

8

8.1   Introduction

Finer details of the solutions of the induced kinetic differential equations of reactions are investigated: we try to find out how the solutions (as functions of time) and their projections onto the space of concentrations—the trajectories—evolve. After establishing the assuring fact that the components of the concentration vector always remain nonnegative, we state which ones remain strictly positive and which ones stay strictly zero throughout their domain of existence. Next, the domain of existence itself becomes the object of study: we look for conditions under which the domain is bounded, meaning that the system "blows up" or has a finite escape time—a common situation in combustion models. Afterward we delimit the smallest set that the trajectories are confined to. It turns out that the trajectories not only remain in the stoichiometric compatibility class corresponding to the initial concentration vector, but they may not even leave the kinetic subspace, which is in general a smaller subspace of the species space. The existence of nonlinear first integrals may further restrict the set in which the trajectories can move. Then three important theorems characterizing the time evolution of the concentrations follow: one is about detailed balanced mechanisms, one is about reactions with acyclic Volpert graphs, and in between comes the celebrated zero-deficiency theorem, a far-reaching generalization of the classical statement on detailed balanced mechanisms ensuring **regular behavior**, the behavior which is usually expected of a mechanism by the chemist. Next, **exotic behavior** follows: conditions ensuring or excluding oscillation, and also a few remarks on oligo-oscillation and chaos occurring in reactions. The chapter closes with special topics such as symmetries of induced kinetic differential equations and the graphs of element fluxes allowing a more detailed insight into the evaluation of concentrations. Finally, as usual, a set of solved and open problems follows.

© Springer Science+Business Media, LLC, part of Springer Nature 2018
J. Tóth et al., *Reaction Kinetics: Exercises, Programs and Theorems*,
https://doi.org/10.1007/978-1-4939-8643-9_8

8.2   Well-Posedness

Before formulating more detailed statements about the solutions of induced kinetic differential equations and their trajectories, we need a few fundamental statements. At the beginning of the twentieth century, Jacques Hadamard formulated three requirements important for problems of applied mathematics. These requirements (altogether called **well-posedness**) are as follows:

- The problem should have a solution.
- The solution should be unique.
- It should continuously depend on parameters, in particular, on initial data.

We formulate below theorems expressing the fact that the initial value problems for induced kinetic differential equations of mechanisms are well-posed problems.

Theorem 8.1 *The induced kinetic differential equation* (6.3) *of any mechanism has a unique solution once the initial value of the concentrations has been specified.*

Proof The right-hand side of Eq. (6.3) is continuously differentiable; thus the statement immediately follows from the Picard–Lindelöf Theorem 13.42.   □

Remark 8.2 The consequence of the theorem may not remain true if one has reaction steps with fractional order, smaller than 1. To remain on the safe side, one usually assumes that the stoichiometric coefficients of the reactant complexes can be any nonnegative real number except those in the open interval $]0, 1[$. In most cases it is not a strict restriction to assume that they are integers.

Remark 8.3 The fact that the right-hand side depends on the parameters (mainly, reaction rate coefficients) in continuously differentiable way together with continuous differentiability of the right-hand side itself implies that the solutions are also continuously differentiable functions of the parameters and the initial conditions (Perko 1996, Section 2.3).

Example 8.4 As to the quantitative dependence on parameters, let us start with the example of the irreversible Lotka–Volterra reaction, and let us write its induced kinetic differential equation in the following form, explicitly expressing the dependence of the variables on the vector of reaction rate coefficients $\mathbf{k} = \begin{bmatrix} k_1 & k_2 & k_3 \end{bmatrix}^\top$:

$$\dot{x}_1(t, \mathbf{k}) = k_1 x_1(t, \mathbf{k}) - k_2 x_1(t, \mathbf{k}) x_2(t, \mathbf{k}), \tag{8.1}$$

$$\dot{x}_2(t, \mathbf{k}) = k_2 x_1(t, \mathbf{k}) x_2(t, \mathbf{k}) - k_3 x_2(t, \mathbf{k}). \tag{8.2}$$

Specifying the sensitivities (see (13.23)) $s_{m,p}(t, \mathbf{k}) := \frac{\partial x_m(t, \mathbf{k})}{\partial k_p}$ for the case of induced kinetic differential equations of reactions with mass-action type kinetics, we have

$$\dot{s}_{1,1} = x_1 + k_1 s_{1,1} - k_2 s_{1,1} x_2 - k_2 x_1 s_{2,1}, \tag{8.3}$$

$$\dot{s}_{2,1} = k_2 s_{1,1} x_2 + k_2 x_1 s_{2,1} - k_3 s_{2,1}, \tag{8.4}$$

$$\dot{s}_{1,2} = k_1 s_{1,2} - x_1 x_2 - k_2 s_{1,2} x_2 - k_2 x_1 s_{2,2}, \tag{8.5}$$

$$\dot{s}_{2,2} = x_1 x_2 + k_2 s_{1,2} x_2 + k_2 x_1 s_{2,2} - k_3 s_{2,2}, \tag{8.6}$$

$$\dot{s}_{1,3} = k_1 s_{1,3} - k_2 s_{1,3} x_2 - k_2 x_1 s_{2,3}, \tag{8.7}$$

$$\dot{s}_{2,3} = k_2 s_{1,3} x_2 + k_2 x_1 s_{2,3} - x_2 - k_3 s_{2,3}. \tag{8.8}$$

(Problem 8.1 asks the reader to derive these equations in full generality.) Solving the system (8.1)–(8.8) with the initial conditions

$$x_1(0) = x_{1,0} \in \mathbb{R}_0^+, \; x_2(0) = x_{2,0} \in \mathbb{R}_0^+,$$

$$s_{1,1} = 0, \; s_{2,1} = 0, \; s_{1,2} = 0, \; s_{2,2} = 0, \; s_{1,3} = 0, \; s_{2,3} = 0$$

provides information on the (time-varying) effect of the individual reaction rate coefficients on the concentrations. Note that this effect is depending on the actual values of the reaction rate coefficients and also on the initial concentrations. Note also that it is enough to solve, e.g., the four variable system (8.1)–(8.2), (8.5)–(8.6) if one is only interested in the effect of k_2.

One can proceed symbolically (without relying on numerical calculations) by focusing on the stationary concentrations $\begin{bmatrix} k_3/k_2 & k_1/k_2 \end{bmatrix}^\top$ of the above reaction. Upon substituting these into (8.3)–(8.8), one can see that at this point the sensitivity equations simplify to

$$\begin{aligned} \dot{s}_{1,1} &= k_3/k_2 - k_3 s_{2,1}, & \dot{s}_{2,1} &= k_1 s_{1,1}, \\ \dot{s}_{1,2} &= -k_1 k_3/k_2^2 - k_3 s_{2,2}, & \dot{s}_{2,2} &= k_1 k_3/k_2^2 + k_1 s_{1,2}, \\ \dot{s}_{1,3} &= -k_3 s_{2,3}, & \dot{s}_{2,3} &= k_1 s_{1,3} - k_1/k_2. \end{aligned} \tag{8.9}$$

Let us concentrate on the effects of k_1. This is expressed by

$$s_{1,1}(t) = \sqrt{k_3/k_1} \sin\left(\sqrt{k_1 k_3} t\right) / k_2, \quad s_{2,1}(t) = \left(1 - \cos\left(\sqrt{k_1 k_3} t\right)\right) / k_2$$

showing that at the beginning both effects are small, then, at time $\pi/(2\sqrt{k_1 k_3})$ the effect of k_1 on the first variable is maximal. The effect on the second variable reaches its maximum at time $\pi/\sqrt{k_1 k_3}$, i.e., at the same time when the effect on the first variable vanishes. Then the absolute value of the effect on the first variable starts

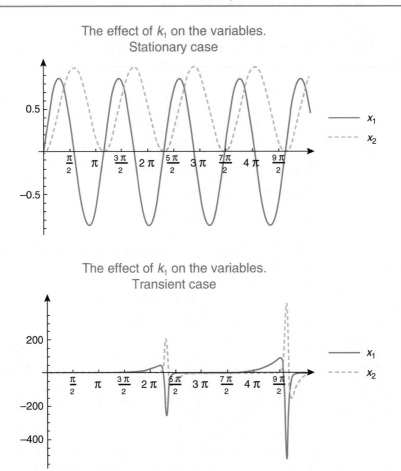

Fig. 8.1 Sensitivities as a function in time in the stationary and transient cases

growing but becomes negative, etc. (see Fig. 8.1). The lower part of the figure shows that the sensitivities in the transient case are growing with time. The treatment of sensitivities in the case of oscillatory reactions needs more care (see Sipos-Szabó et al. 2008; Zak et al. 2005).

The investigation of the trivial stationary concentration $\begin{bmatrix} 0 & 0 \end{bmatrix}^{\top}$ is left to the reader (Problem 8.2).

Remark 8.5 It is worthwhile looking at Turányi and Tomlin (2014) who use a series of different quantities derived from the sensitivity matrix to characterize the parameter dependence of complicated mechanisms. (See also the earlier review paper by Turányi 1990.) There it turns out that the usual way to calculate the effect of the reaction rate coefficients in practice is to numerically approximate the sensitivities. A nice application of sensitivity analysis is that large mechanisms can be reduced by excluding reaction steps with negligible or no effects (see, e.g., Turányi et al. 1993).

8.3 Nonnegativity

The meaning of the dependent variables in an induced kinetic differential equation is concentration (or mass which is a constant multiple of it as volume is fixed); thus it is a natural requirement for the solutions to be nonnegative for their whole domain of existence. If this is a result of the model—as it will turn out to be the case—then our belief in the model is further confirmed. Moreover, there is no need to **assume** the invariance of the first orthant, let alone to **prove** it again and again for special cases, as it so often happens in the literature.

In this section we summarize the results by Volpert (1972) in the special case of mass-action kinetics, but the original paper is worth reading for its far-reaching generalizations using only the qualitative properties of mass-action type kinetics. (However, acknowledgment should be given to Müller-Herold (1975) who stated and proved the statement below in the same generality starting from the Horn and Jackson (1972) paper. Note that this work remained practically unnoticed.)

Theorem 8.6 *Let us consider the reaction* (2.1), *and let the solution of its induced kinetic differential equation* (6.8) *with the initial condition* $\mathbf{c}(0) = \mathbf{c}_0 \in (\mathbb{R}^+)^M$ *be defined on the interval* $J \subset \mathbb{R}_0^+$. *Then, for all* $t \in J : \mathbf{c}(t) \in (\mathbb{R}^+)^M$.

Proof Suppose that the solution is defined on the interval J, but it is not always positive, and let the first time when any of the coordinate functions (let it be c_m) turns zero be $t_0 \in J : c_m(t_0) = 0$. Then we have for all $p \in \mathcal{M}$ and for all $t \in [0, t_0[: c_p(t) > 0$. If for some $r \in \mathcal{R}$ we have $\gamma(m, r) < 0$, then $\alpha(m, r) > 0$ should also hold. Using this fact one can rewrite the mth equation of the induced kinetic differential equation (6.8) in the following way:

$$\dot{c}_m(t) = c_m(t)\varphi(t) + \psi(t), \qquad (8.10)$$

where the continuous functions φ and ψ defined on J are as follows:

$$\varphi(t) := \sum_{r:\gamma(m,r)<0} \gamma(m, r)\frac{w_r(\mathbf{c}(t))}{c_m(t)} \quad \text{and} \quad \psi(t) := \sum_{r:\gamma(m,r)\geq0} \gamma(m, r)w_r(\mathbf{c}(t)).$$

Solving (8.10) as a linear differential equation gives

$$c_m(t) = c_m(0) \exp\left(\int_0^t \varphi\right) + \int_0^t \exp\left(\int_u^t \varphi\right) \psi(u)\, du, \quad (t \in J)$$

showing that $c_m(t_0) > 0$, in contradiction with our assumption. □

Remark 8.7 Obviously, the conclusion remains true if a function $\mathbf{g} \in \mathscr{C}(\mathbb{R}, (\mathbb{R}_0^+)^M)$ (representing **time-dependent input**) is added to the right-hand side of the induced kinetic differential equation.

Corollary 8.8 *Let us consider the reaction* (2.1)*, and let the solution of its induced kinetic differential equation* (6.8) *with the initial condition* $\mathbf{c}(0) = \mathbf{c}_0 \in (\mathbb{R}_0^+)^M$ *be defined on the interval* $J \subset \mathbb{R}_0^+$*. Then, for all* $t \in J : \mathbf{c}(t) \in (\mathbb{R}_0^+)^M$*.*

Proof Continuous dependence of the solutions of differential equations—see, e.g., Perko (1996), Section 2.3—on initial data and Theorem 8.6 implies the statement: add a vector $\boldsymbol{\varepsilon} \in (\mathbb{R}^+)^M$ to the nonnegative initial concentrations, apply the previous theorem, and let $\boldsymbol{\varepsilon}$ tend to $\mathbf{0} \in \mathbb{R}^M$.   □

This corollary also follows from the fact that the velocity vector points into the interior of the first orthant. However, to prove the above stricter theorem, one cannot argue this way.

One can also formulate more precise statements about positivity of the coordinates. The essence of the statements below is that the concentration of species which can be produced from a given set of species initially present is positive for all positive times, whereas the concentration of the other ones is zero for all positive times. This also means that an initially positive concentration cannot turn into zero in a finite time, (as opposed to the non-mass-action case, see Póta 2016), while an initially zero concentration may turn positive. Let us introduce a formal definition first.

Definition 8.9 An acyclic subgraph Γ of the Volpert graph of a reaction is a **reaction path**, if it contains together with its vertices corresponding to the reaction steps also all the edges pointing into the these vertices. A species is an **initial species** if no edge belonging to Γ points into it. (In other words, its **in-degree** is zero.)

Example 8.10 In the reaction $X + Y \longrightarrow 0 \quad U \longrightarrow Z + Y$, a reaction path is the subgraph of the Volpert graph with all the edges except the one pointing from the second reaction step to the species Z, because it contains together with all the reaction steps all the edges pointing into them. The initial species of this reaction path are X, Y, U.

Definition 8.11 Let $\emptyset \neq \mathscr{M}_0 \subset \mathscr{M}$ be a nonempty set of species. Species $X(m)$ is said to be **reachable** from $\mathscr{M}_0$, if there exists a reaction path which contains $X(m)$ and has no other initial species than those in $\mathscr{M}_0$.

Now the reason why it was useful to introduce indices in Chap. 3 will be obvious immediately. The following statement can be proved by induction on indices:

Lemma 8.12 *The index of a species is finite if and only if it is reachable from the set of species* $\mathscr{M}_0$*.*

The connection between indices and positivity follows. From now on let the set of species $\mathcal{M}_0$ be defined as those with positive initial concentration:

$$\mathcal{M}_0 := \{p \in \mathcal{M} \mid c_p(0) > 0\}.$$

Theorem 8.13 *Suppose that the zero complex is not a reactant complex in the reaction. Then for all species $m \in \mathcal{M}$ unreachable from $\mathcal{M}_0$ and for all $t \in J$ in the domain of existence of the solution $c_m(t) = 0$.*

A kind of pair of the above theorem follows.

Theorem 8.14 *For all species $m \in \mathcal{M}$ reachable from $\mathcal{M}_0$ and for all $t \in J$ in the domain of existence of the solution $c_m(t) > 0$.*

Remark 8.15 Obviously, the last theorem remains true if a possibly time-dependent input is added to the right-hand side of the induced kinetic differential equation. Both theorems above are obvious from the point of view of the chemist.

Example 8.16 The usual initial condition for the Michaelis–Menten reaction

$$E + S \rightleftharpoons C \longrightarrow E + P$$

has the property that $e(0) > 0, s(0) > 0, c(0) = p(0) = 0$. Calculating the Volpert indices using the corresponding function of **ReactionKinetics**

```
VolpertIndexing[
{"Michaelis-Menten"}, {"E", "S"}, Verbose -> True]
```

we get that all the species have a finite index; therefore all the concentrations are positive for positive times. More generally, if either C or both of E and S are initially present, this will again be the case; otherwise some of the species will become unavailable. We can formulate this fact that the sets {E, S} and {C} are **minimal initial sets** of species to produce positive concentrations, cf. page 209.

More complicated examples can be found in Kovács et al. (2004).

Remark 8.17 The absolute probabilities of a Markovian pure jump process (with time-independent rates) in general (not only those of a reaction) behave in a similar way: they remain either zero or positive for all positive times in their domain (see, e.g., Remark 2.48 in Liggett 2010). This is a consequence of the above theorems (only) in the case when the state space is finite, because in this case the master equation may be considered the induced kinetic differential equation of a compartmental system.

8.4   Blowing Up

Some experience in the theory of differential equations causes us to be quite careful speaking about a solution defined on the open interval $J \subset \mathbb{R}$. The very simple example of the simple second-order autocatalytic step $2\,\mathrm{X} \xrightarrow{k} 3\,\mathrm{X}$ shows that one cannot neglect the investigation of the **maximal solution** (see Definition 13.43) in the case of induced kinetic differential equations. Namely, the induced kinetic differential equation of this reaction being $\dot{c} = kc^2$ can symbolically be solved (not only by pen and paper but also using **Concentrations**) to give

$$[0, \frac{1}{kc(0)}[\ni t \mapsto c(t) = \frac{c(0)}{1 - ktc(0)}, \tag{8.11}$$

i.e., the concentration vs. time curve is only defined up to the time $\frac{1}{kc(0)}$.

This phenomenon is called **blowup** (or blowing up) in the theory of differential equations, but several other expressions are also used. In the theory of stochastic processes, practically the same phenomenon is called **finite escape time** or **first infinity**. We have to note that the problem is much more popular in the field of partial differential equations and stochastic processes than in the field of ordinary differential equations. If we remain within the framework of applications in chemical kinetics, then we may also use the terms the onset of **thermal runaway** or **ignition**.

The detailed analysis below is mainly based on the paper by Csikja and Tóth (2007) and some references cited therein. Above we had to use a nonlinear example (the model of a second-order reaction), because such a phenomenon cannot occur in linear differential equations with constant coefficients (even less in the case of induced kinetic differential equations of first-order reactions): the maximal solution is always defined on the whole real line. In other words we have what is called **global existence**. In the example above, we have a **movable singularity**, meaning that the location of the singularity depends on the initial value (Fig. 8.2).

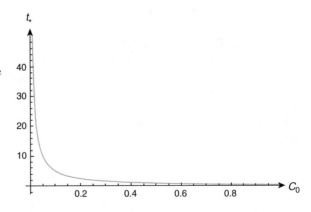

Fig. 8.2 Dependence of the blow-up time on the initial concentration in the reaction $2\,\mathrm{X} \xrightarrow{2} 3\,\mathrm{X}$. The larger the initial concentration c_0 is, the earlier the concentration becomes infinity

Definition 8.18 Let the solution of the initial value problem

$$\dot{\mathbf{c}}(t) = \mathbf{f}(\mathbf{c}(t)) \quad \mathbf{c}(0) = \mathbf{c}_0$$

(with $\mathbf{f} \in \mathscr{C}^1(\mathbb{R}^M, \mathbb{R}^M)$, $\mathbf{c}_0 \in \mathbb{R}^M$) be $J \ni t \mapsto \varphi(t, \mathbf{c}_0)$, where $J \subset \mathbb{R}$ is an open interval containing 0. This solution is said to **blow up** (from the left) at $t_* \in \mathbb{R} \setminus J$ if for all $M \in \mathbb{R}^+$, there exists $\delta \in \mathbb{R}^+$ for which and for all $t \in J$ such that $t_* - t < \delta$ the inequality $\|\varphi(t, \mathbf{c}_0)\| \geq M$ holds.

8.4.1 An Algebraic Method

As a first step, we present a sufficient condition for blowup, given by Getz and Jacobson (1977). (The case when one has a single species can easily be treated (see Problem 8.4).) Their method is based on an estimate using a corresponding scalar equation.

Naturally, if a linear combination of the components blows up, then at least one of the components should also blow up.

Let us consider the problem of blowing up with quadratic right-hand sides. Let $M \in \mathbb{N}$; $\mathbf{A}_1, \mathbf{A}_2, \ldots, \mathbf{A}_M \in \mathbb{R}^{M \times M}$; $\mathbf{b}_1, \mathbf{b}_2, \ldots, \mathbf{b}_M \in \mathbb{R}^M$; $c_1, c_2, \ldots, c_M \in \mathbb{R}$, and consider the initial value problem

$$\dot{x}_m = \mathbf{x}^\top \mathbf{A}_m \mathbf{x} + \mathbf{b}_m^\top \mathbf{x} + c_m \quad (m \in \mathscr{M}), \quad \mathbf{x}(0) = \mathbf{x}_0, \tag{8.12}$$

where one assumes (without the restriction of generality) that the matrices $\mathbf{A}_m$ are symmetric. Let us introduce

$$\mathbf{A}(\boldsymbol{\omega}) := \sum_{m \in \mathscr{M}} \omega_m \mathbf{A}_m \quad \mathbf{b}(\boldsymbol{\omega}) := \sum_{m \in \mathscr{M}} \omega_m \mathbf{b}_m \quad c(\boldsymbol{\omega}) := \sum_{m \in \mathscr{M}} \omega_m c_m \quad (\boldsymbol{\omega} \in \mathbb{R}^M)$$

$$\Delta(\boldsymbol{\omega}) := \frac{\lambda(\boldsymbol{\omega})}{\boldsymbol{\omega}^\top \boldsymbol{\omega}} (\mathbf{b}(\boldsymbol{\omega})^\top \mathbf{A}(\boldsymbol{\omega})^{-1} \mathbf{b}(\boldsymbol{\omega}) - 4c(\boldsymbol{\omega})) \quad (\boldsymbol{\omega} \in \mathbb{R}^M \setminus \{\mathbf{0}\}), \tag{8.13}$$

where $\lambda(\boldsymbol{\omega}) \in \mathbb{R}$ is the smallest eigenvalue of the matrix $\mathbf{A}(\boldsymbol{\omega})$, and $\mathbf{A}(\boldsymbol{\omega})$ will only be used below when it is invertible. With these definitions the (slightly corrected) main result of the mentioned authors is the following:

Theorem 8.19 (Getz and Jacobson (1977)) *If there exists $\boldsymbol{\omega} \in \mathbb{R}^M$ such that $\mathbf{A}(\boldsymbol{\omega})$ is positive definite, then the solution to (8.12) blows up:*

1. *if $\Delta(\boldsymbol{\omega}) < 0$, then for all $\mathbf{x}_0 \in \mathbb{R}^M$;*
2. *if $\Delta(\boldsymbol{\omega}) = 0$, then for all $\mathbf{x}_0 \in \mathbb{R}^M$ fulfilling*

$$\boldsymbol{\omega}^\top \mathbf{x}_0 > -\frac{1}{2} \boldsymbol{\omega}^\top \mathbf{A}(\boldsymbol{\omega})^{-1} \mathbf{b}(\boldsymbol{\omega});$$

3. if $\Delta(\omega) > 0$, then for all $\mathbf{x}_0 \in \mathbb{R}^M$ fulfilling

$$\omega^\top \mathbf{x}_0 > -\frac{1}{2}\omega^\top \mathbf{A}(\omega)^{-1}\mathbf{b}(\omega) + \frac{\sqrt{\Delta(\omega)}\omega^\top \omega}{2\lambda(\omega)}.$$

The upper estimates for the blow-up time t_* in the cases above are as follows:

1.

$$t_* \leq \frac{2}{\sqrt{-\Delta(\omega)}}\left(\frac{\pi}{2} - \frac{\arctan(2\mathbf{z}(\omega)\lambda(\omega))}{\sqrt{-\Delta(\omega)}}\right),$$

2.

$$t_* \leq -\frac{\omega^\top \omega}{\lambda(\omega^\top \mathbf{x}_0 + \omega^\top \mathbf{A}^{-1}\mathbf{b})},$$

3.

$$t_* \leq \frac{1}{\sqrt{\Delta(\omega)}}\ln\left(\frac{\mathbf{z}(\omega) + \frac{\sqrt{\Delta(\omega)}}{2\lambda(\omega)}}{\mathbf{z}(\omega) - \frac{\sqrt{\Delta(\omega)}}{2\lambda(\omega)}}\right),$$

with $\mathbf{z}(\omega) := \frac{\omega^\top(\mathbf{x}_0 + \frac{1}{2}\mathbf{A}(\omega)^{-1}\mathbf{b}(\omega))}{\omega^\top \omega}$.

Example 8.20 Let us try to show that the reaction

$$0 \xrightarrow{\ k_1\ } Y \quad 2X \xrightarrow{\ k_2\ } X + Y \quad 2Y \xrightarrow{\ k_3\ } 3X$$

blows up for some initial concentration vectors. The induced kinetic differential equation of the reaction is $\dot{x} = -k_2 x^2 + 3k_3 y^2$ $\dot{y} = k_2 x^2 - 2k_3 y^2 + k_1$; therefore with the notation of Theorem 8.19, one has $\mathbf{A}_1 = \begin{bmatrix} -k_2 & 0 \\ 0 & 3k_3 \end{bmatrix}$, $\mathbf{A}_2 = \begin{bmatrix} k_2 & 0 \\ 0 & -2k_3 \end{bmatrix}$, $\mathbf{b}_1 = \mathbf{b}_2 = \mathbf{0}$, $c_1 = 0$, and $c_2 = k_1$. It is easy to find an ω satisfying Case 1 of Theorem 8.19: $\omega = \begin{bmatrix} 3 & 4 \end{bmatrix}$ is an example. Let us also fix the reaction rate coefficients as $k_1 = 1, k_2 = 2$, and $k_3 = 3$. Then $\lambda(\omega) = 2, c(\omega) = 4$, and $\Delta(\omega) = -32/25$; thus the solutions blow up with any initial values x_0, y_0. For this fixed ω, one can also apply the upper estimate of the blow-up time if one uses the values $x_0 = 1, y_0 = 2$:

$$t_* < \frac{2}{\sqrt{32/25}}\left(\frac{\pi}{2} - \frac{\arctan(44/25)}{\sqrt{32/25}}\right) \approx 1.12977.$$

Plotting the result provided by

```
ReplaceAll @@ Concentrations[
    {2X -> X + Y, 0 -> Y, 2Y -> 3X},
    {2, 1, 3}, {1, 2}, {0, 2}]
```

shows that the blow-up time is around 0.771.

How to find the best vector ω in general? The matrix

$$\mathbf{A}(\omega) := \begin{bmatrix} 2(\omega_2 - \omega_1) & 0 \\ 0 & 3(3\omega_1 - 2\omega_2) \end{bmatrix}$$

is positive definite if and only if $\frac{3}{2}\omega_1 > \omega_2 > \omega_1$ (implying $\omega_1 > 0$); thus one has to choose ω from the area defined by these two inequalities in such a way that the upper estimate of t_* be as small as possible. (Note that $\Delta(\omega) = -\frac{8}{\omega_1^2 + \omega_1^2}$ will be negative for all such choices.)

A series of natural questions is formulated at the end of the chapter.

A numerical method follows for the detection of blowup in quadratic equations. Case 1 of Theorem 8.19 can be considerably simplified for numerical methods.

Theorem 8.21 *Case 1 of Theorem 8.19 holds if and only if there exists an $\omega \in \mathbb{R}^M$ satisfying*

$$\mathbf{H}(\omega) := \begin{bmatrix} 4c(\omega) & b(\omega)^\top \\ b(\omega) & \mathbf{A}(\omega) \end{bmatrix} \succcurlyeq \mathbf{I}_{(M+1)\times(M+1)}. \tag{8.14}$$

Proof Clearly, $\Delta(\omega) < 0$ if and only if $4c(\omega) - b(\omega)^\top \mathbf{A}(\omega)^{-1} b(\omega) > 0$. Using Haynsworth's theorem (Theorem 13.23), the latter inequality and $\mathbf{A}(\omega) \succ \mathbf{0}$ together are equivalent to $\mathbf{H}(\omega) \succ \mathbf{0}$.

To complete the proof, observe that such an ω^* exists if and only if there exists one that satisfies $\mathbf{H}(\omega^*) \succcurlyeq \mathbf{I}$. Namely, suppose $\omega^* \in \mathbb{R}^M$ satisfies $\mathbf{A}(\omega^*) \succ \mathbf{0}$ and $\mathbf{H}(\omega^*) \succ \mathbf{0}$, and let $\lambda_1 > 0$ be the smallest eigenvalue of $\mathbf{H}(\omega^*)$. Then for every $c > 0 : \mathbf{A}(c\omega^*) \succ \mathbf{0}$, and the smallest eigenvalue of $\mathbf{H}(c\omega^*) - \mathbf{I}$ is $c\lambda_1 - 1$. Hence, for $c \geq 2/\lambda_1$, $\mathbf{H}(c\omega^*) \succcurlyeq \mathbf{I}$ holds. $\square$

If the condition of the above theorem does not hold, we can still find an ω satisfying $\mathbf{A}(\omega) \succ \mathbf{0}$, or $\mathbf{H}(\omega) \succcurlyeq \mathbf{I}$. The minimization or maximization of $\Delta(\omega)$, however, seems to be difficult.

Constraints of the form $\mathbf{A}(\omega) \succcurlyeq \mathbf{B}$, where $\mathbf{A}$ is a fixed linear function of the variable vector ω and $\mathbf{B}$ is a given matrix, are called **linear matrix inequalities** or **semi-definite constraints**. The problem of finding an ω satisfying such inequalities, or, more generally, the problem of finding $\sup\{\mathbf{c}^\top\mathbf{x} \,|\, \mathbf{A}(\mathbf{x}) \succcurlyeq \mathbf{B}, x \in \mathbb{R}^M\}$ is called a **semi-definite optimization problem** or **semi-definite program**. The similarity between semi-definite programs and linear programs is not coincidental; semi-definite programs are generalizations of linear programming problems. There are several efficient numerical algorithms (specialized interior-point methods) to solve semi-definite programs, which we shall not discuss in this book. The reader interested in the topic is encouraged to consult the rather accessible survey Vandenberghe and Boyd (1996) for a good introduction; Wolkowicz et al. (2000) is a considerably more in-depth survey of the (nearly) state-of-the-art.

Example 8.22 As a continuation of Example 8.20, let us check inequality (8.14). This reduces to

$$\omega_2 > \frac{1}{4k_1}, \quad \omega_2 > \omega_1 + \frac{1}{k_2}, \quad \omega_2 < \frac{3}{2}\omega_1 - \frac{1}{2k_3},$$

what can always be fulfilled (as the reader can convince herself/himself), no matter what the value of the reaction rate coefficients are, i.e., there always exists an ω for which $\Delta(\omega) < 0$, and thus the system blows up for every initial vector (including nonnegative ones).

8.4.2 Transformations of Differential Equations

Transformation of kinetic differential equations is a recurrent theme. It turns out that the transformations proposed by Beklemisheva (1978) may throw light on the phenomenon of blowing up.

Every polynomial differential equation (6.39) can obviously be written in the following form (see the special case of Example 8.25 on how the mapping from (6.39) to (8.16) works):

$$\dot{x}_m = x_m \left(\lambda_m + \sum_{r \in \mathcal{R}} A_{mr} \prod_{p=1}^{M} x_p^{B_{pr}} \right) \quad (m \in \mathcal{M}), \tag{8.15}$$

or using the more compact notation reviewed in Chap. 13,

$$\dot{\mathbf{x}} = \mathbf{x} \odot (\lambda + \mathbf{A}\mathbf{x}^{\mathbf{B}}) \tag{8.16}$$

where $\mathbf{A} \in \mathbb{R}^{M \times R}, \mathbf{B} \in \mathbb{R}^{M \times R}, \lambda \in \mathbb{R}^M$. For polynomial equations the entries of $\mathbf{B}$ are integers not smaller than -1; in what follows we relax these assumptions and allow $\mathbf{B}$ to have arbitrary, even non-integer elements; thus Eq. (8.16) is not always of the Kolmogorov form (see Definition 6.16).

Theorem 8.23 *Let* $\mathbf{C} \in \mathbb{R}^{M \times M}$ *be a non-singular (invertible) matrix, and let us introduce new variables with the definition* $\mathbf{y} := \mathbf{x}^{\mathbf{C}^{-1}}$. *Then, Eq. (8.16) is transformed into an equation of the same form but with the following parameters:*

$$\lambda' := (\mathbf{C}^{-1})^{\top}\lambda \quad \mathbf{A}' := (\mathbf{C}^{-1})^{\top}\mathbf{A} \quad \mathbf{B}' := \mathbf{C}\mathbf{B}. \tag{8.17}$$

Proof First of all, we remind the reader that the operations in the theorem and the proof are defined and treated in the Appendix. As $\mathbf{y} := \mathbf{x}^{\mathbf{C}^{-1}}$ implies (see the Appendix) $\mathbf{x} = \mathbf{y}^{\mathbf{C}}$, one can use (14.61) to get

$$\dot{\mathbf{y}} = (\mathbf{x}^{\mathbf{C}^{-1}})^{\cdot} = \mathbf{x}^{\mathbf{C}^{-1}} \odot \left((\mathbf{C}^{-1})^{\top} \frac{\dot{\mathbf{x}}}{\mathbf{x}} \right) = \mathbf{y} \odot \left((\mathbf{C}^{-1})^{\top} \lambda + (\mathbf{C}^{-1})^{\top} \mathbf{A} \mathbf{y}^{\mathbf{CB}} \right)$$

meaning that for $\mathbf{y}$ one has an equation of the form Eq. (8.16) with the parameters as given in Eq. (8.17). $\qquad\square$

Remark 8.24

- An immediate consequence of (8.17) is that the transformation has two invariants: $\mathbf{B}^{\top}\lambda$ and $\mathbf{B}^{\top}\mathbf{A}$, because

$$\mathbf{B}'^{\top}\lambda' = \mathbf{B}^{\top}\lambda, \quad \mathbf{B}'^{\top}\mathbf{A}' = \mathbf{B}^{\top}\mathbf{A}.$$

- A transformation of this form is sometimes called **quasi-monomial transformation**.
- Systems of the form (8.16) together with the given type of transformations have also been used by Szederkényi et al. (2005) and earlier by Hernández-Bermejo and Fairén (1995).

Let us consider special cases when this transformation leads to a simplified form, which may also reveal if blowup occurs or not.

8.4.2.1 Decoupling

Suppose that $M < R$, and the rank of the matrix $\mathbf{B} \in \mathbb{R}^{M \times R}$ in Eq. (8.16) is $M_0 < M$, then there exist independent vectors:

$$\mathbf{v}_{M_0+1}, \mathbf{v}_{M_0+2}, \ldots, \mathbf{v}_M \in \mathbb{R}^M \text{ for which } \mathbf{v}_m^{\top}\mathbf{B} = \mathbf{0}_R^{\top} \quad (m = M_0+1, M_0+2, \ldots, M),$$

and we can choose $\mathbf{C}$ in the following way:

$$\mathbf{C} := \begin{bmatrix} \mathbf{I}_{M_0} & \mathbf{0}_{M_0 \times (M-M_0)} \\ \mathbf{v}_{M_0+1}^{\top} & \\ \cdots & \\ \mathbf{v}_M^{\top} & \end{bmatrix}_{M \times M},$$

where $\mathbf{I}_{M_0}$ is a $M_0 \times M_0$ unit matrix, and $\mathbf{0}_{M_0 \times (M-M_0)}$ is a zero matrix. (Note that the first M_0 rows are partitioned, while the last $M - M_0$ rows are not.) In this case we have $\mathbf{B}' = \mathbf{C}\mathbf{B} = \begin{bmatrix} \mathbf{B}_{M_0 \times R} \\ \mathbf{0}_{(M-M_0) \times R} \end{bmatrix}$, which means that the first M_0 equations will only contain the first M_0 variables, and solving these one can substitute these variables into the last $M - M_0$ equations to get a linear equation (with nonconstant coefficients). This procedure is thus a special case of lumping (Tóth et al. 1997) and has been applied in the special case of second-order reactions earlier (Tóth and Érdi 1978, pp. 290–295).

The explicit form of the transformed equations is

$$\dot{y}_m = y_m \left(\lambda'_m + \sum_{j=1}^{m} A'_{mj} \prod_{p=1}^{M_0} y_p^{B_{pj}} \right) \qquad (m = 1, 2, \ldots, M_0)$$

$$\dot{y}_m = y_m \left(\lambda'_m + \sum_{j=1}^{m} A'_{mj} \right) \qquad (m = M_0 + 1, M_0 + 2, \ldots, M).$$

The obtained system blows up if and only if its nonlinear part does.

8.4.2.2 Lotka–Volterra Form

In the very special case when $M = R$ and $\mathbf{B}$ is invertible, an even more transparent form can be obtained: a quadratic polynomial equation of the **generalized Lotka–Volterra form**; see (6.34). Let us choose $\mathbf{C} := \mathbf{B}^{-1}$; then $\lambda' = \mathbf{B}^\top \lambda$, $\mathbf{A}' = \mathbf{B}^\top \mathbf{A}$, $\mathbf{B}' = \mathbf{I}$ show that these parameters are invariant for any further quasi-monomial transformations, meaning that we have arrived at the simplest possible form in a certain sense: any further transformation can only "distort" the simple exponent of $\mathbf{x}$. Note in particular that this transformation disposes of the negative and non-integral exponents on the right-hand side of the equation. Another important point is that the transformed equation is always a kinetic differential equation. Let us consider an example.

Example 8.25 In the case of the equation

$$\dot{x}_1 = x_1 \left(1 - \frac{x_2^{2/3}}{x_3^2} \right), \quad \dot{x}_2 = x_2 \left(-1 - \frac{3x_2^{2/3}}{x_3^2} + x_2 \right), \quad \dot{x}_3 = x_3 \left(2 + x_2 + \frac{5x_1 x_2^3}{x_3^1} \right)$$

we have

$$\lambda = \begin{bmatrix} 1 \\ -1 \\ 2 \end{bmatrix}, \quad \mathbf{A} = \begin{bmatrix} -1 & 0 & 0 \\ -3 & 1 & 0 \\ 0 & 1 & 5 \end{bmatrix}, \quad \mathbf{B} = \begin{bmatrix} 0 & 0 & 1 \\ \frac{2}{3} & 1 & 3 \\ -2 & 0 & -1 \end{bmatrix}, \qquad (8.18)$$

therefore

$$\mathbf{C} := \mathbf{B}^{-1} = \begin{bmatrix} -\frac{1}{2} & 0 & -\frac{1}{2} \\ -\frac{8}{3} & 1 & \frac{1}{3} \\ 1 & 0 & 0 \end{bmatrix} \quad \mathbf{B}^\top \mathbf{A} = \begin{bmatrix} -2 & -\frac{4}{3} & -10 \\ -3 & 1 & 0 \\ -10 & 2 & -5 \end{bmatrix} \quad \mathbf{B}^\top \lambda = \begin{bmatrix} -5 \\ -1 \\ -4 \end{bmatrix}.$$

$$(8.19)$$

With these we get the Lotka–Volterra form of the equations:

$$\dot{y}_1 = y_1 \left(-\frac{14}{3} - 2y_1 - \frac{4}{3}y_2 - 10y_3 \right)$$
$$\dot{y}_2 = y_2 \left(-1 - 3y_1 + y_2 \right)$$
$$\dot{y}_3 = y_3 \left(-4 - 10y_1 + 2y_2 - 5y_3 \right).$$

Remark 8.26

1. This method enables us to eliminate non-integer exponents. This implies that the original equation may have not fulfilled the Lipschitz condition, whereas the transformed one obeys it.
2. The transformed equation is always a kinetic differential equation, even if the original was not one.
3. The authors Brenig and Goriely (1994) have carried out a systematic investigation on how to find an appropriate **C** matrix.
4. Beklemisheva (1978) always assumes the invertibility of **B** (which in the case of reaction kinetic applications means that the reaction step vectors are independent). Moreover, she only treats the case $\lambda = \mathbf{0}$.

Exclusion of the phenomena of blowing up is also a consequence of general theorems of which we mention a few now. The solution of the induced kinetic differential equation of

- a stoichiometrically subconservative (Problem 8.5), stoichiometrically mass-conserving, or kinetically mass-conserving reaction;
- a detailed balanced mechanism (Theorem 8.45);
- a zero-deficiency reaction (Theorem 8.47);
- a reaction with acyclic Volpert graph (Theorem 8.54);
- a first-order reaction (being a linear differential equation)

does not blow up.

To close the topics of blowing up, we mention that Csikja and Tóth (2007) initiated the use of **Kovalevskaya exponents** and **Painlevé analysis** for finding movable singularities in induced kinetic differential equations.

8.5 First Integrals

The meaning and existence of (positive) linear first integrals have been studied in Chap. 4 in detail: they usually represent mass conservation. The existence of a positive linear first integral together with nonnegativity (Volpert 1972; Volpert and

Hudyaev 1985) of the solutions implies that the complete solution of the kinetic differential equation is defined on the whole real line (i.e., it does not blow up), which is not necessarily the case for systems that are not (stoichiometrically) mass conserving.

However, quadratic first integrals were almost neglected; therefore we dedicate a few lines to them, as well.

We may reformulate the quest of first integrals as wanting to restrict even more that part of the state space where the trajectories of an induced kinetic differential equation can wander.

Definition 8.27 Let $M \in \mathbb{N}, \mathbf{f} \in \mathscr{C}^1(\mathbb{R}^M, \mathbb{R}^M)$, and consider the differential equation $\dot{\mathbf{x}} = \mathbf{f} \circ \mathbf{x}$. The function $\varphi \in \mathscr{C}^1(\mathbb{R}^M, \mathbb{R})$ (or, **functional**, if one wants to emphasize that it is a scalar-valued function) is said to be the **first integral** of the given differential equation if for all solutions $\boldsymbol{\xi}$ of it one has $\varphi(\boldsymbol{\xi}(t)) = $ constant ($t \in \mathscr{D}_{\boldsymbol{\xi}}$).

8.5.1 Linear First Integrals

Let us start with the simplest case and try to find (homogeneous) linear functions which are first integrals. Such a function can be identified with a vector in $\mathbb{R}^M$. Without restriction of generality, one can assume that a linear first integral is a homogeneous linear function.

Recall the general mass-action type induced kinetic differential equation:

$$\dot{\mathbf{c}} = \mathbf{f} \circ \mathbf{c} = (\boldsymbol{\beta} - \boldsymbol{\alpha}) \cdot \mathbf{w} \circ \mathbf{c}. \tag{6.5}$$

By integrating this equation we get

$$\mathbf{c}(t) = \mathbf{c}_0 + (\boldsymbol{\beta} - \boldsymbol{\alpha}) \cdot \int_0^t \mathbf{w}(\mathbf{c}(s)) \, ds \tag{8.20}$$

showing that the trajectories can only wander in the parallels of the stoichiometric space or, more precisely, in its part lying in the first orthant: the nonnegative reaction simplex. This formula also shows that for a mass-conserving reaction one has

$$\boldsymbol{\varrho}^\top \mathbf{c}(t) = \text{const}(= \boldsymbol{\varrho}^\top \mathbf{c}_0). \tag{8.21}$$

In some cases it may happen that an even narrower set may contain the trajectories. Obviously, one can also state that the trajectories remain in the linear space spanned by the range of the right-hand side,

$$\mathscr{S}^* := \text{span} \{\mathscr{R}_{\mathbf{f}}\}, \tag{8.22}$$

which may justifiably have the name **kinetic subspace**. (Note that $\mathscr{R}_{\mathrm{f}}$ itself is in general not a linear space, cf. Problem 8.32.) The natural question (answered by Feinberg and Horn 1977) is although trivially

$$\mathscr{S}^* \subset \mathscr{S}, \tag{8.23}$$

under what condition can we have strict inclusion and when do we have equality in (8.23)?

Theorem 8.28 (Feinberg and Horn (1977)) *Let us consider a reaction endowed with mass-action kinetics, and suppose that the Feinberg–Horn–Jackson graph has L components and T ergodic components, and the deficiency of the mechanism is δ.*

1. *If $T = L$, then the stoichiometric subspace and the kinetic subspace coincide: $\mathscr{S} = \mathscr{S}^*$.*
2. *If $T - L > \delta$, then the stoichiometric and the kinetic subspaces do not coincide: $\mathscr{S} \subsetneq \mathscr{S}^*$.*

Remark 8.29 Let us note that the numbers T and L only depend on the "reacts to" relation and do not depend on the values of the stoichiometric coefficients (even less on reaction rate coefficients).

Note that weakly reversible reactions satisfy $T = L$. (Theorem 3.8.)

Corollary 8.30 *For every weakly reversible reactions (a fortiori for every reversible reactions), the stoichiometric and kinetic subspaces coincide.*

Corollary 8.31 *For reactions of deficiency zero, the stoichiometric and kinetic subspaces coincide if and only if each connected component contains precisely one ergodic component.*

Example 8.32 The reaction in Fig. 8.3 has $\{A + B, C\}$, $\{D + E, F\}$, and $\{G, H, 2\,J\}$ as strong components (Definition 13.30), and $\{D + E, F\}$ and $\{G, H, 2\,J\}$ as ergodic components (or terminal strong linkage classes). Our program calculates this result in the following way:

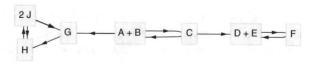

Fig. 8.3 Reaction with more ergodic classes than linkage classes: $2 = T > L = 1$

```
Last /@ ReactionsData[
    {H <=> 2J -> G <- A+B <=> C -> D+E <=> F, G -> H}]
    ["fhjterminalstronglyconnectedcomponents"]
```

Definition 8.33 A mechanism with **f** as the right-hand side of its induced kinetic differential equation is said to be **kinetically mass conserving** if there exists a vector $\varrho \in \mathbb{R}^M$ with positive coordinates so that for all $\mathbf{c} \in (\mathbb{R}_0^+)^M$

$$\varrho^\top \cdot \mathbf{f}(\mathbf{c}) = 0 \tag{8.24}$$

holds.

Remark 8.34

- The concepts **kinetically mass producing** and **kinetically mass consuming** can be defined in an analogous way.
- Stoichiometrically mass-conserving mechanisms are also kinetically mass conserving, and Problem 8.12 shows that the converse is not true.

8.5.2 Nonlinear First Integrals

It may also happen that a differential equation has a nonlinear function as its first integral. The simplest case is that of quadratic functions.

8.5.2.1 Quadratic First Integrals

Here we try to determine classes of mass-action type kinetic differential equations with the property of having a **quadratic first integral**. Since the introduction of the name of first integral by E. Nöther in 1918, it turned out that first integrals may help

- prove that the complete solution of the induced kinetic differential equation is defined for all positive times (Volpert 1972, p. 586, Theorem 9);
- reduce the number of variables either by constructing an appropriate lumping scheme (Li et al. 1994) or by simply eliminating some variables;
- apply the generalization of the Bendixson and Bendixson–Dulac criterion to higher dimensional cases (Tóth 1987; Weber et al. 2010).

Our main tool to find such first integrals is simply the comparison of coefficients of polynomials and the characterization of kinetic differential equations within the class of polynomial ones (see Theorem 6.27). It is a very natural requirement that a numerical method aimed at solving (6.3) should keep the total mass $\sum_{m \in \mathcal{M}} \varrho_m c_m(t)$ constant (independent of time) in case of a kinetically mass-conserving reaction. There are some methods to have this property (see, e.g.,

Bertolazzi 1996). A similar requirement is to keep other, e.g., quadratic first integrals, which has also been shown for some methods by Rosenbaum (1977).

However, not much is known about equations, especially kinetic differential equations with quadratic first integrals. Obviously, equations of mechanics, like that of the standard harmonic oscillator $x' = y \quad y' = -x$, may have quadratic first integrals, $\varphi(p, q) := p^2 + q^2$ in this case, and here the meaning of the quadratic first integral is the total mechanical energy.

We cite some of the statements on the existence and nonexistence of quadratic first integrals from Nagy and Tóth (2014); the reader might consult the paper for proofs and further details.

Let us start with the simplest case: when the candidate first integral is a sum of squares, i.e., when one has a **diagonal quadratic first integral**.

Theorem 8.35 *Let us consider the following system of differential equations*

$$\dot{x}_m = F_m \circ (x_1, x_2, \ldots, x_M), \quad (m \in \mathscr{M}) \tag{8.25}$$

where the functions F_m are quadratic functions of the variables, that is,

$$F_m(x_1, x_2, \ldots, x_M) = \sum_{p \in \mathscr{M}} A_{m,p} x_p^2 + \sum_{\substack{p=1 \\ p \neq m}}^{M} B_{m,p} x_m x_p$$

$$+ \sum_{\substack{p,q=1 \\ p < q \\ p \neq m, q \neq m}}^{M} C_{p,q}^m x_p x_q + \sum_{p \in \mathscr{M}} D_{m,p} x_p + E_m. \tag{8.26}$$

Suppose that the system of differential equations is kinetic. *The function*

$$\varphi(x_1, x_2, \ldots, x_M) = a_1 x_1^2 + a_2 x_2^2 + \cdots + a_M x_M^2$$

(with $a_m > 0$ for $m \in \mathscr{M}$) is a first integral for the above system if and only if the functions F_m have the following form with $K_{m,p} \geq 0$:

$$F_m(x_1, x_2, \ldots, x_M) = \sum_{\substack{p=1 \\ p \neq m}}^{M} a_p K_{m,p} x_p^2 - \sum_{\substack{p=1 \\ p \neq m}}^{M} a_p K_{p,m} x_m x_p. \tag{8.27}$$

Proof The function φ is a first integral for the system (8.25) if and only if its derivative with respect to the system (its **Lie derivative**; see Definition 13.47) is equal to zero. This fact can be expressed by the coefficients of the polynomials on the right-hand side. Investigating the signs of the coefficients and using the absence of negative cross effect, one arrives at the only part of the statement.

The proof of the **if** part is obvious. $\qquad\square$

Example 8.36 Let $M = 2$ and suppose that $\varphi(x, y) = x^2 + y^2$. Then (8.27) specializes to

$$\dot{x} = ay^2 - bxy, \quad \dot{y} = bx^2 - axy \qquad (8.28)$$

which may be considered as the induced kinetic differential equation of the reaction

$$X \xleftarrow{\;a\;} X + Y \xrightarrow{\;b\;} Y \qquad 2\,X \xrightarrow{\;b\;} 2\,X + Y \qquad 2\,Y \xrightarrow{\;a\;} X + 2\,Y \qquad (8.29)$$

as the application of

```
RightHandSide[
    {X <- X+Y -> Y, 2X -> 2X+Y, 2Y -> X+2Y}
    ,{a, b, b, a}, {x,y}]
```

gives: $\{a\ y^2\text{-}b\ x\ y,\ b\ x^2\text{-}a\ x\ y\}$. A typical trajectory is shown in Fig. 8.4. Naturally arises the question if the differential equation (8.28) can be represented with a mechanism only containing three complexes 2 X, 2 Y, and X + Y. It can be easily shown that the answer is negative (see Problem 6.21).

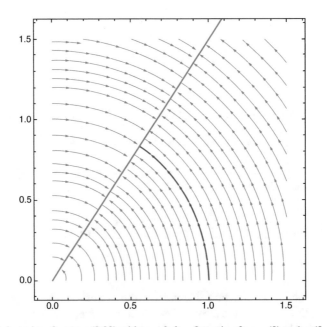

Fig. 8.4 Trajectories of system (8.28) with $a = 2$, $b = 3$ starting from $x(0) = 1$, $y(0) = 0$

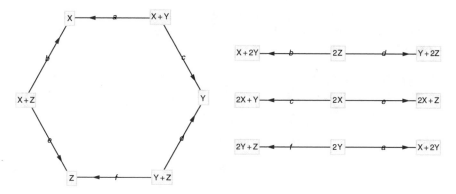

Fig. 8.5 3D system with a quadratic first integral

Example 8.37 Let $M = 3$ and suppose that $\varphi(x, y, z) = x^2 + y^2 + z^2$. Then (8.27) specializes to

$$
\begin{aligned}
\dot{x} &= ay^2 + bz^2 - cxy - exz \\
\dot{y} &= cx^2 + dz^2 - axy - fyz \\
\dot{z} &= ex^2 + fy^2 - bxz - dyz
\end{aligned}
\tag{8.30}
$$

(with nonnegative coefficients a, b, c, d, e, f) which may be considered as the induced kinetic differential equation of the reaction shown in Fig. 8.5 as again the application of **RightHandSide** verifies.

Corollary 8.38 *As the divergence of the system* (8.30) *is* $-ax - bx - cy - dy - ez - fz < 0$ *in the first orthant (if at least one of the coefficients is different from zero) and the system has a first integral, (Tóth 1987, Theorem 3.3) (actually, a version of K. R. Schneider's theorem) implies that it has no periodic orbit in the first orthant.*

A similar proof shows that (even weighted) sum of squares cannot be a first integral if mass is conserved.

Theorem 8.39 *Let us consider the differential equation* (8.25) *where the functions F_m are of the form* (8.26). *Suppose that the differential equation is kinetic and kinetically mass conserving. The function*

$$
\varphi(x_1, x_2, \ldots, x_M) = a_1 x_1^2 + a_2 x_2^2 + \cdots + a_M x_M^2
$$

(where $a_m \neq 0$ for all m) is a first integral for the system (8.25), *if and only if for all m*

$$
F_m(x_1, x_2, \ldots, x_M) = 0.
$$

The paper Nagy and Tóth (2014) contains results also on more general quadratic forms with and without the assumption of (kinetic) mass conservation. Let us only cite a single example.

Example 8.40 If $\varphi(x_1, x_2, y_1, y_2, z) = x_1^2 + x_2^2 - y_1^2 - y_2^2$ is a first integral and the vector of masses to be conserved is $\varrho^{x_1} = \varrho^{x_2} = \varrho^{y_1} = \varrho^{y_2} = \varrho^z = 1$, then the system to have this first integral and conserving mass is $(a, b, c, d \geq 0)$:

$$\dot{x}_1 = ay_1z + by_2z \quad \dot{x}_2 = cy_1z + dy_2z \quad \dot{y}_1 = ax_1z + cx_2z \quad \dot{y}_2 = bx_1z + dx_2z$$
$$\dot{z} = -\dot{x}_1 - \dot{x}_2 - \dot{y}_1 - \dot{y}_2 \tag{8.31}$$

A possible inducing reaction is the following:

$$X_1 + Z \xrightarrow{1} aY_1 + bY_2 + X_1 + (1 - a - b)Z$$
$$X_2 + Z \xrightarrow{1} cY_1 + dY_2 + X_2 + (1 - c - d)Z$$
$$Y_1 + Z \xrightarrow{1} aX_1 + cX_2 + Y_1 + (1 - a - c)Z$$
$$Y_2 + Z \xrightarrow{1} bX_1 + dX_2 + Y_2 + (1 - b - d)Z$$

Another possible reaction can be seen in Fig. 8.6. We have also investigated the case in two dimensions when the first integral is a binary quadratic form.

Finally, let us mention that most of our statements can be obtained using the package **ReactionKinetics** and by simple additional programs. Also, previously (see page 95), we provided simple codes to check if negative cross effect is present in a polynomial or not.

Fig. 8.6 Reaction system for (8.31)

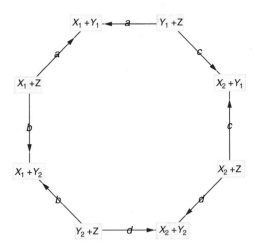

8.5.2.2 Further Forms

We may find to try other types of first integrals. Let us mention one simple, still interesting result.

Theorem 8.41 *Among the polynomial differential equations of the form*

$$\dot{x} = ax^2 + bxy + cy^2 + dx + ey + f$$
$$\dot{y} = Ax^2 + Bxy + Cy^2 + Dx + Ey + F \tag{8.32}$$

(defined in the positive quadrant) the only one having

$$\varphi(p, q) := p + q - \ln(p) - \ln(q)$$

as its first integral is

$$\dot{x} = bxy - bx \quad \dot{y} = -bxy + by. \tag{8.33}$$

Note that it is not assumed that (8.32) is a kinetic differential equation, and even if it is assumed, in the result no restriction is made on the sign of b.

This result is very similar to the results leading uniquely to the Lotka–Volterra model under different circumstances (see, e.g., Morales 1944; Hanusse 1972, 1973; Tyson and Light 1973; Póta 1983; Tóth and Hárs 1986b; Schuman and Tóth 2003). One might try to generalize this result to the multidimensional case.

Another form of interesting first integrals is a free energy-like function:

$$\varphi(\mathbf{c}) := \sum_{m \in \mathcal{M}} c_m \ln \left(\frac{c_m}{c_m^0} \right),$$

which will turn out to be a useful Lyapunov function for broad classes of reactions as we will see below. Gonzales-Gascon and Salas (2000) have systematically found this type of first integrals (and other types, as well) for three-dimensional Lotka–Volterra systems.

8.6 Invariant Sets

Invariance is a key concept; we shall meet it a few times in the book.

Suppose $\varphi \in \mathcal{C}^1(\mathbb{R}^M, \mathbb{R})$ is a first integral of the system

$$\dot{\mathbf{x}} = \mathbf{f} \circ \mathbf{x} \quad \mathbf{x}(0) = \mathbf{x}_0, \tag{8.34}$$

where $\mathbf{f} \in \mathscr{C}^1(\mathbb{R}^M, \mathbb{R}^M)$. Then for all $t \in \mathscr{D}_{\mathbf{x}}$, one has $\varphi(\mathbf{x}(t)) = \varphi(\mathbf{x}_0)$, or $\varphi(\mathbf{x}(t)) - \varphi(\mathbf{x}_0) = 0$, i.e., the set

$$\Phi := \{\overline{\mathbf{x}} \in \mathbb{R}^M \, | \, \varphi(\overline{\mathbf{x}}) - \varphi(\mathbf{x}_0) = 0\} \tag{8.35}$$

is an **invariant set** of the system (8.34), meaning that solutions starting in Φ will remain in Φ as far as they are defined. The fact that in the case of kinetic differential equations solutions starting in $(\mathbb{R}_0^+)^M$ remain in this set shows that the existence of an invariant set may not imply the existence of a nontrivial (time-independent) first integral. Now we show a method to find invariant sets (which may also lead to finding a first integral, if it exists) following Romanovski and Shafer (2009), Section 3.6 and Antonov et al. (2016). Other methods can be found in the papers cited in Tóth et al. (1997).

Definition 8.42 The function $H \in \mathscr{C}^1(\mathbb{R}^M, \mathbb{R})$ is said to be a **generalized eigenfunction** of the equation $\dot{\mathbf{x}} = \mathbf{f} \circ \mathbf{x}$ if there exists a function $\varphi \in \mathscr{C}^1(\mathbb{R}^M, \mathbb{R})$ such that

$$H'(\mathbf{x})\mathbf{f}(\mathbf{x}) = \varphi(H(\mathbf{x})) \quad \mathbf{x} \in \mathbb{R}^M \tag{8.36}$$

holds.

Remark 8.43 If one introduces the linear operator $\mathscr{C}^1(\mathbb{R}^M, \mathbb{R}) \ni H \mapsto \mathscr{A}H := H'\mathbf{f} \in \mathscr{C}(\mathbb{R}^M, \mathbb{R})$, then Eq. (8.36) can be written in the following way: $\mathscr{A}H = \varphi \circ H$; therefore, more precisely, H is a generalized eigenfunction of the operator $\mathscr{A}$ (actually defined using $\mathbf{f}$). We can use the name instead of scalar-valued functions H for vector-valued functions in a similar meaning.

Theorem 8.44 *Suppose that there exists $H \in \mathscr{C}^1(\mathbb{R}^M, \mathbb{R})$ and $\varphi \in \mathscr{C}^1(\mathbb{R}, \mathbb{R})$ with $\varphi(0) = 0$ such that Eq. (8.36) holds. Then the set $\Phi := \{\overline{\mathbf{x}} \in \mathbb{R}^M \, | \, H(\overline{\mathbf{x}}) = 0\}$ is an invariant set of (8.34).*

Proof Let $\mathbf{x}_0 \in \Phi$, i.e., $H(\mathbf{x}_0) = 0$. Consider the function h defined by $h(t) := H(\mathbf{x}(t))$ $(t \in \mathscr{D}_{\mathbf{x}})$, where $\mathbf{x}$ is the solution of (8.34). Then,

$$h(0) = H(\mathbf{x}(0)) = H(\mathbf{x}_0) = 0, \quad h'(t) = H'(\mathbf{x}(t))\mathbf{f}(\mathbf{x}(t)) = \varphi(H(\mathbf{x}(t)) = \varphi(h(t)).$$

However, the zero function is also a solution of the initial value problem

$$h'(t) = \varphi(h(t)) \quad h(0) = 0$$

and because of uniqueness (see Appendix, Theorem 13.42), the function h must be identical with the zero function. $\qquad\square$

This theorem provides a constructive method to find invariant sets. For example, one can specialize the condition in the following way:

$$H'(\overline{\mathbf{x}})\mathbf{f}(\overline{\mathbf{x}}) = K(\overline{\mathbf{x}})H(\overline{\mathbf{x}}) \quad (\overline{\mathbf{x}} \in \mathbb{R}^M) \tag{8.37}$$

with a scalar-valued function K. In the case when $\mathbf{f}$ is a polynomial, we can look for polynomials H and K with unknown coefficients fulfilling the above condition. If H is linear, then the obtained invariant surface is an **invariant plane**. This is the method used in Antonov et al. (2016).

It is easy to show that once we have polynomials H_i and K_i fulfilling (8.37) (called **Darboux polynomials** and **cofactor**s, respectively), then $\prod_i H_i^{\lambda_i}$ is a first integral of (8.34) if the constants λ_i fulfill $\sum_i \lambda_i K_i = 0$. (Check it!)

8.7   Transient Behavior

Our main concern is the characterization of the time evolution of concentrations. Earlier, in Chap. 7 we treated the stationary behavior; here we are interested in what happens in intermediate times as opposed to long times. The expression "transient" is also used here in the strict sense: not the initial, not the long-time behavior, and it does not mean here ephemeral. We cite statements assuring "dull" behavior; when all the concentrations tend to a fixed value, nothing interesting happens. In this case we speak about regular behavior, because this is expected for almost all reactions from the chemist.

It is more interesting when something unusual happens, when there exist more than one stationary state, or periodic or chaotic solutions exist. These phenomena are not only more exciting for the mathematician; they turned out to be extremely important from the point of view of applications, especially in biology, in chemical technology, and in other fields. Without an exact mathematical definition, we shall refer to non-regular behavior as exotic.

8.7.1   Regular Behavior

Most of the chemists, especially those not so much involved in reaction kinetics, have the following picture about the time evolution of reactions. All the concentrations of the species tend to the unique positive stationary point in the stoichiometric compatibility class (or positive reaction simplex) corresponding to the initial concentration vector. The stationary point is (relatively asymptotically globally) stable.

We may say that in these cases the reaction shows **regular** (or using the terminology by Horn and Jackson 1972 **quasi-thermodynamic**) behavior.

Such a widespread belief should either be rigorously proved within the framework of a model or should be disproved by appropriate, chemically acceptable counterexamples.

The last 50 years of formal reaction kinetics has seen results in both directions. Since the late 1960s and early 1970s of the last century, it is a well-formulated theorem that detailed balanced reactions show regular behavior.

Shear (1967) was the first to state the theorem, although he thought it to be more general than it is, as shown by Higgins (1968). A few years later, Volpert and Hudyaev (1985), p. 635, formulated and proved the statement rigorously, mentioning early (1938) ideas by Zeldovich. One could also add the even earlier approach by Boltzmann, although the last two authors used the concept in a stochastic framework. However, detailed balance is similar but not exactly the same in the two models; see, e.g., Joshi (2015) and Chap. 10.

Once we know that detailed balance implies regular behavior, one may wish to have necessary and sufficient conditions for this property. These we have presented in Sect. 7.8.2 and are used in our program as **DetailedBalance** (see the examples in the mentioned subsection).

8.7.1.1 Time Evolution of Detailed Balanced Reactions

Theorem 8.45 *Suppose that the reversible reaction (7.7) endowed with mass-action type kinetics is detailed balanced at some positive stationary concentration* $\mathbf{c}_* \in (\mathbb{R}^+)^M$. *Then:*

1. *The solution* $t \mapsto \mathbf{c}(t)$ *of the induced kinetic differential equation of the reaction with nonnegative initial concentrations* $\mathbf{c}_0 \in (\mathbb{R}_0^+)^M$ *is defined for all nonnegative times and is also bounded.*
2. *The induced kinetic differential equation has no nontrivial nonnegative periodic solutions.*
3. *The mechanism is detailed balanced.*
4. *The ω-limit set of the initial value problem consists of either a single (positive) detailed balanced stationary point or of nonnegative stationary points fulfilling the condition equation (7.9) (of detailed balancing).*
5. *(Positive) detailed balanced stationary points are stable and are relatively asymptotically stable within the stoichiometric compatibility class within which they reside.*

Proof We give only a sketch of the proof. First of all, let us introduce the function:

$$F(\mathbf{c}) := \mathbf{c}^\top \left(\ln(\mathbf{c}/\mathbf{c}_*) - \mathbf{1}_M \right) \tag{8.38}$$

1. It can be shown that $F(\mathbf{c}(t)) \leq F(\mathbf{c}_0)$ and this—together with nonnegativity of the solutions—imply existence for all nonnegative times and boundedness.
2. Suppose a solution $\mathbf{c}$ with period $T \in \mathbb{R}^+$ exists. Then, $F(\mathbf{c}(t)) = F(\mathbf{c}(t + T))$ would hold which contradicts to the fact that F is strictly decreasing along nonconstant solutions.

3. For constant solutions $\mathbf{c}_{**} \in (\mathbb{R}_0^+)^M$ (stationary points), one should have $k_r \mathbf{c}_{**}^{\alpha(\cdot,r)} = k_{-r} \mathbf{c}_{**}^{\beta(\cdot,r)}$ meaning just the condition of detailed balance at the point $\mathbf{c}_{**}$.
4. This follows from the fact that F is bounded from below and decreasing along the solutions; therefore $\lim_{t \to +\infty} F(\mathbf{c}(t))$ exists.
5. Use the level sets of F.

$\square$

Remark 8.46

1. The proof uses the fact that $(a - b)(\ln(a) - \ln(b)) \geq 0$ and also continuous dependence of the solutions on the parameters, cf. the proof of nonnegativity of solutions (see Corollary 8.8 above).
2. F can be thought of as the free energy of the system.
3. Note that in the Volpert graph of a reversible reaction, each of the species is vertices of a cycle, a property reminding to weak reversibility.

A far-reaching generalization of detailed balance is complex balance (see Sect. 7.6). As Theorem 7.15 states, a simple-to-check sufficient condition of complex balancing in the mass-action case is that the reaction is weakly reversible and its deficiency is zero.

8.7.1.2 The Zero-Deficiency Theorem
One of the most important statements in the last 40 years of formal reaction kinetics follows.

Theorem 8.47 *Suppose that the deficiency of a reaction is zero. Then:*

1. *No nontrivial periodic solutions of the induced kinetic differential equation can exist.*
2. *Furthermore,*
   a. *If the reaction is not weakly reversible, then it cannot have a positive stationary point, no matter what form the kinetics has.*
   b. *If the reaction is weakly reversible and the kinetics is of the mass-action form, then with any choice of positive rate constants (i.e., with any mechanism built on the given reaction):*
      i. *There exists in each positive stoichiometric compatibility class exactly one positive stationary concentration.*
      ii. *Each of the positive stationary concentrations is relatively asymptotically stable relative to the stoichiometric compatibility class in which it resides.*

Proof Instead of giving the proof, we provide a few links, because the whole proof is quite long. A short introduction can be read in Feinberg (1980), and a sketch of the proof is given in Feinberg (1977), whereas Feinberg (1979) gives all the technical details. Gunawardena (2003) gives a relatively short proof, which is enlightening from many respects. Bamberger and Billette (1994) also give a short proof, and they also state that the integral $\int_0^{+\infty} (\mathbf{c}(t) - \mathbf{c}_*)^2 \, dt$ converges. Although only a part of the theorem is proved, it has been done in a really short way by Boros (2013c).   □

Horn (1974) realized that complex balancing may only imply local (relative) stability; still for 40 years, we had the **Global Attractor Hypothesis** asserting that global stability also holds. This has been proved for the one linkage class case by Anderson (2011), and by entirely new methods in the general case by Craciun (2016) (see also http://www.sjsu.edu/people/matthew.johnston/GAC_Workshop/). (Here we can only give a loose definition of attractor as a set of states, invariant under the dynamics, toward which neighboring states approach in the long run.)

Uniqueness of the stationary state can be found in a larger class of reactions, Feinberg (1980). We only cite the **deficiency-one theorem** without proof.

Theorem 8.48 *Suppose that a weakly reversible reaction (endowed with mass-action type kinetics) has L linkage classes, and suppose the lth linkage class has a deficiency δ_l ($l \in \{1, 2, \ldots, L\}$), and the deficiency of the full mechanism is δ. If $\delta_l \leq 1$ ($l \in \{1, 2, \ldots, L\}$) and $\sum_{l=1}^{L} \delta_l = \delta$ holds, then for every set of reaction rate coefficients, there exists exactly one positive stationary point in each reaction simplex.*

Remark 8.49

- A short proof of the deficiency-one theorem has been provided by Boros (2013b, 2012). His method was also capable of giving a necessary and sufficient condition for the existence of an interior stationary point for reactions covered by the deficiency-one theorem.
- Consider a zero-deficiency reaction with L linkage classes, containing N_l complex in the lth linkage class which has deficiency δ_l for $l = 1, 2, \ldots, L$. If the deficiency of the reaction is zero, then, since $\sum_{l=1}^{L} S_l \geq S$, via

$$0 = \delta = N - L - S \geq \sum_{l=1}^{L} (N_l - 1 - S_l) = \sum_{l=1}^{L} \delta_l$$

implies that $\delta_l = 0$ for all $l = 1, 2, \ldots, L$. This shows that Theorem 8.48 subsumes zero-deficiency reactions, as well.

As an illustration consider the reaction in Fig. 8.7.

The concept of **complex graph** introduced by Horn (1973) and recalled here from Definition 3.30 of Chap. 3 also contains important information on the time evolution of the solution of the induced kinetic differential equation. Before that one needs a definition.

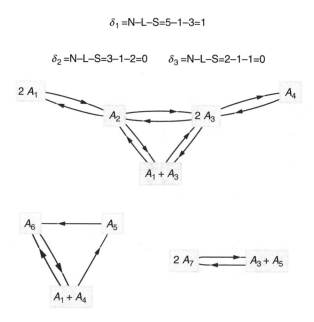

$$\delta_1 = N-L-S = 5-1-3 = 1$$

$$\delta_2 = N-L-S = 3-1-2 = 0 \qquad \delta_3 = N-L-S = 2-1-1 = 0$$

Fig. 8.7 Illustration of the Deficiency One Theorem 8.48. $\delta = N - L - S = 10 - 3 - 6 = 1$

Definition 8.50 A reaction is **quasi-thermodynamic** with respect to the concentration $\mathbf{a} \in (\mathbb{R}^+)^M$ if:

- $\mathbf{c}_* \in (\mathbb{R}^+)^M$ is a stationary concentration if and only if $\ln(\mathbf{c}_*) - \ln(\mathbf{a}) \in \mathscr{S}^\perp$; and
- for all $\mathbf{c} \in (\mathbb{R}^+)^M : (\ln(\mathbf{c}) - \ln(\mathbf{a})) \cdot \mathbf{f}(\mathbf{c}) \le 0$, with equality holding if and only if $\mathbf{c}$ is a stationary point.

Theorem 8.51 *If the complex graph (Definition 3.30) of a reaction only containing three short complexes and endowed with mass-action kinetics does not contain an even cycle or an odd dumbbell (recall Definition 3.31), then:*

1. *weak reversibility implies quasi-thermodynamic behavior;*
2. *violation of weak reversibility implies the nonexistence of positive steady states, and violation of quasi-thermodynamic behavior.*

Corollary 8.52 *The complex graph of a weakly reversible mechanism only containing three short complexes endowed with mass-action kinetics must contain an even cycle or an odd dumbbell if the induced kinetic differential equation of the reaction is to admit one of the following:*

1. *an unstable stationary point,*
2. *two or more stationary points in one and the same positive reaction simplex,*
3. *nontrivial periodic solutions with positive coordinates.*

Example 8.53 The complex graph of the reversible—therefore weakly reversible—reaction A + B $\rightleftharpoons$ C being 0–3 1–2 fulfills the condition of Theorem 8.51; therefore it is quasi-thermodynamic. (The reader may check this property directly from the definition, as well.)

Finally, let us mention that considering reactions as flows in networks in the sense of operations research proved to be useful in a simplified proof of the zero-deficiency theorem and an extension of the deficiency-one theorem (Boros 2013c,a).

8.7.1.3 The Case of Acyclic Volpert Graphs
Theorem 8.54 *Suppose that the Volpert graph of the reaction* (2.1) *is acyclic, and assume that the reaction is endowed with mass-action type kinetics. Then:*

1. *The solution* $t \mapsto \mathbf{c}(t)$ *of the induced kinetic differential equation of the mechanism with nonnegative initial concentrations is defined for all nonnegative times and is also bounded.*
2. *The induced kinetic differential equation has no nontrivial nonnegative periodic solutions.*
3. *There exists*

$$\lim_{t \to +\infty} \mathbf{c}(t) =: \mathbf{c}_*, \tag{8.39}$$

and it is a stationary point of the induced kinetic differential equation.
4. *Each of the nonnegative stationary points of the induced kinetic differential equation is a solution to the equation* $\mathbf{c}^\alpha = \mathbf{0} \in \mathbb{R}^R$.
5. *There exists a constant* $K \in \mathbb{R}^+$, *independent from the initial condition, such that*

$$\int_0^{+\infty} |\dot{c}_m(t)| \, \mathrm{d}t < K \tag{8.40}$$

holds.

Proof

1. Let $\varrho > \mathbf{0} \in \mathbb{R}^M$ be the solution of $\varrho^\top \gamma < \mathbf{0}$ (the existence of which has been formulated as Problem 4.5); then along the solutions $\mathbf{c}$ of the induced kinetic differential equation of the reaction, one has $\frac{\mathrm{d}}{\mathrm{d}t}(\varrho^\top \mathbf{c}(t)) \leq 0$; thus $\varrho^\top \mathbf{c}(t) \leq \varrho^\top \mathbf{c}(0)$ implies the statement.
2. If $\mathbf{c}$ is a nonconstant solution of the induced kinetic differential equation then for some $t \in \mathbb{R}$ and $r \in \mathscr{R} : k_r \mathbf{c}(t)^{\alpha(\cdot, r)} > 0$ (otherwise we had for all $t \in \mathbb{R}$ and $r \in \mathscr{R} : k_r \mathbf{c}(t)^{\alpha(\cdot, r)} = 0$ implying that $\mathbf{c}(t) = \mathbf{c}(0)$). Thus, $\varrho^\top \mathbf{c}(t)$ cannot be constant, but if $\mathbf{c}$ is a periodic solution with period $T \in \mathbb{R}^+$, then $\varrho^\top \mathbf{c}(t) = \varrho^\top \mathbf{c}(t + T)$ would also imply a contradiction.

3. Along the solutions $\varrho^{\top}\mathbf{c}(t)$ is decreasing and being also bounded from below, $\lim_{t\to+\infty} \varrho^{\top}\mathbf{c}(t)$ exists; thus for all $r \in \mathscr{R} : \int_0^{+\infty} k_r \mathbf{c}(s)^{\alpha(\cdot,r)} \, ds < +\infty$. The fact that $c_m(t) = c_m(0) + \sum_{r\in\mathscr{R}} \gamma(m,r) \int_0^t k_r \mathbf{c}(s)^{\alpha(\cdot,r)} \, ds$ (8.39) holds implies that $\mathbf{c}_*$ is a stationary point.

4. If $\mathbf{c}_*$ is a stationary point, then $(\varrho^{\top}\gamma)\mathbf{k} \cdot \mathbf{c}_*^{\alpha} = \mathbf{0}$ implies that $\mathbf{k} \cdot \mathbf{c}_*^{\alpha} = \mathbf{0}$.

5. Let P be the $M \times M$ matrix of independent solutions to $\varrho^{\top}\gamma \leq \mathbf{0}$. Then the components of $\mathbf{d}(t) := P\mathbf{c}(t)$ are non-increasing functions of time; thus

$$\int_0^{+\infty} |\dot{d}_m(s)| \, ds = \left| \int_0^{+\infty} \dot{d}_m(s) \, ds \right| = d_m(0) - d_m(+\infty) \leq K_1,$$

with some positive constant. As P is invertible, one can return to the original variables via $\mathbf{c}(t) = P^{-1}\mathbf{d}(t)$ to get the required result.

$\square$

Remark 8.55

- All the complexes in a weakly reversible reaction are vertices of a cycle of the Feinberg–Horn–Jackson graph; in this sense the Feinberg–Horn–Jackson graph contains many cycles. The theorem by Volpert relates reactions with no cycles at all in the Volpert graph. Although the two graphs are different (but related; see Sect. 3.3), these two theorems cover two extreme cases.
- Many textbook reactions are of deficiency zero; therefore the zero-deficiency theorem has a wide scope. On the other hand, if a single reaction step is reversible, then its Volpert graph is not acyclic; therefore the scope of Volpert's theorem is limited.
- As complex balancing is a far-reaching generalization of detailed balancing, the zero-deficiency theorem represents a giant leap from the folkloristic statement on detailed balanced reactions.
- An absolutely important (aesthetic or didactic, if you wish) property of the main theorems ensuring regular behavior is that the conditions are formulated in chemical terms, which means that they can be applied even by researchers (say, chemists not interested in mathematical details) who would not listen to a statement formulated in purely mathematical terms.

Remark 8.56 Here we mention that under very special circumstances (more precisely, for a small class of reactions), the **Ljapunov exponent** (characterizing the exponential growth or decrease of the concentrations; see Definition 13.49) can be calculated from the structure of the Volpert graph (Volpert and Hudyaev 1985, pp. 645–648). This is similar to the characterization of initial behavior in terms of the Volpert indices (see Chap. 9).

8.7.2   Exotic Behavior

Our main concern up to this point was to collect sufficient conditions to ensure regular behavior, a behavior usually expected from a reaction by the chemist. However, in many respects, it is more interesting to have **exotic behavior**: oscillation, multistability, multistationarity, or chaos. Interesting they are not only from the mathematical point of view, but oscillatory behavior in a reaction may form the basis of periodic behavior in biological systems of which one can find those with different period: minutes, 1 day, 1 year, etc.; see Murray (2002), Winfree (2001), Edelstein-Keshet (2005). Analogous expectations can be expressed in connection with the other phenomena.

8.7.2.1 Oscillation: Absence or Presence of Periodic Solutions

In his famous 1900 lecture, D. Hilbert formulated as the second part of his XVIth problem to find the number of limit cycles of two-variable polynomial differential equations (see, e.g., Gaiko 2013). The last more than 100 hundred years have shown that this is a very hard problem (see, e.g., Ye and Lo 1986). However, from the point of view of applications, one would need even more: find the number of periodic solutions of polynomial differential equations in any number of variables although one would be content with the solution of the kinetic case, i.e., with a subclass of quadratic or cubic polynomials only. Early attacks by Escher (1980, 1981) are worth mentioning.

Let us turn to the history of oscillation from the chemical point of view. In a classical chemical experiment, two colorless solutions are mixed, and at first there is no visible reaction. After a certain time, the solution suddenly turns to dark blue. In some variations the solution will repeatedly turn from colorless to blue and back to colorless, until the reagents are depleted: an **oscillatory reaction** is obtained. This suddenly ending reactions are called **chemical clock**s or **clock reaction**s. Examples of clock reactions are the Belousov–Zhabotinsky reaction, the Briggs–Rauscher reaction, the Bray–Liebhafsky reaction, and the iodine clock reaction. (Cf. Lente et al. 2007.) Chemical clocks are important from the point of view of biological applications, as the chemical basis of **biological clock**s, and also from the point of view of physics, because they are examples when it is not true that all the concentrations tend to an asymptotically stable stationary state as it is generally believed.

Especially when the experiments of Belousov (1958) became known—first, through the work by Zhabotinsky (1964)—the question emerged what are the **structural conditions** of the existence of periodic solutions in a kinetic differential equation. (On the difference between structural and parametric conditions, see, e.g., Beck 1992, 1990.)

Necessary conditions to exclude the existence of periodic solutions can be found in the classical central theorems on regular behavior in Sect. 8.7.1. Given a special mechanism, one may also try to apply the theorem by Bendixson (Theorem 13.51).

Example 8.57 Consider the reversible Wegscheider reaction in Fig. 7.4 with the induced kinetic differential equation:

$$\dot{x} = -k_1 x + k_{-1} y - 2k_2 x^2 + 2k_{-2} y^2 = f \circ (x, y)$$
$$\dot{y} = k_1 x - k_{-1} y + 2k_2 x^2 - 2k_{-2} y^2 = g \circ (x, y).$$

In the first orthant, the divergence of the right-hand side being

$$\partial_1 f(x, y) + \partial_2 g(x, y) = -k_1 - 4k_2 x - k_{-1} - 4k_{-2} y < 0$$

the reaction cannot have a closed trajectory there.

The Bendixson theorem can also be used for other purposes.

Example 8.58 Consider the induced kinetic differential equation of the Lotka–Volterra reaction $\dot{x} = k_1 x - k_2 xy$, $\dot{y} = k_2 xy - k_3 y$, and let us calculate the divergence of the right-hand side: $\text{div}(\begin{bmatrix} k_1 x - k_2 xy & k_2 xy - k_3 y \end{bmatrix}^\top) = k_1 - k_3 + k_2(x - y)$. As this expression is not of the constant sign, we cannot immediately apply Theorem 13.51. However, we can say that if the induced kinetic differential equation of the Lotka–Volterra reaction is to have closed trajectories in the first quadrant, then they should cross the line $\{\begin{bmatrix} \overline{x} & \overline{y} \end{bmatrix}^\top \in \mathbb{R}^2 \mid k_1 - k_3 + k_2(\overline{x} - \overline{y}) = 0\}$. Using several Dulac functions, one can learn more and more about the location of the possible closed trajectories.

A serious disadvantage of the Bendixson–Dulac theorem is that it is about planar systems, and cannot easily and naturally be generalized to higher dimensional cases. It may happen that a system can be reduced to a 2D system via first integrals, and then one can apply the theorem (see, e.g., Tóth 1987).

Let us see a more powerful application of the Bendixson–Dulac theorem. It turned out relatively early that the simplest possible model—admitting chemical interpretation—that leads to oscillations is the **Lotka–Volterra reaction**. The theorem below has first been given a complete proof by Póta (1983). Later, an alternative proof based on the classification of planar vector fields and promising some extensions has been produced by Schuman and Tóth (2003). The earlier discussion by Tyson (1980) and Kádas and Othmer (1979b,a); Kádas and Othmer (1980) is also worth studying.

Theorem 8.59 (Póta–Hanusse–Tyson–Light) *Two-species second-order reactions with short product complexes cannot have a limit cycle in $(\mathbb{R}^+)^2$.*

Proof The induced kinetic differential equation of a second-order reaction among two species is of the form

$$\dot{x} = -a_{11}^1 x^2 + a_{12}^1 xy + a_{22}^1 y^2 + b_1^1 x + b_2^1 y + c^1 \tag{8.41}$$

$$\dot{y} = a_{11}^2 x^2 + a_{12}^2 xy - a_{22}^2 y^2 + b_1^2 x + b_2^2 y + c^2, \tag{8.42}$$

where $a_{22}^1, b_2^1, c^1, a_{11}^2, b_1^2, c^2 \geq 0$ (to exclude nonnegative cross effects) and as no long complexes are present, one also has that $a_{11}^1 \geq 0, a_{22}^2 \geq 0$; furthermore (from the same reason), of a_{12}^1 and a_{12}^2 not more than one can be positive. Let us choose the Dulac function in the following way: $(\mathbb{R}^+)^2 \ni (x, y) \mapsto B(x, y) := \frac{1}{xy}$ and calculate $D := \operatorname{div}(B\mathbf{f})$ (with $\mathbf{f}$ the right-hand side of (8.41)–(8.42)) to get:

$$-\left(\frac{a_{11}^1}{y} + \frac{a_{22}^1 y}{x^2} + \frac{b_2^1}{x^2} + \frac{c^1}{x^2 y} + \frac{a_{11}^2 x}{y^2} + \frac{a_{22}^2}{x} + \frac{b_1^2}{y^2} + \frac{c^2}{xy^2} \right).$$

Assuming that (8.41)–(8.42) has a closed trajectory fully contained in a simply connected open set, $E \subset (\mathbb{R}^+)^2$ implies that for all $(x, y) \in E : D(x, y) = 0$; thus $a_{11}^1 = a_{22}^1 = b_2^1 = c^1 = a_{11}^2 = a_{22}^2 = b_1^2 = c^2 = 0$; therefore our equation simplifies to

$$\dot{x} = a_{12}^1 xy + b_1^1 x \quad \dot{y} = a_{12}^2 xy + b_2^2 y. \tag{8.43}$$

If any of the coefficient here is zero, then one of the derivatives is of the constant sign excluding periodicity. Therefore one may assume that all the coefficients are different from zero. If a_{12}^1 and b_1^1 (or a_{12}^2 and b_2^2) are of the same sign, again the derivative of x (or the derivative of y) is of the constant sign. If $\operatorname{sign}(b_1^1) = \operatorname{sign}(b_2^2) = -\operatorname{sign}(a_{12}^1) = -\operatorname{sign}(a_{12}^2)$, then the (positive!) stationary point would be a saddle (Problem 8.20) which cannot be surrounded by a closed orbit (Andronov et al. 1973, pp. 205–219). Finally, the only systems left are

$$\dot{x} = b_1^1 x - \overline{a_{12}^1} xy \quad \dot{y} = -\overline{b_2^2} y + a_{12}^2 xy \tag{8.44}$$

and

$$\dot{x} = -\overline{b_1^1} x + a_{12}^1 xy \quad \dot{y} = b_2^2 y - \overline{a_{12}^2} xy \tag{8.45}$$

with positive coefficients only. But both of these are Lotka–Volterra models with **conservative oscillation** (i.e., having a first integral; see Definition 8.27), i.e., without limit cycles. To completely agree with the usual form of the Lotka–Volterra equation, one has to have the same coefficient before the (mixed) second-degree terms xy. The substitution $X := x, Y := \frac{a_{12}^1}{a_{12}^2} y$ in (8.45) will achieve this. $\square$

Remark 8.60

1. A consequence of the theorem is that among the two-species second-order reactions, (practically) the only oscillatory reaction is the Lotka–Volterra model. It is interesting that the same result is obtained if one starts from different premises: if the linearized form of the Lotka–Volterra model is given, then the simplest model with this linearized form is again the Lotka–Volterra model (see Tóth and Hárs 1986b).
2. A case simpler to treat is known from the 1950s of the last century (see Problem 8.19).
3. As conservative oscillation may only be stable but not asymptotically stable, it is natural to look for models as simple as possible that have a limit cycle. Such a model, the Brusselator, $0 \rightleftharpoons X \longrightarrow Y \quad 2X + Y \longrightarrow 3X$ with a single three-order reaction step, has been created by Prigogine and Lefever (1968) and has become extremely popular. The existence of a periodic solution (without the proof that it is a limit cycle) is shown in Problems 8.22 and 8.23.
4. Várdai and Tóth (2008) shows an animation how a Hopf bifurcation emerges in the Brusselator model. Enjoy and modify it appropriately.
5. Escher (1981) contains chemical examples with two-species and second-order reactions having even more than one limit cycles, but he allows long product complexes, as well; thus his constructions are not in contradiction with the statement of the theorem above.

As to the presence of periodic solutions or closed trajectories, we start with a simple observation.

Example 8.61 The Lotka–Volterra model is a model with conservative oscillation because it has a first integral. Consider $\dot{x} = k_1 x - k_2 xy$, $\dot{y} = k_2 xy - k_3 y$; then obviously $\varphi(\overline{x}, \overline{y}) := \ln(\overline{x}^{k_3} \overline{y}^{k_1}) - k_2 \overline{x} - k_3 \overline{y}$ is a first integral; therefore the trajectories remain on the level curves of this function which are closed curves. How do we find this first integral? One may use the following (admittedly, ad hoc) method. The equations imply (in the open first quadrant) $\frac{\dot{x}}{x} = k_1 - k_2 y$, $\frac{\dot{y}}{y} = k_2 x - k_3$. Now let us multiply the two equations to get $\frac{\dot{x}}{x}(k_2 x - k_3) = (k_1 - k_2 y)\frac{\dot{y}}{y}$. Taking the integral from 0 to t, one has $k_2 x(t) - k_3 \ln(x(t)) - k_2 x(0) + k_3 \ln(x(0)) = k_1 \ln(y(t)) - k_2 y(t) - k_1 \ln(y(0)) + k_2 y(0)$ showing that φ above is really a first integral of the Lotka–Volterra mechanism.

Let us finish with a final note on the application of Theorem 13.54. Textbooks on differential equations usually apply this theorem to equations with negative cross effects. Problem 8.22 shows a kinetic example.

Another fruitful tool is the Theorem 13.57 as it has been shown, e.g., by Hsü (1976): He has rigorously shown that the Oregonator model of the Belousov–Zhabotinsky mechanism has periodic solutions. Another application is shown in Schneider et al. (1987) containing a kinetic differential equation modeling synaptic slow waves and having periodic solutions as a consequence of Theorem 13.57. The interested reader might study the paper: it contains lengthy calculations. Application of the Theorem 13.57 to the Brusselator is the topic of Problem 8.23.

More complicated systems may need a customized analysis.

Example 8.62 We encourage the reader to consult the paper by Kertész (1984) containing a nice analysis of the **Explodator** defined by Noszticzius et al. (1984)

$$X \longrightarrow \beta_1 X \quad X + Y \longrightarrow Z \longrightarrow \beta_2 Y \quad Y \longrightarrow 0 \quad (\beta_1, \beta_2 > 1) \qquad (8.46)$$

an alternative to describe the Belousov–Zhabotinsky reaction with the following properties:

1. In the open first orthant, it has a single unstable stationary point.
2. Its Jacobian has a negative real eigenvalue and two eigenvalues with positive real parts (and with or without imaginary parts depending on the values of the parameters).
3. A one-dimensional stable manifold corresponds to the negative eigenvalues (two trajectories go into the stationary point).
4. All the other trajectories go to infinity in various ways.

The solution of Problem 8.15 is a good preparation to reading the paper.

It is also useful to know which are the simplest reactions still able to show oscillations (Wilhelm and Heinrich 1995; Smith 2012; Tóth and Hárs 1986b), cf. page 121.

Finally, we cite a necessary condition of periodicity and multistationarity from Schlosser and Feinberg (1994).

Theorem 8.63 *If the reaction* (2.1) *has either a periodic solution or multiple stationary states in* $(\mathbb{R}^+)^M$, *then its S-C-L graph is cyclic.*

Proof Suppose the S-C-L graph is acyclic. Then by Theorem 3.28, the deficiency of the reaction is zero; therefore—according to the zero-deficiency theorem (Theorem 8.47)—the existence of positive multiple stationary states and periodic solutions is excluded. □

8.7.2.2 Oligo-Oscillation or Overshoot–Undershoot Phenomenon
Another interesting phenomenon which has also been observed experimentally, e.g., by Rábai et al. (1979) and Murphy et al. (2005) is **oligo-oscillation**: this happens when some of the concentration vs. time curves show multiple (usually a finite number of) extrema. General statements on transient behavior of solutions to differential equations are hard to obtain. We are in a lucky situation to have such a general statement for a class of reactions.

Theorem 8.64 (Póta–Jost) *If in a closed compartmental system of M compartments with the deterministic model*

$$\dot{c} = Ac \tag{8.47}$$

all the eigenvalues of **A** *are real numbers, then none of the components of the concentration vs. time function can have more than* $M - 2$ *strict local extrema.*

Proof For a closed compartmental system, none of the eigenvalues can be positive, and at least one of them is zero (Problem 8.18). Let us denote the different eigenvalues by $\lambda_0 := 0$, and $\lambda_1, \lambda_2, \ldots, \lambda_K$ with the multiplicities $\mu_1, \ldots, \mu_K$ where $K \in \mathbb{N}$; $\sum_{k=1}^{K} \mu_k = M - 1$. Then the components of the concentration vs. time functions are of the form

$$c_m(t) = c_m^* + \sum_{k=1}^{K} P_{mk}(t) e^{\lambda_k t}, \tag{8.48}$$

where $P_{mk}(t)$ are polynomials of degree $\mu_k - 1$. Now let us apply the generalized Higgins lemma 13.39 to the derivative of c_m to get the result. (Actually, we do not need the fact that the nonzero eigenvalues are negative.) □

A trivial consequence of the theorem is the well-known statement: in the consecutive reaction A $\longrightarrow$ B $\longrightarrow$ C, the concentration of the intermediate species B cannot have more than one extremum. However, the above theorem does not exclude that even compartmental systems might show quite interesting behavior: some of the eigenvalues of the coefficient matrix may be complex, and— as a consequence of this fact—some of the concentration vs. time curves may have an infinite number of local extrema; see Problem 8.25.

Let us consider the reversible triangle reaction of Fig. 8.8. As the reaction is a closed compartmental system, it is stoichiometrically mass conserving with the vector $\varrho := (1, 1, 1)^\top$, or—in plain English—the sum of the concentrations is constant. Therefore, instead of the full induced kinetic differential equation, it is enough to consider the differential equation:

$$\dot{a} = -(k_1 + k_{-3} + k_3)a + (k_{-1} - k_3)b + k_3(a_0 + b_0 + c_0)$$
$$\dot{b} = (k_1 - k_{-2})a - (k_{-1} + k_{-2} + k_2)b + k_{-2}(a_0 + b_0 + c_0). \tag{8.49}$$

Different sets of the reaction rate coefficients may lead to markedly different qualitative behavior of the concentration vs. time curves; see Problems 8.24 and 8.25. However, so far as the eigenvalues are real, the concentration vs. time curves cannot have more than one extrema.

It is much harder to get any result on nonlinear systems; therefore it may be useful to see even a very special statement without proof; for the proof based on the investigation of the trajectories, see Problem 8.17.

Fig. 8.8 The reversible
triangle reaction

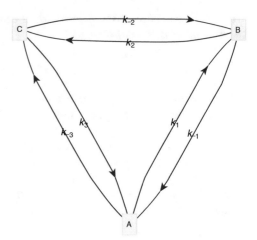

The reader may also be interested in Póta (1992) and Kordylewski et al. (1990).

8.7.2.3 Multistability and Multistationarity

Multistability is used in the more general sense (although not really correctly): it
means that a differential equation has multiple attractors/repellors (see page 176),
some of which may be stationary points, and others may be limit cycles or even
more complicated sets. One obvious example is obtained when one considers Hopf
bifurcation: as the parameter changes, a limit cycle may appear, but still one has a
stationary point (which may however lose its stability).

In case of **multistationarity**, the differential equation has multiple stationary
points. As the literature on these topics is growing intensively, we can only mention
a few results. First, let us see an example showing that although two-species
second-order reaction may not have periodic trajectories except the Lotka–Volterra
reactions, still they may have more than one stationary points, depending on the
values of the reaction rate coefficients.

Example 8.65 The kinetic differential equation

$$\dot{x} = y^2 - 6x + 3y + 2 \quad \dot{y} = x^2 - y^2 + 6x - 6y$$

has $\begin{bmatrix} 1 \\ 1 \end{bmatrix}$ and $\begin{bmatrix} 2 \\ 2 \end{bmatrix}$ as its stationary points and has no other positive stationary points.

An important special class of systems is the homogeneous **continuous-flow stirred-
tank reactor** (CFSTR) in which beyond the reaction steps, some of the species
are present in the feed stream and all of them are present in the effluent stream.
The in- and outflow can obviously be described by (formal) reaction steps of the
form $0 \longrightarrow X$ and $X \longrightarrow 0$. However, Feinberg and his students (see Schlosser
and Feinberg (1994) and the references therein) elaborated a theory using only

the steps describing the chemical reactions. They introduced the S-C-L graph (see Definition 3.27 and Theorem 8.63) and have shown that some characteristics of this graph may provide information on the existence or nonexistence of multiple stationary states. Theorem 8.63 cited above is only their simplest result; the interested reader should consult the above papers for results and the cited references for proofs.

Example 8.66 The S-C-L graph of the reaction in Fig. 8.9 shown in Fig. 8.10 does not contain cycles; therefore no assignment of reaction rate coefficients (and no addition of any kind of in- and outflow) can produce more than one positive stationary states. Note that the reaction itself without in- and outflow has zero deficiency: $\delta = 11 - 4 - 7 - 0$ and is weakly reversible; therefore the reaction itself is not capable to produce more than one stationary points, but the reaction, if in- and outflows are added, has a large deficiency (actually, six). Let us remark that the right-hand side of the induced kinetic differential equation only consists of quadratic terms: it is a homogeneous quadratic polynomial, cf. Halmschlager and

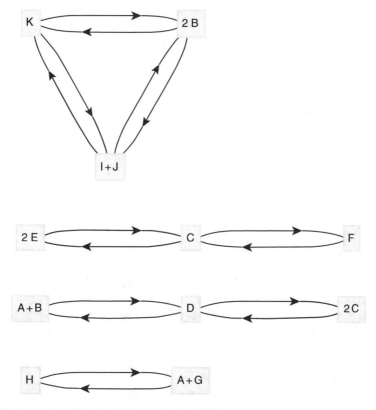

Fig. 8.9 A reaction with no multistationarity in CSFTR

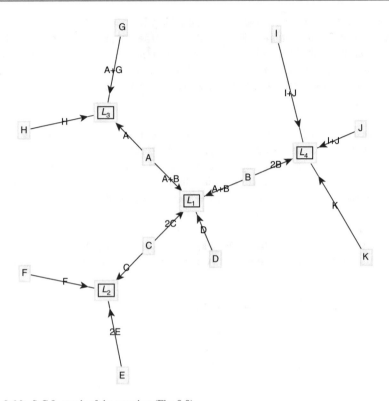

Fig. 8.10   S-C-L graph of the reaction (Fig. 8.9)

Tóth (2004). Note that much simpler reactions may admit more than one stationary states (see, e.g., Problem 8.44).

Continuation of the work can be found in Craciun and Feinberg (2006, 2010).

8.7.2.4 Chaos

There is a lot of experimental and numerical evidence that chaos is present in homogeneous kinetics; see, e.g., Epstein and Pojman (1998), Györgyi and Field (1991), Scott (1991, 1993, 1994), and also the early models by Rössler (1976), Willamowski and Rössler (1980). However, in this field—similar to other ones— there are almost no rigorous results showing that any property characterizing chaos such as the following appear in a given chemical model or in a class of models:

1. extreme **sensitivity to initial data**, or **butterfly effect**, making **long-term prediction** impossible;
2. **topological transitivity** (or **mixing**): the system evolves in time so that any given region or open set of its phase space will eventually be mapped so as to overlap with any other given region;

3. dense **closed trajectories**: every point in the phase space is approached arbitrarily closely by periodic orbits;
4. presence of a **strange attractor**, typically having a **fractal** structure;
5. properties of the recurrence plot or of the **Poincaré map**.

One of the few exceptions is Huang and Yang (2006) who give a computer-assisted proof of the existence of chaotic dynamics in a three-variable model of the Belousov–Zhabotinsky reaction, the bromate–malonic acid–ferroin system. To do this, they investigate the Poincaré map derived from the induced kinetic differential equation of the reaction and show the existence of horseshoes.

Example 8.67 Ivanova (Volpert and Hudyaev 1985, p. 630) constructed a reaction only consisting from bimolecular steps for which one can easily (Problem 8.45) show that the trajectories are bounded, and numerical solutions suggest that the solutions are neither periodic nor tending to a stable stationary point. The reaction is as follows:

$$A + B \xrightarrow{100} 2B \quad B + C \xrightarrow{660} 2C \quad C + A \xrightarrow{600} 2A \qquad (8.50)$$
$$A + D \xrightarrow{100} 2D \quad D + E \xrightarrow{660} 2E \quad E + A \xrightarrow{360} 2A.$$

Let us have a look at the double loop in the Volpert graph of this seemingly chaotic reaction. Note also that the right-hand side is a homogeneous quadratic polynomial (Figs. 8.11 and 8.12).

Let us mention that Rössler (1976) was able to create a polynomial differential equation with a single nonlinear term on the right-hand side showing chaotic behavior numerically.

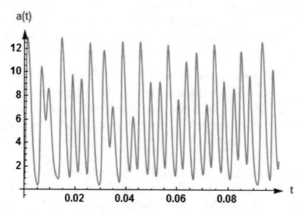

Fig. 8.11 Time evolution of the concentration of species A in the reaction (8.50) with the initial vector (1, 2, 3, 4, 5)

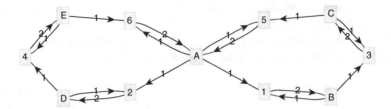

Fig. 8.12 Volpert graph of the reaction (8.50)

What we do have are some negative results. Fu and Heidel (1997) investigated three-dimensional quadratic systems and were able to show for many of them that they are unable to show chaotic behavior in any sense. Heidel and Fu (1999) have also shown that most of the three-dimensional quadratic systems with a total of four terms on the right-hand side of the equations and having a zero divergence (to avoid misunderstanding we are not using the term **conservative** here) do not exhibit chaos.

Attempts have been made to connect the presence of chaos with the absence of negative cross effects (Tóth and Hárs 1986a), also using the theory of algebraic invariants of polynomial differential equations Halmschlager et al. (2003). Finally, let us mention a kind of taming chaos: chaos control by Petrov et al. (1993).

8.8   The Influence Diagram

The effect of species onto each other can also be represented by an influence diagram defined as follows:

Definition 8.68 The **influence diagram** of the mechanism $\langle \mathcal{M}, \mathcal{R}, \boldsymbol{\alpha}, \boldsymbol{\beta}, \mathbf{k} \rangle$ at the concentration $\mathbf{c} \in (\mathbb{R}_0^+)^M$ is a directed graph with the species as vertices and with edges pointing from vertex X_m to the point X_p. The edge has a positive sign at a point $t \in \mathbb{R}_0^+$ in the domain of the solution of the induced kinetic differential equation if $\partial_m f_p(\mathbf{c}(t))$ is positive, and it has a negative sign if $\partial_m f_p(\mathbf{c}(t))$ is negative; otherwise there is no edge between X_m and X_p. We may for short call $\frac{\partial f_p(\mathbf{c})}{\partial c_m}$ the **effect of species** X_m onto the species X_p. The influence diagram is **uniform** in a subset of $(\mathbb{R}_0^+)^M$ if it does not depend on the value of the concetration vector $\mathbf{c}$. **Strong** influence diagrams are those which are connected, uniform in $(\mathbb{R}_0^+)^M$, and in which all the vertices have at least one edge entering and leaving them.

Then, King (1982) and Thomas (1978) discretize both time and the state space and introduce Boolean graphs to describe the dynamics of a reaction in the state space and formulate statements which from the pure mathematical viewpoint can only be considered as conjectures supported by some examples. Recently, Domijan and Pécou (2012) has proved a few theorems in this direction.

Although this approach seems intuitively to be quite attractive and has many biological applications (mainly in molecular genetics), they seem to be of limited value in formal kinetics as there seems to exist no general theory making them applicable to reactions in general.

Klee and van den Driessche (1977) proposed a method (see also Jeffries et al. 1987) to check **sign stability** of matrices, an extremely strong form of stability. To establish this property, they introduced a graph which is actually the same as the influence diagram. Coloring the vertices of this graph provides a tool to check the sign stability of a matrix. This procedure is also useful to investigate linearized forms of nonlinear systems, mainly in biology (see, e.g., Kiss and Kovács 2008; Kiss and Tóth 2009). Here we only give the definition of sign stability because of two reasons: first, this concept is important in many fields of applications; second, its relevance in formal kinetics is a nice open problem.

Definition 8.69 The matrix $\mathbf{A} \in \mathbb{R}^{M \times M}$ is said to be **sign stable**, if all the matrices $\mathbf{B} \in \mathbb{R}^{M \times M}$ with the same sign pattern are asymptotically stable in the sense that all of their eigenvalues have a negative real part.

A characterization of sign stable matrices follows:

Theorem 8.70 *The matrix* $\mathbf{A} = (a_{mp}) \in \mathbb{R}^{M \times M}$ *is sign stable if and only if the following relations hold:*

1. *for all* $m \in \mathcal{M} : a_{mm} \leq 0,$
2. *there exists* $m \in \mathcal{M} : a_{mm} < 0,$
3. *for all* $m, p \in \mathcal{M}$ *so that* $m \neq p : a_{mp} a_{pm} < 0,$
4. *for each sequence of* $k \geq 3$ *different indices one has* $a_{mp} a_{pi} \ldots a_{qr} a_{rm} = 0,$
5. $\det(\mathbf{A}) \neq 0.$

As we have seen above, an important property of differential equations used to model reactions cannot have negative cross effect (see Definition 6.24 in Chap. 6), and this property directly excludes the third condition above in case of first-order reactions.

8.9 Dynamic Symmetries

Earlier in Sect. 7.9, we have shown how different restrictions on the stationary points can uniformly be formulated. Now we formulate restrictions on the right-hand sides of differential equations in a similar way, cf. also Tóth and Érdi (1988) and Ronkin (1977), p. 24.

Definition 8.71

1. The system $\dot{\mathbf{x}} = \mathbf{f} \circ \mathbf{x}$ is a **gradient system**, if there exists a function (called **potential**) $V \in \mathscr{C}^2(\mathbb{R}^M, \mathbb{R})$ such that $\mathbf{f} = V'$, or equivalently, for all $m, p \in \mathscr{M}$ one has $\partial_p f_m = \partial_m f_p$, i.e., $\mathbf{f}' = (\mathbf{f}')^\top$ holds.
2. The system $\dot{\mathbf{x}} = \mathbf{f} \circ \mathbf{x}$ is a **Shahanshani gradient system**, if $(\mathbb{R}^+)^M \ni \overline{\mathbf{x}} \mapsto \frac{\mathbf{f}(\overline{\mathbf{x}})}{\overline{\mathbf{x}}}$ is a gradient system, or equivalently for all $\overline{\mathbf{x}} \in (\mathbb{R}^+)^M$ one has $\left(\frac{\mathbf{f}(\overline{\mathbf{x}})}{\overline{\mathbf{x}}}\right)' = \left(\left(\frac{\mathbf{f}(\overline{\mathbf{x}})}{\overline{\mathbf{x}}}\right)'\right)^\top$.
3. The system $\dot{\mathbf{x}} = \mathbf{u} \circ (\mathbf{x}, \mathbf{y})$, $\dot{\mathbf{y}} = \mathbf{v} \circ (\mathbf{x}, \mathbf{y})$ (with $\mathbf{u}, \mathbf{v} \in \mathscr{C}^1(\mathbb{R}^{2M}, \mathbb{R}^M)$) is a **Hamiltonian system**, if the system with the right-hand side $\begin{bmatrix} -\mathbf{v} \\ \mathbf{u} \end{bmatrix}$ is a gradient system, or equivalently there exists a function called **Hamiltonian** $H \in \mathscr{C}^2(\mathbb{R}^{2M}, \mathbb{R})$ such that $\mathbf{u} = \partial_2 H$, $\mathbf{v} = -\partial_1 H$.
4. The system $\dot{\mathbf{x}} = \mathbf{u} \circ (\mathbf{x}, \mathbf{y})$, $\dot{\mathbf{y}} = \mathbf{v} \circ (\mathbf{x}, \mathbf{y})$ (with $\mathbf{u}, \mathbf{v} \in \mathscr{C}^1(\mathbb{R}^{2M}, \mathbb{R}^M)$) is a **Cauchy–Riemann–Erugin system**, if both the systems with the right-hand side $\begin{bmatrix} \mathbf{v} \\ \mathbf{u} \end{bmatrix}$ and $\begin{bmatrix} \mathbf{u} \\ -\mathbf{v} \end{bmatrix}$ are gradient systems. In this case there exists an analytic function F from $\mathbb{C}^M$ into $\mathbb{C}$ such that its real part is $\mathbf{u}$ and its imaginary part is $\mathbf{v}$, both as functions of $\mathbf{x} + i\mathbf{y}$.

A common formulation of the above definitions can be given as follows. Let $\mathscr{F} : \mathscr{C}^1(\mathbb{R}^M, \mathbb{R}^M) \longrightarrow \mathscr{C}^1(\mathbb{R}^M, \mathbb{R}^M)$ be an operator such that $\mathscr{F}(\mathbf{f})(\overline{\mathbf{x}}) \in \mathbb{R}^{M \times M}$, and let us suppose that

$$\mathscr{F}(\mathbf{f})'(\overline{\mathbf{x}}) = \mathscr{F}(\mathbf{f})'(\overline{\mathbf{x}})^\top \tag{8.51}$$

holds. Then, $\dot{\mathbf{x}} = \mathbf{f} \circ \mathbf{x}$ is a gradient system, if the requirement (8.51) holds with $\mathscr{F} = \mathrm{id}$, and it is a Shahshahani gradient system if it holds with $\mathscr{F}(\mathbf{f}) = \mathbf{f}/\mathrm{id}$. For Hamiltonian systems let us use $\mathscr{F}(\mathbf{f}) := \begin{bmatrix} \mathbf{0} & -\mathrm{id} \\ \mathrm{id} & \mathbf{0} \end{bmatrix} \mathbf{f}$, and for Cauchy–Riemann–Erugin systems, we have two requirements, one with $\mathscr{F}_1(\mathbf{f}) := \begin{bmatrix} \mathbf{0} & \mathrm{id} \\ \mathrm{id} & \mathbf{0} \end{bmatrix} \mathbf{f}$ and one with $\mathscr{F}_2(\mathbf{f}) := \begin{bmatrix} \mathrm{id} & \mathbf{0} \\ \mathbf{0} & -\mathrm{id} \end{bmatrix} \mathbf{f}$.

Why is it good if a differential equation has any of the above forms? Because there exists a lot of results on the qualitative behavior of the solutions of such systems (see, e.g., Guckenheimer and Holmes 1983; Hirsch et al. 2004).

Example 8.72 A short verification shows that the reaction $0 \xrightarrow{f} X \underset{2e}{\overset{2e}{\rightleftharpoons}} Y \xleftarrow{F} 0$
has as its induced kinetic differential equation the gradient system

$$\dot{x} = -2ex + 2ey + f \quad \dot{y} = 2ex - 2ey + F$$

with $V(x, y) := -e(x - y)^2 + fx + Fy$.

How relevant are these concepts from the point of view of formal kinetics? These questions are to be studied and are formulated as open problems (page 209), because there exist only a few simple (negative) results; see Problem 8.29 and one more reassuring one which we cite now.

Theorem 8.73 *Suppose that in a reaction for all reaction steps, either the reactant complex vector is a multiple of the basis vectors or a sum of different basis vectors. If the induced kinetic differential equation of such a reaction is a gradient system, then one can have for no $m \in \mathcal{M}$ a negative term in $\frac{\partial f_m}{\partial c_p}$, if $p \neq m$.*

Proof Suppose on the contrary that $\frac{\partial f_m}{\partial c_p}$ with some $p \neq m$ contains terms with negative sign, then f_m should contain a term with negative sign: $-g(\mathbf{c})c_m c_p$, where $g(\mathbf{c}) \in \mathbb{R}_0^+$ does depend neither on c_p nor on c_p. Thus, $\frac{\partial f_m}{\partial c_p} = -g(\mathbf{c})c_m$, consequently $f_p(\mathbf{c}) = -g(\mathbf{c})c_m^2/2 < 0$ expressing negative cross effect which is impossible.   □

Remark 8.74

1. One may say that in the definition of a gradient system, one has **quantitative** requirements: equality of the corresponding coefficients of a given polynomial differential equation. These requirements have **qualitative** implications concerning the sign of some coefficients, and these are the facts to have immediate kinetic consequences.
2. As the class of the reactions in the theorem subsume all second-order reactions and many more, one may call them **weakly realistic**.
3. The property that none of the species cause the decrease of another one may be formulated as the reaction being **cross catalytic**.

With all these concepts, Theorem 8.73 can be reformulated as follows:

Theorem 8.75 *If the induced kinetic differential equation of a weakly realistic reaction is a gradient system, then the canonical realization of the induced kinetic differential equation is necessarily cross catalytic.*

8.10   Persistence

It is an important property of the induced kinetic differential equation of a reaction if none of the concentration vs. time curves turn to zero during the entire course of the reaction. (It is nonetheless important if a reaction is used as a model of some biological phenomenon: here persistence is just the opposite of extinction, i.e., it means survival.) Therefore we are interested in structural conditions to ensure or rule out that a reaction has this property. An early result on the topics is that by Simon (1995). The authors Angeli (2010), Angeli et al. (2007, 2011) have a number of results in this respect, mainly using the structure of the Volpert graph (which they call Petri nets, similarly to Volpert's papers in the early 1970s of the last century).

Let us consider the reaction (2.1):

$$\sum_{m \in \mathcal{M}} \alpha(m, r) X(m) \longrightarrow \sum_{m \in \mathcal{M}} \beta(m, r) X(m) \quad (r \in \mathcal{R}). \tag{8.52}$$

Definition 8.76 The reaction above is said to be **persistent**, if for all positive initial conditions $\mathbf{c}(0) = \mathbf{c}_0$ the solution $\mathbf{c}$ of the induced kinetic differential equation of the reaction fulfills $\liminf_{t \to +\infty} c_m(t) > 0$ for every m.

Remark 8.77 Instead of positivity of $\mathbf{c}_0$, it is enough to require nonnegativity, if all the species can be reached via reaction path from the initially positive species, cf. Theorem 8.14.

Definition 8.78 A nonnegative vector $\varrho \in (\mathbb{R}_0^+)^M$ for which $\varrho^\top \gamma = 0$ is said to be a **weak mass**.

Obviously, a reaction is mass conserving if it has a strictly positive weak mass vector, in other words, if it has a weak mass vector whose support is the whole index set $\mathcal{M}$. The terminology is ours, and while it may not be perfect, we prefer this to *P*-**semiflow** of Angeli et al. (2007, 2011), as the previous one fits more easily to known concepts.

Definition 8.79 The nonempty subset $\mathcal{M}_0 \subset \mathcal{M}$ of the species is said to be a **siphon** if $\forall r \in \mathcal{R} \, (\exists m \in \mathcal{M}_0 : \beta(m, r) > 0 \longrightarrow \exists p \in \mathcal{M}_0 : \alpha(p, r) > 0)$, i.e., if each reaction step that produces a species in $\mathcal{M}_0$ also has some species in $\mathcal{M}_0$ as one of its reactant species.

The siphon is **minimal** if none of its proper subsets are siphons.

Theorem 8.80 (Angeli et al. (2007)) *Suppose the reaction (2.1) is mass conserving and each siphon contains the support of a weak mass. Then the network is persistent.*

The papers by De Leenheer et al. (2006), Angeli et al. (2011) present sufficient conditions for persistence for special classes of reactions.

8.11 Controllability and Observability

Scientific (as opposed to engineering) approach means to understand what is going on in a given, say, chemical system. The engineering approach is that one wants to achieve something, e.g., maximize the yield, minimize the dangerous byproducts, avoid blowing up, etc. All these mean that one wants to direct the course of a reaction with some kinds of input so as to have an advantageous result. This is the topic of **controllability**, a fundamental property of dynamical systems. There are only a few papers containing results in this field; we give a very short review based on the Introduction of Drexler and Tóth (2016).

The well-founded methods in control engineering are based on linear dynamical systems; however the dynamics of chemical reactions is usually nonlinear. Control and controllability of nonlinear systems is only available for a relatively small class of systems and is more involved than control of systems with linear dynamics (see, e.g., Isidori 1995).

Controllability of chemical reactions is usually analyzed using control theory developed for linear systems, and the linear model is acquired by linearizing the dynamics at an operating point. Reactions with a positive stationary point have been analyzed this way by Farkas (1998a). Yuan et al. (2011b) treated the liquid-phase catalytic oxidation of toluene to benzoic acid based on linearization at five different operating points. Polymerization at several operating points was investigated by Lewin and Bogle (1996). Amand et al. (2014) identified a linear model from measurements to describe the time evolution and control of protein glycosylation. Otero-Muras et al. (2008) analyzed the connection between the controllability and the structure of chemical reactions.

Working at operating points however only gives local results. Controlling chemical reactions is a key issue in chemical engineering science (see, e.g., Maya-Yescas and Aguilar 2003), where control of nonlinear chemical reactors is considered. A review of the topic can be found in Yuan et al. (2011a). To study nonlinear reactions, Ervadi-Radhakrishnan and Voit (2005) suggested the application of Lie algebra rank condition; however no general results were given.

Drexler and Tóth (2016) analyzed the Lie algebra of the vector fields related to the reaction steps and used the Lie algebra rank condition to get global controllability results for chemical reactions. Reaction rate coefficients of the reaction steps are considered as control inputs, and the lowest number of control inputs needed to control the system is determined. The analysis is symbolical, as opposed to the analysis done in the literature that is usually numerical. Chemical reactions are shown to be controllable almost everywhere in this setting. The reaction steps whose reaction rate coefficients need to be control inputs are also identified. Initializer reaction steps are defined and proved that their reaction rate coefficients need to be control inputs, which has already been shown by experiments on a polymerization example. Consecutive reactions can be controlled with a single control input. In a more recent manuscript, (Drexler et al. 2017) the control inputs are the temperature of the reaction and the inflow rates of some species. The chemical reactions are strongly reachable on a subspace that has the same dimension as the dimension of the stoichiometric subspace in every point except where the concentration of reactant species is zero and that this result holds for reactions in continuously stirred tank reactors as well. As an application, the controllability of the anode and cathode reactions of a polymer electrolyte membrane fuel cell with the inflow rates of hydrogen and oxygen being the control inputs is analyzed, and it has been shown that by using the inflow rates as control inputs, the dimension of the subspace on which the system is controllable can be greater than the number of independent reaction steps. It turns out that it is not possible to strictly separate the two approaches mentioned in the beginning of the present subsection. Already in the very early history of control theory (or system theory), it turned out that controllability is very closely related to the following problem of **observability**: Given the present state of a system, is it possible to tell where did it come from? (Note that the question where the system started if it is in a given state now may be called scientific in the strict sense; still, it is treated—because of technical reasons— parallel with controllability.) Here we only mention a few papers on the topics: Farkas (1998b), Horváth (2002–2008), Horváth (2002) which we are expounding later in Sect. 9.5.6.

8.12   Reaction–Diffusion Models

Some experimentalists interested in reaction kinetics claim that homogeneous reaction kinetics is not interesting any more. We hope to have proved just the opposite in the present book, especially with the open problems attached to the individual chapters. One can say that taking into consideration spatial inhomogeneities is more interesting as it means investigation of more complex systems. However, in this book we can only dedicate a few pages to the problem, including links to the relevant literature. First of all, the equation we use to describe diffusion and reaction is

$$\frac{\partial c_m(t,\mathbf{x})}{\partial t} = D_m \Delta c_m(t, \mathbf{x}) + \sum_{r \in \mathscr{R}} \gamma(m, r) k_r \mathbf{c}(t, \mathbf{x})^{\boldsymbol{\alpha}(\cdot, r)} \qquad (8.53)$$

$$\mathbf{x} \in \Omega \subset \mathbb{R}^K, \quad m \in \mathscr{M}$$

where $D_m \in \mathbb{R}^+$ are the diffusion coefficients of the individual species, $K \in \mathbb{N}$ is the number of spatial dimensions, and $\Omega \subset \mathbb{R}^K$ is an open connected set. Note that no cross diffusion is taken into consideration: all the species only affect their own diffusion. Usually one has to assume appropriate initial and boundary conditions to have a well-posed problem; see Sect. 8.2.

A few classical books on the topics are as follows: Frank-Kamenetskii (1947), Fife (1979), Britton (1986), the first one being more applications oriented while the last two is about theory. The most popular book with mathematicians concentrating on the case with a single species and a single spatial dimension has been written by Smoller (1983). Another book for mathematicians on global stability is Rothe (1984). Martin and his coworkers have written a series of mathematical papers

(Hollis et al. 1987; Martin 1987) on reaction–diffusion systems of more complicated type than (8.53). Below we only cite one of his earlier results from Martin (1987) on the global uniform asymptotic stability of the stationary state. A random choice from the numerical literature is Ladics (2007). Hollis (2004) gives the documentation of a *Mathematica* program called `ReactionDiffusionLab`. Note that at the moment the *Mathematica* function `NDSolve` can automatically solve reaction–diffusion systems with first-order reactions only.

A few (again mathematical) papers have been dedicated to systems where electric charges also play a role: Glitzky et al. (1996, 1994), Glitzky and Hünlich (2000, 1997), Gröger (1992). An early result was provided by Volpert (1972) who has shown that spatial discretization usually used to solve reaction–diffusion equations numerically provides a system of ordinary differential equations which can in themselves be considered to be the induced kinetic differential equation of a (very large) reaction; therefore nice properties (such as, e.g., positivity) of kinetic differential equations are automatically inherited. On the other side, results by Volpert on stability for homogeneous kinetics have been generalized for the reaction–diffusion case by Mincheva and Siegel (2004, 2007). The papers by Shapiro and Horn (1979b,a) show how to apply results on homogeneous kinetics to spatially distributed systems; no wonder that it contains results similar to those by Volpert mentioned above. We do not have place here to describe the interesting experimental and theoretical work by Á. Tóth and D. Horváth as seen in the randomly selected papers Bohner et al. (2016), Tóth et al. (1996), and Horváth and Tóth (1998) and the works also connected to nanoscience by Lagzi and coworkers, e.g., Lagzi et al. (2010a,b).

8.12.1 Chemical Waves

In reaction–diffusion systems, concentration **waves** can propagate (see, e.g., Epstein and Pojman 1998). **Concentric** chemical waves were observed in the Belousov–Zhabotinsky reaction. If such waves are broken, **spiral waves** may appear.

Two-dimensional wave fronts can propagate in thicker layers of solution. Breaking such fronts results in formation of three-dimensional **scroll waves**. Concentric, spiral, and scroll waves attracted much attention (Epstein and Pojman 1998). Many other spatiotemporal patterns have been discovered in various Belousov–Zhabotinsky reaction–diffusion systems (Nagy-Ungvárai et al. 1989). The paper by Noszticzius et al. (1987) is the more interesting because it was the gel reactor in which it was possible to find Turing instabilities (see below, page 198) a few years later by Castets et al. (1990).

Here we show an extremely simple model of a chemical wave which can symbolically be treated. More involved models are treated, e.g., in Tyson (1977). Some of our readers might find the lecture notes by Póta (1996) quite useful.

Let us start with the autocatalytic reaction $A + B \xrightarrow{1} 2B$ and assume that it takes place in a one-dimensional vessel where diffusion cannot be neglected. Then the reaction–diffusion equations (with the simplifying assumption that the two chemical species have the same diffusion coefficients D) to describe the model are as follows:

$$\partial_0 a(t, x) = D\partial_1^2 a(t, x) - a(t, x)b(t, x) \tag{8.54}$$

$$\partial_0 b(t, x) = D\partial_1^2 b(t, x) + a(t, x)b(t, x) \tag{8.55}$$

with the boundary conditions

$$a(0, x) = \mathfrak{B}(x \leq 0)a_0 \tag{8.56}$$

$$b(0, x) = \mathfrak{B}(x > 0)a_0. \tag{8.57}$$

These equations imply that $h(t, x) := a(t, x) + b(t, x)$ obeys the equations:

$$\partial_0 h(t, x) = D\partial_1^2 h(t, x) \quad h(0, x) = a_0. \tag{8.58}$$

The solution to this is $h(t, x) = a_0$, and using this fact we get a single equation for the function b:

$$\partial_0 b(t, x) = D\partial_1^2 b(t, x) + (a_0 - b(t, x))b(t, x). \tag{8.59}$$

Now we are looking for solutions of (8.59) of the form $b(t, x)/a_0 = \varphi(x + ct)$, because these solutions have the property that they are constant along lines in the (t, x) plane that means they are wave solutions in a certain sense. Substituting this form into (8.59) one gets

$$D\varphi'' - c\varphi' + a_0(1 - \varphi)\varphi = 0 \tag{8.60}$$

$$\varphi(-\infty) = 0 \quad \varphi(+\infty) = 1. \tag{8.61}$$

It turns out the wave velocities c for which the last boundary problem has a solution are those for which $c \geq 2\sqrt{D}$ and experimentally measurable wave is obtained with the value for which the equality holds. Now the problem has been simplified to the solution (numerically) of the boundary value problem (8.60)–(8.61) and to the comparison with measurements.

8.12.2 Turing Instability

From the large literature on Turing instability (Castets et al. 1990; Cavani and Farkas 1994; Dilao 2005; Herschkowitz-Kaufman 1975) and also (Edelstein-Keshet 2005; Murray 2002; Epstein and Pojman 1998), we only cite a few.

Turing instability is a phenomenon (predicted by the same Turing who was also a pioneer in computer science and a code breaker in the Second World War; see Turing 1952) which is quite strange from the classical viewpoint: Here diffusion is the cause of (spatial) inhomogeneity contrary to the general expectation that it should, in general, equilibrate inhomogeneities. The reason why Turing was interested in this problem was that he tried to give a model of the formation of the (asymmetric) embryo starting from a symmetric state. He constructed a system which he considered a reaction–diffusion system (we are going to return to this point below) in which there exists a stable homogeneous stationary state losing its stability as a result of inhomogeneous perturbations.

The phenomenon has also been shown to exist in an experimental setup almost 40 years after Turing's work. It was putting the CIMA reaction into the gel ring reactor—designed by Noszticzius et al. (1987)—by DeKepper et al. (Castets et al. 1990) which is generally considered to have produced the long-sought-for result first: the emergence of stationary patterns as a result of diffusion-driven instability.

Elementary arguments are enough to show that the presence of cross-inhibition is a necessary condition of Turing instability, at least in the case of systems with one, two, or three chemical species (Szili and Tóth 1997). This result implies that the presence of higher-than-first-order reactions is a necessary condition of Turing instability. The generalization of the statement for an arbitrary number of species needs more refined tools (Szili and Tóth 1993) of which Martin (1987) is one of the most important one.

Consider the reaction–diffusion equation (8.53) with initial conditions

$$c_m(0, \mathbf{x}) = c_m^0(\mathbf{x}) \quad (\mathbf{x} \in \Omega) \tag{8.62}$$

and either with fixed boundary conditions:

$$c_m(t, \mathbf{x}) = c_m^* \quad (\mathbf{x} \in \partial\Omega) \tag{8.63}$$

or with zero flux conditions

$$\partial_{\nu(\mathbf{x})} c_m(t, \mathbf{x}) = 0 \quad (\mathbf{x} \in \partial\Omega) \tag{8.64}$$

where $\nu(\mathbf{x})$ is the outer normal to $\partial\Omega$ at the point $\mathbf{x} \in \Omega$ of the spatial domain Ω where the reaction takes place. Suppose that the reaction without diffusion has a nonnegative stationary state $\mathbf{c}_*$, i.e., $\mathbf{f}(\mathbf{c}_*) = \mathbf{0}$. This stationary state is said to be **Turing unstable** if it is an asymptotically stable stationary state of the reaction without diffusion, but it is an unstable solution of the reaction–diffusion equation with the initial condition (8.62) and either with the fixed boundary condition (8.63) or with the zero flux condition (8.64).

Let $r(\mathbf{A}) := \max\{\Re(\lambda)|\lambda$ is an eigenvalue of $\mathbf{A}\}$ be the **spectral abscissa** of the matrix $\mathbf{A}$. Obviously, if $r(\mathbf{f}'(\mathbf{c}_*)) < 0$, then $\mathbf{c}_*$ is an asymptotically stable stationary point of the homogeneous system. Let $\kappa_k, k = 0, 1, 2, \ldots$ be the eigenvalues of the Laplace operator on the domain Ω, and suppose $r(\mathbf{f}'(\mathbf{c}_*) + \kappa_k D) < 0$ holds for all eigenvalues κ_k of the Laplace operator. Then, according to Martin (1987), $\mathbf{c}_*$ is a globally uniformly asymptotically stable solution of the reaction–diffusion equation.

We need another concept.

Definition 8.81 The chemical species $X(m)$ is said to cross-inhibit $X(p)$ at the concentration $\overline{c}$, if $\partial_m f_p(\overline{c}) < 0$ (Cf. Remark 8.74 and also Definition 8.68).

Obviously, for polynomial differential equations in general, the presence of negative cross effect implies the presence of cross-inhibition, at least if the corresponding term does depend at all on the corresponding variable c_m. Kinetic differential equations, however, are only able to show cross-inhibition. Now we are in the position to formulate our result.

Theorem 8.82 *The presence of cross-inhibition is a necessary condition of Turing instability.*

A simple consequence of this theorem is that Turing instability cannot appear in a reaction with first-order reaction steps if the kinetics is of the mass-action type.

Let us make a remark on the example given by Turing (1952). Here the "reaction terms" are as follows: $\dot{c}_1 = 5c_1 - 6c_2 + 1 \quad \dot{c}_2 = 6c_1 - 7c_2 + 1$. As this system contains negative cross effect, it cannot be the mass-action type induced kinetic differential equation of any reaction; thus Turing's example is nonkinetic and linear (contrary to the view according to which it is kinetic and nonlinear).

Emergence of Turing patterns in fields outside chemistry can be found in the papers by Cavani and Farkas (1994), Farkas (1995).

Finally, let us mention that good sources for the theory of waves and patterns are Edelstein-Keshet (2005), Murray (2002).

8.13 Exercises and Problems

8.1 Calculate the sensitivity equations in full generality for mechanisms with mass-action kinetics.

(Solution: page 411.)

8.2 What is the effect of the individual reaction rate coefficients on the trivial stationary point of the Lotka–Volterra mechanism?

(Solution: page 411.)

8.3 Find a polynomial differential equation containing negative cross effect which still can only have nonnegative solutions if started from the first orthant.

(Solution: page 412.)

8.4 Show that the solution of the induced kinetic differential equation of a single-species mechanism may blow up for all positive initial concentrations if and only if its right-hand side is a polynomial of order larger than or equal to two which has no positive roots.

(Solution: page 412.)

8.5 Show that the solutions of a mass-consuming reaction do not blow up.

(Solution: page 412.)
    The use of the *Mathematica* function **FindInstance** might be very useful to find the vector ω. We used the same function when constructing the examples.

8.6 Show that the solutions of the kinetic differential equation

$$\dot{x} = 1 - x^2/4 + 2y^2 + z^2/2, \quad \dot{y} = x^2 - y^2, \quad \dot{z} = -z^2 \qquad (8.65)$$

blow up for any choice of the initial conditions.

(Solution: page 412.)

8.7 Show that the solutions of the (kinetic) differential equation

$$\dot{x} = -x^2/2 + 2y^2 + z^2, \quad \dot{y} = x^2 - y^2, \quad \dot{z} = -z^2 \qquad (8.66)$$

blow up for some choices of the initial conditions. Show also that the solutions blow up for all x_0, y_0, z_0 for which $x_0 \geq 0$, $y_0 \geq 0$, $z_0 \geq 0$, $x_0 + y_0 + z_0 > 0$ holds.

(Solution: page 413.)

8.8 Show that the solutions of the kinetic differential equation

$$\dot{x} = -x - x^2/2 + 2y^2 + z^2, \quad \dot{y} = x^2 - y^2, \quad \dot{z} = -z^2 \qquad (8.67)$$

blow up for some choices of the initial conditions.

(Solution: page 413.)

8.9 Prove that in the reaction $X \xrightarrow{1} Y \xrightarrow{1} 4X$, the unique stationary point is $\begin{bmatrix} 0 & 0 \end{bmatrix}^T$. Calculate $\lim_{t \to +\infty} \mathbf{c}(t)$.

(Solution: page 413.)

8.10 Find a reaction which is not weakly reversible; still the number of the components of its Feinberg–Horn–Jackson graph is the same as that of its ergodic components.

(Solution: page 413.)

8.11 Show that $T - L > 0$ alone is not enough to ensure that the stoichiometric subspace properly contains the kinetic subspace.

(Solution: page 413.)

8.12 Show that the reaction $X \xrightarrow{k} Y \quad X \xrightarrow{1} Z \quad Y + Z \xrightarrow{1} 2X$ (Feinberg 1987, p. 2266) is stoichiometrically mass conserving for all positive values of the reaction rate coefficient k with the vector of masses $\varrho_1 := \begin{bmatrix} 1 & 1 & 1 \end{bmatrix}^T$. However, if $k = 1$, it is kinetically mass conserving with the vector of masses $\varrho_2 := \begin{bmatrix} 2 & 1 & 3 \end{bmatrix}^T$, i.e.,

$$\varrho_2^T \cdot \begin{bmatrix} \dot{x} \\ \dot{y} \\ \dot{z} \end{bmatrix} = 0 \text{ holds with the solutions of the induced kinetic differential equation}$$

of the reaction.

(Solution: page 414.)

8.13 Find an example showing that a reaction can be complex balanced for some reaction rate coefficients even if it is either not weakly reversible or not of the zero deficiency.

(Solution: page 414.)

8.14 Show that with appropriate parameters the induced kinetic differential equation of the autocatalator (Gray and Scott 1986)

$$P \xrightarrow{k_1} A \quad A + 2B \xrightarrow{k_2} 3B \quad B \xrightarrow{k_3} C, \tag{8.68}$$

—where P and C are external species— has a periodic solution.

(Solution: page 415.)

8.15 Investigate the stationary points of the Explodator model of Example 8.62.

(Solution: page 416.)

8.16 Find the single positive stationary point of the reaction (8.50), and investigate its stability.

(Solution: page 416.)

8.17 The first component of the concentration vs. time function of the induced kinetic differential equation of the reaction

$$Y \xrightarrow{1} 0 \underset{1}{\overset{4}{\rightleftharpoons}} X \xrightarrow{1} X + Y \xrightarrow{7} 2X$$

with the initial condition $x(0) = 0$, $y(0) = 1$ has exactly one strict local maximum.

(Solution: page 416.)

8.18 Consider the induced kinetic differential equation $\dot{c} = Ac$ of a closed compartmental system. Prove that A does not have any positive real eigenvalues but that at least one of its eigenvalues is zero.

(Solution: page 416.)

8.19 Prove a simplified version of Bautin's theorem (Bautin 1954): the (Kolmogorov type or Lotka–Volterra type, if you like) differential equation

$$\dot{x} = x(ax + by + c) \quad \dot{y} = y(Ax + By + C)$$

cannot have a limit cycle in the open first quadrant.

(Solution: page 417.)

8.20 Show that the positive stationary point of the equations $\dot{x} = -bxy + dx$ $\dot{y} = -\beta xy + \delta y$ can only be a saddle if all the coefficients b, d, β, δ are of the same sign.

(Solution: page 417.)

8.21 Show that the modification of the Brusselator

$$0 \underset{1}{\overset{a}{\rightleftharpoons}} X \xrightarrow{b} Y \quad 2X + Y \xrightarrow{1} 3X \quad X + 2Y \xrightarrow{1} Y$$

cannot have periodic solutions in the first orthant.

(Solution: page 417.)

8.22 By constructing a bounded closed positively invariant set and using Theorem 13.54 show that the Brusselator has a periodic solution.

(Solution: page 418.)

8.23 Verifying the conditions of the Theorem 13.57 shows that that the Brusselator has a periodic solution.

(Solution: page 418.)

8.24 Show that the induced kinetic differential equation of the reversible triangle reaction cannot have periodic solutions if any of the reaction rate coefficients is positive.

(Solution: page 418.)

8.25 We follow the manuscript by Nagy et al. (2006). Consider the induced kinetic differential equation of the triangle reaction of Fig. 8.8 with the initial condition [A](0)=1, [B](0)=0, [C](0)=0 with three different sets of the reaction rate coefficients:

I. $k_1 = k_3 = k_5 = 1$ and $k_2 = k_4 = k_6 = 0$;
II. $k_1 = k_2 = k_3 = k_4 = k_5 = k_6 = 1$;
III. $k_1 = 2$ and $k_2 = k_3 = k_4 = k_5 = k_6 = 1$.

Show that in Case I $t \mapsto a(t)$ has an infinite number of strict local extrema. Case II represents a detailed balanced reaction, and in this case $t \mapsto a(t)$ has a single maximum. Case III is not detailed balanced; still here $t \mapsto a(t)$ has a single maximum, as well (contrary to the statement of Alberty (2004) who—based on erroneous numerical calculations—stated that the named function can have multiple extrema).

(Solution: page 420.)

8.26 Show that the induced kinetic differential equation of the reaction in Fig. 8.13 is a gradient system.

(Solution: page 420.)

8.27 Show that the induced kinetic differential equation of the reaction in Fig. 8.14 is a Hamiltonian system.

(Solution: page 420).

8.28 Show that the induced kinetic differential equation of the reaction in Fig. 8.15 is a Cauchy–Riemann–Erugin system, and use this fact to symbolically solve its induced kinetic differential equation.

Fig. 8.13
Feinberg–Horn–Jackson
graph of a reaction having an
induced kinetic differential
equation which is a gradient
system

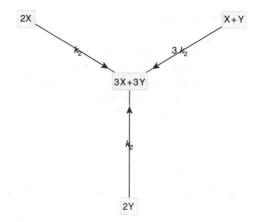

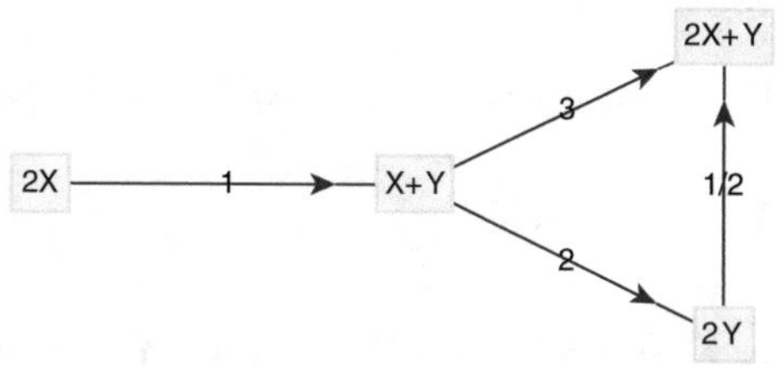

Fig. 8.14
Feinberg–Horn–Jackson
graph of a reaction having an
induced kinetic differential
equation which is a
Hamiltonian system

Fig. 8.15
Feinberg–Horn–Jackson
graph of a reaction having an
induced kinetic differential
equation which is a
Cauchy–Riemann–Erugin
system

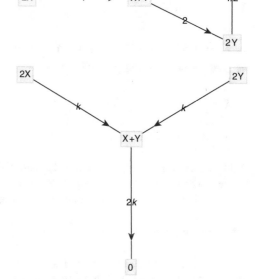

(Solution: page 420.)

8.29 Find all the induced kinetic differential equations of second-order two-species mass-conserving systems which are gradient systems, Hamiltonian sytems, or Cauchy–Riemann–Erugin systems.

(Solution: page 420.)

8.30 ^{210}Bi is known to decay to ^{210}Po with a half-life of 5 days, and ^{210}Po further decays with a half-life of 138 days to ^{206}Pb which is not radioactive. Suppose the decay can be modeled by first-order kinetics. Suppose we are only given ^{210}Bi at the beginning. When does the quantity of ^{210}Po reach its maximum? At what time does the quantity of ^{206}Pb reach half of its final value?

(Solution: page 421.)

8.31 Consult Tomlin et al. (1992), p. 110; or Turányi and Tomlin (2014) to formulate a temperature-dependent model in the general case.

(Solution: page 422.)

8.32 Find $\mathscr{R}_f$ and $\mathscr{S}^*$ (see Eq. (8.22)) for the following reactions:

1. $2\,X \xrightarrow{k} 3\,X,$
2. $X \xrightarrow{k_1} Y \quad X \xrightarrow{k_2} Z,$
3. $X \xrightarrow{k_1} 2\,X \quad X + Y \xrightarrow{k_2} 2\,Y \quad Y \xrightarrow{k_3} 0,$
4. (see Johnston and Siegel 2011) $Y \xrightarrow{k_1} X \xrightarrow{k_2} 0 \longleftarrow 2\,Y \xrightarrow{k_3} 3\,X \quad (k_2 \neq k_3),$
5. $Y \xleftarrow{k_1} X \xleftarrow{k_2} 2\,Y \xrightarrow{k_2} 3\,X.$

(Solution: page 424.)

8.33 Under which conditions can it happen that the maxima of the concentrations in the Lotka–Volterra reaction appear at the same time? What are the conditions that the period of the two concentrations are the same?

(Solution: page 424.)

8.34 Why is it impossible to transform the equation of the harmonic oscillator (6.41) and the Lorenz equation (6.40) into Lotka–Volterra form using the method of Example 8.25? Does the method work on the nonkinetic polynomial equation:

$$\dot{x} = xy^2 \quad \dot{y} = -x?$$

(Solution: page 424.)

8.35 An application of Corollary 8.52 shows that the complex graph of the reversible Lotka–Volterra reaction does contain an odd dumbbell (actually, two): a necessary condition to have nontrivial periodic solutions.

(Solution: page 425.)

8.36 Show that the induced kinetic differential equation of no three-species second-order reaction with a linear first integral can have a limit cycle in $(\mathbb{R}^+)^3$.

(Solution: page 425.)

8.37 Show that the induced kinetic differential equation of all three-species second-order reactions with the first integral $\psi(x, y, z) := xyz$ is a generalized Lotka–Volterra system.

(Solution: page 426.)

8.38 Show that all the coordinate hyperplanes are invariant sets of Kolmogorov type equations.

(Solution: page 426.)

8.39 Show that the Wegscheider reaction has a single positive stationary concentration vector in all the stoichiometric compatibility classes for any set of reaction rate coefficients.

(Solution: page 426.)

8.40 Show that the reversible triangle reaction has a single positive stationary concentration vector in all the stoichiometric compatibility classes for any set of reaction rate coefficients.

(Solution: page 426.)

8.41 Calculate the total net flux for all the atoms in the reaction step:

$$CH_3 + C_3H_7 \longrightarrow C_4H_8 + H_2. \tag{8.69}$$

Turányi and Tomlin (2014)

(Solution: page 427.)

8.42 Calculate the total net flux for all the atoms in the Ogg reaction Ogg (1947) (see Eq. (2.11)):

(Solution: page 427.)

8.43 Find subsets of the first quadrant in which the Lotka–Volterra reaction

$$X \xrightarrow{k_1} 2\,X \quad X + Y \xrightarrow{k_2} 2\,Y \quad Y \xrightarrow{k_3} 0$$

is uniform.

(Solution: page 428.)

8.44 Using direct analysis of the induced kinetic differential equations, show that $A + B \rightleftharpoons 2\,A$ and $A + 2\,B \rightleftharpoons 3\,A$ cannot admit multiple stationary states if put into a CSFTR, whereas $2\,A + B \rightleftharpoons 3\,A$ can (Schlosser and Feinberg 1994).

(Solution: page 429.)

8.45 Show that the trajectories of the reaction (8.50) are always bounded.

(Solution: page 430.)

8.14 Open Problems

1. In the sensitivity equations (8.3)–(8.8), there a few terms expressing negative cross effect. Characterize those reactions for which the sensitivity equations are kinetic.
2. Theorem 8.1 provided a sufficient condition to ensure the existence and uniqueness of solutions of initial value problems for induced kinetic differential equations. Is it possible to formulate a necessary and sufficient condition in terms of the components of the reactant vectors in case of non-integer stoichiometric coefficients (Cf. Remark 8.2)?
3. In connection with Theorem 8.19, the following questions arise.
   (a) The theorem is formulated for polynomial differential equations, which means that the lack of negative cross effect is not utilized.
   (b) The theorem should also be reformulated in structural terms, i.e., in terms understandable by the chemist.
   (c) How to give a lower estimate for the blow-up time?
   (d) How to select those components which tend to infinity?
   (e) Which are the conditions for the matrices $\mathbf{A}_m$ to ensure the existence of ω such that $\mathbf{A}(\omega)$ is positive definite? Certainly, we are mainly interested in structural conditions with immediate chemical meaning, but numerical methods for finding an appropriate vector ω are also of value. A trivial sufficient condition for this is that all the matrices are semi-definite, and at least one of them is definite.
   (f) If there exists more than one such ω, how to choose among them to receive the best estimate for the blow-up time?

(g) From the practical point of view (of firemen, let us say), the most important question is what is the "smallest, simplest" reaction step to be included to prevent blowup?

4. We know which species becomes nonnegative or positive at all times of the domain of existence of the concentration curves (Theorems 8.8 and 8.6), and we also know that in a zero-deficiency reaction, some of the stationary concentrations should be zero if the reaction is not weakly reversible. Which conditions would be sufficient or necessary to ensure that a given species has a zero stationary concentration?

5. Design an algorithm or give an easy-to-treat necessary and sufficient condition to select minimal initial sets (see page 155) to produce positive concentrations.

6. Does (8.16) have any further invariants (except the trivial ones of the form $g(\mathbf{B}\lambda, \mathbf{BA})$? More precisely, determine all the functions $\mathbf{f} \in \mathscr{C}^1(\mathbb{R}^M \times \mathbb{R}^{M \times R} \times \mathbb{R}^{R \times M}, \mathbb{R}^K)$ (with some $K \in \mathbb{N}$) for which the functional equation

$$\mathbf{f}(\lambda, \mathbf{A}, \mathbf{B}) = \mathbf{f}((\mathbf{C}^{-1})^\top \lambda, (\mathbf{C}^{-1})^\top \mathbf{A}, \mathbf{CB}) \qquad (8.70)$$

holds. Even identifying all multilinear invariants $\mathbf{f}$ might be interesting.

7. Find sufficient conditions under which a mass-action type mechanism is kinetically mass conserving.

8. Give a general characterization of induced kinetic differential equations which are also gradient systems, Hamiltonian systems, or Cauchy–Riemann–Erugin systems.

9. How to choose the reaction rate coefficients and the initial concentrations so as to have $N - 2$ local extrema in some/all species concentrations in a compartmental system?

10. Is there any general criterion to ensure or exclude oligo-oscillation in higher-order reactions?

11. Delineate large classes of reactions for which sign stability holds for the linearized form of the induced kinetic differential equation.

References

Alberty RA (2004) Principle of detailed balance in kinetics. J Chem Educ 81(8):1206–1209

Amand MMS, Tran K, Radhakrishnan D, Robinson AS, Ogunnaike BA (2014) Controllability analysis of protein glycosylation in cho cells. Plos One 9:1–16

Anderson DF (2011) A proof of the global attractor conjecture in the single linkage class case. SIAM J Appl Math 71(4):1487–1508

Andronov AA, Leontovich EA, Gordon II, Maier AG (1973) Qualitative theory of second-order dynamic systems. Wiley, New York

Angeli A (2010) A modular criterion for persistence of chemical reaction networks. IEEE Trans Aut Control 55(7):1674–1679

Angeli D, De Leenheer P, Sontag ED (2007) A Petri net approach to study the persistence in chemical reaction networks. Math Biosci 210(2):598–618

Angeli D, De Leenheer P, Sontag ED (2011) Persistence results for chemical reaction networks with time-dependent kinetics and no global conservation laws. SIAM J Appl Math 71(1):128–146

Antonov V, Dolicanin D, Romanovski VG, Tóth J (2016) Invariant planes and periodic oscillations in the May–Leonard asymmetric model. MATCH Commun Math Comput Chem 76(2):455–474

Bamberger A, Billette E (1994) Quelques extensions d'un théorème de Horn et Jackson. Comptes rendus de l'Académie des sciences Série 1, Mathématique 319(12):1257–1262

Bautin NN (1954) On periodic solutions of a system of differential equations. Prikl Mat Mekh 18:128–134

Beck MT (1990) Mechanistic and parametric conditions of exotic chemical kinetics. React Kinet Catal Lett 42(2):317–323

Beck MT (1992) Mechanistic and parametric conditions of exotic chemical kinetics. Why are there so few oscillatory reactions? Acta Chir Hung 129:519–529

Beklemisheva LA (1978) Classification of polynomial systems with respect to birational transformations i. Differentsial'nye Uravneniya 14(5):807–816

Belousov BP (1958) A periodic reaction and its mechanism. Sb Ref Radiats Med Moscow pp 145–147

Bertolazzi E (1996) Positive and conservative schemes for mass action kinetics. Comput Math 32:29–43

Bohner B, Endrődi B, Horváth D, Tóth Á (2016) Flow-driven pattern formation in the calcium-oxalate system. J Chem Phys 144(16):164504

Boros B (2012) Notes on the deficiency one theorem: multiple linkage classes. Math Biosci 235(1):110–122

Boros B (2013a) On the dependence of the existence of the positive steady states on the rate coefficients for deficiency-one mass action systems: single linkage class. J Math Chem 51(9):2455–2490

Boros B (2013b) On the existence of the positive steady states of weakly reversible deficiency-one mass action systems. Math Biosci 245(2):157–170

Boros B (2013c) On the positive steady states of deficiency-one mass action systems. PhD thesis, School of Mathematics. Director: Miklós Laczkovich, Doctoral Program of Applied Mathematics, Director: György Michaletzky, Department of Probability Theory and Statistics, Institute of Mathematics, Faculty of Science, Eötvös Loránd University, Budapest

Brenig L, Goriely A (1994) Painlevé analysis and normal forms. In: Tournier E (ed) Computer algebra and differential equations. London mathematical society lecture note series. Cambridge University Press, Cambridge, pp 211–238

Britton NF (1986) Reaction-diffusion equations and their applications to biology. Academic, London

Castets V, Dulos E, Boissonade J, De Kepper P (1990) Experimental evidence of a sustained standing turing-type nonequilibrium chemical pattern. Phys Rev Lett 64(24):2953

Cavani M, Farkas M (1994) Bifurcations in a predator-prey model with memory and diffusion II: turing bifurcation. Acta Math Hung 63(4):375–393

Craciun G (2016) Toric differential inclusions and a proof of the global attractor conjecture. arXiv:150102860

Craciun G, Feinberg M (2006) Multiple equilibria in complex chemical reaction networks: extensions to entrapped species models. IEE Proc-Syst Biol 153(4):179–186

Craciun G, Feinberg M (2010) Multiple equilibria in complex chemical reaction networks: semiopen mass action systems. SIAM J Appl Math 70(6):1859–1877

Csikja R, Tóth J (2007) Blow up in polynomial differential equations. Enformatika Int J Appl Math Comput Sci 4(2):728–733

De Leenheer P, Angeli D, Sontag ED (2006) Monotone chemical reaction networks. J Math Chem 41(3):295–314

Dilao R (2005) Turing instabilities and patterns near a Hopf bifurcation. Appl Math Comput 164(2):391–414

Domijan M, Pécou E (2012) The interaction graph structure of mass-action reaction networks. J Math Biol 65(2):375–402

Drexler DA, Tóth J (2016) Global controllability of chemical reactions. J Math Chem 54(6):1327–1350

Drexler DA, Virágh E, Tóth J (2017) Controllability and reachability of reactions with temperature and inflow control. Fuel. Published online 7 October 2017

Edelstein-Keshet L (2005) Mathematical models in biology. In: O'Malley RE Jr (ed) Classics in applied mathematics, vol 46. Society for Industrial and Applied Mathematics, Philadelphia

Epstein I, Pojman J (1998) An introduction to nonlinear chemical dynamics: oscillations, waves, patterns, and chaos. Topics in physical chemistry series. Oxford University Press, New York. http://books.google.com/books?id=ci4MNrwSlo4C

Ervadi-Radhakrishnan A, Voit EO (2005) Controllability of non-linear biochemical systems. Math Biosci 196(1):99–123

Escher C (1980) Models of chemical reaction systems with exactly evaluable limit cycle oscillations and their bifurcation behaviour. Berichte der Bunsengesellschaft für physikalische Chemie 84(4):387–391

Escher C (1981) Bifurcation and coexistence of several limit cycles in models of open two-variable quadratic mass-action systems. Chem Phys 63(3):337–348

Farkas M (1995) On the distribution of capital and labour in a closed economy. SE Asian Bull Math 19(2):27–36

Farkas G (1998a) Local controllability of reactions. J Math Chem 24:1–14

Farkas G (1998b) On local observability of chemical systems. J Math Chem 24:15–22

Feinberg M (1977) Mathematical aspects of mass action kinetics. In: Lapidus L, Amundson N (eds) Chemical reactor theory: a review. Prentice-Hall, Englewood Cliffs, pp 1–78

Feinberg M (1979) Lectures on chemical reaction networks. Notes of lectures given at the Mathematics Research Center. University of Wisconsin, Feinberg

Feinberg M (1980) Chemical oscillations, multiple equilibria, and reaction network structure. In: Stewart W, Ray WH, Conley C (eds) Dynamics and modelling of reactive systems. Academic, New York, pp 59–130

Feinberg M (1987) Chemical reaction network structure and the stability of complex isothermal reactors—I. The deficiency zero and deficiency one theorems. Chem Eng Sci 42(10):2229–2268

Feinberg M, Horn FJM (1977) Chemical mechanism structure and the coincidence of the stoichiometric and kinetic subspaces. Arch Ratl Mech Anal 66(1):83–97

Fife PC (1979) Mathematical aspects of reacting and diffusing systems. Springer, Berlin

Frank-Kamenetskii DA (1947) Diffusion and heat transfer in chemical kinetics. USSR Academy of Science Press, Moscow

Fu Z, Heidel J (1997) Non-chaotic behaviour in three-dimensional quadratic systems. Nonlinearity 10:1289–1303

Gaiko V (2013) Global bifurcation theory and Hilbert's sixteenth problem. Mathematics and its applications, vol 562. Springer Science & Business Media, Berlin

Getz WM, Jacobson DH (1977) Sufficiency conditions for finite escape times in systems of quadratic differential equations. J Inst Math Applics 19:377–383

Glitzky A, Hünlich R (1997) Global estimates and asymptotics for electro-reaction-diffusion systems in heterostructures. Appl Anal 66:205–226

Glitzky A, Hünlich R (2000) Electro-reaction-diffusion systems including cluster reactions of higher order. Math Nachr 216:95–118

Glitzky A, Gröger K, Hünlich R (1994) Existence, uniqueness and asymptotic behaviour of solutions to equations modelling transport of dopants in semiconductors. Bonner Mathematische Schriften 258:49–78

Glitzky A, Gröger K, Hünlich R (1996) Free energy and dissipation rate for reaction diffusion processes of electrically charged species. Appl Anal 60:201–217

Gonzales-Gascon F, Salas DP (2000) On the first integrals of Lotka-Volterra systems. Phys Lett A 266(4–6):336–340

Gray P, Scott SK (1986) A new model for oscillatory behaviour in closed systems: the autocatalator. Berichte der Bunsengesellschaft für physikalische Chemie 90(11):985–996. http://dx.doi.org/10.1002/bbpc.19860901112

Gröger K (1992) Free energy estimates and asymptotic behaviour of reaction-diffusion processes. IAAS-Preprint 20

Guckenheimer J, Holmes P (1983) Nonlinear oscillations, dynamical systems, and bifurcations of vector fields. Applied mathematical sciences, vol 42. Springer, New York

Gunawardena J (2003) Chemical reaction network theory for in-silico biologists. http://vcpmedharvardedu/papers/crntpdf

Györgyi L, Field RS (1991) Simple models of deterministic chaos in the Belousov–Zhabotinskii reaction. J Phys Chem 95(17):6594–6602

Halmschlager A, Tóth J (2004) Über Theorie und Anwendung von polynomialen Differentialgleichungen. In: Wissenschaftliche Mitteilungen der 16. Frühlingsakademie, Mai 19–23, 2004, München-Wildbad Kreuth, Deutschland, Technische und Wirtschaftswissenschaftliche Universität Budapest, Institut für Ingenieurweiterbildung, Budapest, pp 35–40

Halmschlager A, Szenthe L, Tóth J (2003) Invariants of kinetic differential equations. Electron J Qual Theory Differ Equ 2003(14):1–14. Proceedings of the 7'th Colloquium on the Qualitative Theory of Differential Equations

Hanusse P (1972) De l'existence d'un cycle limite dans l'évolution des systémes chimiques ouverts. C R Acad Sci Ser C 274:1245–1247

Hanusse P (1973) Simulation des systémes chimiques par une methode de Monte Carlo. C R Acad Sci Ser C 277:93

Heidel J, Fu Z (1999) Non-chaotic behaviour in three-dimensional quadratic systems, II. The conservative case. Nonlinearity 12:617–633

Hernández-Bermejo B, Fairén V (1995) Nonpolynomial vector fields under the Lotka–Volterra normal form. Phys Lett A 206(1):31–37

Herschkowitz-Kaufman M (1975) Bifurcation analysis of nonlinear reaction-diffusion equations ii. Steady state solutions and comparison with numerical simulations. Bull Math Biol 37(6):589–636

Higgins J (1968) Some remarks on Shear's Liapunov function for systems of chemical reactions. J Theor Biol 21:293–304

Hirsch MW, Smale S, Devaney RL (2004) Differential equations, dynamical systems, and an introduction to chaos. Pure and applied mathematics, vol 60. Elsevier—Academic, Amsterdam

Hollis S (2004) Reaction-diffusionlab.m. http://www.math.armstrong.edu/faculty/hollis/mmade/RDL/

Hollis SL, Martin RH Jr, Pierre M (1987) Global existence and boundedness in reaction-diffusion systems. SIAM J Math Anal 18(3):744–761

Horn F (1973) On a connexion between stability and graphs in chemical kinetics. II. Stability and the complex graph. Proc R Soc Lond A 334:313–330

Horn FJM (1974) The dynamics of open reaction systems. In: SIAM-AMS proceedings, vol VIII. SIAM, Philadelphia, pp 125–137

Horn F, Jackson R (1972) General mass action kinetics. Arch Ratl Mech Anal 47:81–116

Horváth Z (2002) Effect of lumping on controllability and observability. Poster presented at the colloquium on differential and difference equations dedicated to Prof. František Neuman on the occasion of his 65th birthday, Brno, September, pp 4–6

Horváth Z (2002–2008) Effect of lumping on controllability and observability. arxivorg http://arxiv.org/PS_cache/arxiv/pdf/0803/0803.3133v1.pdf

Horváth D, Tóth Á (1998) Diffusion-driven front instabilities in the chlorite–tetrathionate reaction. J Chem Phys 108(4):1447–1451

Hsü ID (1976) Existence of periodic solutions for the Belousov–Zaikin–Zhabotinskiĭ reaction by a theorem of Hopf. J Differ Equ 20(2):399–403

Huang Y, Yang XS (2006) Numerical analysis on the complex dynamics in a chemical system. J Math Chem 39(2):377–387

Isidori A (1995) Nonlinear control systems. Springer, Berlin

Jeffries C, Klee V, van den Driessche (1987) Qualitative stability of linear systems. Linear Algebra Appl 87:1–48

Johnston MD, Siegel D (2011) Linear conjugacy of chemical reaction networks. J Math Chem 49(7):1263–1282

Joshi B (2015) A detailed balanced reaction network is sufficient but not necessary for its Markov chain to be detailed balanced. Discrete Contin Dyn Syst Ser B 20(4):1077–1105

Kádas Z, Othmer H (1979a) Erratum: Stable limit cycles in a two-component bimolecular reaction system. J Chem Phys 72(12):1845

Kádas Z, Othmer H (1979b) Stable limit cycles in a two-component bimolecular reaction system. J Chem Phys 70(4):1845–1850

Kádas Z, Othmer H (1980) Reply to comment on stable limit cycles in a two-component bimolecular reaction system. J Chem Phys 72(4):2900–2901

Kertész V (1984) Global mathematical analysis of the Explodator. Nonlinear Anal 8(8):941–961

King RB (1982) The flow topology of chemical reaction networks. J Theor Biol 98(2):347–368

Kiss K, Kovács S (2008) Qualitative behaviour of n-dimensional ratio-dependent predator prey systems. Appl Math Comput 199(2):535–546

Kiss K, Tóth J (2009) n-Dimensional ratio-dependent predator-prey systems with memory. Differ Equ Dyn Syst 17(1–2):17–35

Klee V, van den Driessche P (1977) Linear algorithms for testing the sign stability of a matrix and for finding Z-maximum matchings in acyclic graphs. Numer Math 28(3):273–285

Kordylewski W, Scott SK, Tomlin AS (1990) Development of oscillations in closed systems. J Chem Soc Faraday Trans 86(20):3365–3371

Kovács K, Vizvári B, Riedel M, Tóth J (2004) Computer assisted study of the mechanism of the permanganate/oxalic acid reaction. Phys Chem Chem Phys 6(6):1236–1242

Ladics T (2007) Application of operator splitting in the solution of reaction-diffusion equations. Proc Appl Math Mech 7(1):2020135–2020136. http://dx.doi.org/10.1002/pamm.200701017

Lagzi I, Kowalczyk B, Wang D, Grzybowski B (2010a) Nanoparticle oscillations and fronts. Angew Chem 122(46):8798–8801

Lagzi I, Soh S, Wesson PJ, Browne KP, Grzybowski BA (2010b) Maze solving by chemotactic droplets. J Amer Chem Soc 132(4):1198–1199

Lente G, Bazsa G, Fábián I (2007) What is and what isn't a clock reaction? New J Chem 31:1707–1707

Lewin DR, Bogle D (1996) Controllability analysis of an industrial polymerization reactor. Comp Chem Engng 20:871–876

Li G, Rabitz H, Tóth J (1994) A general analysis of exact nonlinear lumping in chemical kinetics. Chem Eng Sci 49(3):343–361

Liggett TM (2010) Continuous time Markov processes: an introduction, vol 113. American Mathematical Society, Providence

Martin RH Jr (1987) Applications of semigroup theory to reaction-diffusion systems. In: Gill TL, Zachary WW (eds) Nonlinear semigroups, partial differential equations and attractors. Lecture notes in mathematics, vol 1248. Springer, Berlin, pp 108–126

Maya-Yescas R, Aguilar R (2003) Controllability assessment approach for chemical reactors: nonlinear control affine systems. Chem Eng J 92:69–79

Mincheva M, Siegel D (2004) Stability of mass action reaction-diffusion systems. Nonlinear Anal 56(8):1105–1131

Mincheva M, Siegel D (2007) Nonnegativity and positiveness of solutions to reaction-diffusion systems. J Math Chem 42(4):1135–1145

Morales MF (1944) On a possible mechanism for biological periodicity. Bull Math Biophys 6:65–70

Müller-Herold U (1975) General mass-action kinetics. Positiveness of concentrations as structural property of Horn's equation. Chem Phys Lett 33(3):467–470

Murphy MD, Ogle CA, Bertz SH (2005) Opening the "black box": oscillations in organocuprate conjugate addition reactions. Chem Commun (7):854–856

Murray JD (2002) Mathematical biology (I. An introduction), interdisciplinary applied mathematics, vol 17, 3rd edn. Springer, New York

Nagy I, Tóth J (2014) Quadratic first integrals of kinetic differential equations. J Math Chem 52(1):93–114

Nagy I, Póta G, Tóth J (2006) Detailed balanced and unbalanced triangle reactions (manuscript)

Nagy-Ungvárai Z, Tyson JJ, Hess B (1989) Experimental study of the chemical waves in the cerium-catalyzed Belousov–Zhabotinskii reaction. 1. Velocity of trigger waves. J Phys Chem 93(2):707–713

Noszticzius Z, Farkas H, Schelly ZA (1984) Explodator and Oregonator: parallel and serial oscillatory networks. A comparison. React Kinet Catal Lett 25:305–311

Noszticzius Z, Horsthemke W, McCormick WD, Swinney HL, Tam WY (1987) Sustained chemical waves in an annular gel reactor: a chemical pinwheel. Nature 329(6140):619–620

Ogg RAJ (1947) The mechanism of nitrogen pentoxide decomposition. J Chem Phys 15(5):337–338

Otero-Muras I, Szederkényi G, Hangos KM, Alonso AA (2008) Dynamic analysis and control of biochemical reaction networks. Math Comput Simul 79:999–1009

Perko L (1996) Differential equations and dynamical systems. Springer, Berlin

Petrov V, Gaspar V, Masere J, Showalter K (1993) Controlling chaos in the Belousov—Zhabotinsky reaction. Nature 361(6409):240–243

Póta G (1983) Two-component bimolecular systems cannot have limit cycles: a complete proof. J Chem Phys 78:1621–1622

Póta G (1992) Exact necessary conditions for oscillatory behaviour in a class of closed isothermal reaction systems. J Math Chem 9(4):369–372

Póta G (1996) Chemical waves and spatial patterns in reaction-diffusion systems. Kossuth Egyetemi Kiadó, Debrecen (in Hungarian)

Póta G (2016) Solutions defined on finite time intervals in a model of a real kinetic system. J Math Chem 54:1–5

Prigogine I, Lefever R (1968) Symmetry breaking instabilities in dissipative systems, II. J Chem Phys 48:1695–1700

Rábai G, Bazsa G, Beck MT (1979) Design of reaction systems exhibiting overshoot-undershoot kinetics. J Am Chem Soc 101(12):6746–6748

Romanovski V, Shafer D (2009) The center and cyclicity problems: a computational algebra approach. Birkhäuser, Boston

Ronkin LI (1977) Elements of multivariate complex function theory. Naukova Dumka, Kiev

Rosenbaum JS (1977) Conservation properties for numerical integration methods for systems of differential equations. 2. J Phys Chem 81(25):2362–2365

Rössler OE (1976) Chaotic behavior in simple reaction system. Zeitschrift für Naturforsch A 31:259–264

Rössler OE (1976) An equation for continuous chaos. Phys Lett A 57(5):397–398

Rothe F (1984) Global stability of reaction-diffusion systems. Springer, Berlin

Schlosser PM, Feinberg M (1994) A theory of multiple steady states in isothermal homogeneous CFSTRs with many reactions. Chem Eng Sci 49(11):1749–1767

Schneider K, Wegner B, Tóth J (1987) Qualitative analysis of a model for synaptic slow waves. J Math Chem 1:219–234

Schuman B, Tóth J (2003) No limit cycle in two species second order kinetics. Bull Sci Math 127:222–230

Scott SK (1991, 1993, 1994) Chemical chaos. International series of monographs on chemistry, vol 24. Oxford University Press, Oxford

Shapiro A, Horn F (1979a) Erratum. Math Biosci 46(1–2):157

Shapiro A, Horn F (1979b) On the possibility of sustained oscillations, multiple steady states, and asymmetric steady states in multicell reaction systems. Math Biosci 44(1–2):19–39

Shear D (1967) An analog of the Boltzmann H-theorem (a Liapunov function) for systems of coupled chemical reactions. J Theor Biol 16:212–225

Simon LP (1995) Globally attracting domains in two-dimensional reversible chemical dynamical systems. Ann Univ Sci Budapest Sect Comput 15:179–200

Sipos-Szabó E, Pál I, Tóth J, Zsély IG, Turányi T, Csikász-Nagy A (2008) Sensitivity analysis of a generic cell-cycle model. In: 2nd FUNCDYN meeting, Rothenburg oberhalb der Taube. http://www.math.bme.hu/~jtoth/pubtexts/SiposSzaboTothPal.pdf

Smith HL (2012) Global dynamics of the smallest chemical reaction system with Hopf bifurcation. J Math Chem 50(4):989–995

Smoller J (1983) Shock waves and reaction-diffusion equations. Springer, New York

Szederkényi G, Hangos KM, Magyar A (2005) On the time-reparametrization of quasi-polynomial systems. Phys Lett A 334(4):288–294

Szili L, Tóth J (1993) Necessary condition of the Turing instability. Phys Rev E 48(1):183–186

Szili L, Tóth J (1997) On the origin of Turing instability. J Math Chem 22(1):39–53

Thomas R (1978) Logical analysis of systems comprising feedback loops. J Theor Biol 73(4):631–656

Tomlin AS, Pilling MJ, Turányi T, Merkin J, Brindley J (1992) Mechanism reduction for the oscillatory oxidation of hydrogen: sensitivity and quasi-steady-state analyses. Comb Flame 91:107–130

Tóth J (1987) Bendixson-type theorems with applications. Zeitschrift für Angewandte Mathematik und Mechanik 67(1):31–35

Tóth J, Érdi P (1978) Models, problems and applications of formal reaction kinetics. A kémia újabb eredményei 41:227–350

Tóth J, Érdi P (1988) Kinetic symmetries: some hints. In: Moreau M, Turq P (eds) Chemical reactivity in liquids. Fundamental aspects, Paris, Sept. 7–11 1987. Plenum Press, New York, pp 517–522

Tóth J, Hárs V (1986a) Orthogonal transforms of the Lorenz- and Rössler-equations. Physica D 19:135–144

Tóth J, Hárs V (1986b) Specification of oscillating chemical models starting form a given linearized form. Theor Chim Acta 70:143–150

Tóth Á, Lagzi I, Horváth D (1996) Pattern formation in reaction-diffusion systems: cellular acidity fronts. J Phys Chem 100(36):14837–14839

Tóth J, Li G, Rabitz H, Tomlin AS (1997) The effect of lumping and expanding on kinetic differential equations. SIAM J Appl Math 57:1531–1556

Turányi T (1990) Sensitivity analysis of complex kinetic systems. Tools and applications. J Math Chem 5(3):203–248

Turányi T, Tomlin AS (2014) Analysis of kinetic reaction mechanisms. Springer, Berlin

Turányi T, Györgyi L, Field RJ (1993) Analysis and simplification of the GTF model of the Belousov–Zhabotinsky reaction. J Phys Chem 97:1931–1941

Turing AM (1952) The chemical basis of morphogenesis. Philos Trans R Soc Lond Ser B Biol Sci 237(641):37–72

Tyson JJ (1977) Analytic representation of oscillations, excitability, and traveling waves in a realistic model of the Belousov–Zhabotinskii reaction. J Chem Phys 66(3):905–915

Tyson JJ (1980) Comment on stable limit cycles in a two-component bimolecular reaction system. J Chem Phys 72(4):2898–2899

Tyson JJ, Light JC (1973) Properties of two-component bimolecular and trimolecular chemical reaction systems. J Chem Phys 59(8):4164–4273

Vandenberghe L, Boyd S (1996) Semidefinite programming. SIAM Rev 38(1):49–95

Várdai J, Tóth J (2008) Hopf Bifurcation in the Brusselator. http://demonstrations.wolfram.com/HopfBifurcationInTheBrusselator/, from The Wolfram Demonstrations Project

Volpert AI (1972) Differential equations on graphs. Mat Sb 88(130):578–588

Volpert AI, Hudyaev S (1985) Analyses in classes of discontinuous functions and equations of mathematical physics. Martinus Nijhoff Publishers, Dordrecht (Russian original: 1975)

Weber A, Sturm T, Seiler W, Abdel-Rahman EO (2010) Parametric qualitative analysis of ordinary differential equations: computer algebra methods for excluding oscillations. Lect Notes Comput Sci 6244:267–279. Extended Abstract of an Invited Talk, CASC 2010

Wilhelm T, Heinrich R (1995) Smallest chemical reaction system with Hopf bifurcation. J Math Chem 17:1–14

Willamowski KD, Rössler OE (1980) Irregular oscillations in a realistic abstract quadratic mass action system. Zeitschrift für Naturforschung A 35(3):317–318

Winfree AT (2001) The geometry of biological time, vol 12. Springer Science & Business Media, Berlin

Wolkowicz H, Saigal R, Vandenberghe L (2000) Handbook of semidefinite programming: theory, algorithms, and applications, vol 27. Springer Science & Business Media, Berlin

Ye YQ, Lo CY (1986) Theory of limit cycles, vol 66. American Mathematical Society, Providence

Yuan Z, Chen B, Zhao J (2011a) An overview on controllability analysis of chemical processes. AIChE J 57(5):1185–1201

Yuan Z, Chen B, Zhao J (2011b) Controllability analysis for the liquid-phase catalytic oxidation of toluene to benzoic acid. Chem Eng Sci 66:5137–5147

Zak DE, Stelling J, Doyle FJ (2005) Sensitivity analysis of oscillatory (bio) chemical systems. Comput Chem Eng 29(3):663–673

Zhabotinsky AM (1964) Periodic liquid-phase oxidation reactions. Doklady Akademii Nauk SSSR 157:392–395

Approximations of the Models

9

9.1 Introduction

Most of the induced kinetic differential equations of realistic models in kinetics cannot be symbolically solved (if one disregards the Taylor-series solution, see Brenig and Goriely 1994). Therefore one has to rely upon symbolic and numeric approximations. When dealing with stoichiometry, one meets the problems of linear algebra and linear (integer) programming. To determine the stationary points of a reaction, one has to solve large systems of nonlinear equations. The most important model we use is a system of ordinary differential equations which is almost always a stiff one; therefore special numerical methods are needed. It may also be useful to reduce the number of variables by lumping or by other techniques. The induced kinetic differential equation of a reaction also has a series of good properties: it only has nonnegative solutions, it may have linear or nonlinear first integrals, and it may also have periodic solutions. Numerical methods preserving these important properties are preferred. Some problems may best be formulated in terms of differential algebraic equations; therefore special methods to solve such equations are also needed. Some of these areas are shortly reviewed and illustrated by examples. In the case of the stochastic model, some characteristics can only be calculated approximately. Even simulating this model may be carried out faster, if one uses an approximate method.

The approximation methods can be divided into two large groups. In one of them, one is capable to formulate an approximation symbolically, i.e., to provide an approximation applicable also to models which are not fully specialized in the sense that it also contains some parameters (of which one may only know that they are positive). In the other group, one can only approximate models which are specialized up to the very last parameter. In these cases the effect of parameters can only be studied if one also chooses some of the parameters systematically.

© Springer Science+Business Media, LLC, part of Springer Nature 2018
J. Tóth et al., *Reaction Kinetics: Exercises, Programs and Theorems*,
https://doi.org/10.1007/978-1-4939-8643-9_9

9.2   Behavior at the Start

The first symbolic method helps tell the order of magnitude of the concentrations of the species at the beginning of the process as a function of Volpert indices.

Let us consider the reaction $\langle \mathscr{M}, \mathscr{R}, \alpha, \beta, \mathbf{k} \rangle$, and let $\mathscr{M}_0 \subset \mathscr{M}$ be a set of species (to be interpreted below as species with positive initial concentrations). Suppose the zero complex is not a reactant complex. We rely on the definitions introduced in Sect. 3.22.

Theorem 9.1 *For a species* $m \in \mathscr{M} \setminus \mathscr{M}_0$ *reachable from* $\mathscr{M}_0$ *and having the index* κ_m

$$c_m(t) = t^{\kappa_m} d_m(t) \tag{9.1}$$

holds in a neighborhood $[0, \tau]$ *of zero with a continuous function* $d_m \in \mathscr{C}([0, \tau], \mathbb{R})$.

Example 9.2

- In the Robertson reaction

$$A \longrightarrow B \quad 2B \longrightarrow B + C \longrightarrow A + C$$

(if, as usual, $\mathscr{M}_0 := \{A, B\}$) one has that $t \mapsto c(t)/t$ is continuous at zero, where the concentration of C at time t is denoted by $c(t)$, because the Volpert index of the C is 1 as given by

```
First @ VolpertIndexing[
    {A -> B, 2B -> B+C -> A+C}, {A, B}].
```

- Consider the Michaelis–Menten reaction $E + S \rightleftharpoons C \longrightarrow E + P$, and let, as usual, $\mathscr{M}_0 := \{E, S\}$. Then the theorem says that $t \mapsto c(t)/t$ and $t \mapsto p(t)/t^2$ are continuous at zero, where the concentration of C and P at time t is denoted by $c(t)$ and $p(t)$, respectively, because the Volpert indices of the species are 0, 0, 1, 2:

```
VolpertIndexing[{"Michaelis-Menten"}, {"E", "S"}].

{{E -> 0, S -> 0, C -> 1, P -> 2},
 {E+S -> C -> 0, C -> E+S -> 1, C -> E+P -> 1}}
```

Now we received the indices of the reaction steps, as well.

Remark 9.3

1. (9.1) formally is not true for unreachable species.

2. (9.1) can be rephrased that the order of the solution is at least κ_m. The next statement is that it is exactly κ_m in case of first-order reactions.

Theorem 9.4 *Suppose that the reaction is of first order. Then, for a species $m \in \mathcal{M} \setminus \mathcal{M}_0$ reachable from $\mathcal{M}_0$ and having the index κ_m the concentration $c_m(t)$ has order of zero equal to κ_m, i.e., one has $d_m(0) \neq 0$ in this case.*

Example 9.5 Consider the consecutive reaction A $\longrightarrow$ B $\longrightarrow$ C, and let, as usual, $\mathcal{M}_0 := \{A\}$. Then the theorem says that close enough to zero $t \mapsto \frac{b(t)}{t}$ and $t \mapsto \frac{c(t)}{t^2}$ are continuous functions with nonzero values at zero, where the concentration of B and C at time t is denoted by $b(t)$ and $c(t)$, respectively.

9.3   Behavior at the End: Omega-Limit Sets

Volpert and Ivanova (1987) offer the following procedure to determine the ω-limit sets of trajectories of the solutions of induced kinetic differential equations: Starting from the reaction, they construct another (simpler, "smaller") reaction with trajectories having the same ω-limit sets. First, we need a definition.

Definition 9.6 The reaction rate w_r is said to be **summable** if for all bounded solutions $\mathbb{R}_0^+ \ni t \mapsto \mathbf{c}(t)$ of the induced kinetic differential equation

$$\int_0^{+\infty} w_r(\mathbf{c}(s))\, ds < +\infty \tag{9.2}$$

holds. Reaction step r is said to be summable if (9.2) holds for all reaction rates which fulfil Conditions 2 and are constant, if the reactant complex is the zero complex.

A simple sufficient condition of summability can be formulated, the proof of which is given as the solution of Problem 9.2. The above authors also give an algorithm to find all the summable reaction steps. First, consider all the species which occur in at least one reactant complex but do not occur in any product complex. Discard the reaction steps containing such species. Then, repeat the procedure. At the end, the discarded steps will be exactly those which are summable.

The algorithm shows that in a circuit-free compartmental system (and also in a circuit-free generalized compartmental system in the narrow sense), all the steps are summable, whereas none of the reaction steps in the usual Michaelis–Menten reaction, Lotka–Volterra reaction, or the reversible bimolecular reaction A + B $\rightleftharpoons$ C are summable. In the Robertson reaction, the rate of all the reaction steps turns out to be summable.

Definition 9.7 The ω-**limit graph** of a reaction is obtained from the Volpert graph by discarding all the summable reaction steps together with the edges adjacent to these.

Now the main assertion of the authors follows.

Theorem 9.8 *Consider those solutions of the kinetic differential equation which are defined for all nonnegative times and are bounded. The ω-limit set of these solutions is a connected invariant set consisting of the trajectories of the reaction having ω-limit graph as its Volpert graph.*

In the same review paper, the authors also deal with the ω-limit set and stability of stationary points on the boundary of the stoichiometric compatibility classes. Another general statement on **strict Lyapunov exponents** follows.

Theorem 9.9 *If $\lim_{t \to +\infty} \mathbf{c}(t)$ exists for the solution $\mathbf{c}$ of the induced kinetic differential equation, then for all $m \in \mathcal{M}$ the strict Lyapunov exponent $\lim_{t \to +\infty} \frac{1}{t} \ln |c_m(t)|$ also exists.*

The authors also cite an algorithm to calculate the strict Lyapunov exponent.

9.4   Transient Behavior: On the Quasi-Steady-State Hypothesis

Probably the oldest symbolic approximation method is the **quasi-steady-state hypothesis**, or **quasi-steady-state assumption**, or using again another name: the **Bodenstein principle**. The physical basis of this hypothesis is that in many reactions, one or more of the species are produced and consumed much more quickly than the others; therefore it is not a bad idea to consider the concentration of the slowly varying species (the Bodenstein species) to be nearly constant.

   The first reaction to be treated by this assumption was the **Michaelis–Menten reaction** by Briggs and Haldane (1925) and by Michaelis and Menten (1913):

$$E + S \underset{k_{-1}}{\overset{k_1}{\rightleftharpoons}} C \overset{k_2}{\longrightarrow} E + P, \qquad (9.3)$$

where E is an **enzyme** (a catalyst which is protein by composition), S is the material to be transformed by the enzyme named **substrate**, C is a (temporary) **enzyme-substrate complex**, and finally, P is the **product**, the "goal" of the reaction. In this reaction it is the enzyme-substrate complex C which is supposed to be formed and decomposed in a relatively fast way; therefore the concentration of this species is supposed to be approximately constant.

   In 1903, Henri found that enzyme reactions were initiated by a bond between the enzyme and the substrate. His work was taken up by Michaelis and Menten (Michaelis and Menten

1913) who investigated the kinetics of an enzymatic reaction mechanism, invertase, that catalyzes the hydrolysis of sucrose into glucose and fructose. They proposed the above model of the reaction. The specificity of enzyme action is explained in terms of the precise fitting together of enzyme and substrate molecules: the **lock and key hypothesis** by Fischer. Some years later a more precise formulation of the Michaelis–Menten equation was given by Briggs and Haldane (1925). They pointed out that the Michaelis assumption that an equilibrium exists between E, S, and C is not always justified and should be replaced by the assumption that C is present not necessarily at equilibrium, but in a steady state. The resulting equation is of the same form, but the Michaelis constant has a different meaning with respect to the different rate constants. We do not follow the recent developments of the topic; we only show how this argument can be made and has been made rigorous by Heineken et al. (1967), Segel (1988) and Tzafriri and Edelman (2007) based on the singular perturbation theory of Tikhonov (1952).

The usual treatment (actually proposed by Briggs and Haldane 1925) is that one starts from the induced kinetic differential equation of the reaction (9.3):

$$\begin{aligned}
\dot{e} &= -k_1 es + k_{-1}c + k_2 c & \dot{s} &= -k_1 es + k_{-1}c \\
\dot{c} &= k_1 es - k_{-1}c - k_2 c & \dot{p} &= k_2 c.
\end{aligned} \tag{9.4}$$

If one takes the usual initial conditions $e(0) = e_0, s(0) = s_0, c(0) = p(0) = 0$, then the above system implies that

$$e(t) + c(t) = e_0 + c_0 \qquad s(t) + c(t) + p(t) = s_0, \tag{9.5}$$

meaning that the total quantities of enzyme and substrate, respectively (no matter in which form they are present, bounded, or free), are constant. An algorithmic way to arrive at these linear first integrals—which does not rely on chemical intuition—is obtained by calculating the stoichiometric matrix

$$\gamma = \begin{bmatrix} -1 & 1 & 1 \\ -1 & 1 & 0 \\ 1 & -1 & -1 \\ 0 & 0 & 1 \end{bmatrix}$$

and finding its left null space via solving the linear algebraic equation $\varrho^\top \gamma = 0$, or

$$\begin{bmatrix} \varrho_1 & \varrho_2 & \varrho_3 & \varrho_4 \end{bmatrix} \cdot \begin{bmatrix} -1 & 1 & 1 \\ -1 & 1 & 0 \\ 1 & -1 & -1 \\ 0 & 0 & 1 \end{bmatrix} = \begin{bmatrix} 0 \\ 0 \\ 0 \\ 0 \end{bmatrix},$$

and getting $\varrho_1 + \varrho_2 = \varrho_3$ $\varrho_1 + \varrho_4 = \varrho_3$. Two linearly independent solutions can be obtained in many different ways, e.g., they are $\begin{bmatrix} 1 & 0 & 1 & 0 \end{bmatrix}^\top$ and $\begin{bmatrix} 0 & 1 & 1 & 1 \end{bmatrix}^\top$ corresponding to the linear combinations (9.5) above. Note that the generating vectors of linear first integrals might be a larger set than the left null space of γ, see Theorem 8.28.

By issuing the command **GammaLeftNullSpace**, we obtain another, but equivalent, representation of the null space in question.

GammaLeftNullSpace[{mm}, {e0, s0, 0, 0}, {e, s, c, p}]

gives

{-e + p + s == -e0 + s0, c + e == e0}.

With these, (9.4) can be reduced to

$$\dot{s} = -k_1 s(e_0 - c) + k_{-1} c \quad \dot{c} = k_1 s(e_0 - c) - (k_{-1} + k_2)c. \tag{9.6}$$

Using the assumption that $\dot{c} = 0$ one gets

$$c = \frac{k_1 e_0 s}{k_{-1} + k_2 + k_1 s}, \tag{9.7}$$

and for the production rate of the product P:

$$\dot{p} = k_2 \frac{k_1 e_0 s}{k_{-1} + k_2 + k_1 s} = \frac{k_2 e_0 s}{K_M + s}, \quad \text{where } K_M := \frac{k_{-1} + k_2}{k_1}$$

is the **Michaelis constant**.

If the usual assumption $e_0 \ll s_0$ is made, then $e(t) \approx e_0$; therefore the (initial) formation rate of P is approximately $\frac{k_2 e s}{K_M + s}$.

This is the usual approach widespread among the chemists (see, e.g., Keleti 1986). But if one requires the concentration c in Eq. (9.7) to be really constant, i.e., having zero derivative, then one arrives at the constant solution $s = 0, c = 0$ corresponding to the stationary point $\begin{bmatrix} e_0 & 0 & 0 & s_0 \end{bmatrix}^\top$. (Check it.)

A more precise approach formulated in a more general setting is obtained in the following way. Consider the differential equation

$$\dot{x}(t) = f(x(t), y(t)) \quad \dot{y}(t) = g(x(t), y(t)) \tag{9.8}$$

and suppose that the (algebraic) equation $g(x_*, y_*) = 0$ has a single solution for y_* with any x_*, i.e., there exists a function φ such that $g(x_*, \varphi(x_*)) = 0$. The question is if one uses the differential equation

$$\dot{x}(t) = f(x(t), \varphi(x(t))) \tag{9.9}$$

to describe approximately the process obeying (9.8), is it a good approximation, what is its error, etc. To make the thing even more complicated, we ask these types of questions: Does the solution of (9.9) approximate the solution of

$$\dot{x}(t) = f(x(t), y(t)) \quad \varepsilon \dot{y}(t) = g(x(t), y(t)) \tag{9.10}$$

if ε is a small enough positive number? The small parameter ε is to counterbalance the large values of $\dot{y}(t)$, the rate of the **fast variable**.

The mathematical tool called **singular perturbation theory** needed here has been formulated by Tikhonov and his coworkers (Tikhonov 1952) and has first been applied to the Michaelis–Menten reaction by a founding father of mathematical chemistry, Rutherford Aris, and his coworkers (Heineken et al. 1967). Now after such a long verbal introduction, we present the theory very shortly, based partially on the following papers: Klonowski (1983), Segel (1988), Segel and Slemrod (1989), Schnell and Maini (2003), Turányi et al. (1988), Turányi and Tóth (1992) and Zachár (1998). We also mention that recently, combination of algebraic and analytic tools opened a new approach (see, e.g., Goeke and Walcher 2014; Goeke et al. 2015).

Let us start from $\dot{s} = -k_1 s(e_0 - c) + k_{-1} c,\quad \dot{c} = k_1 s(e_0 - c) - (k_{-1} + k_2)c$, and let us introduce new (**dimensionless**) variables by

$$S := \frac{s}{s_0}, \quad C := \frac{c}{\frac{e_0 s_0}{s_0 + K_M}}, \quad T := t k_1 e_0.$$

Now the transformed equation is

$$S' = -S(1 - \alpha C) + \beta(1 - \alpha)C, \tag{9.11}$$

$$\mu C' = S(1 - \alpha C) - (1 - \alpha)C, \tag{9.12}$$

—with $\alpha := s_0/(s_0 + K_M)$, $\beta := k_{-1}/(k_{-1} + k_2)$, $\mu := e_0/(s_0 + K_M)$—where all the terms are of the same magnitude. If μ is small ($e_0 \ll s_0 + K_M$), then (9.12) is close to

$$0 = S(1 - \alpha C) - (1 - \alpha)C. \tag{9.13}$$

Now, instead of the two initial conditions $S(0) = 1$ and $C(0) = 0$, we have only the freedom to choose $S(0)$, and $C(0)$ should be calculated from (9.13) to be 1. Now we can apply Theorem 13.50, because its conditions are fulfilled. Here

$$\varphi(S) = \frac{S}{1 + \alpha(S - 1)} = \frac{S}{\alpha S + (1 - \alpha)}. \tag{9.14}$$

There are a few practical problems with this kind of approach. First, in more complicated reactions, φ cannot be easily calculated. Second, it is far from being trivial which parameter (or parameter combination) can be chosen as small (see, e.g., Segel 1988; Segel and Slemrod 1989 and also Goeke and Walcher 2014; Goeke et al. 2015).

Nevertheless, let us substitute $\varphi(S)$ into (9.11) for C to get

$$S' = -S\left(1 - \alpha\frac{S}{1 + \alpha(S - 1)}\right) + \beta(1 - \alpha)\frac{S}{1 + \alpha(S - 1)} \tag{9.15}$$

$$= -\frac{(\alpha - 1)(\beta - 1)S}{\alpha(S - 1) + 1}, \tag{9.16}$$

a separable equation which can easily be solved (although not explicitly) to get

$$-(1-\alpha)(1-\beta)T = \alpha(S(T)-1)+(1-\alpha)\ln(S(T)). \qquad (9.17)$$

However, *Mathematica* gives the explicit solution in terms of the `ProductLog` function—which is defined as giving the principal solution to the equation $z = we^w$—as

$$S(T) = \frac{1-\alpha}{\alpha}\texttt{ProductLog}\left(\frac{\alpha}{1-\alpha}e^{T(\beta-1)+\frac{\alpha}{1-\alpha}}\right),$$

after some nonautomatic simplifications. The pair $T \mapsto (S(T), C(T))$ is called the **outer solution**. Now we try to approximate the behavior of the solution at the beginning, and this will be the **inner solution**.

As a first step, the time will be rescaled by the parameter μ in such a way that the time variable will be $\vartheta := T/\mu$. Then the equations will be transformed into

$$\dot{S} = \mu(-S(1-\alpha C)+\beta C(1-\alpha))$$
$$\dot{C} = S(1-\alpha C)-C(1-\alpha).$$

If $\mu = 0$, then $\dot{S} = 0$; thus $S(\vartheta) = s_0 = 1$. Putting this into the second equation, one gets $\dot{C} = 1 - C$; thus $C(\vartheta) = 1 - e^{-\vartheta}$. The two approximations together with the "exact" (numerically calculated) functions can be seen in Fig. 9.1.

9.5   Reduction of the Number of Variables by Lumping

Throughout the present chapter, we are showing methods for approximate simplification of models of chemical kinetics. One family of such methods consists of transformations of the models to simpler forms. Exact transformations have been met elsewhere in the book (see, e.g., Sects. 6.5.3 and 8.4.2). A method to approximate the solutions by reducing the number of variables has been shown in the previous section. Here we add another useful approximate method.

A major problem with kinetic differential equations is the very large number (dozens, hundreds, sometimes even thousands) of variables. This is especially so with practical problems in the fields of combustion, metabolism, and atmospheric chemistry. Therefore it may be useful from the numerical point of view if one is able to use a model with much less variables. Furthermore, a model with a few number of variables may also show much more clearly the essence of the phenomenon.

There are a lot of methods to reduce the number of (dependent) variables, the best known being linear and nonlinear **lumping** originally initiated by Wei and Kuo in the 1960s of the twentieth century (Wei and Kuo 1969; Kuo and Wei 1969).

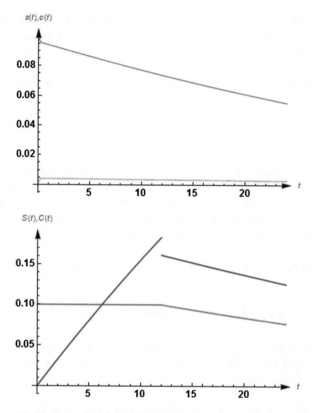

Fig. 9.1 Concentrations in the Michaelis–Menten reactions, exact and approximate. The parameters are as follows: $k_1 = 40$, $k_{-1} = 5$, $k_2 = 0.5$, $e(0) = 0.01$, $s(0) = 0.1$, $c(0) = p(0) = 0$

Before going into details, let us mention that the same concept is known in other fields of science as well, sometimes under different names, such as **aggregation** and **dynamic factor analysis** in econometry, **coarse graining** in physics, etc.

Consider the equation

$$\dot{\mathbf{x}} = \mathbf{f} \circ \mathbf{x} \tag{9.18}$$

with $\mathbf{f} \in \mathscr{C}^1(\mathbb{R}^M, \mathbb{R}^M)$. Let $\hat{M} \leq M$ be a natural number, and let us try to find out if there exists a function $\mathbf{h} \in \mathscr{C}(\mathbb{R}^M \times \mathbb{R}^{\hat{M}})$ and a function $\hat{\mathbf{f}} \in \mathscr{C}(\mathbb{R}^{\hat{M}} \times \mathbb{R}^{\hat{M}})$ so that an autonomous differential equation

$$\frac{d}{dt}\hat{\mathbf{x}} = \hat{\mathbf{f}} \circ \hat{\mathbf{x}} \tag{9.19}$$

describes the time evolution of $\hat{\mathbf{x}} := \mathbf{h} \circ \mathbf{x}$.

Definition 9.10 The function **h** above is said to be a (nonlinear) **lumping function**, the Eq. (9.18) is **(exactly) nonlinearly lumped** to the **lumped equation** (9.19), or Eq. (9.19) is expanded to Eq. (9.18).

Let us see an example why the problem is less trivial than it seems to be.

Example 9.11 Consider the differential equation

$$\dot{x} = ax + by \quad \dot{y} = cx + dy. \tag{9.20}$$

What can we say about the evolution of $x + y$? Is the time derivative of this function a function of the sum itself, or, does there exist a function $\hat{f}$ so that $(x + y)' = \hat{f} \circ (x + y)$ holds? What we know is that $(x + y)' = (a + c)x + (b + d)y$, and the right-hand side of this equation is only under very special circumstances a function of $x + y$, namely, if and only if $a + c = b + d$ (see Problem 9.9).

Now let us turn to general results.

9.5.1 Exact Linear Lumping of Linear Equations

The case when all of $\mathbf{f}, \mathbf{h}$ and $\hat{\mathbf{f}}$ are linear functions (i.e., $\mathbf{f}(\mathbf{x}) = \mathbf{Kx}, \mathbf{h}(\mathbf{x}) = \mathbf{Mx}, \hat{\mathbf{f}}(\hat{\mathbf{x}}) = \hat{\mathbf{K}}\hat{\mathbf{x}}$) was treated by Wei and Kuo (1969), pioneers of the topic. We mention their basic results.

Theorem 9.12 *The necessary and sufficient condition of exact linear lumpability of the linear equation*

$$\dot{\mathbf{x}} = \mathbf{Kx} \tag{9.21}$$

with $\mathbf{M} \in \mathbb{R}^{\hat{M} \times M}$ *(where* $\hat{M} \leq M$*) into the linear equation*

$$\frac{d}{dt}\hat{\mathbf{x}} = \hat{\mathbf{K}}\hat{\mathbf{x}} \tag{9.22}$$

is that

$$\mathbf{MK} = \hat{\mathbf{K}}\mathbf{M}. \tag{9.23}$$

holds.

Proof Let us calculate the time derivative of the transformed vector $\hat{\mathbf{x}}$ in two different ways. As

$$\frac{d}{dt}\hat{\mathbf{x}} = \hat{\mathbf{K}}\hat{\mathbf{x}} = \hat{\mathbf{K}}\mathbf{Mx} \quad \text{and} \quad \frac{d}{dt}\hat{\mathbf{x}} = \mathbf{M}\dot{\mathbf{x}} = \mathbf{MKx}, \tag{9.24}$$

the statement follows.   □

Theorem 9.12 helps us check whether a given pair $\mathbf{M}$ and $\hat{\mathbf{K}}$ defines exact linear lumping or not.

Example 9.13 In the case of Eq. (9.20)—assuming $a + c = b + d$—one has

$$\mathbf{K} = \begin{bmatrix} a & b \\ c & d \end{bmatrix}, \quad \mathbf{M} = \begin{bmatrix} 1 & 1 \end{bmatrix}, \quad \hat{\mathbf{K}} = \begin{bmatrix} a + c \end{bmatrix},$$

and really, $\mathbf{MK} = \hat{\mathbf{K}}\mathbf{M}$.

Example 9.14 The question arises if it is possible at all to linearly lump Eq. (9.20) to a single equation. In formulas this means that we are looking for numbers $m_1, m_2, \hat{k} \in \mathbb{R}$ so that

$$\begin{bmatrix} m_1 & m_2 \end{bmatrix} \begin{bmatrix} a & b \\ c & d \end{bmatrix} = \hat{k} \begin{bmatrix} m_1 & m_2 \end{bmatrix} \tag{9.25}$$

holds. But this means nothing else that the vector $\begin{bmatrix} m_1 & m_2 \end{bmatrix}$ should be a left eigenvalue of the coefficient matrix and $\hat{k}$ should be the corresponding eigenvalue.

One would like to find explicit representations of these linear mappings.

Theorem 9.15 *Suppose that $\mathbf{M}$ is of the full rank, and the necessary and sufficient condition (9.23) holds. Then $\hat{\mathbf{K}}$ can be calculated by the formula*

$$\hat{\mathbf{K}} = \mathbf{M}\mathbf{K}\mathbf{M}^\top (\mathbf{M}\mathbf{M}^\top)^{-1}. \tag{9.26}$$

Proof A simple reformulation of (9.23) gives the statement.   □

Example 9.16 To continue our example, we have

$$\hat{\mathbf{K}} = \mathbf{M}\mathbf{K}\mathbf{M}^\top (\mathbf{M}\mathbf{M}^\top)^{-1} = \begin{bmatrix} 1 & 1 \end{bmatrix} \begin{bmatrix} a & b \\ c & d \end{bmatrix} \begin{bmatrix} 1 & 1 \end{bmatrix}^\top \left(\begin{bmatrix} 1 & 1 \end{bmatrix} \begin{bmatrix} 1 & 1 \end{bmatrix}^\top \right)^{-1}$$

$$= \frac{1}{2} \begin{bmatrix} a + b + c + d \end{bmatrix} = \begin{bmatrix} a + c \end{bmatrix} \in \mathbb{R}^{1 \times 1}.$$

Note that Theorem 9.15 above is valid in full generality, not only for deterministic models of first-order reactions. The same is true for the next result which is a generalization of Example 9.14.

Theorem 9.17 (Li and Rabitz (1989)) *Linear differential equations can always be exactly linearly lumped. Appropriate lumping matrices can be obtained via taking some of the left eigenvectors of the coefficient matrix of the original differential equation as row vectors.*

Proof The idea of the proof can be seen in Example 9.14.   □

It may easily happen (see Problem 9.14) that an induced kinetic differential equation when exactly lumped does not remain kinetic. Therefore, it is useful to have conditions to ensure that the lumped system is kinetic as well, in this case we say that the lumping matrix is **kinetic**.

Theorem 9.18 (Farkas (1999)) *A nonnegative lumping matrix is kinetic if and only if it has a nonnegative generalized inverse. (See Definition 13.16.)*

Example 9.19 The matrix $\mathbf{M} = \begin{bmatrix} 1 & 1 \end{bmatrix}$ has a nonnegative generalized inverse $\overline{\mathbf{M}} = \begin{bmatrix} 1/2 & 1/2 \end{bmatrix}^{\top}$ (which is also a Moore–Penrose inverse); therefore $\mathbf{M}$ is a kinetic lumping matrix (in case it is a lumping matrix at all).

The case when all the old species are transformed into exactly one new species (and all the new species contains at least one old species) is called **proper lumping**. This means that the columns of $\mathbf{M}$ are vectors of the standard basis, usually repeated a few times.

Corollary 9.20 *Proper lumping matrices are kinetic.*

Proof If $\mathbf{M}$ is a proper lumping matrix, then $\mathbf{M}\mathbf{M}^{\top}$ is a diagonal matrix with positive integers on its main diagonal; therefore its inverse is of the same form, a positive definite diagonal matrix, and thus it fulfills the condition in Theorem 9.18.   □

Theorem 9.21 *Suppose we have a closed compartmental system which is also detailed balanced. Then, exactly lumping it with a proper lumping matrix, a differential equation is obtained, which can be considered to be the induced kinetic differential equation of a closed compartmental system which is detailed balanced.*

Proof As in this case $\mathbf{M}$ is of the full rank, one can use the representation (9.26). Direct calculation shows that the off-diagonal elements of $\hat{\mathbf{K}}$ are nonnegative, and the diagonal elements are equal to the column sums. For the proof of inheritance of detailed balance, see Wei and Kuo (1969).   □

Example 9.22 The reversible triangle reaction can exactly be lumped by $\mathbf{M} = \begin{bmatrix} 1 & 0 & 0 \\ 0 & 1 & 1 \end{bmatrix}$ if and only if $k_{-1} = k_3$ holds, the lumped variables then are $\hat{x} = x, \hat{y} = y + z$, and the lumped reaction is $\hat{X} \underset{k_3}{\overset{k_1 + k_{-3}}{\rightleftharpoons}} \hat{Y}$. Detailed balancing in the original reaction means $k_1 k_2 k_3 = k_{-1} k_{-2} k_{-3}$, which implies for the lumped reaction $k_1 k_2 = k_{-2} k_{-3}$, a condition which is vacuous here, because the lumped reaction is independent of k_2 and k_{-2} and is detailed balanced no matter what the value of the reaction rate coefficients are.

Fig. 9.2 A square reaction

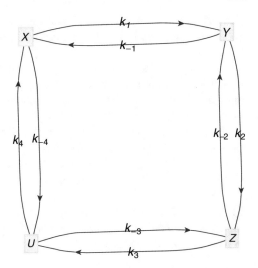

Astonishing it may seem, a detailed balanced reaction can in some cases not be properly lumped.

Example 9.23 Consider the square in Fig. 9.2, and let us try to apply $\mathbf{M} :=$
$\begin{bmatrix} 1\,0\,0\,0 \\ 0\,1\,0\,0 \\ 0\,0\,1\,1 \end{bmatrix}$ as lumping matrix. Then the condition (9.23) implies that $k_{-2} = k_4 = 0$;
thus the reaction is not even weakly reversible. Neither will the obtained three species triangle reaction be a weakly reversible reaction, furthermore the lumped reaction will not depend on the reaction rate coefficients k_3 and k_{-3}; this is quite natural, because if one is going to lump Z and U into one species, then the traffic between these to species is not relevant.

9.5.2 Exact Linear Lumping of Nonlinear Equations

Let us define exact linear lumping for autonomous differential equations in general. Consider the Eq. (9.18) with $\mathbf{f} \in \mathscr{C}(\mathbb{R}^M, \mathbb{R}^M)$. Let $\hat{M} \leq M$ be a natural number, and let us try to find out, if there exists a matrix $\mathbf{M} \in \mathbb{R}^{\hat{M} \times M}$ of full rank so that an autonomous differential equation

$$\frac{\mathrm{d}}{\mathrm{d}t}\hat{\mathbf{x}} = \hat{\mathbf{f}} \circ \hat{\mathbf{x}} \tag{9.27}$$

describes the time evolution of $\hat{\mathbf{x}} := \mathbf{Mx}$.

Definition 9.24 The matrix $\mathbf{M}$ above is said to be a **lumping matrix**, the Eq. (9.18) is **(exactly) linearly lumped** to the **lumped equation** of (9.27).

One can say that exact linear lumping can also serve as a model for the situations in which instead of the individual concentrations of species, we can only measure the sum, or more generally, the linear combination of the concentrations. Such is the case, e.g., with methods of spectroscopy.

Exact linear lumping has been investigated by Li (1984) for the case of first- and second-order reactions and extended by Li and Rabitz (1989) to the general case. Practically, it is the derivative of the right-hand side what takes over the role of the coefficient matrix of the linear case.

Theorem 9.25 (Li and Rabitz (1989)) *With the above notations, the necessary and sufficient condition of exact linear lumpability of (9.18) by the matrix $\mathbf{M}$ is that any of the following equalities hold:*

$$\mathbf{M}\mathbf{f}(\mathbf{x}) = \mathbf{M}\mathbf{f}(\overline{\mathbf{M}}\mathbf{M}\mathbf{x}) \tag{9.28}$$

$$\mathbf{M}\mathbf{f}'(\mathbf{x}) = \mathbf{M}\mathbf{f}'(\overline{\mathbf{M}}\mathbf{M}\mathbf{x})\overline{\mathbf{M}}\mathbf{M} \tag{9.29}$$

$$\mathbf{M}\mathbf{f}'(\mathbf{x}) = \mathbf{M}\mathbf{f}'(\overline{\mathbf{M}}\mathbf{M}\mathbf{x}) \tag{9.30}$$

with all the generalized inverses $\overline{\mathbf{M}}$ of the matrix $\mathbf{M}$ for which $\mathbf{M}\overline{\mathbf{M}} = \mathrm{id}_{\hat{M}}$ holds.

Theorem 9.26 (Li and Rabitz (1989)) *Suppose $\mathbf{f}'(\mathbf{x})$ has a nontrivial invariant subspace $\mathcal{M}$, independent of $\mathbf{x}$, and let the rows of $\mathbf{M}$ be the basis vectors of $\mathcal{M}$. If the matrix $\mathbf{M}$ obtained this way fulfils any of the necessary and sufficient conditions (9.28), (9.29) and (9.30) and the eigenvalues of $\mathbf{f}'(\mathbf{x})^\top$ and $\mathbf{f}'(\overline{\mathbf{M}}\mathbf{M}\mathbf{x})^\top$ are the same, then $\mathbf{M}$ is an exact lumping matrix.*

Let us turn to exact nonlinear lumping, an obvious generalization of exact linear lumping.

9.5.3 Exact Nonlinear Lumping of Nonlinear Equations

We make a few regularity assumptions to make the treatment simpler. Suppose that all the functions are differentiable as many times as needed, they are defined on the whole space, $\mathbf{f}(\mathbf{0}) = \mathbf{0}$, $\mathbf{h}(\mathbf{0}) = \mathbf{0}$, the function $\mathbf{h}$ is nondegenerate in the sense that its coordinate functions are independent, and the solutions to all the initial value problems below are defined for all nonnegative times (what is certainly the case for the important class of mass conserving reactions).

Again, we start with a series of necessary and sufficient conditions of lumpability, and then we formulate the connection between lumpability and the existence of first integrals. Then the effect of lumping on the qualitative properties of the solutions is treated. To describe the attempts to find nonlinear lumping functions would take too much space; thus the interested reader should consult the papers by Li, Rabitz, and their coworkers.

Now we cite the most important statements in the form as they presented by Tóth et al. (1997).

Theorem 9.27 *Equation* (9.18) *can be lumped into* (9.19) *with the lumping function* $\mathbf{h} \in \mathscr{C}^1(\mathbb{R}^M, \mathbb{R}^{\hat{M}})$ *if and only if*

$$\mathbf{h'f} = \hat{\mathbf{f}} \circ \mathbf{h} \tag{9.31}$$

is fulfilled. Furthermore, the representation

$$\hat{\mathbf{f}} = (\mathbf{h'f}) \circ \overline{\mathbf{h}} \tag{9.32}$$

is also valid with any generalized (right) inverse $\overline{\mathbf{h}}$ *of the function* $\mathbf{h}$ *(i.e.,* $\mathbf{h} \circ \overline{\mathbf{h}} = \mathrm{id}_{\mathbb{R}^{\hat{M}}}$*).*

Proof Let us calculate the derivative of the transformed quantity $\hat{\mathbf{x}} := \mathbf{h} \circ \mathbf{x}$ in two different ways.

$$\frac{\mathrm{d}}{\mathrm{d}t}\hat{\mathbf{x}} = \frac{\mathrm{d}}{\mathrm{d}t}(\mathbf{h} \circ \mathbf{x}) = (\mathbf{h'} \circ \mathbf{x}) \cdot \dot{\mathbf{x}} = (\mathbf{h'} \circ \mathbf{x}) \cdot (\mathbf{f} \circ \mathbf{x}) = (\mathbf{h'f}) \circ \mathbf{x}$$

$$\frac{\mathrm{d}}{\mathrm{d}t}\hat{\mathbf{x}} = \hat{\mathbf{f}} \circ \hat{\mathbf{x}} = \hat{\mathbf{f}} \circ (\mathbf{h} \circ \mathbf{x}) = (\hat{\mathbf{f}} \circ \mathbf{h}) \circ \mathbf{x}$$

Applying Lemma 13.34 shows that both formulas (9.31) and (9.32) follow. Independence of the representation from the choice of the generalized inverse is shown in Problem 9.8.   □

Now let us formulate a necessary and sufficient condition without using the transformed right-hand side.

Theorem 9.28 *Equation* (9.18) *can be lumped into* (9.19) *with the lumping function* $\mathbf{h} \in \mathscr{C}^1(\mathbb{R}^M, \mathbb{R}^{\hat{M}})$ *if and only if*

$$\mathbf{h'f} = \mathbf{h'} \circ (\overline{\mathbf{h}} \circ \mathbf{h}) \cdot \mathbf{f} \circ (\overline{\mathbf{h}} \circ \mathbf{h}) \tag{9.33}$$

is fulfilled.

A necessary condition follows, showing that invariant sets (see Sect. 8.6) play an important role here, as well.

Theorem 9.29 *If Eq.* (9.18) *can be lumped into* (9.19) *with the lumping function* $\mathbf{h} \in \mathscr{C}^1(\mathbb{R}^M, \mathbb{R}^{\hat{M}})$, *then the set* $\mathbf{h}^{-1}(\mathbf{0}) = \{\bar{\mathbf{x}} \in \mathbb{R}^M \,|\, \mathbf{h}(\bar{\mathbf{x}}) = \mathbf{0}\}$ *is an* **invariant set** *of* (9.18).

Proof The proof is the same as the proof of Theorem 8.44 but uses (9.31).   □

The next (necessary and sufficient) condition does not contain the generalized inverse of $\mathbf{h}$.

Theorem 9.30 *Equation* (9.18) *can be lumped into* (9.19) *with the lumping function* $\mathbf{h} \in \mathscr{C}^1(\mathbb{R}^M, \mathbb{R}^{\hat{M}})$ *if and only if there exists a matrix valued function* $\mathbf{X} : \mathbb{R}^M \longrightarrow \mathbb{R}^{M \times L}$ $(L \geq M - \hat{M})$ *such that its value is of the rank* $M - \hat{M}$ *at each argument and*

$$\mathbf{h}'(\mathbf{x})\mathbf{X}(\mathbf{x}) = \mathbf{0} \quad (\mathbf{h}'\mathbf{f})'(\mathbf{x})\mathbf{X}(\mathbf{x}) = \mathbf{0}. \tag{9.34}$$

In order to obtain the exact lumping functions, it is enough to find all the invariant functions of (9.18) and choose those which also fulfil any of the necessary and sufficient conditions (9.31), (9.33), or (9.34).

9.5.4 Construction of the Lumping Function

The problem with the exact construction is that it is equivalent to finding global nonlinear first integrals, as we will see below.

Now we learn that an exact lumping function with values in the space $\mathbb{R}^{\hat{M}}$ is the function of $\hat{M}$ independent generalized eigenfunctions. (See Definition 8.42.) In other words, to construct exact lumping functions means to construct generalized eigenfunctions; this last problem however leads to the determination of global first integrals, not an easy task to solve. Before that, we need another definition.

Definition 9.31 The generalized eigenfunction $H \in \mathscr{C}^1(\mathbb{R}^M, \mathbb{R})$ is a **normed generalized eigenfunction** of the Eq. (9.18) (or, of the linear operator $\mathscr{A}$ defined by $\mathscr{A}H := H'\mathbf{f}$) if $\mathscr{A}H = 1$ also holds.

In order to find the normed generalized eigenfunctions of the operator $\mathscr{A}$, one has to determine the M independent solutions to the quasilinear partial differential equation $\mathscr{A}H = 1$. To get those solutions, one has to find such a nontrivial solution of the system of partial differential equations

$$AH = 1 \quad A\phi_m = 0 \quad (m = 1, 2, \ldots, M - 1) \tag{9.35}$$

in which the functions ϕ_m $(m = 1, 2, \ldots, M - 1)$ are independent.

Theorem 9.32 (Li et al. (1994))

1. *All the exact lumping functions* $\mathbf{h} : \mathbb{R}^M \longrightarrow \mathbb{R}^{\hat{M}}$ *of* (9.18) *can be obtained as* $\hat{M}$ *functions of* $\hat{M}$ *independent normed generalized eigenfunctions of the operator* $\mathscr{A}$.
2. *Any* $\hat{M}$ *functions of* $\hat{M}$ *independent normed generalized eigenfunctions of the operator* $\mathscr{A}$ *defines an exact lumping function of the Eq.* (9.18).

If (9.18) can be lumped by the function $\mathbf{h} \in \mathscr{C}^1(\mathbb{R}^M, \mathbb{R}^{\hat{M}})$ to the Eq. (9.19), then, obviously $\mathscr{A}\mathbf{h} = \hat{\mathbf{f}} \circ \mathbf{h}$. We can use this relation to determine $\mathbf{h}$, since this means that the generalized eigenfunctions are scalar-valued lumping functions with the corresponding function $\hat{\mathbf{f}}$ providing the lumped right-hand side. The real problem is that to know M normed generalized eigenfunctions is the same as to know one normed generalized eigenfunction and $M - 1$ global first integrals, and no general methods are known to find these in the general case. What we can do is that we look for normed generalized eigenfunctions (i.e., lumping functions) within a class of functions, e.g., polynomials. Choosing the class of linear functions reduces the problem to that of linear lumping.

We do not cite here methods to find **approximate nonlinear lumping**; we only mention that such a method can be based, e.g., on singular perturbation (Li et al. 1993) and has been applied to the irreversible Michaelis–Menten reaction with the numerical data taken from the trypsin-catalyzed hydrolysis of the benzoyl-L-arginine ethyl ester. This was an application of approximate lumping without constraints. One may wish to leave some species unlumped, and this may be expressed as a constraint. This method of constrained approximate nonlinear lumping has been applied to the a combustion model of hydrogen in a closed vessel. Another application of lumping to a physically based pharmacokinetic model (PBPK) can be found in Brochot et al. (2005).

In Tóth et al. (1997) we have formulated a series of open problems. One of them has been solved by the late Gyula Farkas (Farkas 1999); others are repeated below in the section of Open Problem.

Let us also mention the papers on the connection between local and global controllability and lumping by Farkas (1998a,b) and by Horváth (2002–2008).

9.5.5 Lumping and the Qualitative Properties of Solutions

Now let us investigate what an effect lumping has on the qualitative properties of solutions. First, following Tóth et al. (1997), we formulate a few obvious statements.

Theorem 9.33 *If* $\mathbf{f}$ *has the Lipschitz property and* $\mathbf{h}$ *is a nondegenerate lumping function (Tóth et al. 1997, p. 1534), then the right-hand side defined by* (9.32) *also has the Lipschitz property.*

Theorem 9.34 *Exact linear lumping does not increase the degree of a polynomial right-hand side.*

Theorem 9.35 *Under lumping, the image of*

- *a (positively) invariant set of the state space is a (positively) invariant set,*
- *a stationary point is a stationary point,*
- *a closed trajectory is a closed trajectory.*

Let us see some unpleasant counterexamples.

Example 9.36

- Adding up the concentrations in the Ivanova reaction

$$X + Y \longrightarrow 2\,Y \quad Y + Z \longrightarrow 2\,Z \quad Z + X \longrightarrow 2\,X$$

  an induced kinetic differential equation is obtained having constant solutions only.
- The induced kinetic differential equation of the reaction

$$X \longrightarrow 2\,X \quad Y \longrightarrow X + 2\,Y \quad X \longrightarrow X + 2\,Z \quad Z \longrightarrow Y + Z$$

  has solutions with monotonous coordinate functions, but lumping by the function $h(p, q, r) := (p-q, p-r)$ the (nonkinetic) differential equation of the harmonic oscillator is obtained having only periodic solutions.

An easy consequence of the Lemma 13.18 is that at a stationary point which has been obtained by lumping, all the eigenvalues of the Jacobian are eigenvalues of the Jacobian of the original right-hand side. In case of linear lumping, the statement remains valid for nonstationary points too, but this may not be so with nonlinear lumping. Thus we have the following statements:

Theorem 9.37
- *Sources and sinks are lumped into sources and sinks, respectively.*
- *If the stationary point of the original equation as a function of some parameter is always hyperbolic in the sense that the corresponding eigenvalue never has zero real part, then one cannot have Andronov–Hopf bifurcation in the lumped equation.*
- *If one finds Andronov–Hopf bifurcation in the lumped equation when changing a certain parameter, then the situation is the same in the original equation at the corresponding value of the parameter.*

An asymptotically stable stationary point can be lumped into an unstable one. What is more, lumping can produce blow-up (see Problem 9.16). We also have a positive result.

Theorem 9.38 *A solution of* (9.18) *which does not blow up is lumped into a solution of* (9.19) *which does not blow up either. If a solution of the lumped system blows, then it is the image of a blowing up solution.*

Since our aim is to investigate the smaller system instead of the larger one, the most important statements are those which start from the properties of the lumped system and make conclusions about the properties of the original system. Such a statement follows.

Theorem 9.39 *If the lumped stationary point is (asymptotically) stable, then the corresponding stationary point is relatively (asymptotically) stable.*

The original concept of lumping has been extended to reaction-diffusion models (Li and Rabitz 1991; Rózsa and Tóth 2004) and also time discrete models, both deterministic and stochastic (Tóth et al. 1996; Iordache and Corbu 1987). Lumping stochastic models is closely related to **dynamic factor analysis** (see, e.g., Bolla 2013; Bolla and Kurdyukova 2010; Forni et al. 2015).

9.5.6   Observability and Controllability and Lumping

One may also wish to know how controllability and observability change as a result of lumping. The statements below by Horváth (2002–2008) and Horváth (2002) solves this problem.

Theorem 9.40 *Let us assume that* (9.21) *is exactly lumpable to* (9.22) *by* **M** *and the linear system* (9.21) *is completely controllable. Then the lumped system* (9.22) *is also completely controllable.*

Problem 9.17 shows that complete controllability of the lumped system does not imply complete controllability of the original one.

Theorem 9.41 *Let us assume that* (9.21) *is exactly lumpable to* (9.22) *by* **M** *and that for every eigenvalue* λ *of* **K** *holds that all partial multiplicities of* λ *are equal to unity. Furthermore let us consider system* (9.22) *with the observation function*

$$\hat{y}(t) = \hat{C}\hat{x}(t). \tag{9.36}$$

Then if the linear system (9.21) *is completely observable with the observation function* $y(t) = Cx(t)$, *then the linear system* (9.22), (9.36) *is also completely observable. (Here* $\hat{C} \in \mathbb{R}^{p \times k}$ *and* $\hat{C} := C\tilde{M}^\top$.)

The main goal of Farkas (1998b) is to provide sufficient conditions to guarantee local observability. Similarly, Farkas (1998a) gives sufficient conditions for local controllability of reactions. Coxson (1984) establishes a relationship between lumpability and observability of the lumped observation system for both finite and infinite dimensional linear systems (continuous species).

9.6   Numerical Approximations

Problems of reaction kinetics are usually nontrivial from the computational point of view. In the present chapter, we try to collect and present some of those we have met during writing and using the package **ReactionKinetics**. What is left are the problems of linear algebra and linear programming, the solution of nonlinear equations, and finally, the most important one: the solution of systems of nonlinear differential equations. The different methods of stochastic simulation will be treated in Sect. 10.7.

One of the main advantages of implementing **ReactionKinetics** in a computer algebra system such as *Mathematica* is that the underlying system provides us with a usually reliable collection of numerical and symbolic computational toolset, effectively hiding most of the computational difficulties. Therefore what we have done, and propose to our readers to do, is to rely on the built-in numerical and symbolic methods of *Mathematica* as long as they are efficient enough for your purposes and accept the answers given by built-in functions as far as they are not suspicious. (*Mathematica* also does some self-testing before returning solutions, e.g., of ill-conditioned systems of linear equations, and gives a warning if there are doubts about the correctness of the solutions.) If one thinks that something might be wrong, then one might try to apply options, such as algorithm selection, of the functions. If this does not help either, then one should turn to the literature and to experts of numerical mathematics—but not before. To help this itinerary, we formulated our code in such a way that the built-in options of the used *Mathematica* functions are transferred to our functions. Let us see a graphical example. (A numerical example will be shown in Sect. 9.6.4.)

We can use **ShowFHJGraph[{"Triangle"}, {k_1, k_2, k_3}]** with only the required arguments of our function **ShowFHJGraph** (see Fig. 9.3). However, if we want to have a more informative figure, then we can utilize the fact that **ShowFHJGraph** is based on **GraphPlot** and is able to use its options, as in the next example:

```
ShowFHJGraph[{"Triangle"}, {k1, k2, k3},
    DirectedEdges -> True, VertexLabeling -> All,
    ImageSize->400,
    StronglyConnectedComponentsColors->{Pink}],
```

(see Fig. 9.3). (Each optional argument but the last one is a built-in option of the function **GraphPlot**. The last row contains an option which was introduced by us. The options can be given in any order.)

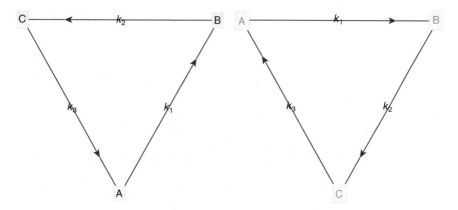

Fig. 9.3 Feinberg–Horn–Jackson graph of the triangle reaction without and with built-in options

After this short introduction, let us analyze the problems of the mentioned areas as they arise in reaction kinetics in a more systematic way.

9.6.1 Numerical and Symbolic Linear Algebra

Symbolic and numerical problems of linear algebra recurrently arise; here we give an example of calculating the stationary point of a compartmental model abbreviated as **jr** and investigated by Ross (2008), p. 2136:

$$0 \longrightarrow X_1 \rightleftharpoons X_2 \rightleftharpoons X_3 \rightleftharpoons X_4 \rightleftharpoons X_5 \rightleftharpoons X_6 \rightleftharpoons X_7 \rightleftharpoons X_8 \longrightarrow 0.$$

Expression **StationaryPoint[jr]** gives the result symbolically (based upon **LinearSolve**). We do not reproduce the result *verbatim* because of the length of the formulae, but here is an equivalent symbolic form:

$$c_i^* = k_0 \sum_{j=1}^{9-i} \frac{\left(\prod_{l=10-j}^{8} k_l\right) \left(\prod_{l=i}^{8-j} k_{-l}\right)}{\prod_{l=i}^{8} k_l} \quad (i = 1, 2, \ldots, 8).$$

The verification of this result can be done in the following way:

```
rhs=RightHandSide[
    {0 -> X1 <=> X2 <=> X3 <=> X4 <=>
        X5 <=> X6 <=> X7 <=> X8 -> 0},
    Join[{k[0]}, Flatten[Transpose[{k /@ Range[7],
        k /@ -Range[7]}]], {k[8]}], x /@ Range[8]];
rhs /. {x[i_] -> k[0] Sum[Product[k[l], {l, 10-j, 8}]
    Product[k[-l], {l, i, 8-j}] / Product[k[l],
    {l, i, 8}], {j, 1, 9-i}]} // Simplify
```

The generalization of this claim is left to the reader. Let us note that the result does not depend on the initial concentrations, i.e., this reaction shows absolute concentration robustness (see Sect. 7.10), a fact which does not follow from Theorem in Shinar and Feinberg (2010, p. 1390); because the deficiency of the reaction is zero. A moral of this example may be that some symbolic results obtained by computer may be useless.

For numerical linear algebra, *Mathematica* largely depends on extensions of standard open-source libraries. In particular, an extended precision extension of LAPACK is used for solving systems of linear equations, finding eigenvalues, and computing matrix decompositions. At the time of writing this book, these libraries are continuously maintained and incorporate state-of-the-art algorithms which also exploit the structure and certain properties (such as definiteness) of the underlying matrices. Therefore we use the built-in routines as a black box and expect that we shall not encounter computational difficulties that originate in problems with the numerical linear algebra routines.

9.6.2 Linear Programming

Some functions in the stoichiometric part of the package rely on linear programming. Mass conservation relations can be found by solving linear programs (recall Chap. 4), and some of the algorithms for the decomposition of overall reactions into elementary steps also use linear programming, too (Chap. 5). We have not encountered any serious problems during the development and use of our package, hence only mention a couple of points briefly.

The main linear programming solver in *Mathematica* is an implementation of the well-known **simplex method**. It is capable of both (extended precision) numerical and symbolic calculations using rational arithmetic. This algorithm, with default options, is expected to be sufficient for obtaining mass conservation relations and also to be used in the algorithms of Sects. 5.3.4 and 5.3.5 when searching for decompositions.

A particularly useful property of the simplex method is that when the linear program has multiple optimal solutions, the simplex method returns a "sparser" one (i.e., one with more zero components), than other algorithms, such as interior point methods. This is particularly desirable for the partial enumeration of decompositions discussed in Sect. 5.3.5, as it leads to both simpler decompositions and to a larger number of decompositions. An interior point method for linear programming is also implemented in *Mathematica*, but it is not used in `ReactionKinetics`.

For the linear programming-based enumeration of every decomposition presented in Sect. 5.3.3, another feature of the simplex method is very useful. In this decomposition algorithm, a multitude of nearly identical linear programs are solved; each is obtained by adding one constraint to one of the previously solved linear programs. Linear programs obtained this way need to be solved "from scratch," but their solution can be accelerated using information gained during the solution of the previous linear program, using an algorithm called the **dual simplex method**.

At present, the linear programming solver in *Mathematica* does not support this feature, although most industry standard linear programming solvers do.

It should also be mentioned that recently *Mathematica*'s built-in `Reduce` function has been extended to handle the solution of linear Diophantine equations and may be used to decompose reactions in the case when there are only finitely many solutions. This can be done using the `Method->"Reduce"` option in `Decompositions`. (Note that this method cannot be used to find only minimal decompositions.)

9.6.3 Nonlinear Equations

As we mentioned in Chap. 7, the first step in finding the stationary points of an induced kinetic differential equation is solving the equation $f(c^*) = 0$, where f is the right-hand side of the induced kinetic differential equation of the given reaction. Here f is a multivariate polynomial; thus we are interested in the roots of a polynomial. Knowing the facts that

- finding the roots of two quadratic two variable polynomials leads to the solution of a sixth-degree polynomial equation in one variable
- no **general** formula exists to express the roots of higher than fourth order one variable polynomials,

suggests that the symbolic solution of this problem is formidable.

Example 9.42 Suppose that we have a (large) system of linear algebraic equations, i.e., an equation for a multivariate linear polynomial, then the usual method proposed by Gauss consists in transforming our equation into triangular form and solving recursively each of the resulting equations, always for a single variable. Thus, instead of the equation

$$-x + 3y + 3z = 2$$
$$3x + y + z = 4$$
$$2x - 2y + 3z = 10$$

one solves the equivalent triangular system

$$-x + 3y + 3z = 2$$
$$y + z = 1$$
$$5z = 10$$

by first calculating $z = 2$ from the last equation, then using this knowledge to calculate $y = 1 - z = -1$ from the penultimate one, and finally, one gets $x = -2 + 3y + 3z = -2 - 3 + 6 = 1$.

An important recent development initiated by Buchberger (2001) is that **systems of polynomial equations with finitely many solutions can also be triangularized.** (See also Lichtblau 1996.) This does not contradict the fact that the roots of higher than fourth-order polynomials cannot be found symbolically in general. But one can find the (finitely many) roots of the emerging one variable polynomial either symbolically or numerically and substitute these roots one by one into the penultimate, two-variable polynomial to get the possible values of the next variable and continue in the same manner working backward until all solutions are enumerated.

The key concept in obtaining the triangular form is that of the **Gröbner basis**. Very loosely speaking, a Gröbner basis of a system of polynomials is another, "equivalent" system with some favorable properties that make the solution of several algebraic problems involving these polynomials (solving the corresponding system of equations is only one of them) easier to solve. In *Mathematica*, Gröbner bases can be obtained using the function **GroebnerBasis**, which is used also by other *Mathematica* functions such as **Solve, Reduce, Resolve, Eliminate,** and **FindInstance** for computations involving multivariate polynomials.

Let us see an example.

Example 9.43 The induced kinetic differential equation of the reaction

$$X \longrightarrow 2X \quad X \longrightarrow X + Y \longrightarrow Y \quad 2Y \longrightarrow Y$$

is $\dot{x} = x - xy$, $\dot{y} = x - y^2$; therefore to find the stationary points, one has to solve the system of polynomial equations $0 = x - xy$, $0 = x - y^2$. As

```
GroebnerBasis[{x - x y, x - y^2}, {x, y}]
```

gives

```
{-y^2 + y^3, x - y^2},
```

we can easily find the stationary points by first finding the (two) solutions of $-y^2 + y^3 = 0$ and substituting them into $x - y^2 = 0$.

Gröbner bases are not unique and are not necessarily "triangular," which is shown by the fact that just by changing the order of the two variables and computing **GroebnerBasis[{x-x y, x-y²}, {y, x}]**, we get an entirely different basis—in fact, even the *number of polynomials* of the bases and the degrees of the polynomials are different. Of course, we get the same stationary points either way, and these are the same as those what we would obtain by simply invoking

```
Reduce[{x - x y == 0, x - y^2 == 0}].
```

Gröbner bases can sometimes be very difficult and time-consuming to compute even for a small number of polynomials of degrees as low as three or four. See Problem 9.7 for an example. The interested reader may find more information on Gröbner bases in algebraic geometry textbooks, such as Cox et al. (2004).

9.6.4 Ordinary Differential Equations

We are usually interested in the solutions of the induced kinetic differential equation of a reaction which can rarely be calculated symbolically using **DSolve** and more often than not, numerically, using **NDSolve**.

9.6.4.1 Black Box Use

Take as an example, to illustrate the first case, the triangle reaction.

```
Concentrations[{Triangle},
    {k1, k2, k3}, {a0, b0, c0}]
```

uses only **DSolve** (note that neither the time interval is given, and nor numerical values are provided for the parameters). We do not care how the result is obtained; we only use it.

9.6.4.2 Use of Built-in Options of Mathematica

When solving induced kinetic differential equations, first we try to use **NDSolve** without any external intervention. If the result is not convincing or appropriate, then we may try to use some of the built-in options of *Mathematica* similarly as we have shown above. Let us consider a version of the Oregonator, a model for the Belousov–Zhabotinsky reaction (Deuflhard and Bornemann 2002, p. 17–18):

$$BrO_3^- + Br^- \xrightarrow{134/100} HBrO_2 \quad HBrO_2 + Br^- \xrightarrow{16 \times 10^8} P \quad Ce(IV) \xrightarrow{1} Br^-$$

$$HBrO_2 + BrO_3^- \xrightarrow{8 \times 10^3} Ce(IV) + 2\,HBrO_2 \quad 2\,HBrO_2 \xrightarrow{4 \times 10^7} P.$$

Figure 9.4 shows the concentration vs. time curve of $HBrO_2$ without additional options, only using the built-in function **NDSolve**:

```
cin = Concentrations[or={"A"+"Y" -> "X", "X"+"Y" -> "P",
        "A"+"X" -> 2"X"+"Z", 2"X"->"P", "Z"->"Y"},
    {134/100, 16*10^8, 8*10^3, 4*10^7, 1},
    {5*10^(-1), 10^(-6), 10^(-8), 6*10^(-2), 10^(-4)},
    {0, 100}, {a, y, x, p, z}];
Plot[Log[x[t]/.Last[cin]], {t, 0, 100}, PlotRange->All,
    AxesLabel ->
    (Style[#, Bold, 12]& /@ {"t", "x(t)"})]
```

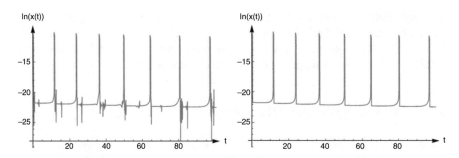

Fig. 9.4 Logarithm of the concentration of HBrO$_2$ vs. time without and with options $[BrO_3^-](0) = 5 \times 10^{-1}, [Br^-](0) = 10^{-6}, [HBrO_2](0) = 10^{-8}, [P](0) = 6 \times 10^{-2}, [Ce(IV)](0) = 10^{-4}$

and also using a few options and changing the default values of some parameters

```
con = Concentrations[or,
    {134/100, 16*10^8, 8*10^3, 4*10^7, 1},
    {5*10}^(-1), 10^(-6), 10^(-8), 6*10^(-2), 10^(-4)},
    {0,100}, {a, y, x, p, z},
    Method -> "BDF", WorkingPrecision -> 32,
    MaxSteps->10^6];
Plot[Log[x[t] /. Last[con]], {t, 0, 100},
    PlotRange -> All,
    AxesLabel ->
    (Style[#, Bold, 12]& /@ {"t", "x(t)"})]
```

give a much more reliable result.

Let us note that Deuflhard and Bornemann (2002) had to be supplemented with initial concentrations and time to be reproducible.

An early review on the requirements toward numerical methods and codes to be applied in chemical kinetics has been presented by Dahlquist et al. (1982).

9.6.4.3 Stiffness
The phenomenon causing a numerical problem most often when solving induced kinetic differential equations of reactions is **stiffness**. This means that there are orders of magnitude differences between the rate of changes of different concentrations. If this is the case, then a simple numerical method does not know how to choose the step size. If it is too large, then fine changes in the fast variable will be missed; if it is chosen to be too small, then the process of integration may take very long time. This is the case especially with oscillatory reactions: when one of the species concentrations is around its extremum with almost zero derivative, the other one goes through the largest change (Fig. 9.5).

Stiffness originating in chemical kinetics urged mathematicians, first of them Gear (1992) to devise numerical methods for solving such problems. Since then many other methods have been introduced and are in use; one of them is the

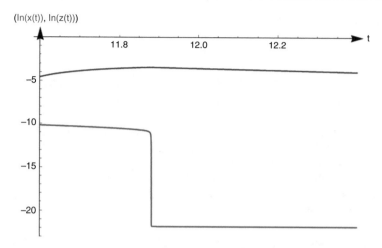

Fig. 9.5 Logarithm of the concentrations of $HBrO_2$ (red) and $Ce(IV)$ (line) vs. time (Deuflhard and Bornemann 2002, pp. 17–18)

Backward Differentiation Formula (BDF) used above, which is particularly useful for stiff differential equations and differential algebraic equations.

Darvishi et al. (2007) propose the variational iteration method to solve stiff systems, and they also apply the method to solve the Rosenbrock model.

9.6.4.4 Numerical Methods Conserving Important Qualitative Properties

Even the simplest kinetic differential equations can only be solved by numerical methods; therefore the question if such a method is able to keep important qualitative properties of the models arouse very early. It is always true that starting from a nonnegative concentration vector, one can never have negative concentrations later (see Theorem 8.6). In many cases the investigated reaction is mass conserving, i.e., the induced kinetic differential equation of the reaction has a (positive) linear first integral (see, e.g., Chap. 4). It may also happen that it has a first integral with not necessarily positive coefficients (see Problem 9.18) or that it has a nonlinear (e.g., quadratic or more complicated) first integral (see Sect. 8.5).

An obvious requirement from a numerical method is to keep as many of these and similar properties as possible. Now we review some of the basic results.

First Integrals

As to keeping first integrals, numerical methods were constructed to keep linear (Robertson and McCann 1969) and quadratic (Rosenbaum 1977; LaBudde and Greenspan 1976) first integrals. A more recent review on general first integrals has been given by Shampine (1986).

The first results in this field come from the 1970s of the last century, when computers began to be applied in reaction kinetics.

Definition 9.44 Let us consider the initial value problem

$$\dot{\mathbf{x}} = \mathbf{f} \circ \mathbf{x} \quad \mathbf{x}(0) = \mathbf{x}_0, \tag{9.37}$$

with $\mathbf{f} \in \mathscr{C}(\mathbb{R}^M, \mathbb{R}^M)$ and $\mathbf{x}_0 \in \mathbb{R}^M$; and suppose a numerical method provides a sequence of approximations $\mathbf{x}^1, \mathbf{x}^2, \ldots, \mathbf{x}^n, \ldots$ of the solution.

1. If $\overline{\mathbf{x}} \mapsto \boldsymbol{\omega}^\top \overline{\mathbf{x}}$ is a linear first integral of the initial value problem (9.37) (where $\boldsymbol{\omega} \in \mathbb{R}^M$), and for all $n \in \mathbb{N} : \boldsymbol{\omega}^\top \mathbf{x}^n = \boldsymbol{\omega}^\top \mathbf{x}_0$, then the given numerical method is said to be **linearly conservative**.
2. If $\overline{\mathbf{x}} \mapsto \overline{\mathbf{x}}^\top \mathbf{B} \overline{\mathbf{x}}$ is a quadratic first integral (where $\mathbf{B} \in \mathbb{R}^{M \times M}$ is a positive definite symmetric matrix), and for all $n \in \mathbb{N} : \mathbf{x}^{n\top} \mathbf{B} \mathbf{x}^n = \mathbf{x}_0^\top \mathbf{B} \mathbf{x}_0$, then the given numerical method is said to be **quadratically conservative**.

Problem 9.21 shows that the requirement of, say, linear conservativity in itself is not enough to ensure that the approximations given by the numeric method are good in any other sense.

The two fundamental statements were formulated by Rosenbaum (1977).

Theorem 9.45

1. (Variable step) linear multistep methods are linearly conservative.
2. (Variable step) Runge–Kutta methods are linearly conservative.

Let us recapitulate that the concept we met in Chap. 4 might be more precisely called **stoichiometrically mass conserving**. This is different from the property that a reaction is kinetically mass conserving (see Definition 8.33).

Remark 9.46

- A stoichiometrically mass conserving reaction is kinetically mass conserving, as well.
- The induced kinetic differential equation of a kinetically mass conserving reaction has a linear first integral.

None of the conversions is true; this fact is trivial in the second case; Problem 9.20 shows that it is neither true in the first one.

Let us remark in passing that although the existence of a first integral makes it possible to reduce the number of variables by elimination, this may not necessarily be a good idea from the numerical point of view. Also, elimination of a variable using the quasi-steady-state hypothesis is not a good idea either: it may destroy linear first integrals.

You may not think that all the numerical methods are linearly conservative: Rosenbaum (1977) mentions a few nonconservative methods, as well.

Positivity

Volpert's Theorem 8.6 reassures us that none of the coordinate functions of the solution to a kinetic differential equation can be negative at any positive time. Interestingly, this important property has been started to be investigated not so early in spite of its importance (Faragó and Komáromi 1990; Shampine 1986; Shampine et al. 2005), but since then the investigations have been extended to partial differential equations (of which parabolic ones are especially important as reaction diffusion equations belong to this category) too (see Faragó 1996; Faragó and Horváth 2001; Faragó and Komáromi 1990), but the first result again comes from Volpert (1972) who has shown that the method of finite differences when applied to the reaction diffusion equation leads to a formal reaction of higher dimension, and thus one can again apply the abovementioned theorem to get nonnegativity of the numerical approximation.

Horváth (1998, 2005) asked the question how large the step size can be in Runge–Kutta methods so that the method still conserves positivity of the solutions. Other authors like Bertolazzi (1996), Faragó (1996), and Karátson and Korotov (2009) treat positivity and mass conservation of numerical methods, also applied for mass action kinetics. Antonelli et al. (2009) analyze the impact of positivity and mass conservation properties of a recent version of VODE—a stiff ordinary differential equation solver using backward differentiation formulas—on the prediction of ignition in numerical simulation in combustion. The property that solutions starting in the first orthant do remain in the first orthant can be rephrased as the (positive) invariance of the first orthant. Cheuh et al. (1977) put this question in a wider perspective.

Further Relevant Properties to Keep

Another important qualitative property is the presence of a periodic solution, be it a conservative oscillation as in the case of the Lotka–Volterra reaction or a limit cycle as in the case of the Brusselator.

Some differential equations have the property that the volume of the phase space is constant along the solutions (Quispel and Dyt 1998).

Ketcheson (2009) and Ketcheson et al. (2009) discuss the preservation of monotonicity (here strong stability or monotonicity mean that some convex functional of the solution is nonincreasing in time, dissimilarly from the meaning we have used in Sect. 6.4.3). Hadjimichael et al. (2016) proves keeping a kind of stability by numerical methods and Lóczi and Chávez (2009) is about bifurcation preserving methods.

Skeel and Gear (1992) treat the question if the Hamiltonian structure of a differential equation is kept by a numerical method or not. Such a method is termed as **symplectic**. The problem how often an induced kinetic differential equation can have a Hamiltonian structure has been raised by Tóth and Érdi (1988).

9.6.5   Differential Algebraic Equations

The time evolution of a reaction is most often described by systems of ordinary differential equations. However, it is often the case that the model consists of differential equations and also some algebraic relations between the concentrations. Then we have a **differential algebraic equation system**. We show two examples how one may arrive at such a model.

Example 9.47 One of our favorite models, the Robertson reaction (see, e.g., Eq. (2.6)), is obviously mass conserving; one can take $\varrho = \begin{bmatrix} 1 & 1 & 1 \end{bmatrix}^{\mathsf{T}}$. Therefore, instead of solving the induced kinetic differential equation in the usual way, one can add the equation $a(t) + b(t) + c(t) = a(0) + b(0) + c(0)$ expressing mass conservation and solve the emerging differential algebraic equation to get Fig. 9.6.

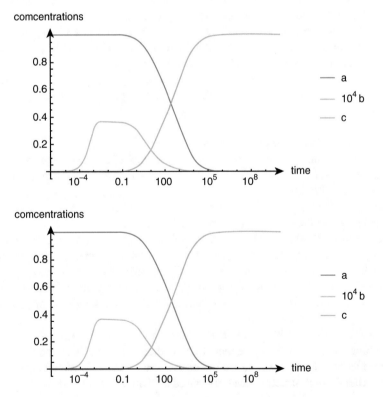

Fig. 9.6 Exact solution of the induced kinetic differential equation of the Robertson reaction, and its solution as an algebraic differential equation. Parameters: $k_1 = 0.04, k_{-1} = 3 \times 10^7, k_2 = 10^4, a_0 = 1, b_0 = 0, c_0 = 0$

Example 9.48 A usual simplification method to reduce the number of variables in kinetic differential equations is obtained when one realizes that one of the variables is changing much slowly and assumes its rate of change is zero. Let us consider the Michaelis–Menten reaction: $E + S \underset{k_{-1}}{\overset{k_1}{\rightleftharpoons}} C \xrightarrow{k_{-2}} E + P$ having the induced kinetic differential equation

$$\dot{e} = -k_1 es + k_{-1}c + k_{-2}c$$
$$\dot{s} = -k_1 es + k_{-1}c$$
$$\dot{c} = k_1 es - k_{-1}c - k_{-2}c$$
$$\dot{p} = k_{-2}c$$

with the usual initial condition

$$e(0) = e_0, \ s(0) = s_0, \ c(0) = 0, \ p(0) = 0.$$

According to the usual assumption, the rate of change of the intermediate complex C is so small that it can be considered to be zero (see more details in Sect. 9.4). Here we only concentrate on the possibility that one can solve instead of the above induced kinetic differential equation the following differential algebraic equations:

$$\dot{e} = -k_1 es + k_{-1}c + k_{-2}c \tag{9.38}$$
$$\dot{s} = -k_1 es + k_{-1}c \tag{9.39}$$
$$0 = k_1 es - k_{-1}c - k_{-2}c \tag{9.40}$$
$$\dot{p} = k_{-2}c, \tag{9.41}$$

and the results show similarity with the usual approximation. (The initial concentration of c should be chosen carefully when using (9.41).) All the data are taken from (Heineken et al. 1967, Figure 1) (Fig. 9.7).

Finally, let us mention that the case when either the number of species or the number of reaction steps is infinite may have either direct importance or can be used as an approximation method. This is the case of **continuous species** or **continuous components** and reaction steps, which may be useful in systems where the number of species is extremely large (Aris 1989; Aris and Gavalas 1966) or if they can really be only parameterized by a continuously changing parameter, e.g., by an angle characterizing the form of the molecule (Érdi and Tóth 1989, pp. 78–79).

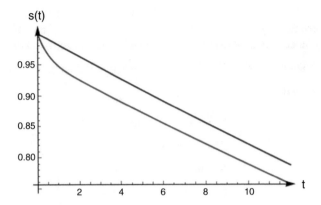

Fig. 9.7 Solution of the induced kinetic differential equation of the Michaelis–Menten reaction without approximation: red line, as an algebraic differential equation: dashed blue line. Parameters: $k_1 = 1, k_{-1} = 0.625, k_{-2} = 0.375, e_0 = 0.1, s_0 = 1.0$

9.7   Exercises and Problems

9.1 How can you estimate the order of the concentrations at the beginning for the reversible bimolecular reaction?

(Solution: page 430)

9.2 Suppose that there exists $\varrho \in (\mathbb{R}_0^+)^M$ for which $\varrho^\top \gamma \leq 0$ holds. Then all reaction steps for which $\sum_{m \in \mathcal{M}} \varrho_m \gamma(m, r) < 0$ holds is a summable reaction step.

(Solution: page 430)

9.3 Apply, if possible, the sufficient condition of summability in Problem 9.2 to the Lotka–Volterra reaction, the simple reversible bimolecular reaction, the Michaelis–Menten reaction, and the Robertson reaction.

(Solution: page 431)

9.4 Show that the stationary point $\begin{bmatrix} 0 & 0 \end{bmatrix}^\top$ of the Eq. (9.6) is an asymptotically stable node.

(Solution: page 431)

9.5 Carry out the calculations shown for the case of the Michaelis–Menten model for the much simpler reaction $S \rightleftharpoons C \longrightarrow P$ assuming that $c(t) \ll s(t)$.

(Solution: page 431)

9.6 Find the symbolic solution of the Eq. (9.11) after substituting C from the Eq. (9.14).

(Solution: page 431)

9.7 Determine the number of stationary points of the (kinetic) differential equation

$$\dot{x} = 2 + x - x^2 + y^2 \quad \dot{y} = 3 + xy - y^3 + xz \quad \dot{z} = 2 + x^3 - yz - z^2$$

using **Reduce**, **Solve**, or **NSolve**, and also "manually", using Gröbner bases. Experiment with different options of the **GroebnerBasis** function.

(Solution: page 432)

9.8 Show that the representation (9.32) is independent of the choice of the generalized inverse $\bar{\mathbf{h}}$.

(Solution: page 432)

9.9 Let $a, b, c, d \in \mathbb{R}$. Show that the function $\mathbb{R}^2 \ni (x, y) \mapsto w(x, y) := (a + c)x + (b + d)y \in \mathbb{R}$ can be represented as a function $\mathbb{R}^2 \ni (x, y) \mapsto u(x, y) := v(x + y) \in \mathbb{R}$ if and only if $a + c = b + d$.

(Solution: page 432)

9.10 Show that the constant eigenvalue of the matrix-valued function

$$\mathbb{R}^2 \ni (x_1, x_2) \mapsto \begin{bmatrix} x_1 + 2 & x_2 \\ x_1 & x_2 + 2 \end{bmatrix} \in \mathbb{R}^{2 \times 2}$$

has a nonconstant eigenvector, while its nonconstant eigenvalue corresponds to a fixed, constant eigenvector.

(Solution: page 432)

9.11 Do the calculations supporting the statements in Examples 9.22 and 9.23.

(Solution: page 433)

Fig. 9.8 A reversible
triangle reaction to lump

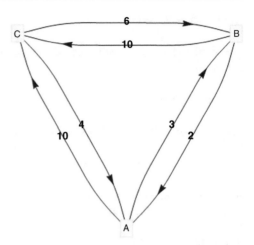

9.12 Lump properly the reversible triangle reaction with the reaction rate coeffi-
cients as shown in Fig. 9.8 using $\mathbf{M} := \begin{bmatrix} 1 & 1 & 0 \\ 0 & 0 & 1 \end{bmatrix}$, and calculate the coefficient matrix
of the lumped equation.

(Solution: page 434)

9.13 Consider the reaction

$$X_i \underset{k_{ij}}{\overset{k_{ji}}{\rightleftharpoons}} X_j \quad (i, j = 1, 2, 3, 4) \tag{9.42}$$

with the assumption $k_{41} = k_{42} = k_{43}$, and show that $\mathbf{M} := \begin{bmatrix} 1 & 1 & 1 & 0 \\ 0 & 0 & 0 & 1 \end{bmatrix}$ is an exact
lumping matrix, i.e., there exists $\kappa, \lambda \in \mathbb{R}^+$ such that

$$\dot{y}_1 = -\kappa y_1 + \lambda y_2 \quad \dot{y}_2 = \kappa y_1 - \lambda y_2; \tag{9.43}$$

or, to put it another way, the reaction (9.42) can linearly be lumped into the reaction

$$Y_1 \underset{\lambda}{\overset{\kappa}{\rightleftharpoons}} Y_2. \tag{9.44}$$

(Solution: page 434)

9.14 Find an induced kinetic differential equation which is transformed by linear
lumping into a differential equation which is not kinetic.

(Solution: page 434)

9.15 Apply proper lumping to a first-order reaction with five species into one having two species. Show that if the starting reaction was a closed, half-open, or open compartmental system, so is the lumped system.

(Solution: page 435)

9.16 Show that an asymptotically stable stationary point may be lumped into an unstable one, and what is more, a lumped system can blow up even if the original did not have this property.

(Solution: page 435)

9.17 Show that although the system

$$\dot{x} = -kx + ky + u + v \quad \dot{y} = kx - 2ky + kz + u \quad \dot{z} = ky - kz + u - v$$

is not completely controllable, still the lumped system

$$\dot{\hat{x}} = -\frac{k}{2}\hat{x} + \frac{k}{2}\hat{y} + 3u + 2v \quad \dot{\hat{y}} = \frac{k}{2}\hat{x} - \frac{k}{2}\hat{y} + 3u - 2v$$

is completely controllable. (What is the lumping matrix?)

(Solution: page 436)

9.18 Construct a reaction with a nonpositive linear first integral.

(Solution: page 436)

9.19 The class of **Runge–Kutta methods** for solving the equation

$$\dot{x} = f \circ x \tag{9.45}$$

can be defined as follows (Frank 2008, Chapter 8):

$$Y_i = x_n + \sum_{j=1}^{s-1} a_{ij} f(Y_j) \quad (i = 1, 2, \ldots, s); \quad x_{n+1} = x_n + h \sum_{i=1}^{s} b_i f(Y_i),$$

where s is the **number of stages**, b_i are the **weights**, and a_{ij} are the **internal coefficients**. Show that, if for all $i, j = 1, 2, \ldots, s$ $b_i b_j - b_i a_{ij} - b_j a_{ji} = 0$ holds, then the method preserves the quadratic first integrals of (9.45).

(Solution: page 436)

9.20 Find a positive vector with which the induced kinetic differential equation of the reaction (Feinberg and Horn 1977, p. 89)

$$F \rightleftharpoons D + E \longleftarrow C \rightleftharpoons A + B \longrightarrow G \longrightarrow H \rightleftharpoons 2J \longrightarrow G \qquad (9.46)$$

can be shown to be kinetically mass conserving; still with this vector the reaction cannot be shown to be stoichiometrically mass conserving.

(Solution: page 437)

9.21 Show that the "numerical methods" for the initial value Problem (9.37) defined below

1. $\mathbf{x}^{n+1} = \mathbf{x}^n$, $\quad \mathbf{x}^0 = \mathbf{x}_0$
2. $\mathbf{x}^{n+1} = \mathbf{x}^n - n \cdot 10^6 \cdot \mathbf{f}(\mathbf{x}^n)$, $\quad \mathbf{x}^0 = \mathbf{x}_0$

are linearly conservative.

(Solution: page 437)

9.22 Show that the Euler method when applied with a small enough step size keeps the invariance of the first orthant.

(Solution: page 437)

9.23 Show that the stationary points of an autonomous differential equation are the same as those obtained by the Euler method.

(Solution: page 437)

9.24 What happens with the oscillatory character of the harmonic oscillator

$$\dot{x} = y \quad \dot{y} = -x$$

if one applies the Euler method?

(Solution: page 438)

9.25 The initial derivatives of the concentrations are six orders of magnitude different from each other in the mass action type induced kinetic differential equation of the reaction $0 \xrightarrow{10^6} X, Y \xrightarrow{1} 0$ if the units used are mol, dm^3, and s. What happens if one uses micromoles instead?

(Solution: page 438)

9.26 Write down the mass action type induced kinetic differential equation of the reaction

$$Br_2 \xrightarrow{k_1} 2\,Br, \text{ initialization} \tag{9.47}$$

$$H_2 + Br \xrightarrow{k_2} H + HBr, \text{ propagation} \tag{9.48}$$

$$H + Br_2 \xrightarrow{k_3} Br + HBr, \text{ propagation} \tag{9.49}$$

$$H + HBr \xrightarrow{k_4} H_2 + Br, \text{ retardation} \tag{9.50}$$

$$2\,Br \xrightarrow{k_5} Br_2, \text{ termination} \tag{9.51}$$

then assuming that the derivative of [Br] and [H] is zero, express the approximate time derivative of [HBr]. This procedure is a kind of application of the quasi-steady-state approximation (QSSA) to the rate of the overall reaction $H_2 + Br_2 \longrightarrow 2\,HBR$, which can formally be obtained as the linear combination of the reaction steps in (9.49) with the coefficients 1, 1, 1, 2 and 1 after eliminating the H and Br radicals and keeping only one H_2 and Br_2 molecule on the left side and two molecules of HBr on the right side.

(Solution: page 438)

9.8 Open Problems

1. It would be desirable to have a description of the behavior at infinite time in chemical terms or in terms of the structure of the reaction.
2. How do you determine the minimal sets of species with initially positive concentrations so that all the concentrations of the species become positive during the course of the reaction?
3. How is the stability on the boundaries explained by Volpert and Ivanova (1987) connected to the Global Attractor Hypothesis?
4. The first assumption of Theorem 9.4 is that the zero complex is not a reactant complex. Is it essential? Can you give an example where this condition and the consequence of the theorem equally fails?

References

Antonelli L, Briani M, D'Ambra P, Fraioli V (2009) Positivity issues in adaptive solutions of detailed chemical schemes for engine suimulations. Commun SIMAI Congr 3:303–314. http://cab.unime.it/journals/index.php/congress/article/download/303/248

Aris R (1989) Reactions in continuous mixtures. AIChE J 35(4):539–548

Aris R, Gavalas GR (1966) On the theory of reactions in continuous mixtures. Philos Trans R Soc Lond Ser A, Math Phys Sci 260(1112):351–393

Bertolazzi E (1996) Positive and conservative schemes for mass action kinetics. Comput Math 32:29–43

Bolla M (2013) Factor analysis, dynamic. Wiley StatsRef: Statistics Reference Online

Bolla M, Kurdyukova A (2010) Dynamic factors of macroeconomic data. Ann Univ Craiova— Math Comput Sci Ser 37(4):18–28

Brenig L, Goriely A (1994) Painlevé analysis and normal forms. In: Tournier E (ed) Computer algebra and differential equations. London Mathematical Society Lecture Note Series. Cambridge University Press, Cambridge, pp 211–238

Briggs GE, Haldane JBS (1925) A note on the kinetics of enyzme action. Biochem J 19:338–339

Brochot C, Tóth J, Bois FY (2005) Lumping in pharmacokinetics. J Pharmacokinet Pharmacodyn 32(5-6):719–736

Buchberger B (2001) Gröbner bases: a short introduction for systems theorists. In: EUROCAST, pp 1–19. https://doi.org/10.1007/3-540-45654-6_1

Cheuh K, Conley C, Smoller J (1977) Positively invariant regions for systems of nonlinear diffusion equations. Indiana Univ Math J 26(2):373–392

Cox DA, Little J, O'Shea D (2004) Using algebraic geometry, 2nd edn. Springer, New York

Coxson PG (1984) Lumpability and observability of linear systems. J Math Anal Appl 99(2):435–446

Dahlquist G, Edsberg L, Skollermo G, Soderlind G (1982) Are the numerical methods and software satisfactory for chemical kinetics? In: Hinze J (ed) Numerical integration of differential equations and large linear systems. Lecture Notes in Mathematics, vol 968. Springer, New York, pp 149–164

Darvishi MT, Khani F, Soliman AA (2007) The numerical simulation for stiff systems of ordinary differential equations. Comput Math Appl 54:1055–1063

Deuflhard P, Bornemann F (2002) Scientific computing with ordinary differential equations. In: Marsden JE, Sirovich L Golubitsky M, Antmann SS (eds) Texts in applied mathematics, vol 42. Springer, New York

Érdi P, Tóth J (1989) Mathematical models of chemical reactions. Theory and applications of deterministic and stochastic models. Princeton University Press, Princeton

Faragó I (1996) Nonnegativity of the difference schemes. Pure Math Appl 6:147–159

Faragó I, Horváth R (2001) On the nonnegativity conservation of finite element solutions of parabolic problems. In: Proceedings of conference on finite element methods: three-dimensional problems, Gakkotosho, Tokyo. GAKUTO international series. Mathematical sciences and applications, vol 15, pp 76–84

Faragó I, Komáromi N (1990) Nonnegativity of the numerical solution of parabolic problems. In: Proceedings of conference on numerical methods. Colloquia Mathematica Societatis János Bolyai, vol 59, pp 173–179

Farkas G (1998a) Local controllability of reactions. J Math Chem 24:1–14

Farkas G (1998b) On local observability of chemical systems. J Math Chem 24:15–22

Farkas G (1999) Kinetic lumping schemes. Chem Eng Sci 54:3909–3915

Feinberg M, Horn FJM (1977) Chemical mechanism structure and the coincidence of the stoichiometric and kinetic subspaces. Arch Ratl Mech Anal 66(1):83–97

Forni M, Hallin M, Lippi M, Zaffaroni P (2015) Dynamic factor models with infinite-dimensional factor spaces: one-sided representations. J Econ 185(2):359–371

Frank J (2008) Numerical modelling of dynamical systems. Lecture notes. Author. http://homepages.cwi.nl/~jason/Classes/numwisk/index.html

Gear CW (1992) Invariants and numerical methods for ODEs. Phys D 60:303–310

Goeke A, Walcher S (2014) A constructive approach to quasi-steady state reductions. J Math Chem 52(10):2596–2626. https://doi.org/10.1007/s10910-014-0402-5

Goeke A, Walcher S, Zerz E (2015) Quasi-steady state–intuition, perturbation theory and algorithmic algebra. In: Gerdt VP, Koepf W, Seiler WM, Vorozhtsov EV (eds) Computer algebra in scientific computing. Lecture notes in computer science, vol 9301. Springer, Cham, pp 135–151

Hadjimichael Y, Ketcheson D, Lóczi L, Németh A (2016) Strong stability preserving explicit linear multistep methods with variable step size. SIAM J Numer Anal 54(5):2799–2832

Heineken FG, Tshuchiya HM, Aris R (1967) On the mathematical status of pseudo-steady state hypothesis. Math Biosci 1:95–113

Horváth Z (1998) Positivity of Runge–Kutta and diagonally split Runge–Kutta methods. Appl Numer Math 28:309–326

Horváth Z (2002) Effect of lumping on controllability and observability. In: Poster presented at the colloquium on differential and difference equations dedicated to Prof. František Neuman on the occasion of his 65th birthday, Brno, Czech Republic, September, pp 4–6

Horváth Z (2002–2008) Effect of lumping on controllability and observability. arxivorg http://arxiv.org/PS_cache/arxiv/pdf/0803/0803.3133v1.pdf

Horváth Z (2005) On the positivity step size threshold of Runge-Kutta methods. Appl Numer Math 53:341–356

Iordache O, Corbu S (1987) A stochastic model of lumping. Chem Eng Sci 42(1):125–132

Karátson J, Korotov S (2009) An algebraic discrete maximum principle in Hilbert space with applications to nonlinear cooperative elliptic systems. SIAM J Numer Anal 47(4):2518–2549

Keleti T (1986) Basic enzyme kinetics. Akadémiai Kiadó, Budapest

Ketcheson D (2009) Computation of optimal monotonicity preserving general linear methods. Math Comp 78:1497–1513

Ketcheson DI, Macdonald CB, Gottlieb S (2009) Optimal implicit strong stability preserving Runge-Kutta methods. Appl Numer Math 59(2):373–392

Klonowski W (1983) Simplifying principles for chemical and enzyme reaction kinetics. Biophys Chem 18(2):73–87

Kuo JCW, Wei J (1969) Lumping analysis in monomolecular reaction systems. Analysis of approximately lumpable system. Ind Eng Chem Fundam 8(1):124–133. https://doi.org/10.1021/i160029a020, http://pubs.acs.org/doi/abs/10.1021/i160029a020, http://pubs.acs.org/doi/pdf/10.1021/i160029a020

LaBudde RA, Greenspan D (1976) Energy and momentum conserving methods of arbitrary order for the numerical integration of equations of motion II. Motion of a system of particles. Numer Math 26(1):1–16

Li G (1984) A lumping analysis in mono- or/and bimolecular reaction systems. Chem Eng Sci 29:1261–1270

Li G, Rabitz H (1989) A general analysis of exact lumping in chemical kinetics. Chem Eng Sci 44(6):1413–1430

Li G, Rabitz H (1991) A general lumping analysis of a reaction system coupled with diffusion. Chem Eng Sci 46(8):2041–2053

Li G, Tomlin AS, Rabitz H, Tóth J (1993) Determination of approximate lumping schemes by a singular perturbation method. J Chem Phys 99(5):3562–3574

Li G, Rabitz H, Tóth J (1994) A general analysis of exact nonlinear lumping in chemical kinetics. Chem Eng Sci 49(3):343–361

Lichtblau D (1996) Gröbner bases in *Mathematica* 3.0. Math J 6:81–88

Lóczi, L, Chávez, J P (2009) Preservation of bifurcations under Runge-Kutta methods. Int J Qual Theory Differ Equ Appl 3(1–2):81–98

Michaelis L, Menten ML (1913) Die Kinetik der Invertinwirkung. Biochem Z 49:333–369

Quispel GRW, Dyt CP (1998) Volume-preserving integrators have linear error growth. Phys Lett A 242(1):25–30

Robertson HH, McCann MJ (1969) A note on the numerical integration of conservative systems of first-order ordinary differential equations. Comput J 12(1):81

Rosenbaum JS (1977) Conservation properties for numerical integration methods for systems of differential equations. 2. J Phys Chem 81(25):2362–2365

Ross J (2008) Determination of complex reaction mechanisms. Analysis of chemical, biological and genetic networks. J Phys Chem A 112:2134–2143

Rózsa Z, Tóth J (2004) Exact linear lumping in abstract spaces. In: Proceedings of 7th Colloquia QTDE, vol 7

Schnell S, Maini PK (2003) A century of enzyme kinetics: reliability of the K_M and v_{max} estimates. Commun Theor Biol 8:169–187

Segel LA (1988) On the validity of the steady state assumption of enzyme kinetics. Bull Math Biol 50(6):579–593

Segel LA, Slemrod M (1989) The quasi-steady-state assumption: a case study in perturbation. SIAM review 31(3):446–477

Shampine LF (1986) Conservation laws and the numerical solution of ODEs. Comp Math Appl 12(5–6):1287–1296

Shampine LF, Thompson S, Kierzenka JA, Byrne GD (2005) Non-negative solutions of ODEs. Appl Math Comput 170(1):556–569

Shinar G, Feinberg M (2010) Structural sources of robustness in biochemical reaction networks. Science 327(5971):1389–1391

Skeel R, Gear C (1992) Does variable step size ruin a symplectic integrator? Phys D 60:311–313

Tikhonov AN (1952) Systems of differential equations containing a small parameter at the derivatives. Mat Sb 31(73):575–585

Tóth J, Érdi P (1988) Kinetic symmetries: some hints. In: Moreau M, Turq P (eds) Chemical reactivity in liquids. Fundamental aspects. Plenum Press, New York, pp 517–522 (Paris, Sept. 7-11, 1987)

Tóth J, Li G, Rabitz H, Tomlin AS (1996) Reduction of the number of variables in dynamic models. In: Martinás K, Moreau M (eds) Complex systems in natural and economics sciences, Eötvös Loránd Fizikai Társulat, Budapest, pp 17–34 (Mátrafüred, 19–22 September, 1995)

Tóth J, Li G, Rabitz H, Tomlin AS (1997) The effect of lumping and expanding on kinetic differential equations. SIAM J Appl Math 57:1531–1556

Turányi T, Tóth J (1992) Comments to an article of Frank–Kamenetskii on the quasi-steady-state approximation. Acta Chim Hung—Model Chem 129:903–914

Turányi T, Bérces T, Tóth J (1988) The method of quasi-stationary sensitivity analysis. J Math Chem 2(4):401–409

Tzafriri AR, Edelman ER (2007) Quasi-steady-state kinetics at enzyme and substrate concentrations in excess of the Michaelis–Menten constant. J Theor Biol 245:737–748

Volpert AI (1972) Differential equations on graphs. Mat Sb 88(130):578–588

Volpert AI, Ivanova AN (1987) Mathematical models in chemical kinetics. In: Samarskii AA, Kurdyumov SP, Mazhukn VI (eds) Mathematical modelling. Nonlinear differential equations of mathematical physics. Nauka, Moscow, pp 57–102

Wei J, Kuo JCW (1969) A lumping analysis in monomolecular reaction systems. Analysis of the exactly lumpable system. Ind Eng Chem Fundam 8(1):114–123

Zachár A (1998) Comparison of transformations from nonkinetic to kinetic models. ACH—Models Chem 135(3):425–434

Part III

The Continuous Time Discrete State Stochastic Model

The most often used continuous time discrete state stochastic model of reactions will be discussed here. The treatment is—as far as possible—parallel with that of the deterministic model.

Stochastic Models

10

10.1 The Necessity of Stochastic Models

As we have seen in the previous chapters (see Chap. 6), the essence of the usual deterministic model is that the effects of different reaction steps are translated into the change of **concentrations** in time of species reacting with one another. With the assumption that the species are in a well-stirred environment, this approach gave a spatially "averaged" description, which proved to be adequate, e.g., when the number of species is large. However, **small systems** (see Arányi and Tóth 1977; Érdi and Tóth 1989, Chapter 5; Gadgil 2008; Grima et al. 2014; Lente 2010; Turner et al. 2004), or systems operating around an **unstable stationary point** (e.g., models of **chirality** such as in Barabás et al. 2010; Lente 2004) are better described by models which take into account:

- discrete state space for species, i.e., number of molecules, atoms, etc. are used instead of concentrations,
- stochastic dynamics for the reaction steps.

In other words, we are going to take into account both the discrete and the random characters of the reactions by keeping track of the number of reacting molecules. This sort of approach has had sufficient attention in the recent decades as the experiments and analytical techniques are being much more developed, e.g., there are cases where it has become possible to measure individual molecules (Juette et al. 2014; Tóth and Érdi 1992; Arakelyan et al. 2004, 2005; Grima et al. 2014; Qian and Elson 2002; Edman and Rigler 2000; Lee et al. 2010; Sakmann and Neher 1995; English et al. 2006; Stoner 1993; Weiss 1999; English et al. 2006; Velonia et al. 2005; Turner et al. 2004).

The induced kinetic Markov process (the usual stochastic model for short) to be introduced in the next section is expected to be valid in the **mesoscopic** scale, i.e., it forms a bridge between the **macroscopic** and **microscopic** models of reaction

© Springer Science+Business Media, LLC, part of Springer Nature 2018
J. Tóth et al., *Reaction Kinetics: Exercises, Programs and Theorems*,
https://doi.org/10.1007/978-1-4939-8643-9_10

kinetics. On the observable mesoscopic level the—internal or external—fluctuations can force us to reconsider the notion of acceptable agreements with experiments. Randomness implies that along exactly the same circumstances, one can have different results at the same time. It is also worth mentioning that this stochastic behavior may or may not mean **reproducibility**: the experimenter measures different results with **seemingly** the same circumstances (cf. Singer 1953; Nagypál and Epstein 1986, 1988; Lente 2010). If the distribution of the measurements (repeated many times) is the same, then the experiment is reproducible in a more general sense. It was rigorously proved that on a large spatial scale, the average behavior of the usual stochastic model is close to that of the usual deterministic model as described in Chap. 6. Nevertheless it can happen that

- the amplification of fluctuations measured on a small scale or
- the effect of compartmentalization, i.e., when a large system consists of many small components that are weakly reacting with each other,

leads to observable macroscopic phenomena **not** captured by the induced kinetic differential equation (6.3). It is also worth noting that if the experimental results are interpreted within a stochastic framework, then **statistical inference** not possible in the deterministic model may also be drawn from data (see Érdi and Ropolyi 1979; Yan and Hsu 2013; Érdi and Lente 2016, Section 3.9; Érdi and Tóth 1989, Section 5.4 and further references therein).

Nowadays, deterministic models are still the most widespread ones in kinetics, but there are an increasing number of well-established arguments both from theoretical and practical sides why it is of importance to consider stochastic reaction kinetic models. It should however be noticed that if one has an acceptable solution, e.g., for the induced kinetic differential equation with micro- or even nanomoles of different chemical species, then in many cases one does not need to worry about stochastic models. In these cases the more complicated stochastic model would lead to essentially the same results as the deterministic one, but at the price of much more effort, not a good deal. However, there are arguably real, physicochemical situations in which a deterministic model is insufficient to describe what is going on. For illustration we outline examples from three research fields below. We refer to the review Gadgil (2008) for further relevant examples.

In this chapter mainly **direct problems** are considered, i.e., given the components, interactions, and parameters, the behavior of the corresponding model is investigated. Modeling approaches for the inference of mechanisms from experimental data, i.e., **inverse problems**, are tackled in Chap. 11.

We remind the reader that more recently Érdi and Lente (2016) and Anderson and Kurtz (2015) have dedicated separate books to stochastic models of reactions. Since the present book is aimed at capturing a wider audience, it is then hoped to be more rigorous than that of the former one (at the price of less applications), while it is able to treat much more applications than the latter one (although at a less rigorous level). We propose the reader to consult these books, as well. Let us also mention from the earlier literature the monographs by Gardiner (2010) and Van Kampen (2006).

10.1.1 Biological Models, Enzyme Kinetics

The best thing for the reader interested in biological applications is to study Érdi and Lente (2016) and Anderson and Kurtz (2015). The earlier book by Iosifescu and Tăutu (1973) also contains a rich material. Here we only give a teaser.

The stochastic model of compartmental systems is often applied to describe the movement of populations between different areas. It is equally suitable for modeling the distribution of drugs or of compounds labeled with isotopes in the human or animal body. The model itself has been solved in the sense that its absolute probabilities have been calculated many times in the last 80 years; in every decade a new author emerges stating that (s)he found the solution (see Érdi and Lente 2016, p. 78, or Leontovich 1935; Siegert 1949; Gans 1960; Krieger and Gans 1960; Bartholomay 1958; Šolc 2002; Darvey and Staff 2004; Gadgil et al. 2005; Jahnke and Huisinga 2007).

The stochastic model of the Michaelis–Menten reaction is not easy to deal with. We shall see below that the problem is caused by the fact that one of the reaction steps is of the order two. However, as one is usually interested in the case when at the beginning of the reaction the number of enzyme molecules is much smaller than that of the substrate molecules, it turned out to be a fruitful idea to calculate all the quantities in the extreme case when one has a single enzyme molecule at the beginning (Arányi and Tóth 1977). Many years later, the experimental techniques made it possible to carry out measurements in this extreme case.

Signal processing in general and in particular in the olfactory system is a favorite topic (Lánský and Rospars 1995; Pokora and Lánský 2008; Zhang et al. 2005). A deterministic counterpart has numerically been investigated by Tóth and Rospars (2005).

Gene expression is also frequently studied using models of stochastic kinetics (see Li and Rabitz 2014; McAdams and Arkin 1997; Wadhwa et al. 2017; Samad et al. 2005 and also Problem 10.1).

10.1.2 Chirality and the Asymmetric Autocatalysis

A chiral molecule is one that cannot be superimposed onto its mirror image. The source of this effect is usually an asymmetric carbon atom. The molecule and its mirror image are **enantiomers**. Enantiomers of the same composition rotate the plane of the polarized light into opposite direction. Originally the direction of rotation is denoted by the name right (R) and left (L) enantiomers, respectively. An interesting and biologically relevant fact is that the amino acids present in living beings are always of the L-type, whereas sugars are always of the R-type. The major questions of the theory are where this asymmetry came from and if it is once present how it is amplified. The biological significance of chirality modeling has been realized as early as in the middle of the last century starting from Frank (1953), though both modeling and experimental research have started to be very vivid since the discovery of the **Soai reaction** (Soai et al. 1995). This reaction shows the possibility of finding circumstances under which asymmetric molecules can emerge without the presence of any external source of asymmetry (as, e.g., asymmetric crystals). As to the theoretical work: it turned out that the usual stochastic model is much more appropriate to describe the spontaneous synthesis of asymmetric molecules (Lente 2004, 2005), showing good agreement with the experimental results. Barabás et al. (2010) presented a review on the deterministic and stochastic models of emergence and amplification of chirality by mechanisms such as asymmetric autocatalysis and absolute asymmetric synthesis.

10.1.3 Combustion

Randomness plays an important role in combustion processes, as well. Lai et al. (2014) applied the usual stochastic model of chemical kinetics to the formation of **nanoparticles** in combustion in order to characterize the growth of **polycyclic aromatic hydrocarbons** (PAHs), important precursors of carbonaceous nanoparticles and soot, in a premixed laminar benzene flame, using a concurrently developed PAH growth chemical reaction mechanism, as well as an existing benzene oxidation mechanism. The authors hope that the proposed method will benefit engineering of novel combustion technologies to mitigate harmful emissions. Urzay et al. (2014) address the influences of residual radical impurities on the computation and experimental determination of ignition times in H_2/O_2 mixtures, in particular the presence of H-atoms in the initial composition of the mixtures in shock tubes. A **stochastic Arrhenius model** that describes the amount of H-radical impurities in shock tubes is proposed to yield a probability density function for the residual concentration of hydrogen radicals in standard shock tubes. The authors use a short mechanism consisting of five reaction steps. The influence of uncertainties on the ignition time is typically negligible compared to the effects of the uncertainties induced by H-impurities when the short mechanism is used.

Global **sensitivity analysis** shows an increasing importance of the recombining kinetics (see Turányi 1990; Turányi and Tomlin 2014). Simulations of homogeneous ignition subject to Monte Carlo sampling of the concentration of impurities show that the variabilities produced in ignition delays by the uncertainties in H-impurities are comparable to the experimental data scatter and to the effects of typical uncertainties of the test temperature when the Stanford chemical mechanism (Hong et al. 2011) is used. The utilization of two other different chemical mechanisms, namely, San Diego (CombustionResearch 2011) and GRI v3.0 (Smith et al. 2000), yields variations in the ignition delays which are within the range of the uncertainties induced by the H-impurities. Chibbaro and Minier (2014) describes stochastic methods (mainly **stochastic differential equations**) to be applied in fluid mechanics including applications in combustion.

Some authors investigate combustion processes from a viewpoint different from the physicochemical one, e.g., Abbasi and Diwekar (2014) quantify the inherent uncertainties in the **biodiesel** production process arising out of feedstock composition, operating and design parameters in the form of a probability distribution function. Simulation results are evaluated to determine impact of the above uncertainties on process efficiency and quality of biodiesel.

10.2 The Induced Kinetic Markov Process

In what follows we first give a heuristic derivation of the master equation of the induced kinetic Markov process for a specific reaction. Next, in Sect. 10.2.2, we precisely formulate what we mean by the induced kinetic Markov process. The master equation can be thought of as the direct analogue of the induced kinetic differential equation, but now, as opposed to Eq. (6.3), where we relied on the theory of ordinary differential equations, we are going to deal with stochastic processes.

Roughly speaking, we consider a consistent set of time-dependent random variables, a stochastic process, whose possible states are represented by the points of the lattice $\mathbb{N}_0^M$, where $M \in \mathbb{N}$ denotes the total number of species (e.g., molecules, atoms, particles, or charges). Throughout this chapter, if we do not say otherwise, we use the column vector $\mathbf{X}(t) := \begin{bmatrix} X_1(t) & X_2(t) & \cdots & X_M(t) \end{bmatrix}^\top \in \mathbb{N}_0^M$ to denote the state of the induced kinetic Markov process at time t. In particular, for $m \in \mathcal{M}$: $X_m(t)$ is the number of the mth species that is present at time $t \geq 0$. Often it is assumed that the initial state $\mathbf{X}(0)$ is fixed (or deterministically chosen), say $\mathbf{X}(0) = \mathbf{x}_0 \in \mathbb{N}_0^M$.

10.2.1 Heuristic Derivation of the Master Equation

Let us recall the Robertson reaction from Robertson (1966), i.e.,

$$A \xrightarrow{k_1^{\text{sto}}} B \quad 2B \xrightarrow{k_2^{\text{sto}}} B + C \xrightarrow{k_3^{\text{sto}}} A + C,$$

where $k_1^{\text{sto}}, k_2^{\text{sto}}, k_3^{\text{sto}}$ are positive real numbers. Assume now that we are capable of keeping track of the number of individual species occurring in the previous reaction. So let $\mathbf{X}(t) := \begin{bmatrix} X_1(t) & X_2(t) & X_3(t) \end{bmatrix}^\top$ be the number of species present in the system at time $t \geq 0$ of the species A, B, and C, respectively, where we initially drop $\mathbf{x}_0 := \begin{bmatrix} x_1^0 & x_2^0 & x_3^0 \end{bmatrix}^\top \in \mathbb{N}_0^3$ molecules of A, B, and C into the reaction vessel. Furthermore let us denote by

$$p_{\mathbf{x}_0, \mathbf{x}}(t) := \mathbf{P}\{\mathbf{X}(t) = \mathbf{x} \mid \mathbf{X}(0) = \mathbf{x}_0\}$$

the **transition probability**, i.e., the probability that starting $\mathbf{X}$ from $\mathbf{x}_0$, we have $\mathbf{x} \in \mathbb{N}_0^3$ molecules of A, B, and C by time $t \geq 0$. We are going to describe models for the temporal evolution of these probabilities under certain restrictions.

Recalling Sect. 6.2, it is natural to assume that in the "small" time interval $[t, t + h[$, the probability that the process $\mathbf{X}$ will jump to another state, i.e., one or more reaction steps will occur, is given by

$$\mathbf{P}\{\mathbf{X}(t+h) = \mathbf{y} \,|\, \mathbf{X}(t) = \mathbf{x}, (\mathbf{X}(s))_{0 \le s < t}\}$$

$$= \mathbf{P}\{\mathbf{X}(t+h) = \mathbf{y}, \text{one reaction step occurs in } [t, t+h[\,|\, \mathbf{X}(t) = \mathbf{x}, (\mathbf{X}(s))_{0 \le s < t}\}$$

$$+ \mathbf{P}\{\mathbf{X}(t+h) = \mathbf{y}, \text{more than one step occurs in } [t, t+h[\,|\, \mathbf{X}(t) = \mathbf{x}, (\mathbf{X}(s))_{0 \le s < t}\}$$

$$= \sum_{r=1}^{R} \mathfrak{B}(\mathbf{y} - \mathbf{x} = \boldsymbol{\gamma}(\cdot, r)) \cdot \varphi_r(\mathbf{x}) h + \varepsilon(h) h, \tag{10.1}$$

where $\mathbf{x} \ne \mathbf{y} \in \mathbb{N}_0^M$, $\varphi_r : \mathbb{N}_0^M \to \mathbb{R}_0^+$ are given functions and the number of reaction steps is now $R = 3$. Hence for $r \in \mathscr{R}$, φ_r is assigned to the rth reaction step which can be thought of as the "rate" at which the reaction step in question is going to be executed. Note that the probability of occurring two or more reactions in a small time interval of length h is of order $\varepsilon(h)h$, that is, negligible.

To be in full compliance with the law of total probability, we have that

$$\mathbf{P}\{\mathbf{X}(t+h) = \mathbf{x} \,|\, \mathbf{X}(t) = \mathbf{x}, (\mathbf{X}(s))_{0 \le s < t}\} = 1 - \sum_{r=1}^{R} \varphi_r(\mathbf{x}) h + \varepsilon(h) h. \tag{10.2}$$

This reflects the case when no reaction has occurred in the time interval $[t, t+h[$. The previous equations for the (transition) probabilities are supposed to be valid for all sufficiently small but positive h's. Let us make some remarks.

1. As opposed to the first remark after Eq. (6.2), now the state variable $\mathbf{X}(t)$ is **not** continuous at every point; it is indeed a **step function**. The **probability** of $\mathbf{X}(t)$ being at a particular state varies continuously with t by the assumptions from above.
2. The right-hand sides of the above relations do **not** explicitly depend on t.
3. All the information for the propagation of the process $\mathbf{X}$ after some time t is determined by its current state $\mathbf{X}(t) = \mathbf{x}$.
4. The technical term $\varepsilon(h)h$ measures the error of approximation which is considered to be negligibly small even if $h \downarrow 0$, that is, $\lim_0 \varepsilon = 0$.

At this point one may wish to specify the functions φ_r to be as simple as possible. We make the choice

$$\varphi_1(\mathbf{x}) = k_1^{\text{sto}} x_1 \mathfrak{B}(x_1 > 0),$$

$$\varphi_2(\mathbf{x}) = k_2^{\text{sto}} x_2 (x_2 - 1) \mathfrak{B}(x_2 > 1),$$

$$\varphi_3(\mathbf{x}) = k_3^{\text{sto}} x_2 x_3 \mathfrak{B}(x_2 > 0, x_3 > 0),$$

where $\mathbf{x} = \begin{bmatrix} x_1 & x_2 & x_3 \end{bmatrix}^\top \in \mathbb{N}_0^3$. Recall that $\mathfrak{B}(A)$ is 1 if statement A is true and is 0 otherwise.

At the first and third reaction steps, φ_r coincide with their deterministic counterparts aside from the different units of quantities involved. One of the differences came along at the second reaction step, where up to a scaling factor φ_2 is proportional to the binomial coefficient $\binom{x_2}{2}$ rather than to x_2^2 which latter was used in the case of the usual deterministic model. To put it another way, the "speed" of the second reaction step, up to a constant factor, equals to the number of ways two molecules of B can form up.

We also note that the units hence the magnitudes of proportionality factors $k_1^{sto}, k_2^{sto}, k_3^{sto}$ are different from that of deterministic counterparts. These are discussed in Sect. 10.2.3 in more detail.

To obtain a system of ordinary differential equations for the **probabilities**, we follow the lines of Sect. 6.2. By the law of total probability, we arrive at

$$p_{\mathbf{x}_0,\mathbf{x}}(t+h)$$

$$= \sum_{r=1}^{R} \mathbf{P}\{\mathbf{X}(t+h) = \mathbf{x} \mid \mathbf{X}(t) = \mathbf{x} - \boldsymbol{\gamma}(\cdot, r)\}\mathbf{P}\{\mathbf{X}(t) = \mathbf{x} - \boldsymbol{\gamma}(\cdot, r) \mid \mathbf{X}(0) = \mathbf{x}_0\}$$

$$+ \mathbf{P}\{\mathbf{X}(t+h) = \mathbf{x} \mid \mathbf{X}(t) = \mathbf{x}\}\mathbf{P}\{\mathbf{X}(t) = \mathbf{x} \mid \mathbf{X}(0) = \mathbf{x}_0\} + \varepsilon(h)h$$

$$= \sum_{r=1}^{R} \varphi_r(\mathbf{x} - \boldsymbol{\gamma}(\cdot, r))hp_{\mathbf{x}_0,\mathbf{x}-\boldsymbol{\gamma}(\cdot,r)}(t) + \left(1 - \sum_{r=1}^{R} \varphi_r(\mathbf{x})h\right)p_{\mathbf{x}_0,\mathbf{x}}(t) + \varepsilon(h)h.$$

Regrouping and dividing the previous equation by h, we obtain

$$\frac{p_{\mathbf{x}_0,\mathbf{x}}(t+h) - p_{\mathbf{x}_0,\mathbf{x}}(t)}{h}$$

$$= \sum_{r=1}^{R} \left[\frac{1}{h}\varphi_r(\mathbf{x} - \boldsymbol{\gamma}(\cdot, r))hp_{\mathbf{x}_0,\mathbf{x}-\boldsymbol{\gamma}(\cdot,r)}(t) - \frac{1}{h}\varphi_r(\mathbf{x})hp_{\mathbf{x}_0,\mathbf{x}}(t)\right] + \varepsilon(h).$$

Since φ_r does not depend on h, the right-hand side of the previous display has a limit as $h \downarrow 0$ so does the left-hand side implying the differentiability of $p_{\mathbf{x}_0,\mathbf{x}}(\cdot)$ for all $\mathbf{x}_0, \mathbf{x} \in \mathbb{N}_0^3$. Thus

$$\dot{p}_{\mathbf{x}_0,\mathbf{x}}(t) = k_1^{sto}(x_1 + 1)\mathfrak{B}(x_1 \geq 0)p_{\mathbf{x}_0,\mathbf{x}-\boldsymbol{\gamma}(\cdot,1)}(t)$$

$$+ k_2^{sto}(x_2 + 1)x_2\mathfrak{B}(x_2 \geq 1)p_{\mathbf{x}_0,\mathbf{x}-\boldsymbol{\gamma}(\cdot,2)}(t)$$

$$+ k_3^{sto}(x_2 + 1)x_3\mathfrak{B}(x_2 \geq 0, x_3 \geq 1)p_{\mathbf{x}_0,\mathbf{x}-\boldsymbol{\gamma}(\cdot,3)}(t)$$

$$- k_1^{sto}x_1\mathfrak{B}(x_1 > 0)p_{\mathbf{x}_0,\mathbf{x}}(t)$$

$$- k_2^{sto}x_2(x_2 - 1)\mathfrak{B}(x_2 > 1)p_{\mathbf{x}_0,\mathbf{x}}(t)$$

$$- k_3^{sto}x_2x_3\mathfrak{B}(x_2 > 0, x_3 > 0)p_{\mathbf{x}_0,\mathbf{x}}(t),$$

where $\boldsymbol{\gamma}(\cdot, 1) = \begin{bmatrix} -1 \ 1 \ 0 \end{bmatrix}^\top$, $\boldsymbol{\gamma}(\cdot, 2) = \begin{bmatrix} -1 \ 0 \ 1 \end{bmatrix}^\top$, and $\boldsymbol{\gamma}(\cdot, 3) = \begin{bmatrix} 1 \ -1 \ 0 \end{bmatrix}^\top$ are the reaction step vectors. This is called the **master equation** of the usual stochastic model of the Robertson reaction (see the general form in Sect. 10.2.5). Notice that for each given $\mathbf{x}_0$, the above is a **finite system** of constant coefficient linear ordinary differential equations, but—as the number of equations can be very large—it is much more complicated than the induced kinetic differential equation.

10.2.2 Jump Markov Process with Stochastic Kinetics

Let us recapitulate the standard setting from (2.1), that is, a reaction is given by

$$\sum_{m \in \mathcal{M}} \alpha(m, r) X(m) \longrightarrow \sum_{m \in \mathcal{M}} \beta(m, r) X(m) \quad (r \in \mathcal{R}), \tag{10.3}$$

where $X(m)$ denotes the mth species and the stoichiometric matrix $\boldsymbol{\gamma}$ is defined by $\boldsymbol{\gamma} := \boldsymbol{\beta} - \boldsymbol{\alpha}$. Suppose that a nonnegative function $\lambda_r : \mathbb{Z}^M \to \mathbb{R}_0^+$ is assigned to each reaction step $r \in \mathcal{R}$. We require the following property to hold.

$$\boxed{\lambda_r(\mathbf{x}) > 0 \quad \text{if and only if} \quad \mathbf{x} - \boldsymbol{\alpha}(\cdot, r) \in \mathbb{N}_0^M,} \tag{10.4}$$

that is, in plain words, λ_r is positive if and only if there is enough species for the rth reaction step to take place. This condition naturally guarantees for the process we introduce below to keep the state space $\mathbb{N}_0^M$.

The functions $(\lambda_r)_{r \in \mathcal{R}}$ fulfilling (10.4) are called the **stochastic kinetics**, not to be confused with the same name used for the branch of the science. In the chemical literature, λ_r are often called the **propensity** or the **intensity** functions.

Researchers on different fields often consider specific choices for the stochastic kinetics that fit their scenarios. Now, let us discuss the most often used choice. First, we call the $k_r^{\text{sto}} \in \mathbb{R}^+$ the **stochastic reaction rate coefficient** corresponding to the rth reaction step. Then (using the definitions of vectorial operations in the Appendix) we define the stochastic kinetics $(\lambda_r)_{r \in \mathcal{R}}$ as

$$\lambda_r(\mathbf{x}) := k_r^{\text{sto}} \boldsymbol{\alpha}(\cdot, r)! \binom{\mathbf{x}}{\boldsymbol{\alpha}(\cdot, r)} = \kappa_r \binom{\mathbf{x}}{\boldsymbol{\alpha}(\cdot, r)}, \tag{10.5}$$

where $r \in \mathcal{R}$, $\mathbf{x} \in \mathbb{N}_0^M$, and $\kappa_r = k_r^{\text{sto}} \boldsymbol{\alpha}(\cdot, r)!$. With the Definition 13.7 of the falling factorials, one can reformulate (10.5) as

$$\boxed{\lambda_r(\mathbf{x}) = k_r^{\text{sto}} [\mathbf{x}]_{\boldsymbol{\alpha}(\cdot, r)} = k_r^{\text{sto}} \prod_{m=1}^M \frac{x_m!}{(x_m - \alpha(m, r))!} \mathfrak{B}(x_m \geq \alpha(m, r)).} \tag{10.6}$$

One can easily check that condition (10.4) is fulfilled with this choice. We often refer to the above-defined λ_r as the **stochastic mass action type kinetics**. It is also called the **Kurtz-type kinetics** or **combinatorial kinetics**. Combinatorial kinetics is easily accessible within *Mathematica* as

$$\lambda_r(\mathbf{x}) = k_r^{\mathrm{sto}}\, \texttt{Apply[Times, FactorialPower[x, }\alpha(\cdot, r)\texttt{]]}.$$

Let us make some comments.

- The stochastic mass action type kinetics bear quite natural interpretations: for $r \in \mathscr{R}$, the right-hand side of (10.5) is proportional to the number of those combinations of species in state $\mathbf{x}$ that can feed the rth reaction step. This reflects the idea to be discussed in Sect. 10.2.4 that the species in the vessel are well-stirred in the sense that they are equally likely to be at any location at any time.
- If the order of the rth reaction step is at most one, then the form of the stochastic mass action type kinetics λ_r coincides with that of the deterministic one w_r (recall Eq. (6.6)).
- In general, the unit of the stochastic reaction rate coefficients can differ from that of the deterministic counterparts used in the induced kinetic differential equation (see Sect. 10.2.3).

Now, the Markov process assigned to (10.3) can be defined as follows.

Definition 10.1 The **induced kinetic Markov process** of reaction (10.3) endowed with the stochastic kinetics $(\lambda_r)_{r \in \mathscr{R}}$ is defined to be the continuous time, time-homogeneous pure jump Markov process usually denoted as $\mathbf{X}$, where

$$\mathbf{X}(t) = \begin{bmatrix} X_1(t) & X_2(t) & \cdots & X_M(t) \end{bmatrix}^{\mathsf{T}} \qquad (t \in \mathbb{R}_0^+)$$

with state space $\mathbb{N}_0^M$ and **infinitesimal generator** $\mathbf{G}$ acting on a function $f : \mathbb{Z}^M \to \mathbb{R}$ in the following way:

$$(\mathbf{G}f)(\mathbf{x}) := \sum_{r=1}^{R} \lambda_r(\mathbf{x})\big(f(\mathbf{x} + \boldsymbol{\gamma}(\cdot, r)) - f(\mathbf{x})\big). \qquad (10.7)$$

In particular, the **induced kinetic Markov process with stochastic mass action type kinetics** is defined to be the one with stochastic kinetics given in (10.6).

Using different approaches we are going to construct the above-defined process in Sect. 10.2.6. We underline that for some stochastic kinetics, $\mathbf{X}$ might **blow-up**, which phenomenon is going to be discussed in Sect. 10.3.6 below. In these cases there exists a random time, which is finite with positive probability, at which one of the coordinates of $\mathbf{X}$ reaches the "cemetery" state, "∞", and after the explosion, the process ceases to exist by default.

The induced kinetic Markov process is a continuous time, discrete state space stochastic model, a **CDS**) model (cf. Érdi and Tóth 1989, p. 19). Note that the class of these does not coincide with major, well-known classes of stochastic processes (see further details in Sect. 10.4).

The infinitesimal generator $\mathbf{G}$ can heuristically be viewed as a(n infinite) matrix acting from $\mathbb{R}^{\mathbb{N}_0^M}$ to $\mathbb{R}^{\mathbb{N}_0^M}$. Hence, with a slight overload of notation, the generator $\mathbf{G}$ can be identified by the matrix $\left[g_{\mathbf{xy}} \right]_{\mathbf{x},\mathbf{y} \in \mathbb{N}_0^M}$, where

$$g_{\mathbf{xy}} := \begin{cases} \sum_{r \in \mathscr{R}, \mathbf{y}-\mathbf{x} = \gamma(\cdot, r)} \lambda_r(\mathbf{x}), & \text{if } \mathbf{x} \neq \mathbf{y}; \\ -\sum_{\mathbf{z} \in \mathbb{N}_0^M, \mathbf{z} \neq \mathbf{x}} g_{\mathbf{xz}}, & \text{if } \mathbf{x} = \mathbf{y}. \end{cases} \tag{10.8}$$

Roughly speaking, $\mathbf{G}$ describes how the probability of a potential change in states should locally look like, that is,

$$\mathbf{P}\{\mathbf{X}(t+h) = \mathbf{y} \mid \mathbf{X}(t) = \mathbf{x}, (\mathbf{X}(s))_{0 \leq s < t}\} = g_{\mathbf{xy}}h + \varepsilon(h)h$$

$$= \sum_{r=1}^{R} \mathfrak{B}(\mathbf{y} - \mathbf{x} = \gamma(\cdot, r))\lambda_r(\mathbf{x})h + \varepsilon(h)h, \tag{10.9}$$

where h is sufficiently small and $\mathbf{x} \neq \mathbf{y} \in \mathbb{N}_0^M$. On the other hand, we also have that

$$\mathbf{P}\{\mathbf{X}(t+h) = \mathbf{x} \mid \mathbf{X}(t) = \mathbf{x}, (\mathbf{X}(s))_{0 \leq s < t}\} = 1 + g_{\mathbf{xx}}h + \varepsilon(h)h$$

$$= 1 - \sum_{r=1}^{R} \lambda_r(\mathbf{x})h + \varepsilon(h)h. \tag{10.10}$$

These local laws turn out to be quite useful in deriving the master equation. The master equation is a general evolution equation for the probability law of $\mathbf{X}$ (see later in Sect. 10.2.5), which has already been obtained from Eqs. (10.1) and (10.2) for the Robertson reaction (cf. (10.9) and (10.10)).

Let us define the process $(\Upsilon_r(t))_{t \geq 0}$ for each $r \in \mathscr{R}$ that counts how many times the rth reaction step has taken place until time t. Hence, the vector process $\Upsilon(t) := \left[\Upsilon_1(t) \ \Upsilon_2(t) \ \cdots \ \Upsilon_M(t) \right]^\top$ $(t \geq 0)$ is called the **(stochastic) reaction extent** of the induced kinetic Markov process.

Equations (10.9) and (10.10) can be reformulated in terms of Υ_r as

$$\mathbf{P}\{\Upsilon_r(t+h) - \Upsilon_r(t) = 1 \mid \mathbf{X}(t) = \mathbf{x}, (\mathbf{X}(s))_{0 \leq s < t}\} = \lambda_r(\mathbf{x})h + \varepsilon(h)h, \tag{10.11}$$

$$\mathbf{P}\big\{\varUpsilon_r(t+h) - \varUpsilon_r(t) = 0 \,|\, \mathbf{X}(t) = \mathbf{x}, (\mathbf{X}(s))_{0 \le s < t}\big\} = 1 - \lambda_r(\mathbf{x})h + \varepsilon(h)h,$$

$$(10.12)$$

and the probability that $\varUpsilon_r$ increases by more than 1 in a small time interval of length h given $\mathbf{X}(t) = \mathbf{x}$ is of order $\varepsilon(h)h$. Note that for each $r \in \mathscr{R}$: $\varUpsilon_r$ is a nonnegative valued, nondecreasing jump process, a **counting process**. The deterministic analogue $\mathbf{U}$ of $\varUpsilon$ was introduced in Sect. 6.3.5.

For stochastic mass action type kinetics, an analogue of the right-hand side of the induced kinetic differential equation can be introduced (recall Eqs. (6.14) and (6.6)) with the help of the infinitesimal generator $\mathbf{G}$, namely,

$$\mathbf{f}^{\text{sto}}(\mathbf{x}) := (\mathbf{G}\, \text{id})(\mathbf{x}) = \sum_{r=1}^{R} (\boldsymbol{\beta}(\cdot, r) - \boldsymbol{\alpha}(\cdot, r)) \lambda_r(\mathbf{x})$$

$$= \sum_{r=1}^{R} (\boldsymbol{\beta}(\cdot, r) - \boldsymbol{\alpha}(\cdot, r)) k_r^{\text{sto}}[\mathbf{x}]_{\boldsymbol{\alpha}(\cdot, r)}, \qquad (10.13)$$

where $\mathbf{G}$ acts on id : $\mathbb{N}_0^M \to \mathbb{R}^M$ componentwise. The function $\mathbf{f}^{\text{sto}} : \mathbb{N}_0^M \to \mathbb{R}^M$ gets its meaning later at the derivation of the master equation (see below in Sect. 10.2.5).

Finally, we assume that the sample paths of the induced kinetic Markov process are right-continuous having finite limits from the left. In particular, when the rth reaction step takes place at time t, then $\mathbf{X}$ is updated so that $\mathbf{X}(t) = \mathbf{X}(t-) + \boldsymbol{\gamma}(\cdot, r)$.

10.2.3 Conversion of Units

This short section explores the connection between the stochastic and deterministic reaction rate coefficients. To differentiate the two, we have denoted the stochastic ones by k_r^{sto} as opposed to the deterministic ones being $k_r = k_r^{\text{det}}$ $(r \in \mathscr{R})$.

Assume that among M different species, the reaction steps of (10.3) take place in a vessel having a **constant volume** $V \in \mathbb{R}^+$ [dm^3]. In the usual deterministic setting, the solution $t \mapsto \mathbf{c}(t) = \big[c_1(t)\, c_2(t) \cdots c_M(t)\big]^\top \in (\mathbb{R}_0^+)^M$ of the induced kinetic differential equation gives the concentration vs. time curves of each species, that is, $c_m(t)$ is measured in mol dm^{-3}, while the induced kinetic Markov process $\mathbf{X}(t) = \big[X_1(t)\, X_2(t) \cdots X_M(t)\big]^\top \in \mathbb{N}_0^M$ is dimensionless. This latter gives the number of species $X(m)$ for $m \in \mathscr{M}$ that is present at time t in the volume V. Hence

$$[\lambda_r(\mathbf{X}(t))] = \sec^{-1} = [\dot{\mathbf{c}}(t) N_A V] = [w_r(\mathbf{c}(t)) N_A V]$$

holds in units, where N_A is the Avogadro constant ($\approx 6.022 \times 10^{23} \text{mol}^{-1}$). Note that [$\mathbf{u}$] is the **unit** of quantity $\mathbf{u}$.

If the induced kinetic differential equation is endowed with mass action type kinetics (6.7) and in a similar way the induced kinetic Markov process is considered with stochastic mass action type kinetics (10.5), then the following equality holds for the **units** of reaction rate coefficients

$$\boxed{[k_r^{\text{sto}}] = [k_r^{\text{det}}(N_A V)^{1-o(r)}],}\qquad (10.14)$$

where $o(r)$ is the order of the rth reaction step (see Definition 2.4 and Eq. (6.12)). It follows that k_r^{sto} is measured in $\sec^{-1}$ regardless of the order of the reaction steps. In particular, for first-order reactions Eq. (10.14) implies coincidence in units, i.e., $[k_r^{\text{sto}}] = [k_r^{\text{det}}]$.

Hereinafter, with a slight abuse of notation, we will make no distinction in notation between the deterministic and stochastic reaction rate coefficients. Hence, if it does not cause any confusion or ambiguity, the superscript of the stochastic reaction rate coefficients is neglected.

10.2.4 More on the Underlying Assumptions

Moving one step backward, it can be investigated why the above-defined induced kinetic Markov process is so natural to deal with. When one tries to understand complex (bio)chemical or physical phenomena, experiments, which possibly exhibit random behavior, may have several nice characteristics one may wish to build into a mathematical model. Generally speaking, it is up to the modeler to take into account all the relevant features his/her model should obey in order to explain the experiments in good agreement. Let us formulate some of the most natural ones, also pointing out the connections with the induced kinetic Markov process and the induced kinetic differential equation.

Dynamics of the Process

1. The probability of the next reaction step to take place can be predicted from the current information, and no information is needed about the process from the past. That is, the process should have **no memory**.
2. If we know the state of the process at a particular time, then what we will see thereafter is statistically the same as if we have just started the process. In other words, the process should be **time homogeneous**.
3. There is no accumulation of reaction steps at any time. That is, the probability that infinitely many reaction steps take place in a finite time interval is zero. The number of reaction steps taking place in a finite time window can however be arbitrarily large.

How are These Properties Reflected in the Induced Kinetic Markov Process?
The first requirement simply follows from the fact that stochastic kinetics λ_r ($r \in \mathscr{R}$) only depend on the current state of the process. **No** delay or previous state dependence in the kinetics λ_r. Although the usual deterministic model has no memory either, there exist efforts to introduce this effect into models (see Atlan and Weisbuch 1973; Lipták et al. 2017). Time homogeneity is explicitly stated in Definition 10.1, but it also follows from the fact that λ_r are independent of t for $r \in \mathscr{R}$. Note that this is the direct analogue of the **autonomousness** of the induced kinetic differential equation. Assumption on the accumulation does not easily come. We may hope that in physically relevant cases, this property can easily be verified in the mathematical model. However, we cannot avoid the effects of blow-up for some choices of λ_r, even in the case of stochastic mass action type kinetics. We further investigate this topic in Sect. 10.3.6.

State Space

1. The amounts of different species can be measured by discrete or continuous variables.
2. The mixture of the species can be **well-stirred** (space homogeneity) or can be **spatially inhomogeneous**.

How Are These Properties Reflected in the Induced Kinetic Markov Process?
The induced kinetic Markov process deals with discrete quantities by definition. Spatial homogeneity is much simpler to start with since spatial inhomogeneity would require the "spatial derivatives" of quantities involved; spatial inhomogeneous models are out of the scope of the present book.

10.2.5 Master Equation

The **master equation** as we have seen in a specific example (see Sect. 10.2.1) describes the time-dependent probability of the process being in a state $\mathbf{x} \in \mathbb{N}_0^M$ starting from some $\mathbf{x}_0 \in \mathbb{N}_0^M$. Let us introduce the **transition probabilities** associated with an induced kinetic Markov process $\mathbf{X}$ as

$$p_{\mathbf{x}_0, \mathbf{x}}(t) := \mathbf{P}\{\mathbf{X}(t) = \mathbf{x} \mid \mathbf{X}(0) = \mathbf{x}_0\} \quad (\mathbf{x} \in \mathbb{N}_0^M), \tag{10.15}$$

where $\mathbf{x}_0 \in \mathbb{N}_0^M$ is the initial state.

It is easy to see that p satisfies the following (Liggett 2010, Theorem 2.12):

1. for every $t \geq 0$ and $\mathbf{x}_0, \mathbf{x} \in \mathbb{N}_0^M : 0 \leq p_{\mathbf{x}_0, \mathbf{x}}(t) \leq 1$;
2. for all $t \geq 0$ and $\mathbf{x}_0 \in \mathbb{N}_0^M : \sum_{\mathbf{x} \in \mathbb{N}_0^M} p_{\mathbf{x}_0, \mathbf{x}}(t) \leq 1$; and
3. for all $t, s \geq 0$, the **Chapman–Kolmogorov equations** hold, i.e.,

$$p_{\mathbf{x}_0, \mathbf{x}}(t + s) = \sum_{\mathbf{y} \in \mathbb{N}_0^M} p_{\mathbf{x}_0, \mathbf{y}}(t) p_{\mathbf{y}, \mathbf{x}}(s) \quad (\mathbf{x}_0, \mathbf{x} \in \mathbb{N}_0^M).$$

In the sufficiently "small" time interval $[t, t + h[$, the two only ways for $\mathbf{X}$ to reach $\mathbf{x} \in \mathbb{N}_0^M$ are the following:

1. the system was in a preceding state and then it jumps to $\mathbf{x}$ by a reaction step (cf. (10.9));
2. the system stays in the same state; no reaction steps have taken place (cf. (10.10)).

This verbally formulated idea is embodied in a system of differential equations.

Theorem 10.2 (Master Equation) *Let $\mathbf{X}$ be an induced kinetic Markov process with stochastic kinetics λ_r. Then*

$$\dot{p}_{\mathbf{x}_0, \mathbf{x}}(t) = \sum_{r=1}^{R} \left(\lambda_r(\mathbf{x} - \boldsymbol{\gamma}(\cdot, r)) p_{\mathbf{x}_0, \mathbf{x} - \boldsymbol{\gamma}(\cdot, r)}(t) - \lambda_r(\mathbf{x}) p_{\mathbf{x}_0, \mathbf{x}}(t) \right), \tag{10.16}$$

for all $\mathbf{x} \in \mathbb{N}_0^M$, so that $\mathbf{X}(0) = \mathbf{x}_0$ is assumed to be deterministic.

Notice that Eq. (10.16) is a linear system of differential equations with constant coefficients which possibly consist of **infinitely many equations**. Equation (10.16) can also be written in matrix form: $\dot{\mathbf{p}}_{\mathbf{x}_0}(t) = \mathbf{p}_{\mathbf{x}_0}(t) \mathbf{G}$, where $\mathbf{p}_{\mathbf{x}_0}(t) := [p_{\mathbf{x}_0, \mathbf{x}}(t)]_{\mathbf{x} \in \mathbb{N}_0^M}$ is a row vector, while $\mathbf{G}$ is the matrix representation of the infinitesimal generator given in (10.8). Now, the formal solution to the master equation can be written as $\mathbf{p}_{\mathbf{x}_0}(t) = $"$\mathbf{x}_0 \exp(t \, \mathbf{G})$". This is why the generator $\mathbf{G}$ deserved its name: it generates a **one-parameter operator semigroup** (for a comprehensive monograph, see Ethier and Kurtz 2009, Chapters 1 and 4). If the process $\mathbf{X}$, starting from some $\mathbf{x}_0 \in \mathbb{N}_0^M$, can only patrol a finite set of $\mathbb{N}_0^M$ by the reaction (10.3), then the number of equations involved in (10.16) is finite; hence the matrix exponential is straightforward (see Sects. 10.2.7 and 10.3.1). In general, however, much care is needed to be able to define at all the "exponential" of an infinite matrix or more precisely of an operator; see the chapters mentioned in Ethier and Kurtz (2009) and Chapter 3 of Liggett (2010) for further reading.

Proof Pick a small $h > 0$ and then write

$$p_{\mathbf{x}_0,\mathbf{x}}(t+h) = \sum_{\mathbf{y} \in \mathbb{N}_0^M} \mathbf{P}\{\mathbf{X}(t+h) = \mathbf{x} \mid \mathbf{X}(t) = \mathbf{y}\}\mathbf{P}\{\mathbf{X}(t) = \mathbf{y} \mid \mathbf{X}(0) = \mathbf{x}_0\}$$

$$= \sum_{r=1}^{R} \mathbf{P}\{\mathbf{X}(h) = \mathbf{x} \mid \mathbf{X}(0) = \mathbf{x} - \boldsymbol{\gamma}(\cdot, r)\}p_{\mathbf{x}_0,\mathbf{x}-\boldsymbol{\gamma}(\cdot,r)}(t)$$

$$+ \mathbf{P}\{\mathbf{X}(t+h) = \mathbf{x} \mid \mathbf{X}(t) = \mathbf{x}\}p_{\mathbf{x}_0,\mathbf{x}}(t) + \varepsilon(h)h$$

$$= h \sum_{r=1}^{R} \lambda_r(\mathbf{x} - \boldsymbol{\gamma}(\cdot, r))p_{\mathbf{x}_0,\mathbf{x}-\boldsymbol{\gamma}(\cdot,r)}(t)$$

$$+ \left(1 - h \sum_{r=1}^{R} \lambda_r(\mathbf{x})\right)p_{\mathbf{x}_0,\mathbf{x}}(t) + \varepsilon(h)h,$$

where we used the law of total probability, the memorylessness, and time homogeneity of $\mathbf{X}$ and (10.9) and (10.10) (ε is so small that $\lim_0 \varepsilon = 0$). Rearranging the above equation, we arrive at

$$\frac{p_{\mathbf{x}_0,\mathbf{x}}(t+h) - p_{\mathbf{x}_0,\mathbf{x}}(t)}{h} = \sum_{r=1}^{R} \left(\lambda_r(\mathbf{x} - \boldsymbol{\gamma}(\cdot, r))p_{\mathbf{x}_0,\mathbf{x}-\boldsymbol{\gamma}(\cdot,r)}(t) - \lambda_r(\mathbf{x})p_{\mathbf{x}_0,\mathbf{x}}(t)\right) + \varepsilon(h).$$

Finally, taking the limit $h \downarrow 0$, we obtain the desired Eq. (10.16).   □

In many cases it is not possible to **symbolically** solve the master equation even for relatively small systems as the number of possible states and that of coupled differential equations easily becomes very large. Various computational techniques including symbolical and numerical methods have been developed to calculate or approximate the transition probabilities of the species as a function of time. For further details, see Sect. 10.3.1. In **ReactionKinetics MasterEquation** provides the master equation of a given reaction.

Let us summarize the general evolution equations for continuous time jump Markov processes specialized for the case of the induced kinetic Markov process.

Theorem 10.3 (Kolmogorov's Forward and Backward Equations)   *Consider an induced kinetic Markov process* $\mathbf{X}$ *with stochastic kinetics* $(\lambda_r)_{r \in \mathscr{R}}$, *and as usual let* $p_{\mathbf{x},\mathbf{y}}(t) = \mathbf{P}\{\mathbf{X}(t) = \mathbf{y} \mid \mathbf{X}(0) = \mathbf{x}\}$ *be the transition probability. Then we have*

$$\dot{p}_{\mathbf{x},\mathbf{y}}(t) = \sum_{r=1}^{R} \left(\lambda_r(\mathbf{y} - \boldsymbol{\gamma}(\cdot, r))p_{\mathbf{x},\mathbf{y}-\boldsymbol{\gamma}(\cdot,r)}(t) - \lambda_r(\mathbf{y})p_{\mathbf{x},\mathbf{y}}(t)\right), \tag{10.17}$$

$$\dot{p}_{\mathbf{x},\mathbf{y}}(t) = \sum_{r=1}^{R} \lambda_r(\mathbf{x})\big(p_{\mathbf{x}+\boldsymbol{\gamma}(\cdot,r),\mathbf{y}}(t) - p_{\mathbf{x},\mathbf{y}}(t)\big), \tag{10.18}$$

where $p_{\mathbf{x},\mathbf{y}}(0) = \mathfrak{B}(\mathbf{x} = \mathbf{y})$ and $\mathbf{x}, \mathbf{y} \in \mathbb{N}_0^M$.

Proof The system of differential equations simply follows from the application of the law of total probability taking into account (10.9) and (10.10) (for more details see Liggett 2010, Chapter 2). $\qquad\square$

Notice that the master equation (10.16) is indeed Kolmogorov's forward equation (10.17). What follows is a more general system of differential equations for the expectation of an arbitrary function of **X**. This is also called the Dynkin's formula (cf. Øksendal 2003, Section 7.4).

Theorem 10.4 (Dynkin) *Let $a : \mathbb{N}_0^M \to \mathbb{R}$ be any function for which*

$$\mathbf{E}\big|\lambda_r(\mathbf{X}(t))a(\mathbf{X}(t))\big| < +\infty$$

holds for all $r \in \mathcal{R}$ and $t \in [0, T[$ where $T > 0$ and $\mathbf{X}$ is an induced kinetic Markov process with (general) stochastic kinetics $(\lambda_r)_{r\in\mathcal{R}}$. Then we have

$$\frac{\mathrm{d}}{\mathrm{d}t}\mathbf{E}\big\{a(\mathbf{X}(t)) \mid \mathbf{X}(0) = \mathbf{x}\big\}$$

$$= \sum_{r=1}^{R} \mathbf{E}\big\{\lambda_r(\mathbf{X}(t))\big(a(\mathbf{X}(t) + \boldsymbol{\gamma}(\cdot, r)) - a(\mathbf{X}(t))\big) \mid \mathbf{X}(0) = \mathbf{x}\big\}. \tag{10.19}$$

Proof Let $h > 0$ be sufficiently small and start with

$$\mathbf{E}\big\{a(\mathbf{X}(t+h)) - a(\mathbf{X}(t)) \mid \mathbf{X}(0) = \mathbf{x}\big\}$$

$$= \sum_{\mathbf{y}\in\mathbb{N}_0^M} \mathbf{E}\big\{a(\mathbf{X}(t+h)) - a(\mathbf{y}) \mid \mathbf{X}(t) = \mathbf{y}\big\}\mathbf{P}\{\mathbf{X}(t) = \mathbf{y} \mid \mathbf{X}(0) = \mathbf{x}\}$$

$$= \sum_{\mathbf{y}\in\mathbb{N}_0^M} \sum_{r=1}^{R} \lambda_r(\mathbf{y})h \cdot (a(\mathbf{y} + \boldsymbol{\gamma}(\cdot, r)) - a(\mathbf{y}))\mathbf{P}\{\mathbf{X}(t) = \mathbf{y} \mid \mathbf{X}(0) = \mathbf{x}\} + \varepsilon(h)h$$

$$= h\sum_{r=1}^{R} \mathbf{E}\big\{\lambda_r(\mathbf{X}(t))(a(\mathbf{X}(t) + \boldsymbol{\gamma}(\cdot, r)) - a(\mathbf{X}(t))) \mid \mathbf{X}(0) = \mathbf{x}\big\} + \varepsilon(h)h,$$

where we used the memorylessness of **X**, the law of total probability twice, and (10.9). Dividing the previous display by h and sending h to 0, we obtain the desired Eq. (10.19). $\qquad\square$

In particular, if we choose a to be the indicator function, i.e., $a(\mathbf{x}) = \mathcal{B}(\mathbf{x} = \mathbf{y})$ ($\mathbf{x} \in \mathbb{N}_0^M$) for some fixed $\mathbf{y} \in \mathbb{N}_0^M$, then Eq. (10.19) returns Kolmogorov's forward equation or, in other words, the master equation (10.16).

Theorem 10.4 enables us to deduce a system of differential equations for the first moment or, more generally, any kind of moments of the induced kinetic Markov process $\mathbf{X}$. For explicit choices and further details, see Sect. 10.3.3.

10.2.6 Equivalent Representations of the Induced Kinetic Markov Process

Recall the definition of induced kinetic Markov process from Sect. 10.2.2. The present section focuses on different representations resulting in the same process.

10.2.6.1 Memorylessness Emphasized

Since the desired jump process should have no memory, the candidate for the waiting time distribution is the exponential one. An exponentially distributed random variable, say ξ, (and only this among continuous distributions) has the property that $\mathbf{P}\{\xi > s + t \mid \xi > t\} = \mathbf{P}\{\xi > s\}$ for all $t, s > 0$. This formula expresses what is needed: the probability that one will have to wait at least (an additional) time s for an event to happen ($\xi > s$) is the same as the conditional probability given that one has already waited time t without anything happened (i.e., $\xi > s + t$ assuming that $\xi > t$). For further properties of the exponential distribution, see Norris (1998, Section 2.3).

Now, consider a double sequence of nonnegative real numbers called the **transition rates,** $Q(\mathbf{x}, \mathbf{y})$ ($\mathbf{x}, \mathbf{y} \in \mathbb{N}_0^M$). Using the kinetics $(\lambda_r)_{r \in \mathcal{R}}$, one can define Q as

$$Q(\mathbf{x}, \mathbf{y}) := \sum_{r=1}^{R} \mathcal{B}(\mathbf{y} - \mathbf{x} = \boldsymbol{\gamma}(\cdot, r)) \lambda_r(\mathbf{x}).$$

Notice that if the reaction step vectors $(\boldsymbol{\gamma}(\cdot, r))_{r \in \mathcal{R}}$ are all different, then $Q(\mathbf{x}, \mathbf{y})$ equals to $\lambda_r(\mathbf{x})$ whenever $\mathbf{x}, \mathbf{y}$ are chosen so that $\mathbf{y} - \mathbf{x} = \boldsymbol{\gamma}(\cdot, r)$. In other words, $Q(\mathbf{x}, \mathbf{y})$ expresses the **aggregate rate** at which the process will possibly jump from state $\mathbf{x} \in \mathbb{N}_0^M$ to $\mathbf{y} \in \mathbb{N}_0^M$. By using the infinitesimal generator $\mathbf{G}$ (see Eq. (10.8)), one can write that $Q(\mathbf{x}, \mathbf{y}) = g_{\mathbf{xy}}$ if $\mathbf{x} \neq \mathbf{y} \in \mathbb{N}_0^M$. It follows that the **total rate** of leaving state $\mathbf{x} \in \mathbb{N}_0^M$ is

$$\Lambda(\mathbf{x}) := \sum_{r=1}^{R} \lambda_r(\mathbf{x}) = \sum_{\mathbf{y} \in \mathbb{N}_0^M} Q(\mathbf{x}, \mathbf{y}) \tag{10.20}$$

for all $\mathbf{x} \in \mathbb{N}_0^M$.

How should the process X evolve in time?

Take a mutually independent collection of exponentially distributed random variables $(\xi_i)_{i\in\mathbb{N}}$ with unit mean. Assume that $\mathbf{X}$ initially is in state $\mathbf{x} \in \mathbb{N}_0^M$, then the first jump is going to occur at time $\tau_1 := \frac{1}{\Lambda(\mathbf{x})}\xi_1$, and at that time the process will jump to state $\mathbb{N}_0^M \ni \mathbf{z}_1 \neq \mathbf{x}$ with probability $Q(\mathbf{x}, \mathbf{z}_1)/\Lambda(\mathbf{x})$.

Assume now that after $n - 1$ steps $(n > 1)$ $\mathbf{X}$ is in state $\mathbf{z}_{n-1}$ at time τ_{n-1}. Then the next jump is going to occur at time $\tau_n := \tau_{n-1} + \frac{1}{\Lambda(\mathbf{z}_{n-1})}\xi_n$, and $\mathbf{X}$ will jump to state $\mathbb{N}_0^M \ni \mathbf{z}_n \neq \mathbf{z}_{n-1}$ with probability $Q(\mathbf{z}_{n-1}, \mathbf{z}_n)/\Lambda(\mathbf{z}_{n-1})$.

From this construction it follows that $\mathbf{X}$ is a time-homogeneous pure jump Markov process. Indeed, the construction suggests that one could separately handle the jumps of $\mathbf{X}$ from the **waiting times** (or **holding times**) $(\tau_n - \tau_{n-1})_{n\in\mathbb{N}}$ $(\tau_0 := 0)$. The process $(\hat{\mathbf{X}}(n) := \mathbf{X}(\tau_n))_{n\in\mathbb{N}_0}$ is a discrete-time Markov chain; in plain words, if $\hat{\mathbf{X}}$ is in state $\mathbf{x}$ at n, it is then updated to $(\mathbf{x}+\boldsymbol{\gamma}(\cdot, r))_{r\in\mathscr{R}}$ in step $n+1$ according to the probability distribution $(\lambda_r(\mathbf{x})/\Lambda(\mathbf{x}))_{r\in\mathscr{R}}$. Note that the transitions of $\hat{\mathbf{X}}$ do **not** depend on the length of the waiting time, but the waiting times **do** depend on which state the process is in. The process $\hat{\mathbf{X}}$ is also called the **embedded Markov chain** of $\mathbf{X}$. We notice that this approach is the basis of **direct simulation methods** (see those later in Sect. 10.7.1).

In the following we prove that the above construction is in correspondence with the definition of the induced kinetic Markov proces. Indeed, we determine the joint density function of the waiting time and jump distribution from Eqs. (10.9) and (10.10). Start $\mathbf{X}$ from $\mathbf{x} \in \mathbb{N}_0^M$, and let

$$q_{\mathbf{x}}(s) := \mathbf{P}\{\text{first waiting time} > s \mid \mathbf{X}(0) = \mathbf{x}\}$$
$$= \mathbf{P}\{\text{no reaction step takes place in } [0, s] \mid \mathbf{X}(0) = \mathbf{x}\}.$$

Let $[s, s + h]$ be a "small" time interval and then

$$q_{\mathbf{x}}(s + h) = q_{\mathbf{x}}(s)\mathbf{P}\{\text{no reaction step takes place in } [s, s + h] \mid \mathbf{X}(s) = \mathbf{x}\}$$
$$= q_{\mathbf{x}}(s)q_{\mathbf{x}}(h),$$

using the memorylessness and time homogeneity of $\mathbf{X}$. Since $q_{\mathbf{x}}(0) = 1$, we have

$$\frac{q_{\mathbf{x}}(s + h) - q_{\mathbf{x}}(s)}{h} = q_{\mathbf{x}}(s)\frac{q_{\mathbf{x}}(h) - q_{\mathbf{x}}(0)}{h}.$$

Taking the limit $h \downarrow 0$ and recalling (10.10), we get the following initial value problem for a separable differential equation

$$\dot{q}_{\mathbf{x}}(s) = -q_{\mathbf{x}}(s)\Lambda(\mathbf{x}) \quad q_{\mathbf{x}}(0) = 1.$$

It is not hard to see that the solution is $q_{\mathbf{x}}(s) = \exp\{-\Lambda(\mathbf{x})s\}$; hence the (first) waiting time is exponentially distributed with mean $1/\Lambda(\mathbf{x})$ provided that the process was in state $\mathbf{x} \in \mathbb{N}_0^M$. Because of time homogeneity, the only relevant information for the next jump is the current state of the process; hence the density function v of a reaction step occurring first at time $s \geq 0$ is

$$\dot{v}_{\mathbf{x}}(s) = \frac{\mathrm{d}}{\mathrm{d}s}(1 - q_{\mathbf{x}}(s)) = \Lambda(\mathbf{x})\exp\{-\Lambda(\mathbf{x})s\}.$$

Putting the above together, the joint probability density function of the waiting time and first jump is

$$\mathbf{P}\{\mathbf{X} \text{ jumps first at } t \text{ by the } r\text{th reaction step} \mid \mathbf{X}(0) = \mathbf{x}\} = \frac{\lambda_r(\mathbf{x})}{\Lambda(\mathbf{x})}v_{\mathbf{x}}(t)$$

$$= \lambda_r(\mathbf{x})\exp\{-\Lambda(\mathbf{x})t\},$$

where $t \geq 0$ and $r \in \mathcal{R}$, using the fact that the jump from $\mathbf{x}$ is independent of the waiting time.

10.2.6.2 Reaction Steps Emphasized

Another approach becomes apparent when one focuses on how many and what types of jumps (i.e., reaction steps) have taken place until t. It is obvious that the waiting time is a continuous random variable (the distribution of which we calculated in the previous section), but the number of reaction steps having taken place in $[0, t]$ has to be a discrete random variable.

Now, let us again build up the process $\mathbf{X}$ from exponential clocks using the following approach. Take mutually independent sets of exponentially distributed random variables $(\xi_r^{(i)})_{i \in \mathbb{N}}$ for each $r \in \mathcal{R}$. Define recursively the sequence of random reaction steps $(r^{(n)})_{n \in \mathbb{N}}$ as

$$r^{(n)} := \arg\min_{r \in \mathcal{R}}\left\{\frac{\xi_r^{(n)}}{\lambda_r\left(\mathbf{x}_0 + \sum_{j=1}^{n-1}\gamma(\cdot, r^{(j)})\right)}\right\},$$

where $n \in \mathbb{N}$ and $\mathbf{X}(0) = \mathbf{x}_0 \in \mathbb{N}_0^M$ is fixed. In plain words, $r^{(n)}$ is the reaction step executed in the nth step. Now, the total elapsed time until the nth step of $\mathbf{X}$ can be defined as

$$\tau_n := \sum_{i=1}^{n}\frac{\xi_{r^{(i)}}^{(i)}}{\lambda_{r_i}\left(\mathbf{x}_0 + \sum_{j=1}^{i-1}\gamma(\cdot, r^{(j)})\right)}.$$

Thus, the evolution of $\mathbf{X}$ is described by the following equation:

$$\mathbf{X}(t) = \mathbf{x}_0 + \sum_{n=0}^{+\infty}\gamma(\cdot, r^{(n)})\mathcal{B}(\tau_n \leq t) = \mathbf{x}_0 + \sum_{r=1}^{R}\gamma(\cdot, r)\Upsilon_r(t).$$

In the previous display, $\Upsilon_r(t)$ counts the occurrences of the rth reaction step up to time t (see Eqs. (10.11) and (10.12)). The fact that Υ_r behaves as a counting process with local intensity "$\lambda_r(\mathbf{x})h$" suggests that by using random time change arguments, the induced kinetic Markov process $\mathbf{X}$ can be written in the following form:

$$\mathbf{X}(t) = \mathbf{x}_0 + \sum_{r=1}^{R} \boldsymbol{\gamma}(\cdot, r) \mathscr{P}_r\left(\int_0^t \lambda_r(\mathbf{X}(s))\,ds \right), \qquad (10.21)$$

where $(\mathscr{P}_r)_{r \in \mathscr{R}}$ is a set of mutually independent unit rate Poisson processes. Equation (10.21) is referred to as the **Poisson representation** of the induced kinetic Markov process. Moreover, Eq. (10.21) for an unknown $\mathbf{X}$ uniquely determines an induced kinetic Markov process up to time $\sup\{t : \sum_{r \in \mathscr{R}} \Upsilon_r(t) < +\infty\}$.

We underline that the distribution of $\mathscr{P}_r\left(\int_0^t \lambda_r(\mathbf{X}(s))\,ds \right)$ is not necessarily Poissonian. Let us consider two simple examples. For the simple linear inflow $0 \xrightarrow{k} \mathbf{X}$, Eq. (10.21) tells that $X(t) = x_0 + \Upsilon(t) = x_0 + \mathscr{P}(k \cdot t)$, where stochastic mass action type kinetics is assumed. Hence, Υ is a constant rate Poisson process; it has a Poisson distribution with mean $k \cdot t$. Second, consider the somewhat similar reaction $\mathbf{X} \xrightarrow{k} 2\mathbf{X}$ and the corresponding induced kinetic Markov process with stochastic mass action type kinetics. This latter one gives that $X(t) = x_0 + \Upsilon(t) = x_0 + \mathscr{P}\left(\int_0^t kX(s)\,ds \right)$. In particular,

$$\mathbf{P}\{\Upsilon(t) \geq n\} = \mathbf{P}\left\{ \sum_{j=1}^{n} \frac{1}{kj}\xi_j \leq t \right\} = (1 - \exp(-kt))^n,$$

where $t \geq 0$, $n \in \mathbb{N}_0$, and ξ's are mutually independent exponentially distributed random variables with unit mean. So in this case, $\Upsilon(t)$ has a geometric distribution with mean $\exp(kt)$.

The striking similarity between Eq. (10.21) and its deterministic counterpart Eq. (8.20) is quite straightforward. The Poisson representation comes from Gardiner and Chaturvedi (1977) and Kurtz (1978); see also Anderson and Kurtz (2015) and further references therein for a rigorous treatment. We also notice that this approach is the starting point of some **approximate simulation methods** to be discussed in Sect. 10.7.2.

10.2.7 Structure of the State Space

Let us start with some notation. We define the **discrete stoichiometric subspace** as

$$\mathscr{S}_d := \mathscr{S} \cap \mathbb{Z}^M = \operatorname{span}\{\boldsymbol{\gamma}(\cdot, r) \mid r \in \mathscr{R}\} \cap \mathbb{Z}^M.$$

The set of integer coordinate points of the reaction simplex (i.e., the stoichiometric compatibility class) for an $\mathbf{x}_0 \in \mathbb{N}_0^M$ is then $(\mathbf{x}_0 + \mathscr{S}_d) \cap \mathbb{N}_0^M$. Shortly we also call this set the **discrete reaction simplex** or equivalently the **discrete stoichiometric compatibility class**.

Based on the Poisson representation (10.21), one can state the following.

Theorem 10.5 *The trajectory of the induced kinetic Markov process* $\mathbf{X}$ *with stochastic kinetics* λ_r *wanders in the nonnegative integer coordinate points of the reaction simplex, i.e.,* $\mathbf{X}(t) \in (\mathbf{x}_0 + \mathscr{S}_d) \cap \mathbb{N}_0^M$ *holds for every* $t \geq 0$ *if* $\mathbf{X}(0) = \mathbf{x}_0 \in \mathbb{N}_0^M$.

Note again the similarity between the previous theorem and the integrated form of the induced kinetic differential equation (cf. Eq. (8.20)). Let us remark that for mass-consuming and mass-conserving reactions, Theorem 10.5 implies that $\mathbf{X}$ stays in a bounded domain of $\mathbb{N}_0^M$. Furthermore, it is easy to see that for $\mathbf{x}_0, \mathbf{y}_0 \in \mathbb{N}_0^M$: either $(\mathbf{x}_0 + \mathscr{S}_d) \cap \mathbb{N}_0^M = (\mathbf{y}_0 + \mathscr{S}_d) \cap \mathbb{N}_0^M$ or $(\mathbf{x}_0 + \mathscr{S}_d) \cap (\mathbf{y}_0 + \mathscr{S}_d) = \emptyset$ holds. Hence, there exists a countable set of initial vectors $\mathbf{x}_0^{(1)}, \mathbf{x}_0^{(2)}, \ldots$ such that $(\mathbf{x}_0^{(i)} + \mathscr{S}_d) \cap (\mathbf{x}_0^{(j)} + \mathscr{S}_d) = \emptyset$ holds for $i \neq j \in \mathbb{N}$ and $\mathbb{N}_0^M = \cup_{i \in \mathbb{N}} (\mathbf{x}_0^{(i)} + \mathscr{S}_d) \cap \mathbb{N}_0^M$. That is, for any initial vector $\mathbf{x}_0 \in \mathbb{N}_0^M$, there is a unique index $i \in \mathbb{N}$ such that $\mathbf{X}$ will wander in $(\mathbf{x}_0^{(i)} + \mathscr{S}_d) \cap \mathbb{N}_0^M \ni \mathbf{x}_0$.

Now, let us recapitulate the standard classifications of the states of a countable state space Markov process. We say that $\mathbf{x} \in \mathbb{N}_0^M$ is

- **recurrent**, if $\mathbf{P}\{\{t \geq 0 : \mathbf{X}(t) = \mathbf{x}\} \text{ is unbounded} \mid \mathbf{X}(0) = \mathbf{x}\} = 1$;
- **transient**, if $\mathbf{P}\{\{t \geq 0 : \mathbf{X}(t) = \mathbf{x}\} \text{ is unbounded} \mid \mathbf{X}(0) = \mathbf{x}\} = 0$.

The process $\mathbf{X}$ visits infinitely often its recurrent states, while the transient states are those which are never visited after some time. Note that every state is either recurrent or transient.

A state $\mathbf{y} \in \mathbb{N}_0^M$ is **accessible** from $\mathbf{x} \in \mathbb{N}_0^M$ if there exists a sequence of reaction steps $r_1, r_2, \ldots, r_b$ ($b \in \mathbb{N}$) such that $\mathbf{y} = \mathbf{x} + \sum_{k=1}^{b} \gamma(\cdot, r_k)$ and $\lambda_{r_a}(\mathbf{x} + \sum_{k=1}^{a-1} \gamma(\cdot, r_k)) > 0$ holds for all $a \in \{1, 2, \ldots, b\}$. Equivalently, if $p_{\mathbf{x}, \mathbf{y}}(t) > 0$ holds for some $t > 0$, then $\mathbf{y}$ is accessible from $\mathbf{x}$. Two states $\mathbf{x} \neq \mathbf{y} \in \mathbb{N}_0^M$ are said to be **communicating** if they are accessible from one another. Now, the **maximal sets** of states $U \subset \mathbb{N}_0^M$ whose states are communicating with each other are the **communicating classes**. Hence, if U is a communicating class, then for any two states $\mathbf{x}, \mathbf{y}$ of U, there is a positive probability that starting $\mathbf{X}$ from $\mathbf{x}$ it will reach $\mathbf{y}$ at some time. We say that the communicating class U is **closed** if there is no $\mathbf{y} \notin U$ which would be accessible from U. Otherwise, it is said to be **non-closed**. In some contexts closed communicating classes are also referred to as "irreducible components."

It easily follows from the previous concepts that there exists a countable collection $(U_j)_{j \in \mathbb{N}}$ of (distinct) communicating classes for which:

- $\cup_{j \in \mathbb{N}} U_j = \mathbb{N}_0^M$ such that $U_j \cap U_k = \emptyset$ for $j \neq k \in \mathbb{N}$, i.e., $(U_j)_{j \in \mathbb{N}}$ covers the whole state space $\mathbb{N}_0^M$; and
- for each $j \in \mathbb{N}$: U_j consists of only one kind of states, i.e., the recurrence and transience are class properties.

In other words, the communicating classes define a partition of the state space. Let us name some extreme cases:

- When U_j consists of a single recurrent state, it is also called an **absorbing state**.
- If there exists only a single (necessarily closed) communicating class, then the state space is called **irreducible**; otherwise it is said to be **reducible**.

Definition 10.6 We say that $r \in \mathcal{R}$ is an **active reaction step** on a closed communicating class U of an induced kinetic Markov process if there is an $\mathbf{x} \in U$ for which $\lambda_r(\mathbf{x}) > 0$. The set of active reaction steps with respect to a closed communicating class U of the stochastic model is denoted by $\mathcal{R}_U$. The complex vectors of the reaction steps of $\mathcal{R}_U$ are denoted by $\mathcal{C}_U$. $\mathcal{R}_U$ can be empty, e.g., when U consists of a single state.

We say that a closed communicating class U of an induced kinetic Markov process is **positive** if $\mathcal{R}_U = \mathcal{R}$, i.e., when all the reaction steps of (10.3) are active on U.

In connection with the stochastic model of reactions, let us discuss some important consequences. Clearly, the identification of the communicating classes of each set of the collection $\left((\mathbf{x}_0^{(i)} + \mathcal{S}_d) \cap \mathbb{N}_0^M \right)_{i \in \mathbb{N}}$ classifies the whole state space $\mathbb{N}_0^M$ of $\mathbf{X}$. It is of relevance when each of these sets $(\mathbf{x}_0^{(i)} + \mathcal{S}_d) \cap \mathbb{N}_0^M$ can be decomposed into closed communicating classes. When $(\mathbf{x}_0^{(i)} + \mathcal{S}_d) \cap \mathbb{N}_0^M$ is a single (closed) communicating class, one can say that $\mathbf{X}$ is irreducible relative to its discrete reaction simplex. Recall Definition 3.7.

Theorem 10.7 *If the reaction (10.3) is weakly reversible, then all the communicating classes are closed. In particular, for each $\mathbf{x}_0 \in \mathbb{N}_0^M$, $(\mathbf{x}_0 + \mathcal{S}_d) \cap \mathbb{N}_0^M$ is a collection of closed communicating classes.*

Proof Let $\mathbf{x}, \mathbf{y}$ be two states of the induced kinetic Markov process $\mathbf{X}$. Assume that $\mathbf{y}$ is accessible from $\mathbf{x}$. It means that there is a chain of reaction steps $r_1, r_2, \ldots, r_k$ for some $k \in \mathbb{N}$ such that $\mathbf{x} + \sum_{i=1}^{k} \boldsymbol{\gamma}(\cdot, r_i) = \mathbf{y}$, and along the path of states $\mathbf{x} + \sum_{i=1}^{j} \boldsymbol{\gamma}(\cdot, r_i) \in \mathbb{N}_0^M$, the intensity function $\lambda_{r_{j+1}}$ takes on positive value for every $k - 1 \geq j \in \mathbb{N}_0$. Since the reaction (10.3) is weakly reversible, it follows that there exists another sequence of reaction steps $r_1', r_2', \ldots, r_{k'}'$ for some $k' \in \mathbb{N}$ such that $\mathbf{y} + \sum_{i=1}^{k'} \boldsymbol{\gamma}(\cdot, r_i') = \mathbf{x}$ and $\mathbf{y} + \sum_{i=1}^{j} \boldsymbol{\gamma}(\cdot, r_i') \geq \boldsymbol{\alpha}(\cdot, r_{j+1}')$ holds for every $k - 1 \geq j \in \mathbb{N}_0$. Assumption (10.4) ensures that along this latter path, the intensities be positive, as well. This implies that $\mathbf{x}$ is accessible from $\mathbf{y}$. We conclude that the whole state space $\mathbb{N}_0^M$ can be decomposed into closed communicating classes. □

Based on the previous theorem, let us introduce the following notions:

Definition 10.8 We say that the state space $\mathbb{N}_0^M$ of an induced kinetic Markov process with some stochastic kinetics is **essential** if $\mathbb{N}_0^M$ is the union of closed communicating classes. If, apart from a finite number of states, $\mathbb{N}_0^M$ is covered by the union of closed communicating classes, then it is said to be **almost essential**.

As we have seen, the state space of induced kinetic Markov processes of weakly reversible reactions is essential. Now, let us show some examples. First, consider the triangle reaction $A \longrightarrow B \longrightarrow C \longrightarrow A$. The sets of integer coordinate points of the reaction simplexes are the planes $\mathbf{x}_0 + \mathscr{S} = \{(a, b, c) \in \mathbb{R}^3 \mid a + b + c = a_0 + b_0 + c_0\}$ for initial vectors $\mathbf{x}_0 = \begin{bmatrix} a_0 & b_0 & c_0 \end{bmatrix}^\top \in \mathbb{N}_0^3$ intersected by $\mathbb{N}_0^3$. They are disjoint once the initial vectors have different total sums. Hence, using the previous notations, $\mathbf{x}_0^{(i)}$ can be chosen as $\begin{bmatrix} i - 1 & 0 & 0 \end{bmatrix}^\top$ ($i \in \mathbb{N}$). Since the reaction is weakly reversible, Theorem 10.7 applies. In particular, in this case each set $(\mathbf{x}_0^{(i)} + \mathscr{S}_d) \cap \mathbb{N}_0^M$ is a single (closed) communicating class.

Next, consider the reaction $A \longrightarrow 2\,B \longrightarrow 4\,C \longrightarrow A$. Then

$$\mathbf{x}_0 + \mathscr{S} = \{(a, b, c) \in \mathbb{R}^3 \mid a + b + c = a_0 + b_0/2 + c_0/4\},$$

where $\mathbf{x}_0 = \begin{bmatrix} a_0 & b_0 & c_0 \end{bmatrix}^\top \in \mathbb{N}_0^3$. The discrete stoichiometric compatibility classes are the sets $(\mathbf{x}_0^{(i)} + \mathscr{S}_d) \cap \mathbb{N}_0^M$, where $\mathbf{x}_0^{(i)}$ is chosen to be $\begin{bmatrix} 0 & 0 & (i - 1)/4 \end{bmatrix}^\top$ ($i \in \mathbb{N}$). In this case there exist discrete reaction simplexes which are decomposed into more than one closed communicating classes. For instance the relation $a_0 + b_0/2 + c_0/4 = 5/4$ includes four possible states: $\begin{bmatrix} 1 & 0 & 1 \end{bmatrix}^\top$, $\begin{bmatrix} 0 & 2 & 1 \end{bmatrix}^\top$, $\begin{bmatrix} 0 & 0 & 5 \end{bmatrix}^\top$, and $\begin{bmatrix} 0 & 1 & 3 \end{bmatrix}^\top$. The latter one defines a single communicating class, while the former three form the other closed communicating class. The first five (discrete) reaction simplexes are shown in Fig. 10.1.

Finally, in Fig. 10.2 we show the state space and possible transitions of the Lotka–Volterra reaction. The only absorbing state of the process $(X(t), Y(t))_{t \geq 0}$ is $\begin{bmatrix} 0 & 0 \end{bmatrix}^\top$. The state $\begin{bmatrix} 0 & 0 \end{bmatrix}^\top$ is reached whenever Y first consumes X via the reaction step $X + Y \longrightarrow 2\,Y$, and then Y dies out via $Y \longrightarrow 0$. So let $U_1 = \{\begin{bmatrix} 0 & 0 \end{bmatrix}^\top\}$ be the first (closed) communicating class which is recurrent. The positive orthant is another class, i.e., $U_2 = \{\begin{bmatrix} i & j \end{bmatrix}^\top \mid i, j \in \mathbb{N}\}$. The remaining classes consist of single transient states, namely, $U_{1+2j} = \{\begin{bmatrix} 0 & j \end{bmatrix}^\top\}$ and $U_{2+2j} = \{\begin{bmatrix} j & 0 \end{bmatrix}^\top\}$ ($j \in \mathbb{N}$).

Tóth (1988a) and Tóth (1988b) started to investigate the structure of the state space. More recently Paulevé et al. (2014) gave a thorough description of the structure of the state space.

Fig. 10.1 Discrete reaction simplexes of the weakly reversible reaction A $\longrightarrow$ 2 B $\longrightarrow$ 4 C $\longrightarrow$ A

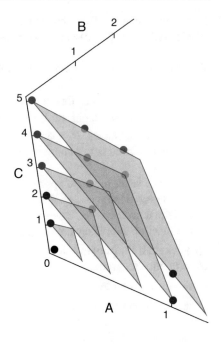

10.3   Transient Behavior

In the following, we investigate the short-term behavior of the induced kinetic Markov process.

10.3.1   Well-Posedness and Solutions of the Master Equation

Consider the set of first-order differential equations of (10.16) with initial condition $p_{\mathbf{x}_0, \mathbf{x}}(0) = \mathfrak{B}(\mathbf{x} = \mathbf{x}_0)$ for $\mathbf{x}_0, \mathbf{x} \in \mathbb{N}_0^M$. The induced kinetic Markov process $\mathbf{X}$ we have constructed from scratch in Sect. 10.2.6 satisfies these equations. It is then natural to ask whether there exists another solution to the initial value problem. The next assertion is formulated in the case of induced kinetic Markov process based on some general theorems from Norris (1998) and Liggett (2010).

Theorem 10.9 *Assume that the process* $\mathbf{X}$ *we have constructed in Sect. 10.2.6 does fulfill the following condition:*

$$\mathbf{P}\left\{ \sum_{n \in \mathbb{N}_0} \frac{1}{\Lambda(\mathbf{X}(\tau_n))} = +\infty \,\middle|\, \mathbf{X}(0) = \mathbf{x}_0 \right\} = 1$$

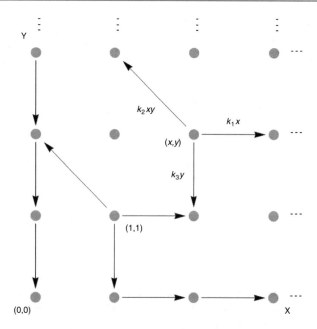

Fig. 10.2 State space and the transitions of the stochastic model of the Lotka–Volterra reaction

for all $\mathbf{x}_0 \in \mathbb{N}_0^M$, *where* τ_n *denote the successive times at which* $\mathbf{X}$ *changes:* $\mathbf{X}(\tau_n-) \neq \mathbf{X}(\tau_n)$ *for* $n \in \mathbb{N}$. *Then the following system of differential equations*

$$\frac{\mathrm{d}\,\widetilde{p}_{\mathbf{x}_0,\mathbf{x}}}{\mathrm{d}t}(t) = \sum_{r=1}^{R} \left(\lambda_r(\mathbf{x} - \boldsymbol{\gamma}(\cdot, r))\widetilde{p}_{\mathbf{x}_0,\mathbf{x}-\boldsymbol{\gamma}(\cdot,r)}(t) - \lambda_r(\mathbf{x})\widetilde{p}_{\mathbf{x}_0,\mathbf{x}}(t)\right) \qquad (10.22)$$

for the unknown $\widetilde{p}_{\mathbf{x}_0,\mathbf{x}}(\cdot)$ *($\mathbf{x}_0, \mathbf{x} \in \mathbb{N}_0$) with initial condition* $\widetilde{p}_{\mathbf{x}_0,\mathbf{x}}(0) = \mathfrak{B}(\mathbf{x} = \mathbf{x}_0)$ *has a unique bounded solution that takes values in* $[0, 1]$. *This unique solution is the transition probability function of* $\mathbf{X}$ *given by* (10.15).

Note that if the discrete stoichiometric compatibility classes are bounded, then Eq. (10.22) consists of finitely many first-order linear ordinary differential equations. Hence, its unique solution can be represented as a matrix exponential (see (10.8)). However, when the discrete stoichiometric compatibility classes and λ_r are both unbounded, then the solution to (10.22) may be not unique. Nonuniqueness can be caused by the blow-up phenomenon which is discussed in Sect. 10.3.6.

Once a system (10.16) is given, how can it be **symbolically** solved? One of the available general tools is the method of **Laplace transform**. For given states $x_0, x \in \mathbb{N}_0^M$, the **Laplace transform of the transition probability function** $p_{x_0,x}(\cdot)$ can be written as

$$L_{x_0,x}(s) = \int_0^{+\infty} \exp(-s \cdot t) p_{x_0,x}(t) \, dt, \tag{10.23}$$

where $s > 0$. Notice that $L_{x_0,x}(\cdot)$ is a bounded continuous function on $]0, +\infty[$ for each $x_0, x \in \mathbb{N}_0^M$. Using L, the master equation (10.16) can be transformed into an algebraic system of equations, namely,

$$L_{x_0,x}(s) = sp_{x_0,x}(0) + \sum_{r=1}^R (\lambda_r(x - \gamma(\cdot, r)) L_{x_0,x-\gamma(\cdot,r)}(s) - \lambda_r(x) L_{x_0,x}(s))$$

$$\tag{10.24}$$

for $s > 0$. Since $p_{x_0,x}(\cdot)$ is a bounded, piecewise continuous function, it follows from classical uniqueness theorems of the Laplace transforms that $L_{x_0,x}(\cdot)$ uniquely identifies $p_{x_0,x}(\cdot)$. It also follows that if (10.16) has a unique solution (e.g., the conditions of Theorem 10.9 hold), then so does the system (10.24). Hence the solution can be obtained by first solving (10.24) and then applying an inverse Laplace transform to get the transition probabilities.

The first exact solutions to these kinds of problems date back to Leontovich (1935), where the general solutions to the master equation of first-order reactions were given. Later, this result has recurrently appeared (see Gans 1960; Krieger and Gans 1960; Matis and Hartley 1971; Bartis and Widom 1974; Darvey and Staff 2004; Gadgil et al. 2005).

Rényi (1954) applied the method of Laplace transform to obtain the explicit form of transition probabilities of the induced kinetic Markov process with stochastic mass action type kinetics of some specific second-order reactions. For a wider range of reactions, Becker (1970, 1973a,b) also calculated the transition probabilities of the stochastic models using, for instance, probability generating functions, which we discuss in the upcoming Sect. 10.3.2. We mention that all these papers were dealing with specific reactions and methods which rely on some symmetries and subtle properties of the underlying models.

For the numerical solutions to the master equations, several strategies were developed during the past decades. The basis of all of them is to somehow truncate the state space and then to solve the truncated (hence finite) system of differential equations by some numerical methods (see, e.g., Munsky and Khammash 2006).

10.3.2 Probability Generating Function

The aim of this section is to transform the master equation into a more transparent
form which might possibly lead to symbolic solutions. Define

$$G(t, \mathbf{z}) := \mathbf{E}\{\mathbf{z}^{\mathbf{X}(t)} \mid \mathbf{X}(0) = \mathbf{x}_0\} = \sum_{\mathbf{x} \in \mathbb{N}_0^M} p_{\mathbf{x}_0, \mathbf{x}}(t) \mathbf{z}^{\mathbf{x}} \quad (t \in \mathbb{R}_0^+,\ \mathbf{z} \in D \subset \mathbb{R}^M)$$

$$(10.25)$$

as the **probability generating function** of the induced kinetic Markov process $\mathbf{X}$
at time t. Recall the definition of p from (10.15). The right-hand side of (10.25) is
uniformly convergent and defines an infinitely differentiable function for all $t \geq 0$
inside the M-dimensional open set $D :=]0, 1[^M$. This can be easily verified using
the Cauchy–Hadamard theorem (see the Appendix on page 371).

Now, one can deduce a partial differential equation for G as a straightforward
consequence of the master equation (10.16) (Tóth 1981, p. 90–91):

$$\dot{G}(t, \mathbf{z}) = \sum_{\mathbf{x} \in \mathbb{N}_0^M} \dot{p}_{\mathbf{x}_0, \mathbf{x}}(t) \mathbf{z}^{\mathbf{x}}$$

$$= \sum_{\mathbf{x} \in \mathbb{N}_0^M} \sum_{r=1}^{R} \big(\lambda_r(\mathbf{x} - \boldsymbol{\gamma}(\cdot, r)) p_{\mathbf{x}_0, \mathbf{x} - \boldsymbol{\gamma}(\cdot, r)}(t) - \lambda_r(\mathbf{x}) p_{\mathbf{x}_0, \mathbf{x}}(t)\big) \mathbf{z}^{\mathbf{x}}$$

$$= \sum_{r=1}^{R} \sum_{\mathbf{y} \in \mathbb{N}_0^M} \big(\lambda_r(\mathbf{y} + \boldsymbol{\alpha}(\cdot, r)) p_{\mathbf{x}_0, \mathbf{y} + \boldsymbol{\alpha}(\cdot, r)}(t)\big) \mathbf{z}^{\mathbf{y}} \big(\mathbf{z}^{\boldsymbol{\beta}(\cdot, r)} - \mathbf{z}^{\boldsymbol{\alpha}(\cdot, r)}\big)$$

$$= \sum_{r=1}^{R} \big(\mathbf{z}^{\boldsymbol{\beta}(\cdot, r)} - \mathbf{z}^{\boldsymbol{\alpha}(\cdot, r)}\big) \sum_{\mathbf{y} \in \mathbb{N}_0^M} \lambda_r(\mathbf{y} + \boldsymbol{\alpha}(\cdot, r)) \mathbf{z}^{\mathbf{y}} p_{\mathbf{x}_0, \mathbf{y} + \boldsymbol{\alpha}(\cdot, r)}(t).$$

The combinatorial form of stochastic mass action type kinetics λ_r, given in (10.6),
implies that

$$\frac{\partial^{o(r)} G}{\partial \mathbf{z}^{\boldsymbol{\alpha}(\cdot, r)}}(t, \mathbf{z}) := \frac{\partial^{o(r)} G}{\partial z_1^{\alpha(1, r)} \partial z_2^{\alpha(2, r)} \cdots \partial z_m^{\alpha(m, r)}}(t, \mathbf{z})$$

$$= \sum_{\mathbf{y} \in \mathbb{N}_0^M} [\mathbf{y} + \boldsymbol{\alpha}(\cdot, r)]_{\boldsymbol{\alpha}(\cdot, r)} \mathbf{z}^{\mathbf{y}} p_{\mathbf{x}_0, \mathbf{y} + \boldsymbol{\alpha}(\cdot, r)}(t)$$

$$= \frac{1}{k_r} \sum_{\mathbf{y} \in \mathbb{N}_0^M} \lambda_r(\mathbf{y} + \boldsymbol{\alpha}(\cdot, r)) p_{\mathbf{x}_0, \mathbf{y} + \boldsymbol{\alpha}(\cdot, r)}(t),$$

where $o(r)$ was defined in (6.12). Therefore G satisfies the following **linear partial differential equation**

$$\dot{G}(t, \mathbf{z}) = \sum_{r=1}^{R} k_r \left(\mathbf{z}^{\beta(\cdot,r)} - \mathbf{z}^{\alpha(\cdot,r)} \right) \frac{\partial^{o(r)} G}{\partial \mathbf{z}^{\alpha(\cdot,r)}}(t, \mathbf{z}), \qquad (10.26)$$

with the **initial condition** $G(0, \mathbf{z}) = \mathbf{z}^{\mathbf{x}_0}$ ($\mathbf{X}(0) = \mathbf{x}_0$, $\mathbf{z} \in [0, 1]^M$) and **boundary condition** $G(t, \mathbf{1}) = 1$ ($t \in \mathbb{R}_0^+$).

The order of Eq. (10.26) is always at least one because of the time derivative on the right-hand side of it. For at least first-order reactions, the order of Eq. (10.26) coincides with that of the considered reaction (10.3).

For the simple linear inflow $0 \xrightarrow{k} X$, we get $\dot{G}(t, z) = k(z - 1)G(t, z)$, the solution of which is $G(t, z) = x_0 \exp(k \cdot t(z - 1))$. This probability generating function uniquely identifies the Poisson distribution with mean $k \cdot t$; nothing weird happened.

Now, if we are given a single second-order reaction step, then (10.26) becomes a second-order partial differential equation. Even in this particular case, it is hopeless to find the general solution to (10.26), or at least it is only known in some very specific cases. When the reaction is of first order, the solution to (10.26) can be obtained by using the method of characteristics. In particular, closed compartmental systems are investigated in Problem 10.14. The functions

ProbabilityGeneratingFunctionEquation and

SolveProbabilityGeneratingFunctionEquation

of **ReactionKinetics** returns and tries to solve Eq. (10.26) of a given reaction, respectively.

At the end of this section, we state an existence and uniqueness theorem for the initial-boundary value problem of (10.26).

Theorem 10.10 *Let α' and β' be $\mathbb{N}_0^{M' \times R'}$ matrices for some $M', R' \in \mathbb{N}$, and set $\gamma' := \beta' - \alpha'$. Assume that*

1. *for each integer $1 \leq m' \leq M'$: if $\gamma'(m', r') < 0$, then $\alpha'(m', r') > 0$ for $1 \leq r' \leq R'$;*
2. *there exists a $\varrho \in (\mathbb{R}^+)^{M'}$ such that $\varrho^\top \gamma' \leq \mathbf{0}^\top$.*

Now, consider the initial-boundary value problem

$$\frac{\partial \widetilde{G}}{\partial t}(t, \mathbf{z}) = \sum_{r=1}^{R'} k_r' \left(\mathbf{z}^{\beta'(\cdot,r)} - \mathbf{z}^{\alpha'(\cdot,r)} \right) \frac{\partial^{o(r)} \widetilde{G}}{\partial \mathbf{z}^{\alpha'(\cdot,r)}}(t, \mathbf{z}), \qquad (10.27)$$

with positive numbers k_r' $(r = 1, 2, \ldots, R')$, where $o(r) = \sum_{m=1}^{M'} \alpha'(m, r)$, $g(0, \mathbf{z}) = \mathbf{z}^{\mathbf{x}_0}$ for some $\mathbf{x}_0 \in \mathbb{N}_0^M$ $(\mathbf{z} \in [0, 1]^M)$ and $g(t, \mathbf{1}) = 1$ $(t \in \mathbb{R}_0^+)$. Then there is a solution to the above problem which is unique among analytic solutions.

Proof Observe that the first condition ensures that there is no negative cross effect (see also Definition 6.24 and Theorem 6.27). So one can set up a reaction which has the appropriate matrices: $\boldsymbol{\alpha}'$, $\boldsymbol{\beta}'$, and $\boldsymbol{\gamma}'$. To this reaction one can assign the induced kinetic Markov process $\mathbf{X}'$ with stochastic mass action type kinetics and stochastic reaction rate coefficients k_r'. Define the probability generating function G of $\mathbf{X}'$ as in (10.25). Clearly, G is an analytic function and solves the initial-boundary value problem of (10.27) as we have already proved it. So we are done with the existence part. For uniqueness notice that by substituting the analytic form of G into (10.27), we arrive at the master equation of $\mathbf{X}'$. Finally, from Theorem 10.17 it follows that there is a unique solution to the master equation provided that $\boldsymbol{\varrho}^\top \boldsymbol{\gamma}' \leq 0$, which proves the uniqueness of G among analytic solutions. □

Note that the previous theorem relied on the existence and uniqueness theorem of the master equation. Hence, any conditions of Theorem 10.17 below imply uniqueness for G once the nonexplosive behavior of $\mathbf{X}'$ is ensured.

10.3.3 Moment Equations

In the following, we obtain differential equations for all the moments of the induced kinetic Markov process $\mathbf{X}$. By moment we mean the expectation of the monomial (power products) or some simple polynomial formed from the components of the induced kinetic Markov process $\mathbf{X}$. We start with the first moments as these are the simplest ones.

10.3.3.1 First Moment
The generating function G was a useful tool to obtain a partial differential equation (see (10.26)) in which all the relevant information for the induced kinetic Markov process $\mathbf{X}$ is contained. The aim of the present section is to find a system of ordinary differential equations for the first moment of $\mathbf{X}$.

The relation between the first moment of the components of $\mathbf{X}$ and the partial derivatives of G is quite straightforward as

$$\mathbf{E}X_m(t) = \frac{\partial G}{\partial z_m}(t, \mathbf{1}), \tag{10.28}$$

where $\mathbf{e}_m$ is the mth standard basis vector. Notice that

$$\frac{\partial G}{\partial z_m}(t, \mathbf{z}) = \sum_{\mathbf{x} \in \mathbb{N}_0^M} x_m \mathbf{z}^{\mathbf{x} - \mathbf{e}_m} p_{\mathbf{x}_0, \mathbf{x}}(t).$$

Taking the limit $\mathbf{z} \uparrow \mathbf{1}$, we obtain (10.28). We then arrive at the following system of differential equations:

$$\frac{\mathrm{d}}{\mathrm{d}t}\mathbf{E}X_m(t) = \sum_{r=1}^{R} k_r(\beta(m,r) - \alpha(m,r))\frac{\partial^{o(r)}G}{\partial \mathbf{z}^{\alpha(\cdot,r)}}(t,\mathbf{1}) = \mathbf{E}\, f_m^{\mathrm{sto}}(\mathbf{X}(t)),$$

(10.29)

where $m \in \mathcal{M}$ (recall (10.13)).

Theorem 10.11 *Assume that the reaction* (10.3) *only consists of first-order reaction steps. Then the equations for the first moments, i.e., Eq.* (10.29), *coincide with the induced kinetic differential equation endowed with mass action type kinetics.*

Proof If the reaction is of first order, then every reactant complex has a unit length; hence the partial derivatives of G on the right-hand side of Eq. (10.29) simplify to the expectation of the only reactant species that changes in the rth reaction step. Hence, for first-order reactions, $\mathbf{f}$ of (6.14) and $\mathbf{f}^{\mathrm{sto}}$ of (10.13) are the same linear functions but measured in different units.   □

Definition 10.12 Assuming mass action type kinetics, we say that the induced kinetic Markov process $\mathbf{X}$ of some reaction is **consistent in mean** with respect to a set of deterministic and stochastic reaction rate coefficients $(k_r^{\mathrm{det}})_{r \in \mathcal{R}}$ and $(k_r^{\mathrm{sto}})_{r \in \mathcal{R}}$, respectively, if the form of Eq. (10.29), i.e., the first moment equations of the process $\mathbf{X}$, coincides with that of the induced kinetic differential equation.

As we have just seen, first-order reactions are consistent in mean regardless of the actual choices for reaction rate coefficients. For second- and higher-order reactions, Eq. (10.29) involves higher-order moments and correlations between the number of different species which is the topic of the next section. In these cases we can only hope that some specific instances for reaction rate coefficients show consistency (see Problem 10.7).

10.3.3.2 Higher-Order Moments
The technique we applied to the first moment in the previous section can also be applied to get differential equations for higher-order moments as well. First, we deduce identities for the **combinatorial moments**.

Let $\mathbf{a} = \begin{bmatrix} a_1 & a_2 & \cdots & a_M \end{bmatrix}^{\mathsf{T}} \in \mathbb{N}_0^M$ and $\bar{a} = \sum_{m \in \mathcal{M}} a_m$, and then the **combinatorial moment** with respect to $\mathbf{a}$ is defined by

$$\mathbf{E}[\mathbf{X}(t)]_{\mathbf{a}} = \frac{\partial^{\bar{a}}G}{\partial \mathbf{z}^{\mathbf{a}}}(t,\mathbf{1}).$$

(10.30)

The previous formula comes from the componentwise differentiation of the generating function G (10.25) (recall that $[\cdot]_{\mathbf{a}}$ denotes the falling factorial function; see in (13.7)). Let us rewrite (10.29) in the following equivalent form

$$\frac{d}{dt}EX_m(t) = \sum_{r=1}^{R} k_r\big(\beta(m,r) - \alpha(m,r)\big)E[X(t)]_{\boldsymbol{\alpha}(\cdot,r)}. \tag{10.31}$$

The previous display also shows that higher-order moments may appear even in the first moment equation. Using Eq. (10.26) we arrive at

$$\frac{d}{dt}E[X(t)]_{\mathbf{a}} = \sum_{r=1}^{R} k_r \frac{\partial^a}{\partial \mathbf{z}^{\mathbf{a}}}\left[\big(\mathbf{z}^{\boldsymbol{\beta}(\cdot,r)} - \mathbf{z}^{\boldsymbol{\alpha}(\cdot,r)}\big)\frac{\partial^{o(r)}G}{\partial \mathbf{z}^{\boldsymbol{\alpha}(\cdot,r)}}\right](t,\mathbf{1}). \tag{10.32}$$

This is often called the **combinatorial moment equation** corresponding to $\mathbf{a} \in \mathbb{N}_0^M$. Some instances follow. In order to get the **covariance equations** corresponding to different pairs of species, we can set $\mathbf{a} = \mathbf{e}_m + \mathbf{e}_p$ ($m \neq p$) and

$$\frac{d}{dt}\mathrm{Cov}(X_m(t), X_p(t)) = \frac{d}{dt}E\big(X_m(t)X_p(t)\big) - \frac{d}{dt}\big(EX_m(t)EX_p(t)\big)$$

$$= \frac{d}{dt}E[X(t)]_{\mathbf{a}} - EX_m(t)\frac{d}{dt}EX_p(t) - EX_p(t)\frac{d}{dt}EX_m(t) =$$

$$= \sum_{r=1}^{R} k_r(\beta(m,r)\beta(p,r) - \alpha(m,r)\alpha(p,r))E[X(t)]_{\boldsymbol{\alpha}(\cdot,r)}$$

$$+ k_r(\beta(m,r) - \alpha(m,r))E[X(t)]_{\boldsymbol{\alpha}(\cdot,r)+\mathbf{e}_p}$$

$$+ k_r(\beta(p,r) - \alpha(p,r))E[X(t)]_{\boldsymbol{\alpha}(\cdot,r)+\mathbf{e}_m}$$

$$- EX_m(t)k_r(\beta(p,r) - \alpha(p,r))E[X(t)]_{\boldsymbol{\alpha}(\cdot,r)}$$

$$- EX_p(t)k_r(\beta(m,r) - \alpha(m,r))E[X(t)]_{\boldsymbol{\alpha}(\cdot,r)}.$$

For the nth **moment equation** of the mth species, we can consider the combinatorial moment equations corresponding to the vectors $n\mathbf{e}_m, (n-1)\mathbf{e}_m, \ldots, 2\mathbf{e}_m, \mathbf{e}_m$. It is not hard to see that a finite linear combination of these equations gives us the desired formula, namely,

$$\frac{d}{dt}EX_m(t)^n = \sum_{k=0}^{n} S(n,k)\frac{d}{dt}E[X(t)]_{k\,\mathbf{e}_m},$$

where $S(n, k) := \left\{\begin{smallmatrix} n \\ k \end{smallmatrix}\right\} = \frac{1}{k!} \sum_{j=0}^{k} (-1)^{k-j} \binom{k}{j} j^n$ denote the **Stirling numbers of the second kind** ($k = 0, 1, \ldots, n$ and $m \in \mathcal{M}$). For instance,

$$X_m(t)^2 = [\mathbf{X}(t)]_{2\mathbf{e}_m} + [\mathbf{X}(t)]_{\mathbf{e}_m},$$

$$X_m(t)^3 = [\mathbf{X}(t)]_{3\mathbf{e}_m} + 3[\mathbf{X}(t)]_{2\mathbf{e}_m} + [\mathbf{X}(t)]_{\mathbf{e}_m},$$

$$X_m(t)^3 = [\mathbf{X}(t)]_{4\mathbf{e}_m} + 6[\mathbf{X}(t)]_{3\mathbf{e}_m} + 7[\mathbf{X}(t)]_{2\mathbf{e}_m} + [\mathbf{X}(t)]_{\mathbf{e}_m}$$

and so on. One can deduce similar formulas for the central moments, as well.

10.3.3.3 Moment Closure Methods

The idea of moment closure techniques is to "close" the moment equations by assuming some relations for the (higher order) moments of the underlying induced kinetic Markov process $\mathbf{X}$. The aim of these methods is to get a finite system of differential equations hoping that the solutions to these provide good approximations to the moments of $\mathbf{X}$. We list some of the most commonly used choices. One can assume that no correlations are present among the species numbers, i.e., $EX_m X_p = EX_m EX_p$ are assumed for every $m \neq p \in \mathcal{M}$.

We can also suppose that either $EX^3 = 3EX^2 EX - 2(EX)^3$, or $EX^3 = -2EX + 3EX^2 + (EX)^3$, or the relation $EX^3 = (EX^2/EX)^3$ holds. A random variable having Gaussian, Poisson, or log-normal distribution fulfills these relations, respectively. Any of the assumptions mentioned may or may not lead to a finite system of differential equations for the moments. If one of the methods results in a finite system, then it might be solved symbolically or by using numerical methods (see Gillespie 2009 and further references therein). The reader is asked to apply one of these methods in Problem 10.13. These methods often involve ad hoc techniques; therefore in each case, one has to prove that the approximate moments are close to the exact ones in some sense.

10.3.4 Conditional Velocity

First, we define certain local quantities, and we calculate them for the stochastic model of reaction (10.3) following Érdi and Tóth (1976).

Definition 10.13 Let $\mathbf{X}$ be an induced kinetic Markov process, and let the nth **moment velocity** $\mathbf{D}_n$ of $\mathbf{X}$ be defined as

$$\mathbf{D}_n(\mathbf{x}) = \lim_{h \downarrow 0} \frac{1}{h} \mathbf{E}\big[(\mathbf{X}(h) - \mathbf{X}(0))^n \mid \mathbf{X}(0) = \mathbf{x}\big]. \qquad (10.33)$$

Note that the power of vectors in this section is understood as a **tensorial product**; hence if $\mathbf{a}$ is a vector, then $\mathbf{a}^2$ is a matrix, $\mathbf{a}^3$ is a three-dimensional array, and so on. The conditional velocity can be computed and one has the following.

Theorem 10.14 *If* **X** *is the induced kinetic Markov process of reaction* (10.3) *with stochastic kinetics* λ_r, *then*

$$\mathbf{D}_n(\mathbf{x}) = \sum_{r=1}^{R} \boldsymbol{\gamma}(\cdot, r)^n \lambda_r(\mathbf{x}). \qquad (10.34)$$

Proof Let h be a "small" number. Then, by Eq. (10.9), it follows that

$$\mathbf{E}\big[(\mathbf{X}(h) - \mathbf{X}(0))^n \mid \mathbf{X}(0) = \mathbf{x}\big] = \sum_{r=1}^{R} \boldsymbol{\gamma}(\cdot, r)^n \lambda_r(\mathbf{x})h + \varepsilon(h)h.$$

Dividing by h and sending it to 0, we get the desired formula. $\qquad \square$

Some remarks follow. It is easy to see that $\mathbf{D}_2(\mathbf{x})$ is a **positive semidefinite** matrix. It is not positive definite when the dimension of the stoichiometric subspace is smaller than the number of species (e.g., in the case of stoichiometrically mass-conserving reactions). Using the conditional velocities, approximations to the master equation as well as to the induced kinetic Markov process can be obtained (see further reading in Gardiner (2010, Section 5–7), Van Kampen (2006), particularly the **Kramers–Moyal expansion** and the **Fokker–Planck equation**).

10.3.5 Martingale Property

In this short section, we investigate the conditions for an induced kinetic Markov process to be a martingale. For a time-homogeneous Markov process

$$\mathbf{Y} = \begin{bmatrix} Y_1 & Y_2 & \cdots & Y_M \end{bmatrix}^{\top} \in \mathbb{N}_0^M$$

to be a **supermartingale**, **martingale**, or **submartingale** in the mth coordinate variable, the following

$$\mathbf{E}\{Y_m(t) \mid \mathbf{Y}(0) = \mathbf{y}\} \le y_m, \ \mathbf{E}\{Y_m(t) \mid \mathbf{Y} = \mathbf{y}\} = y_m, \ \text{or} \ \mathbf{E}\{Y_m(t) \mid \mathbf{Y}(0) = \mathbf{y}\} \ge y_m$$

$$(10.35)$$

is required, respectively, to hold for all $y_m \in \mathbb{N}_0$ and $t \ge 0$ provided that $\mathbf{E}|Y_m(t)| < +\infty$, where $\mathbf{y} = \begin{bmatrix} y_1 & y_2 & \cdots & y_M \end{bmatrix}^{\top}$. The definition and properties of martingales for more general processes can be found, e.g., in Liggett (2010, Section 1.9).

Theorem 10.15 *The mth coordinate process X_m of an induced kinetic Markov process* **X** *is a supermartingale, martingale, or submartingale, respectively, in a* *discrete stoichiometric compatibility class* $(\mathbf{x}_0 + \mathscr{S}_d) \cap \mathbb{N}_0^M$ *for some* $\mathbf{x}_0 \in \mathbb{N}_0^M$ *if*

$$\boxed{\sum_{r=1}^{R} \gamma(m, r)\lambda_r(\mathbf{x}) \leq 0, \ = 0 \ or \ \geq 0} \tag{10.36}$$

holds for all $\mathbf{x} = \begin{bmatrix} x_1 \ x_2 \ \cdots \ x_M \end{bmatrix}^{\top} \in (\mathbf{x}_0 + \mathscr{S}_d) \cap \mathbb{N}_0^M$.

Proof For $\mathbf{y} \in (\mathbf{x}_0 + \mathscr{S}_d) \cap \mathbb{N}_0^M$, let

$$\mu_m(\mathbf{y}) := \sum_{r=1}^{R} \gamma(m, r)\lambda_r(\mathbf{y}).$$

Now, using the total law of probability, the memorylessness property of $\mathbf{X}$, and Eqs. (10.9) and (10.10) in that order, it follows that

$$\mathbf{E}\{X_m(s) \mid \mathbf{X}(0) = \mathbf{x}\}$$

$$= \sum_{\mathbf{y} \in (\mathbf{x}_0 + \mathscr{S}_d) \cap \mathbb{N}_0^M} \mathbf{E}\{X_m(s) \mid \mathbf{X}(s - h) = \mathbf{y}\}\mathbf{P}\{\mathbf{X}(s - h) = \mathbf{y} \mid \mathbf{X}(0) = \mathbf{x}\}$$

$$= \sum_{\mathbf{y} \in (\mathbf{x}_0 + \mathscr{S}_d) \cap \mathbb{N}_0^M} \left(y_m + \sum_{r=1}^{R} \gamma(m, r)\lambda_r(\mathbf{y})h \right) \mathbf{P}\{\mathbf{X}(s - h) = \mathbf{y} \mid \mathbf{X}(0) = \mathbf{x}\} + \varepsilon(h)h$$

$$= \mathbf{E}\{X_m(s - h) \mid \mathbf{X}(0) = \mathbf{x}\}$$

$$+ \sum_{\mathbf{y} \in (\mathbf{x}_0 + \mathscr{S}_d) \cap \mathbb{N}_0^M} \mu_m(\mathbf{y})h\mathbf{P}\{\mathbf{X}(s - h) = \mathbf{y} \mid \mathbf{X}(0) = \mathbf{x}\} + \varepsilon(h)h,$$

where $0 < h$ is sufficiently small and $\mathbf{x} \in (\mathbf{x}_0 + \mathscr{S}_d) \cap \mathbb{N}_0^M$. By rearranging the previous display, we get that

$$\frac{1}{h}\mathbf{E}\{X_m(s) - X_m(s - h) \mid \mathbf{X}(0) = \mathbf{x}\}$$

$$= \sum_{\mathbf{y} \in (\mathbf{x}_0 + \mathscr{S}_d) \cap \mathbb{N}_0^M} \mu_m(\mathbf{y})\mathbf{P}\{\mathbf{X}(s - h) = \mathbf{y} \mid \mathbf{X}(0) = \mathbf{x}\} + \varepsilon(h).$$

Taking the limit as $h \to 0$, we finally arrive at the equality

$$\frac{d}{ds}\mathbf{E}\{X_m(s) \mid \mathbf{X}(0) = \mathbf{x}\} = \sum_{\mathbf{y} \in (\mathbf{x}_0 + \mathscr{S}_d) \cap \mathbb{N}_0^M} \mu_m(\mathbf{y})\mathbf{P}\{\mathbf{X}(s) = \mathbf{y} \mid \mathbf{X}(0) = \mathbf{x}\}.$$

In the cases of $\mu_m \leq 0$, $\mu_m = 0$, and $\mu_m \geq 0$, it implies that

$$\frac{d}{ds}\mathbf{E}\{X_m(s) \mid \mathbf{X}(0) = \mathbf{x}\} \leq 0, \ = 0, \ \text{ and } \geq 0,$$

for all $\mathbf{x} \in (\mathbf{x}_0 + \mathscr{S}_d) \cap \mathbb{N}_0^M$, respectively. We can conclude that

$$-x_m + \mathbf{E}\{X_m(t) \mid \mathbf{X}(0) = \mathbf{x}\} = \int_0^t \frac{d}{ds}\mathbf{E}\{X_m(s) \mid \mathbf{X}(0) = \mathbf{x}\}\, ds \leq 0, \ = 0 \ \geq 0;$$

hence X_m is a supermartingale, martingale, or submartingale in the respective cases.

As an example, consider the induced kinetic Markov process X of the reaction

$$0 \underset{k_{-1}}{\overset{k_1}{\rightleftharpoons}} X \xrightarrow{k_2} 2X$$

endowed with stochastic mass action type kinetics. Note that for any choice of stochastic reaction rate coefficients k_1, k_{-1}, k_2, where at least one of them is nonzero, it holds that $(x_0 + \mathscr{S}_d) \cap \mathbb{N}_0 = \mathbb{N}_0$ for any $x_0 \in \mathbb{N}_0$. Then X is a

- supermartingale (in the strict sense), if $k_1 = 0$ and $k_{-1} > k_2$;
- martingale, if $k_1 = 0$ and $k_{-1} = k_2$;
- submartingale (in the strict sense), if $k_1 > 0$ and $k_{-1} \leq k_2$.

10.3.6 Blowing Up

The blow-up phenomenon can also show up in induced kinetic Markov processes. In the deterministic setting, blow-up can happen at a certain time when at least one of the species concentration becomes "infinite" (Definition 8.18). Since we are now dealing with stochastic processes, if blow-up is about to happen, its time might be random. As usual, let $(\tau_n)_{n \in \mathbb{N}_0}$ with $\tau_0 = 0$ be the jump times, i.e, when $\mathbf{X}$ changes: $\mathbf{X}(\tau_n-) \neq \mathbf{X}(\tau_n)$ for $n \in \mathbb{N}$.

Definition 10.16 Let $\mathbf{X}$ be an induced kinetic Markov process endowed with some stochastic kinetics, and let $\mathbf{X}(0) = \mathbf{x}_0 \in \mathbb{N}_0^M$. We say that $\mathbf{X}$ **blows up** if

$$\mathbf{P}\left\{\tau_\infty := \lim_{n \to +\infty} \sum_{i=1}^n (\tau_i - \tau_{i-1}) < +\infty\right\} > 0. \tag{10.37}$$

If (10.37) holds we often say that $\mathbf{X}$ is **explosive**; otherwise, when $\mathbf{P}\{\tau_\infty < +\infty\} = 0$, it is said to be **nonexplosive**. The random variable τ_∞ is called the **explosion time** of $\mathbf{X}$.

The next theorem from Gikhman and Skorokhod (2004b, p. 210) is articulated in the case of induced kinetic Markov processes.

Theorem 10.17 *The induced kinetic Markov process* $\mathbf{X}$ *started from* $\mathbf{X}(0) = \mathbf{x}_0 \in \mathbb{N}_0^M$ *is* ***nonexplosive*** *if and only if*

$$\mathbf{P}\left\{ \sum_{n=0}^{+\infty} \frac{1}{\Lambda(\mathbf{X}(\tau_n))} = +\infty \,\middle|\, \mathbf{X}(0) = \mathbf{x}_0 \right\} = 1. \tag{10.38}$$

In particular, if

- *the reaction* (10.3) *is mass consuming or mass conserving;* ***or***
- *for each* $r \in \mathscr{R}$: *the (general) stochastic kinetic function* λ_r *is bounded in* $\mathbb{N}_0^M$; ***or***
- $\mathbf{X}$ *is recurrent for each* $\mathbf{x}_0$,

then condition (10.38) *holds; hence* $\mathbf{X}$ *is nonexplosive for every* $\mathbf{X}(0) = \mathbf{x}_0 \in \mathbb{N}_0^M$.

Let us start with a statement that excludes the blow-up for some reactions.

Theorem 10.18 *The induced kinetic Markov process* $\mathbf{X}$ *of a first-order reaction endowed with stochastic mass action type kinetics does not blow up for any* $\mathbf{X}(0) = \mathbf{x}_0 \in \mathbb{N}_0^M$.

Proof Set $\gamma_{\max} := \max_{m \in \mathscr{M}, r \in \mathscr{R}} |\gamma(m, r)|$ and $k_{\max}^{\text{sto}} := \max_{r \in \mathscr{R}} \{k_r^{\text{sto}}\}$. As usual let $(\xi_i)_{i \in \mathbb{N}}$ be a mutually independent and exponentially distributed collection of random variables. Define the jump times $\bar{\tau}_n$ of the process $\bar{X}$ as

$$\bar{\tau}_n = \sum_{i=1}^{n} \frac{1}{k_{\max}^{\text{sto}} \left(\bar{x}_0 + M \gamma_{\max} n \right)} \xi_i,$$

where $\bar{x}_0 = M \max_{m \in \mathscr{M}} \{x_0^m\}$, x_0^m is the mth coordinate of the vector $\mathbf{x}_0$. Then $\bar{X}$ is defined as

$$\bar{X}(t) = \bar{x}_0 + \sum_{n=0}^{+\infty} \gamma_{\max} \, n \, \mathfrak{B}(\bar{\tau}_n \leq t) \qquad (t \geq 0).$$

It is not hard to see that $\mathbf{P}\{\bar{\tau}_\infty < \infty\} = 0$ from which it follows by Theorem 10.17 that $\bar{\mathbf{X}}$ does not blow up.

Recall Sect. 10.2.6 showing that the jump times of $\mathbf{X}$ can be given by $\tau_n = \sum_{i=1}^{n} \frac{1}{\Lambda(\mathbf{X}(\tau_{i-1}))} \xi_i$. Note that from the assumption that the reaction is of first order, it follows that $\Lambda(\mathbf{x}) \leq k_{\max}^{\text{sto}} M \max_{m \in \mathscr{M}} \{x_m\}$ for all $\mathbf{x} \in \mathbb{N}_0^M$.

Since each species increases by at most $\gamma_{\max}$ in each step of $\mathbf{X}$, we get that $\bar{\tau}_n \leq \tau_n$ for all $n \in \mathbb{N}_0^M$ and $\max_{m \in \mathcal{M}}\{X_m(t)\} \leq \bar{X}(t)$ for all $t \geq 0$. One can conclude that $\mathbf{X}$ is defined for all times for every $\mathbf{X}(0) = \mathbf{x}_0 \in \mathbb{N}_0^M$. $\qquad\qquad\square$

Before the main theorem, a simple but relevant example is shown to blow up which enlightens the phenomenon in the stochastic setting. Let us consider the single reaction step

$$2\,\mathrm{X} \xrightarrow{k^{\mathrm{sto}}} 3\,\mathrm{X}, \qquad\qquad (10.39)$$

and let X be the induced kinetic Markov process of it with stochastic mass action type kinetics, i.e., $\lambda_1(x) = k^{\mathrm{sto}}x(x-1)\mathfrak{B}(x > 1)$. We show below that X blows up provided that $x_0 > 1$; in this case the expected explosion time is $\mathbf{E}\tau_{+\infty} = \frac{1}{k^{\mathrm{sto}}(x_0-1)}$; hence $\mathbf{P}\{\tau_\infty < +\infty\} = 1$.

Let us recall the first representation of X outlined in Sect. 10.2.6. Take a mutually independent sequence of exponentially distributed random variables $(\xi_i)_{i \in \mathbb{N}}$ with unit mean. Since there is only one reaction step that can take place, X jumps into state x_0+n at time $\tau_n = \sum_{i=1}^{n} \frac{1}{\lambda_1(x_0+i-1)}\xi_j$ after n steps, provided that $1 < x_0 \in \mathbb{N}_0$ and $n \in \mathbb{N}_0$. Note that for each $i \in \mathbb{N}$:

$$\mathbf{E}\frac{1}{\lambda_1(x_0+i-1)}\xi_i = \sqrt{\mathbf{Var}\left(\frac{1}{\lambda_1(x_0+i-1)}\xi_i\right)} = \frac{1}{k^{\mathrm{sto}}(x_0+i-1)(x_0+i-2)};$$

hence

$$\mathbf{E}\tau_n = \frac{1}{k^{\mathrm{sto}}}\sum_{i=1}^{n}\left(\frac{1}{x_0+i-2} - \frac{1}{x_0+i-1}\right) = \frac{1}{k^{\mathrm{sto}}(x_0-1)} - \frac{1}{k^{\mathrm{sto}}(x_0+n-1)}.$$

Taking the limit as $n \to +\infty$ and using Kolmogorov's two-series theorem (Feller 2008, Sections VIII.5, IX.9), one can conclude that $\lim_{n \to +\infty} \mathbf{E}\tau_n = \mathbf{E}\lim_{n \to +\infty} \tau_n = \mathbf{E}\tau_\infty = \frac{1}{k^{\mathrm{sto}}(x_0-1)} < +\infty$, that is, τ_∞ is almost surely finite.

It is worth comparing this result with the solution (8.11) of the induced kinetic differential equation, also exhibiting the blow-up phenomenon. Moreover, the blow-up time of the induced kinetic differential equation coincides with the **expected time of blow-up** of the induced kinetic Markov process provided that the initial concentration c_0 and the (deterministic) reaction rate coefficient $k = k^{\mathrm{det}}$ are chosen so that $c_0 = \frac{x_0}{N_A V}$ (mol dm^{-3}) and $k^{\mathrm{det}} = k^{\mathrm{sto}}N_A V(1 - 1/x_0)$ (mol^{-1} dm^3 sec^{-1}), where V is the volume of the vessel and N_A is Avogadro's number.

In general, it can be shown that the induced kinetic Markov process endowed with stochastic mass action type kinetics of a single reaction step $a\mathrm{X} \longrightarrow b\mathrm{X}$, where $2 \leq a < b \in \mathbb{N}$, also blows up, if $x_0 \geq a$.

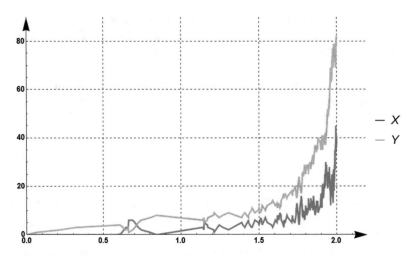

Fig. 10.3 The induced kinetic Markov process of reaction $0 \longrightarrow Y, 2X \longrightarrow X + Y \longrightarrow 2Y \longrightarrow 3X$ with stochastic mass action type kinetics starting from $(0, 0)$ with stochastic reaction rate coefficients $k_1^{sto} = 5$, $k_2^{sto} = k_3^{sto} = 1$, and $k_4^{sto} = 0.2$

Next, let us look at a much less trivial example. Consider the induced kinetic Markov process with stochastic mass action type kinetics of the reaction

$$0 \xrightarrow{k_1^{sto}} Y \quad 2X \xrightarrow{k_2^{sto}} X + Y \xrightarrow{k_3^{sto}} 2Y \xrightarrow{k_4^{sto}} 3X.$$

Simulation suggests that this process also blows up. Figure 10.3 shows a sample path of the induced kinetic Markov process $\begin{bmatrix} X(t) & Y(t) \end{bmatrix}^\top$ for $t \geq 0$. This was generated by using the **Simulation** and **SimulationPlot** of **ReactionKinetics** (see later in Sect. 10.7):

```
SimulationPlot[{0 -> Y, 2X -> X + Y -> 2Y -> 3X},
    {5, 1, 1, 0.2}, {0, 0}, 2]
```

10.4   Usual Categorial Beliefs

Below, we go through on a selection of major classes of stochastic processes showing examples and counterexamples of reactions (if necessary) for which the corresponding induced kinetic Markov processes happen to **belong** or **not belong** to that particular class, respectively. See Fig. 10.4 and for a similar characterization Érdi and Tóth (1989), p. 143.

- **Poisson process**. The Poisson process $(X(t))_{t \geq 0}$ with rate $\lambda > 0$ can be defined as $X(t) := \sup\{n \geq 0 : \sum_{i=1}^{n} \xi_i \leq \lambda t\}$ for $t \geq 0$, where $(\xi_i)_{i \in \mathbb{N}}$ is again

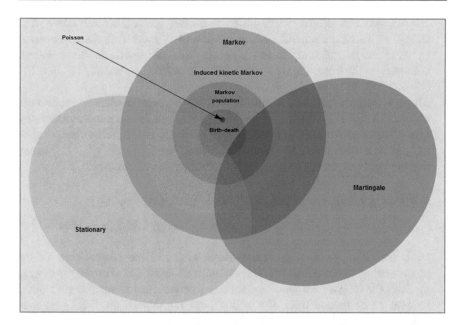

Fig. 10.4 Classes of stochastic processes

a set of mutually independent and exponentially distributed random variables with unit mean. It can be easily verified that in any finite interval, the points of $X(t)$ have Poisson distribution with mean $\lambda \times$ length of the interval and these numbers are independent of each other for nonoverlapping intervals. We have previously seen that in the chemical language, the Poisson process shows up as the induced kinetic Markov process of the simple inflow $0 \xrightarrow{k} X$ with stochastic mass action type kinetics. In this case one can easily verify that the transition probabilities describe a Poisson distribution. However, we underline that the transition probabilities need not be Poissonian. For instance, considering the outflow $X \longrightarrow 0$, the induced kinetic Markov process stays in a bounded domain; hence the distribution of $(p_{x_0,x}(t))_{x \in \mathbb{N}_0}$ cannot be Poissonian.

- **Birth–death process**. The birth–death processes correspond to those jump processes on $\mathbb{N}_0^M$ where the possible transitions are of only two types, "births" or "deaths," i.e., when one of the components of the vectorial process is increased or decreased by one at a time, respectively. In chemical language this means that the induced kinetic Markov processes of reactions that are made of the reaction steps of type $aX(m) \longrightarrow bX(m)$ $(a, b \in \mathbb{N}_0, b - a| \in \{-1, 0, 1\}$ and $m \in \mathcal{M})$ correspond exactly to the birth–death processes. When two or more reacting species are present in the system which undergo mutual changes by a reaction step, then the induced kinetic Markov process is essentially different from a birth–death process.

- **Markov population process**. This concept was introduced by Kingman (1969) as a generalization of the birth–death processes in a way that any of the species can undergo a transformation resulting in another species, but still one unit of change in the whole population size is allowed at once. This class is covered by those induced kinetic Markov processes that correspond to reaction steps consisting of the following type of reaction steps $aX(m) \longrightarrow bX(p)$ $(a, b \in \mathbb{N}_0$, $b - a \in \{-1, 0, 1\}$ and $m, p \in \mathcal{M})$.

- **Markov renewal process**. While the waiting times of a Markov process are always exponentially distributed, the Markov renewal process may have arbitrary type of waiting time distribution. Let $\mathbf{M}$ be a discrete-time Markov chain, and let $(\xi_n)_{n \in \mathbb{N}}$ be a set of (arbitrary) random variables. Setting $\tau_n := \sum_{j=1}^{n} \xi_j$, we say that the joint process $(\mathbf{M}_n, \tau_n)_{n \in \mathbb{N}_0}$ is a **Markov renewal process** if

$$\mathbf{P}\{\xi_{n+1} \leq t, \mathbf{M}_{n+1} = \mathbf{y} \mid (\mathbf{M}_0, \tau_0), (\mathbf{M}_1, \tau_1), \ldots, (\mathbf{M}_n = \mathbf{x}, \tau_n)\}$$
$$= \mathbf{P}\{\xi_{n+1} \leq t, \mathbf{M}_{n+1} = \mathbf{y} \mid \mathbf{M}_n = \mathbf{x}\}$$

holds for all $n \in \mathbb{N}$, $t \geq 0$ and states $\mathbf{x}, \mathbf{y}$; i.e., the current state of the process being $\mathbf{x}$ contains all the relevant information for $(\mathbf{M}, S)$ from the past. In particular, if $(\xi_n)_{n \in \mathbb{N}}$ is a set of mutually independent and identically distributed random variables and their distribution does not depend on $(\mathbf{M}_n)_{n \in \mathbb{N}_0}$, then $(\mathbf{M}_n, \tau_n)_{n \in \mathbb{N}_0}$ is said to be a **renewal process**.

- **Stationary process**. We say that a stochastic process $\mathbf{X}$ is stationary if for all $s, t_1, t_2, \ldots, t_k \geq 0$ and $k \in \mathbb{N}$ the joint distributions of $\left[\mathbf{X}(t_1) \; \mathbf{X}(t_2) \; \cdots \; \mathbf{X}(t_k)\right]^{\top}$ and $\left[\mathbf{X}(t_1 + s) \; \mathbf{X}(t_2 + s) \; \cdots \; \mathbf{X}(t_k + s)\right]^{\top}$ are the same, i.e., the time shifts do not alter the law of the process. In particular, $\mathbf{X}(t)$ has the same distribution for all $t \geq 0$.

An induced kinetic Markov process $\mathbf{X}$ started from some fixed $\mathbf{x}_0 \in \mathbb{N}_0^M$ is almost never a stationary process since the previous condition would imply the relation $\mathbf{X}(t) \equiv \mathbf{x}_0$ to hold for all $t \geq 0$. If $\mathbf{X}$ started from one of its **stationary distributions** (see Sect. 10.5), if any exists, then it would result in a stationary process.

There also exists a weaker form of stationarity (wide-sense or covariance stationarity), when we only require the expectation and autocovariance function of the process (i.e., the covariance of $\mathbf{X}(t_1)$ and $\mathbf{X}(t_2)$) to be invariant with respect to time shifts (Gikhman and Skorokhod 2004a, p. 199). These weaker properties are also strong enough to include only a minor subclass of induced kinetic Markov processes.

- **Martingale**. We investigated the martingale property in Sect. 10.3.5. Now, let us consider an example. The induced kinetic Markov process of reaction

$$X \xrightarrow{k_1} Y \quad X \xrightarrow{k_2} 2X + Y \quad Y \xrightarrow{k_3} 0$$

endowed with stochastic mass action type kinetics satisfies conditions (10.36) for the coordinate process $(X(t))_{t \geq 0}$ if $k_1 \geq k_2$, $k_1 = k_2$ or $k_1 \leq k_2$. In these cases X is a supermartingale, martingale, or submartingale, respectively. The coordinate process in species Y does not fall in either category. Finally, note that a single reaction step is never a martingale.

- **Queueing process**. The main parameters of a queueing system are the following: type of the arrival process (A) and that of the service time distribution (S), the number of servers or channels (c), the number of places in the system (K), the calling population (N), and the queue's discipline (D). The processes A and S can be general processes (even non-Markovian ones are allowed), c denotes the number of servers to which customers can go, while K is the maximum number of customers allowed in the system (including those that are being served). Furthermore, N is the population from which customers arrive at the servers, which can be infinite, as well, and D describes the priority how a server works, e.g., in the first in—first out **FIFO** or the last in—first out **LIFO** cases, customers are to be served in the same order or the reverse order how they arrived in, respectively. Though much more general evolution rules can be attached to a queueing system, one still has that the number of customers under service can change by one at once. Hence the striking difference from the induced kinetic Markov process is clear.

- **Branching process**. A branching process generally speaking describes a layered process where starting from an individual, each ancestor from the previous generation produces offsprings. The random number of descendants then continues to give birth to new offsprings. We can keep track of the size of the most recent population or individuals can live through more than one generation (size-dependent branching processes). More general, e.g., spatial, interactions among different types of individuals can also be taken into account (see Mode 1971; Athreya and Ney 2004). We note that one of the crucial questions in this area is whether the individuals become extinct or they survive with positive probability.

10.5 Stationary Distributions

Long-term behavior is interesting from the applications point of view but usually easier to treat from the mathematical point of view (see e.g. Kolmogoroff 1935; Siegert 1949; Dambrine and Moreau 1981; Vellela and Qian 2007, 2009). Let us start with some crucial definitions. Suppose we are given an induced kinetic Markov process **X** with stochastic kinetics λ_r (see (10.6)).

Definition 10.19 The function $\pi : \mathbb{N}_0^M \to [0, 1]$ is said to be a proper **stationary distribution** of the induced kinetic Markov process **X** if the two conditions below are satisfied:

1. $\sum_{\mathbf{x} \in \mathbb{N}_0^M} \pi(\mathbf{x}) = 1$, and
2. The balance equation

$$\boxed{\sum_{r=1}^{R} \lambda_r(\mathbf{x} - \boldsymbol{\gamma}(\cdot, r)) \, \pi(\mathbf{x} - \boldsymbol{\gamma}(\cdot, r)) = \pi(\mathbf{x}) \sum_{r=1}^{R} \lambda_r(\mathbf{x}),} \qquad (10.40)$$

holds true for every $\mathbf{x} \in \mathbb{N}_0^M$.

Note that the stationary distributions are exactly the constant (i.e., time independent) solutions of the master equation (10.16). To determine all or some of the stationary distributions of an induced kinetic Markov process is a formidable task in general.

In what follows some general facts are summarized about the stationary distributions based on Norris (1998, Chapter 3).

Theorem 10.20 *Assume that* $\mathbf{X}$ *is an induced kinetic Markov process and* U *is one of its closed communicating classes. Then the following statements are equivalent:*

- $\mathbf{X}$ *is nonexplosive and has a stationary distribution* π *concentrated on* U,
- *the states of* U *are positive recurrent, i.e., they are recurrent and the expected return time from* $\mathbf{x}$ *to* $\mathbf{x} \in U$ *is finite.*

In addition, if π *is a stationary distribution, then*

$$\lim_{t \to +\infty} \mathbf{P}\{\mathbf{X}(t) = \mathbf{x} \mid \mathbf{X}(0) = \mathbf{y}\} = \pi(\mathbf{x}) \qquad (10.41)$$

holds for all $\mathbf{x}, \mathbf{y} \in U$, *which is independent of the initial state* $\mathbf{y} \in U$.

The structure of stationary distributions for induced kinetic Markov processes can drastically vary from reaction to reaction. In the case of the celebrated Lotka–Volterra reaction, Reddy (1975) has showed that the only stationary distribution is concentrated on the state $\begin{bmatrix} 0 & 0 \end{bmatrix}^\top$, i.e., when both species die out. However, as the next examples and theorems show, this can be considered as a rather degenerate case. Let us continue with the first-order reaction

$$0 \underset{k_1}{\overset{k_2}{\rightleftharpoons}} X \xrightarrow{k_2} 2X$$

having specially chosen reaction rate coefficients k_1 and k_2. The induced kinetic Markov process X with stochastic mass action type kinetics is nonexplosive for any $k_1, k_2 > 0$ and $x_0 \geq 0$ by Theorem 10.18. The state space of the process, being $\mathbb{N}_0$, is a single closed communicating class; hence X is irreducible. Note that the reaction is **not** weakly reversible and has a deficiency $\delta = 3 - 1 - 1 = 1$.

The stationary balance equation (10.40) results in the following recurrence relation:

$$k_1(x+1)\pi(x+1) + k_2 x \pi(x-1) = (k_1 x + k_2 x + k_2)\pi(x),$$

where $\pi(1)/\pi(0) = k_2/k_1$ and $x \in \mathbb{N}_0$. The solution is given by

$$\pi(x) = \left(1 - \frac{k_2}{k_1}\right)\left(\frac{k_2}{k_1}\right)^x \qquad (x \in \mathbb{N}_0),$$

which is a proper stationary distribution if and only if $k_1 > k_2$. That is, π is a geometric distribution for which the limit (10.41) also holds. The first moment equation is $\frac{d}{dt}\mathbf{E}X(t) = k_2 + (k_2 - k_1)\mathbf{E}X(t)$; hence for $k_1 \neq k_2$,

$$\mathbf{E}X(t) = x_0 \exp(-(k_1 - k_2)t) + \frac{k_2}{k_1 - k_2}\big(1 - \exp(-(k_1 - k_2)t)\big),$$

while for $k_1 = k_2$, $\mathbf{E}X(t) = x_0 + k_2 t$. That is, $\lim_{t \to +\infty} \mathbf{E}X(t) = +\infty$ for $k_1 \leq k_2$.

This, rather dummy, example also provides a counterexample for the usual categorial belief that the Poisson-type distributions should (always) be stationary. Some conditions were given in Tóth and Török (1980), Tóth (1981), and Tóth et al. (1983) under which the Poisson distribution is indeed stationary for the induced kinetic Markov processes of a class of reactions. The next theorem shows in a way the importance of complex balanced stationary points of the induced kinetic differential equation and also gives the stationary distributions for the stochastic model of reactions with mass action type kinetics.

Theorem 10.21 (Anderson et al. (2010)) *Assume that the induced kinetic differential equation with mass action type kinetics of the reaction (10.3) is complex balanced at $\mathbf{c}_* \in \mathbb{R}_0^M$. Then the induced kinetic Markov process of (10.3) with stochastic mass action type kinetics has a product-form stationary distribution.*

If the whole state space $\mathbb{N}_0^M$ is irreducible, then the unique stationary distribution is given by

$$\pi(\mathbf{x}) = \frac{\mathbf{c}_*^{\mathbf{x}}}{\mathbf{x}!}\exp(-\mathbf{c}_*) = \prod_{m=1}^{M}\frac{(c_m)_*^{x_m}}{x_m!}\exp(-(c_m)_*), \qquad (10.42)$$

where $\mathbf{x} \in \mathbb{N}_0^M$.

If $\mathbb{N}_0^M$ is reducible, then the unique stationary distributions π_U concentrated on the closed communicating classes is of the following form:

$$\pi_U(\mathbf{x}) = \frac{1}{Z_U}\frac{\mathbf{c}_*^{\mathbf{x}}}{\mathbf{x}!} = \frac{1}{Z_U}\prod_{m=1}^{M}\frac{(c_m)_*^{x_m}}{x_m!}, \qquad (10.43)$$

where $\mathbf{x} \in U \subset \mathbb{N}_0^M$, $\pi_U(\mathbf{x}) = 0$ otherwise, and Z_U is the normalizing constant.

Let us make some comments.

- The existence of a complex balanced stationary point in the case of deterministic model is ensured by Theorem 7.15. Hence, if the reaction (10.3) has a deficiency zero, then the assumptions of the previous theorem hold if and only if the reaction is weakly reversible. Note that if the reaction is weakly reversible, then by Theorem 10.7 the state space only consists of closed communicating classes.
- Theorem 10.21 does not say that the stationary distribution should be Poisson. For instance, in the case of weakly reversible closed compartmental systems with a single linkage class, the stationary distribution is multinomial (see also Problem 10.17).
- The previous theorem can slightly be generalized by considering more general kinetics (cf. Anderson et al. 2010, Section 6).

Proof First, we check that π given in (10.42) satisfies Eq. (10.40). Taking into account (10.6), it follows that

$$\sum_{r \in \mathcal{R}} k_r \mathbf{c}_*^{-\gamma(\cdot, r)} \prod_{m=1}^{M} \frac{1}{(x_m - \beta(m, r))!} \mathcal{B}(x_m \geq \beta(m, r))$$

$$= \sum_{r \in \mathcal{R}} k_r \prod_{m=1}^{M} \frac{1}{(x_m - \alpha(m, r))!} \mathcal{B}(x_m \geq \alpha(m, r)).$$

This equation is satisfied if for each complex vector $\mathbf{v} = \begin{bmatrix} v_1 & v_2 & \cdots & v_M \end{bmatrix}^\top \in \mathcal{C}$

$$\sum_{\{r \in \mathcal{R}: \boldsymbol{\beta}(\cdot, r) = \mathbf{v}\}} k_r \mathbf{c}_*^{\boldsymbol{\alpha}(\cdot, r) - \mathbf{v}} \prod_{m=1}^{M} \frac{1}{(x_m - v_m)!} \mathcal{B}(x_m \geq v_m)$$

$$= \sum_{\{r \in \mathcal{R}: \boldsymbol{\alpha}(\cdot, r) = \mathbf{v}\}} k_r \prod_{m=1}^{M} \frac{1}{(x_m - v_m)!} \mathcal{B}(x_m \geq v_m), \qquad (10.44)$$

where the summations on the left-hand side and the right-hand side of (10.44) are over all the reaction steps where the reactant complexes and product complexes are $\mathbf{v}$, respectively. Since $\mathbf{x}$ and $\mathbf{v}$ are fixed in the previous display, Eq. (10.44) is equivalent to the complex balance equation of (7.5) (see Sect. 7.6).

The second part of the theorem, when $\mathbb{N}_0^M$ is reducible, can be carried out along the same line as the normalization of π_U does not influence the corresponding equations. $\qquad \Box$

Cappelletti and Wiuf (2016) asked whether the reverse of the previous statement held. Under some restrictions on the state space, they managed to prove the following.

Theorem 10.22 (Cappelletti and Wiuf (2016)) *Let the state space $\mathbb{N}_0^M$ of an induced kinetic Markov process with some stochastic kinetics be almost essential. Furthermore, let π_U be given by (10.43) with some $(k_r)_{r \in \mathscr{R}}$ and some $\mathbf{c}_* \in \mathbb{R}_+^M$ for every closed communicating class U. Now, π_U is a stationary distribution for the stochastic mass action type kinetics for all closed communicating class $U \subset \mathbb{N}_0^M$ if and only if $\mathbf{c}_*$ is a complex balanced stationary point of the induced kinetic differential equation.*

10.5.1 Stochastic Complex Balance

This section defines a concept for complex balance in the stochastic setting, which can be considered as the stochastic analogue of the notion we have defined in Sect. 7.6.

Definition 10.23 A stationary distribution π_U of the induced kinetic Markov process of reaction (10.3) is **stochastically complex balanced** on a closed communicating class $U \subset \mathbb{N}_0^M$ if for every $\mathbf{x} \in U$ and complex vector $\mathbf{v} \in \mathscr{C}_U$, it holds that

$$\sum_{r \in \mathscr{R}_U} \lambda(\mathbf{x} - \gamma(\cdot, r)) \pi_U(\mathbf{x} - \gamma(\cdot, r)) \mathfrak{B}(\beta(\cdot, r) = \mathbf{v}) = \sum_{r \in \mathscr{R}_U} \lambda_r(\mathbf{x}) \pi_U(\mathbf{x}) \mathfrak{B}(\alpha(\cdot, r) = \mathbf{v}),$$

(10.45)

where $\mathscr{R}_U$ denotes the active reaction steps on U and $\mathscr{C}_U$ the corresponding complexes (Definition 10.6).

Definition 10.24 The induced kinetic Markov process is said to be **stochastically complex balanced** if there exists a complex balanced stationary distribution on a **positive** closed communicating class.

Note that the positivity of closed communicating classes in the stochastic setting plays the role of that of complex balance stationary points in the deterministic setting. It turns out below that the above definitions tackle the right concept for complex balance of the stochastic model as a statement similar to Theorem 7.15 can be proved. First, a sort of consistency theorem follows.

Theorem 10.25 (Cappelletti and Wiuf (2016)) *If the induced kinetic Markov process endowed with stochastic mass action type kinetics is stochastically complex balanced, then (10.3) is weakly reversible. Moreover, the induced kinetic differential equation with mass action type kinetics is complex balanced in the sense of (7.5)*

*if and only if the induced kinetic Markov process with stochastic mass action type
kinetics is stochastically complex balanced. If this is the case, then there is a unique
stationary distribution π_U on every closed communicating class U of the state
space, where π_U has the form* (10.43).

This is an extension of the results by Anderson et al. (2010). Finally, the assertion
below is a direct consequence of Theorems 10.25 and 7.15.

Theorem 10.26 (Cappelletti and Wiuf (2016)) *The induced kinetic Markov process of reaction* (10.3) *endowed with stochastic mass action type kinetics is
stochastically complex balanced for any choice of* $(k_r^{sto})_{r \in \mathscr{R}}$ *if and only if the
reaction in question is weakly reversible and its deficiency is zero.*

10.5.2 Stochastic Detailed Balance

The meaning of "detailed balance" can vary from context to context. First, let
us outline the "usual" definition of detailed balance that frequently appears in
probability.

Definition 10.27 An induced kinetic Markov process **X** with transition rate function Q is said to be **detailed balanced** (or **Markov chain detailed balanced**) with
respect to a probability distribution π if

$$\pi(\mathbf{x}) Q(\mathbf{x}, \mathbf{y}) = \pi(\mathbf{y}) Q(\mathbf{y}, \mathbf{x}) \tag{10.46}$$

holds for all $\mathbf{x}, \mathbf{y} \in \mathbb{N}_0^M$.

Note that the above definition is quite general; for instance, it does not assume
anything special for the underlying reaction (10.3). It is an easy consequence that if
X is Markov chain detailed balanced with respect to π, then π is also a stationary
distribution of **X**. Let us also mention that Kolmogorov's criterion gives checkable
conditions for (general) Markov processes to be Markov chain detailed balanced
that can be found in Kelly (1979, Section 1.5) and Whittle (1986, Chapter 4). For
some reason, when (10.46) is satisfied, **X** is also said to be **reversible** with respect
to π in the probability literature.
    A slightly more restrictive concept is due to Whittle (1975).

Definition 10.28 Let us assume that the reaction (10.3) is reversible. Then the
induced kinetic Markov process **X** is said to be **microscopically reversible** (or
stochastically detailed balanced in Whittle sense) if the rate of each forward
reaction step is equal to the rate of each backward reaction steps, that is, for all
$\mathbf{x} \in \mathbb{N}_0^M$, it holds that

$$k_r \lambda_r(\mathbf{x}) = k_{-r} \lambda_{-r}(\mathbf{x}), \tag{10.47}$$

where the corresponding forward and backward reaction steps are now indexed by r and $-r$, where $r \in \mathscr{R}$, making the total number of reactions $2R$.

At first glance the just-defined notion for detailed balance seems somehow different from what we have defined in the deterministic setting (see Sect. 7.8). Indeed, the (deterministic) detailed balance is equivalent to microscopic reversibility (Whittle 1986, Chapter 4), which is stated in the following.

Theorem 10.29 (Whittle) *Assume that the reaction (10.3) is reversible. Then the induced kinetic differential equation with mass action type kinetics (6.7) is detailed balanced (see in Sect. 7.8.2) if and only if the induced kinetic Markov process with stochastic mass action type kinetics is stochastically detailed balanced in Whittle sense.*

It is only left to explore the connection between Definitions 10.27 and 10.28. In this direction Joshi (2015) proved the following.

Theorem 10.30 (Joshi (2015)) *Let the reaction (10.3) be reversible and assume mass action type kinetics for all the models. Then Whittle stochastic detailed balance implies Markov chain detailed balance, but the converse is not true in general. If the columns of the matrix $\boldsymbol{\gamma}$ are all different, then Whittle stochastic detailed balance is equivalent to Markov chain detailed balance.*

In Joshi (2015) a bunch of examples and counterexamples are also given for demonstration. We pick only one showing that Markov chain detailed balance does **not** imply microscopic reversibility in general. The $\boldsymbol{\gamma}$ of reaction $2\,\mathrm{A} \underset{k_{-1}}{\overset{k_1}{\rightleftharpoons}} \mathrm{A} + \mathrm{B} \underset{k_{-2}}{\overset{k_2}{\rightleftharpoons}} 2\,\mathrm{B}$ has two pairs of identical column vectors. In this case Markov chain detailed balance holds regardless of how the reaction rate coefficients are chosen, but for Whittle stochastic detailed balance, the condition $k_1 k_{-2} = k_2 k_{-1}$ is needed.

10.6   Comparison of the Deterministic and Stochastic Models

In the previous sections, we dealt with the intrinsic properties of the usual stochastic model of reactions. Here, it is shown that the properly scaled induced kinetic Markov process converges in some sense to the solution of the induced kinetic differential equation considering mass action type kinetics in both cases. Indeed, a more general treatment of the convergence of Markov processes was done by Kurtz (1970, 1972, 1976), which apply for much more Arnold and Theodosopulu (1980) elaborated and proved several stochastic models for reactions in which diffusion, that is, space inhomogeneity, is also present. More recently Mozgunov et al. (2017) and Bibbona and Sirovich (2017) elaborated approximations to the induced kinetic Markov process under some scaling.

10.6.1 Classical Scaling

Let us consider the induced kinetic Markov process $\mathbf{X}$ with **scaled** stochastic mass action type kinetics given as

$$\lambda_r^{(N)}(\mathbf{x}) := k_r^{\mathrm{sto}}(N)\big[\mathbf{x}\big]_{\alpha(\cdot,r)},$$

where $N = N_A V$, N_A is the Avogadro constant and V is the volume of the vessel in which the considered reaction takes place (see (10.6) and Sect. 10.2.3). Now, set $k_r^{\mathrm{sto}}(N)$ to be $k_r^{\mathrm{det}} N^{1-o(r)}$, and introduce the scaled quantity

$$\mathbf{C}^{(N)}(t) := \frac{1}{N}\mathbf{X}(t) \quad (t \geq 0),$$

the unit of which is now $\mathrm{mol\,dm}^{-3}$. The initial state $\mathbf{C}^{(N)}(0) = \mathbf{C}_0^{(N)}$ is the integer part of $\mathbf{c}_0 N$ divided by N, where $\mathbf{c}_0 \in \mathbb{R}$ is fixed. The next theorem is due to Th. G. Kurtz.

Theorem 10.31 (Kurtz) *Let $\mathbf{X}$ be an induced kinetic Markov process with stochastic mass action type kinetics with the integer part of $\mathbf{c}_0 N$ as initial condition. Then for every $\delta > 0$, it holds that*

$$\lim_{N \to +\infty} \mathbf{P}\big\{ \sup_{0 \leq s \leq t} \|\mathbf{C}^{(N)}(s) - \mathbf{c}(s)\| > \delta \big\} = 0, \tag{10.48}$$

where $\mathbf{c}$ is the solution of the induced kinetic differential equation, where $\mathbf{c}(0) = \mathbf{c}_0$.

Let us make some comments.

- Theorem 10.31 expresses a law of large numbers. In the scaling of the induced kinetic Markov process $\mathbf{X}$, both the number of species and volume tend to infinity such that its ratio, the molarity, is constant. The limit of this machinery is also called the **thermodynamic limit** (see Atkins and Paula 2013, Chapter 15).
- The theorem can be stated for more general processes as well (see for further reading Ethier and Kurtz 2009; Anderson and Kurtz 2015).

Now, let us sketch the main idea behind Theorem 10.31 without going into the technical details of proofs.

First, by using (6.6), observe that

$$\lambda_r^{(N)}(\mathbf{x}) = N w_r\big(\mathbf{c}^{(N)}\big) + \varepsilon\left(\frac{1}{N}\right) \tag{10.49}$$

holds for all $\mathbf{x} \in \mathbb{N}_0^M$, where $\mathbf{c}^{(N)} = \frac{1}{N}\mathbf{x}$ and $\varepsilon(1/N)$ tends to 0 as $N \to +\infty$. Note for first-order reactions that $\lambda_r^{(N)}(\mathbf{x}) = N w_r(\mathbf{c}^{(N)})$ ($\mathbf{c}^{(N)} N = \mathbf{x}$).

It follows from the Poisson representation (10.21) that

$$\mathbf{C}^{(N)}(t) = \mathbf{C}_0^{(N)} + \sum_{r=1}^{R} \frac{1}{N} \boldsymbol{\gamma}(\cdot, r) \mathscr{P}_r \left(\int_0^t \lambda_r^{(N)}(\mathbf{X}(s))\, ds \right)$$

$$= \mathbf{C}_0^{(N)} + \sum_{r=1}^{R} \frac{1}{N} \boldsymbol{\gamma}(\cdot, r) \mathscr{P}_r \left(N \int_0^t w_r(\mathbf{C}^{(N)}(s))\, ds + t\varepsilon\left(\frac{1}{N}\right) \right)$$

$$\approx \mathbf{c}_0 + \sum_{r=1}^{R} \frac{1}{N} \boldsymbol{\gamma}(\cdot, r) \mathscr{P}_r \left(N \int_0^t w_r(\mathbf{C}^{(N)}(s))\, ds \right).$$

By centering the appropriate Poisson processes, i.e., setting

$$\hat{\mathscr{P}}_r(t) := \mathscr{P}_r - t,$$

we get that

$$\sum_{r=1}^{R} \frac{1}{N} \boldsymbol{\gamma}(\cdot, r) \mathscr{P}_r \left(N \int_0^t w_r(\mathbf{C}^{(N)}(s))\, ds \right)$$

$$\sum_{r=1}^{R} \boldsymbol{\gamma}(\cdot, r) \int_0^t w_r(\mathbf{C}^{(N)}(s))\, ds$$

$$+ \sum_{r=1}^{R} \boldsymbol{\gamma}(\cdot, r) \left(\frac{1}{N} \hat{\mathscr{P}}_r \left(N \int_0^t w_r(\mathbf{C}^{(N)}(s))\, ds \right) \right).$$

The law of large numbers for Poisson processes implies that $\frac{1}{N}\hat{\mathscr{P}}_r(Nt)$ tends to zero as $N \to +\infty$; thus

$$\mathbf{C}^{(N)}(t) \approx \mathbf{c}_0 + \int_0^t f(\mathbf{C}^{(N)}(s))\, ds = \mathbf{c}_0 + \sum_{r=1}^{R} \int_0^t k_r^{\text{det}} \boldsymbol{\gamma}(\cdot, r) \mathbf{C}^{(N)}(s)^{\alpha(\cdot, r)}\, ds,$$

which is roughly speaking the integrated form of the induced kinetic differential equation (6.3), where f is the right-hand side of the induced kinetic differential equation (Eq. (6.14)) and w_r is the kinetic function in the deterministic setting (Eq. (6.6)). In plain English, the limit $\lim_{N\to+\infty} \mathbf{C}^{(N)}(t)$ is deterministic, and it satisfies the ordinary differential equation $\dot{\mathbf{c}} = \mathbf{f} \circ \mathbf{c}$ with $\mathbf{c}(0) = \mathbf{c}_0$.

So the deterministic limit of $\mathbf{C}^N$, being the solution of the induced kinetic differential equation was heuristically deduced. One can as well investigate the fluctuations of $\mathbf{C}^{(N)}$ around its average behavior as $N \to +\infty$. Since

$$\frac{1}{\sqrt{N}} \hat{\mathscr{P}}_r(Nt) = \frac{\mathscr{P}_r(Nt) - Nt}{\sqrt{N}}$$

can be approximated by **standard Brownian motions** $(\mathscr{B}_r(t))_{r \in \mathscr{R}}$, the fluctuation around $\mathbf{c}$ is given by the following formula:

$$\mathbf{C}^{(N)}(t) \approx \mathbf{c}(t) + \frac{1}{\sqrt{N}} \mathbf{V}(t),$$

for $t \geq 0$ as $N \to +\infty$, where $\mathbf{V}$ is a **Gaussian process** satisfying

$$\mathbf{V}(t) = \mathbf{V}(0) + \int_0^t \nabla \mathbf{f}(\mathbf{c}(s)) \mathbf{V}(s) \, ds + \sum_{r=1}^R \boldsymbol{\gamma}(\cdot, r) \mathscr{B}_r \left(\int_0^t \lambda_r(\mathbf{c}(s)) \, ds \right),$$

provided that $\mathbf{V}(0)$ is normally distributed. For further details consult (Anderson and Kurtz 2015, Chapter 4) and references therein.

10.6.2 Reaction Extent

Let us briefly discuss the reaction extent $\mathbf{U}$ defined in Sect. 6.3.5 and its connection with $\boldsymbol{\Upsilon}$ in the case of the induced kinetic Markov process (see Sect. 10.2.2). Define the scaled (stochastic) reaction extent as

$$\Upsilon_r^{(N)}(t) = \frac{1}{N} \Upsilon(t) = \frac{1}{N} \mathscr{P}_r \left(\int_0^t \lambda_r^{(N)}(\mathbf{X}(s)) \, ds \right) \quad (t \geq 0),$$

Let $[t, t+h]$ be a "small" time interval, and then

$$\Upsilon_r^{(N)}(t+h) - \Upsilon_r^{(N)}(t) = \frac{1}{N} \mathscr{P}_r \left(\int_t^{t+h} \lambda_r^{(N)}(\mathbf{X}(s)) \, ds \right)$$

$$\approx \frac{1}{N} \mathscr{P}_r \left(N \int_t^{t+h} w_r(\mathbf{C}^{(N)}(s)) \, ds + h\varepsilon \left(\frac{1}{N} \right) \right)$$

$$\approx \int_t^{t+h} w_r(\mathbf{C}^{(N)}(s)) \, ds$$

$$\approx h \cdot w_r(\mathbf{C}^{(N)}(t)) \, ds$$

$$\approx h \cdot w_r \left(\mathbf{c}_0 + \sum_{r=1}^R \boldsymbol{\gamma}(\cdot, r) \Upsilon_r^{(N)}(t) \right),$$

where we first used (10.49), and then the fact that h is small hence with high probability, none of the intensities will change in $[t, t + h]$.

This heuristics suggests, which can be rigorously proved, that for every $r \in \mathscr{R}$: $\Upsilon^{(N)}(t)$ has a limit as $N \rightarrow +\infty$, which is the (deterministic) reaction extent U, defined in Sect. 6.3.5, and solves the initial value problem $\dot{U} = w_r(c_0 + \gamma U)$, $U(0) = 0$ (see Kurtz 1972).

10.7 Stochastic Simulation

In what follows some simple and basic methods are going to be presented which generate sample paths for the induced kinetic Markov process X endowed with some stochastic kinetics. The basis of all these methods is the random (re)generation of subsequent times when certain reaction steps are to be executed. Some simulation methods were selected which can be casted into two groups: exact and approximate ones.

In **ReactionKinetics** the following functions can be used to get the results of a stochastic simulation for a given reaction: **Simulation**, **SimulationPlot**, **SimulationPlot2D** etc. using various simulation methods.

10.7.1 Direct Methods

Based on the constructions in the first part of Sect. 10.2.6, some simulation algorithms can be designed.

10.7.1.1 Direct and First Reaction Methods

One of the simplest methods is the direct reaction method, first published in full generality in Hárs (1976) and then in Gillespie (1977) (see also Érdi et al. 1973; Sipos et al. 1974a,b; Weber and Celardin 1976; Goss and Peccoud 1998). The methods in the rest of this section can be used to make this algorithm more efficient. The steps of the **direct reaction method** (also called the SSA as the "Stochastic Simulation Algorithm" or the Doob–Gillespie algorithm) are the following:

1. **Initialization.** Set $x := x_0 \in \mathbb{N}_0^M$, $t := 0$, and let $t_{max} > 0$ be a fixed time.
2. **while** $t \leq t_{max}$:
   a. Let ξ be an exponentially distributed random variable with mean $1/\Lambda(x)$, then
   b. the rth reaction step is chosen with probability $\lambda_r(x)/\Lambda(x)$.
   c. **Update.** Set $x := x + \gamma(\cdot, r)$, $t := t + \xi$, and update the intensities.

If at some time the total intensity, being $\Lambda(x)$, happens to be zero, then the simulation stops: no more reaction step can be executed. Note that the above method uses two random numbers per iteration, one is for the waiting time and another is for the selection of the next reaction step.

A variant of the previous method is called the first reaction method (see Gillespie 1977). As opposed to the direct reaction method, the following algorithm generates R random times in each iteration from which the smallest one is chosen and the corresponding reaction step is executed. Formally, the algorithm works as follows:

1. **Initialization.** Set $\mathbf{x} := \mathbf{x}_0 \in \mathbb{R}^M$, $t := 0$ and $t_{\max} > 0$ be a fixed time.
2. **while** $t \leq t_{\max}$:
   a. Let ξ_r be exponentially distributed random variables with mean $1/\lambda_r(\mathbf{x})$ for $r \in \mathcal{R}$.
   b. Let $r_{\min} := \arg\min_{r \in \mathcal{R}} \xi_r$.
   c. **Update.** Set $\mathbf{x} := \mathbf{x} + \boldsymbol{\gamma}(\cdot, r_{\min})$, $t := t + \xi_{r_{\min}}$, and update the intensities.

At first glance, the above two algorithms may seem different, but they provably result in the same process (in distribution). We note that one of the bottlenecks of the methods is that "too many" random numbers are being generated in each step which can be cumbersome when the number of reaction steps is large.

10.7.1.2 Next Reaction Method

Let us first define the **dependency graph** attached to a reaction (10.3).

Definition 10.32 The dependency graph $G := (V, E)$ is a directed graph on the set of reaction steps $\mathcal{R}$, i.e., $V = \mathcal{R}$. A directed edge goes from $r_i \in \mathcal{R}$ to $r_j \in \mathcal{R}$ (i.e., $(r_i, r_j) \in E$) if and only if there is an $m' \in \mathcal{M}$ such that $\gamma(m', r_i) \neq 0$ and $\lambda_{r_j}(\mathbf{x})$ depends on the value of $x_{m'}$ (for stochastic mass action type kinetics, it means that $\alpha(m', r_j) > 0$). By default all the loops are included in E, that is, for $r \in \mathcal{R}$, $(r, r) \in E$.

So, in plain words, in a dependency graph, all those reaction step pairs (r_i, r_j) are connected with one another for which there is a species that changes in the r_ith reaction step and affects at the same time the intensity λ_{r_j} of the r_jth reaction step. It is frequent that there are much more reaction steps than different species. In this case it is typical that a reactant species changing by a reaction step only affects a few other reaction steps; hence the dependencies result in a **sparse graph**.

The following algorithm is due to Gibson and Bruck (2000).

1. **Initialization.** Set $\mathbf{x} := \mathbf{x}_0 \in \mathbb{R}^M$, $t := 0$, and let $t_{\max} > 0$ be a fixed time.
2. Generate the dependency graph G (see Definition 10.32).
3. Generate exponentially distributed random numbers ξ_r with mean $1/\lambda_r(\mathbf{x})$ for $r \in \mathcal{R}$, and let $s_r = \xi_r$.
4. **while** $t \leq t_{\max}$:
   (a) Set $r_{\min} := \arg\min_{r \in \mathcal{R}} s_r$, and let $\lambda_r^{\text{old}} := \lambda_r(\mathbf{x})$ for $r \in \mathcal{R}$.
   (b) **update**
      (i) Set $\mathbf{x} := \mathbf{x} + \boldsymbol{\gamma}(\cdot, r_{\min})$ and $t := s_{r_{\min}}$.
      (ii) For each edge $(r_{\min}, j)$ of the dependency graph G, do the following:

(A) Update the λ_j, i.e., $\lambda_j^{\text{new}} := \lambda_j(\mathbf{x})$.

(B) If $j \neq r_{\min}$, then set $s_j := t + \frac{\lambda_j^{\text{old}}}{\lambda_j^{\text{new}}}(s_j - t)$.

(C) If $j = r_{\min}$, then generate ξ having exponential distribution with mean $1/\lambda_{r_{\min}}^{\text{new}}$, and set $s_{r_{\min}} := t + \xi$.

It may happen that $\lambda_{j*} = 0$ for some $j^* \neq r_{\min}$. In this case as long as $\lambda_{j*} = 0$, s_{j*} is set to be ∞. Let t_1 be the first time at which λ_{j*} becomes 0, and let t_2 be the first time at which λ_{j*} ceases to be 0. Let $\lambda_{j*}^{\text{old}}$ be the last pre-0 intensity, and let $\lambda_{j*}^{\text{new}}$ be the first post-0 intensity. Then set $s_{j*} := t_2 + (s_{j*} - t_1)\lambda_{j*}^{\text{old}}/\lambda_{j*}^{\text{new}}$.

The significant improvement made by Gibson and Bruck (2000) is that the above method reduces the random number generations compared to the previous ones in each iteration. Indeed, only the exponentially distributed ξ is regenerated at each step. Note that the next reaction method works with absolute times as opposed to the direct and first reaction methods, where relative times are being generated and compared in each iteration. We also refer to Anderson (2007), in which a modified version of the next reaction method was discussed.

Notice that in all the previous simulation methods, one can make use of the dependency graph. Initially, all the intensities λ_r must be calculated, but in the iterative steps, one can cut down on calculations that result in the same intensities. After executing say the rth reaction step, only those intensities λ_j need to be updated for which (r, j) is an edge of the dependency graph G of the corresponding reaction.

10.7.2 Approximate Methods

This section discusses the "tau-leaping" methods. In the past decade, the investigation of τ-leaping methods has become very vivid after D. T. Gillespie introduced the first version of (explicit) τ-leaping methods in 2001 (see Gillespie 2001; Rathinam et al. 2005). The only goal of this section is to outline the backbone of the τ-leaping methods. At the heart of all of these methods, the leap condition lies, which, roughly speaking, tells us when to execute the next reaction step.

Definition 10.33 (Leap Condition) Assume that the system is in state $\mathbf{x}$ at time t. We say that a $\tau > 0$ satisfies the **leap condition** if in the time interval $[t, t + \tau[$, no intensity function λ_r is likely to change its value by a significant amount, that is, $\lambda_r(\mathbf{x})$ remains roughly unchanged during $[t, t + \tau[$ for all $r \in \mathscr{R}$.

The Poisson representation (10.21) implies that for all $h > 0$ it holds that

$$\mathbf{X}(t + h) = \mathbf{X}(t) + \sum_{r=1}^{R} \boldsymbol{\gamma}(\cdot, r) \mathscr{P}_r \left(\int_t^{t+h} \lambda_r(\mathbf{X}(s)) \, ds \right)$$

provided that $\mathbf{X}(t) = \mathbf{x} \in \mathbb{N}_0^M$. Now, if h satisfies the leap condition, then the number of times the rth reaction step occurs in $[t, t+h[$ follows a Poisson distribution with mean and variance $\lambda_r(\mathbf{x})h$ given that $\mathbf{X}(t) = \mathbf{x}$. Therefore one can propagate the system as

$$\mathbf{X}(t + h) = \mathbf{x} + \sum_{r=1}^{R} \boldsymbol{\gamma}(\cdot, r)\zeta_r,$$

given that $\mathbf{X}(t) = \mathbf{x}$ and ζ_r are mutually independent Poisson random variables with mean $\lambda_r(\mathbf{x})h$. This is the explicit τ-leaping method. The general form of the algorithm follows:

1. **Initialization**. Set $\mathbf{x} := \mathbf{x}_0 \in \mathbb{R}^M$, $t := 0$, and let $t_{\max} > 0$ be a fixed time.
2. **while** $t \geq t_{\max}$:
    (a) Choose τ so that it satisfies the **leap condition**.
    (b) Generate random samples ζ_r according to Poisson distribution with mean $\lambda_r(\mathbf{x})\tau$ for $r \in \mathscr{R}$.
    (c) Consider the following system of equations for the unknown $\mathbf{y}$:

$$\mathbf{y} = \mathbf{x} + \sum_{r=1}^{R} \boldsymbol{\gamma}(\cdot, r)T_r(\mathbf{x}, \mathbf{y}, \zeta_r). \tag{10.50}$$

   where $T_r : \mathbb{N}_0^M \times \mathbb{N}_0^M \times \mathbb{N} \to \mathbb{R}^+$ is a predefined function. Let the solution of (10.50) be $\mathbf{y}^*$ with the components being rounded to the nearest integer value.
    (d) **Update**. Set $\mathbf{x} := \mathbf{y}^*$ and $t := t + \tau$.

The main issue regarding the above method is the selection of an appropriate τ (step size) in each iteration such that the leap condition holds; efficient step-size selection was given by Gillespie and Petzold (2003) and Cao et al. (2006) (see also Anderson 2008). Another problem is that it might happen that "$\mathbf{x}$" reaches negative population; in this direction (Cao et al. 2005; Chatterjee et al. 2005) proposed methods to avoid negative population during the simulation. By now there is an endless literature on the methods and their various improvements.

We can specify the functions T_r in several ways. Let us mention a few:

- if $T_r(\mathbf{x}, \mathbf{y}, \zeta) = \zeta_r$, we get the **explicit τ-leaping method**,
- if $T_r(\mathbf{x}, \mathbf{y}, \zeta) = \zeta_r + \lambda_r(\mathbf{y})\tau - \lambda_r(\mathbf{x})\tau$, we get the **implicit τ-leaping method**, finally
- if $T_r(\mathbf{x}, \mathbf{y}, \zeta) = \zeta_r + \lambda_r(\mathbf{y})\frac{\tau}{2} - \lambda_r(\mathbf{x})\frac{\tau}{2}$, we get the **trapezoidal τ-leaping method**,

where $\mathbf{x}, \mathbf{y} \in \mathbb{N}_0^M$, $\zeta_r \in \mathbb{N}_0$ and $r \in \mathscr{R}$. The implicit method of Rathinam et al. (2003) is quite useful and is an improvement of the explicit one when, e.g., the system appears to be stiff. The numerical results of the trapezoidal scheme proposed by Cao and Petzold (2005) show better results than the explicit and implicit τ-leaping methods.

10.8 Exercises and Problems

10.1 Formulate the master equation and the generating function equation for the stochastic model of the reaction

$$\text{inactive gene} \rightleftharpoons \text{active gene} \longrightarrow \text{messenger} \longrightarrow \text{protein} \longrightarrow 0$$

$$\text{messenger} \longrightarrow 0.$$

Try to symbolically solve the generating function equation by using *Mathematica*.

(Solution: page 439.)

10.2 Consider the **radioactive decay model** (linear outflow) $X \xrightarrow{k} 0$. Find solutions to its master equation and determine the moment equations. Find all the stationary distributions.

(Solution: page 439.)

10.3 Consider the reaction $X + Y \xrightarrow{k_1} 2X \quad X + Y \xrightarrow{k_2} 2Y$. Write down its master equation and the partial differential equation for the generating function. Find all the stationary distributions. Show that the induced kinetic Markov process is a submartingale (supermartingale) in species X if $k_1 \geq k_2$ ($k_2 \geq k_1$). (What about Y?) The process is a martingale for both species X and Y if and only if $k_1 = k_2$. Compare the results with the one obtained from the induced kinetic differential equation.

(Solution: page 440.)

10.4 Write a one-line *Mathematica* code for the conversion of units based on Sect. 10.2.3. Determine the units of the (deterministic and stochastic) reaction rate coefficients for the Michaelis–Menten, Lotka–Volterra, Robertson reactions.

(Solution: page 441.)

10.5 Assume that the induced kinetic Markov processes of the following reactions are endowed with stochastic mass action type kinetics: $0 \longrightarrow X; 0 \rightleftharpoons X; X \rightleftharpoons 2X$ and $X + Y \rightleftharpoons 2X \quad X + Y \rightleftharpoons 2Y$. Find all the communicating classes and classify the states. What are the stationary distributions?

(Solution: page 441.)

10.6 Based on Sect. 10.3.2, write a *Mathematica* program that automatically constructs the partial differential equation for the generating function and is trying to solve it.

(Solution: page 441.)

10.7 Find a second-order reaction which is consistent in mean with respect to a set of reaction rate coefficients.

(Solution: page 441.)

10.8 Show a stoichiometrically **not** mass-conserving reaction for which $\mathbf{D}_2$ is **not** positive definite.

(Solution: page 442.)

10.9 Find stationary distributions for the following reactions:

- $0 \xrightarrow{k_1} X \quad a\,X \underset{k_{-2}}{\overset{k_2}{\rightleftharpoons}} (a-1)\,X,$
- $X \underset{k_{-1}}{\overset{k_1}{\rightleftharpoons}} 2\,X \quad 0 \underset{k_{-2}}{\overset{k_2}{\rightleftharpoons}} Y.$

where stochastic mass action type kinetics are assumed.

(Solution: page 442.)

10.10 Write a simple simulation program in *Mathematica*, based on the direct reaction method.

(Solution: page 443.)

10.11 Assuming mass action type kinetics, write a short program that computes the numeric solution of the induced kinetic differential equation and also simulates the induced kinetic Markov process which is considered to be taking place in a certain volume V (specified by the user).

(Solution: page 443.)

10.12 Consider the induced kinetic Markov process of the reaction

$$X + Y \xrightarrow{k_1} 2\,X + Y \quad Y \xrightarrow{k_2} 2\,Y.$$

(Becker 1970).

Using the generating function, determine the combinatorial moments $E(X(X + 1) \cdots (X+a-1))$ $(a \in \mathbb{N})$. For a given $a \in \mathbb{N}$, prove that the corresponding moment blows up if and only if $a \neq k_2/k_1$. Particularly for

$$t \geq \frac{1}{k_2 - ak_1} \ln \left(\frac{k_2}{ak_1} \right) \tag{10.51}$$

$E(X(X+1) \cdots (X+a-1)) = +\infty$. On the other hand, show that the induced kinetic differential equation has a solution on the whole $\mathbb{R}_0^+$. This is a simple second-order reaction where the deterministic and stochastic models show significant differences.

(Solution: page 443.)

10.13 Consider the induced kinetic Markov process with stochastic mass action type kinetics of the second-order reaction $X \underset{k_{-1}}{\overset{k_1}{\rightleftharpoons}} 2X$. Show that moment equations are not closed. Write down the first two moment equations, and close the equations by substituting the higher moments with one of the formulas of Sect. 10.3.3.3.

(Solution: page 444.)

10.14 Write up the partial differential equation for the generating function in the case of the closed compartmental system: $X(i) \overset{k_{ji}}{\longrightarrow} X(j)$ where $1 \leq i \neq j \leq M$. Can the general solution be given in a compact form?

(Solution: page 445.)

10.15 Consider the reaction $X_1 \underset{k_{12}}{\overset{k_{21}}{\rightleftharpoons}} X_2$. Can the transition probabilities of its induced kinetic Markov process with stochastic mass action type kinetics be given in explicit form? Outline these in the case of $k_{12} = k_{21} = 1$ and $\mathbf{x}_0 = \begin{bmatrix} 1 & 1 \end{bmatrix}^\top$.

(Solution: page 445.)

10.16 Calculate $\mathbf{D}_1, \mathbf{D}_2$ for the closed compartmental system (also called the simple consecutive reaction) $X \overset{k_1}{\longrightarrow} Y \overset{k_2}{\longrightarrow} Z$.

(Solution: page 446.)

10.17 Find all the stationary distributions of the induced kinetic Markov process endowed with mass action type kinetics of a weakly reversible closed compartmental system with a single linkage class.

(Solution: page 446.)

10.18 Write a short *Mathematica* code that symbolically solves Eq. (10.26) with initial value and boundary conditions of the induced kinetic Markov process endowed with mass action type kinetics of a closed compartmental system.

(Solution: page 446.)

10.9 Open Problems

1. Find necessary and sufficient conditions for an induced kinetic Markov process endowed with stochastic mass action type kinetics to be nonexplosive (or explosive) in terms of the α, β and the stochastic reaction rate coefficients $(k_r^{sto})_{r \in \mathscr{R}}$.
2. Classify those induced kinetic Markov processes (with stochastic mass action type kinetics) for which the supermartingale, martingale, or the submartingale property holds.
3. Classify the stationary distributions of the induced kinetic Markov process of mass-conserving second-order, third-order, etc. reactions.

References

Abbasi S, Diwekar UM (2014) Characterization and stochastic modeling of uncertainties in the biodiesel production. Clean Techn Environ Policy 16(1):79–94

Anderson DF (2007) A modified next reaction method for simulating chemical systems with time dependent propensities and delays. J Chem Phys 127(21):214, 107

Anderson DF (2008) Incorporating postleap checks in tau-leaping. J Chem Phys 128(5):054, 103

Anderson DF, Kurtz TG (2015) Stochastic analysis of biochemical systems. Mathematical Biosciences Institute and Springer, Columbus and Berlin

Anderson DF, Craciun G, Kurtz TG (2010) Product-form stationary distributions for deficiency zero chemical reaction networks. Bull Math Biol 72:1947–1970

Arakelyan VB, Simonyan AL, Gevorgyan AE, Sukiasyan TS, Arakelyan AV, Grigoryan BA, Gevorgyan ES (2004) Fluctuations of the enzymatic reaction rate. Electron J Nat Sci 1(2):43–45

Arakelyan VB, Simonyan AL, Kintzios S, Gevorgyan AE, Sukiasyan TS, Arakelyan AV, Gevorgyan ES (2005) Correlation fluctuations and spectral density of the enzymatic reaction rate. Electron J Nat Sci 2(5):3–7

Arányi P, Tóth J (1977) A full stochastic description of the Michaelis–Menten reaction for small systems. Acta Biochim Biophys Hung 12(4):375–388

Arnold L (1980) On the consistency of the mathematical models of chemical reactions. In: Haken H (ed) Dynamics of synergetic systems. Springer, Berlin, pp 107–118

Arnold L, Theodosopulu M (1980) Deterministic limit of the stochastic model of chemical reactions with diffusion. Adv Appl Probab 12(2):367–379

Athreya KB, Ney PE (2004) Branching processes. Courier Corporation, Chelmsford

Atkins P, Paula JD (2013) Elements of physical chemistry. Oxford University Press, Oxford

Atlan H, Weisbuch G (1973) Resistance and inductance-like effects in chemical reactions: influence of time delays. Isr J Chem 11(2-3):479–488

Barabás B, Tóth J, Pályi G (2010) Stochastic aspects of asymmetric autocatalysis and absolute asymmetric synthesis. J Math Chem 48(2):457–489

Bartholomay AF (1958) Stochastic models for chemical reactions: I. Theory of the unimolecular reaction process. Bull Math Biol 20(3):175–190

Bartis JT, Widom B (1974) Stochastic models of the interconversion of three or more chemical species. J Chem Phys 60(9):3474–3482

Becker N (1973a) Carrier-borne epidemics in a community consisting of different groups. J Appl Prob 10(3):491–501

Becker NG (1970) A stochastic model for two interacting populations. J Appl Prob 7(3):544–564

Becker NG (1973b) Interactions between species: some comparisons between deterministic and stochastic models. Rocky Mountain J Math 3(1):53–68

Bibbona E, Sirovich R (2017) Strong approximation of density dependent markov chains on bounded domains. arXiv preprint arXiv:170407481

Cao Y, Petzold L (2005) Trapezoidal τ-leaping formula for the stochastic simulation of biochemical systems. Proceedings of foundations of systems biology in engineering, pp 149–152

Cao Y, Gillespie DT, Petzold L (2005) Avoiding negative populations in explicit Poisson τ-leaping. J Chem Phys 123(5):054, 104, 8

Cao Y, Gillespie DT, Petzold LR (2006) Efficient step size selection for the τ-leaping simulation method. J Chem Phys 124(4):044, 109, 11 pp

Cappelletti D, Wiuf C (2016) Product-form Poisson-like distributions and complex balanced reaction systems. SIAM J Appl Math 76(1):411–432

Chatterjee A, Vlachos DG, Katsoulakis MA (2005) Binomial distribution based τ-leap accelerated stochastic simulation. J Chem Phys 122(2):024, 112

Chibbaro S, Minier JP (2014) Stochastic methods in fluid mechanics. Springer, Wien

CombustionResearch (2011) Chemical-kinetic mechanisms for combustion applications. http://combustion.ucsd.edu, San Diego Mechanism web page, version 2011-11-22

Dambrine S, Moreau M (1981) On the stationary distribution of a chemical process without detailed balance. J Stat Phys 26(1):137–148

Darvey IG, Staff PJ (2004) Stochastic approach to first-order chemical reaction kinetics. J Chem Phys 44(3):990–997

Edman L, Rigler R (2000) Memory landscapes of single-enzyme molecules. Proc Natl Acad Sci USA 97(15):8266–8271

English BP, Min W, van Oijen AM, Lee KT, Luo G, Sun H, Cherayil BJ, Kou SC, Xie XS (2006) Ever-fluctuating single enzyme molecules: Michaelis–Menten equation revisited. Nat Chem Biol 2:87–94

Érdi P, Lente G (2016) Stochastic chemical kinetics. Theory and (mostly) systems biological applications. Springer series in synergetics. Springer, New York

Érdi P, Ropolyi L (1979) Investigation of transmitter-receptor interactions by analyzing postsynaptic membrane noise using stochastic kinetics. Biol Cybern 32(1):41–45

Érdi P, Tóth J (1976, in Hungarian) Stochastic reaction kinetics "nonequilibrium thermodynamics" of the state space? React Kinet Catal Lett 4(1):81–85

Érdi P, Tóth J (1989) Mathematical models of chemical reactions. Theory and applications of deterministic and stochastic models. Princeton University Press, Princeton

Érdi P, Sipos T, Tóth J (1973) Stochastic simulation of complex chemical reactions by computer. Magy Kém Foly 79(3):97–108

Ethier SN, Kurtz TG (2009) Markov processes: characterization and convergence. Wiley, Hoboken

Feller W (2008) An introduction to probability theory and its applications, vol 2. Wiley, Hoboken

Frank FC (1953) On spontaneous asymmetric synthesis. Biochim Biophys Acta 11:459–463

Gadgil C (2008) Stochastic modeling of biological reactions. J Indian Inst Sci 88(1):45–55

Gadgil C, Lee CH, Othmer HG (2005) A stochastic analysis of first-order reaction networks. Bull Math Biol 67(5):901–946

Gans PJ (1960) Open first-order stochastic processes. J Chem Phys 33(3):691–694

Gardiner CW (2010) Stochastic methods: a handbook for the natural and social sciences, 4th edn. Springer series in synergetics. Springer, Berlin

Gardiner CW, Chaturvedi S (1977) The Poisson representation. I. A new technique for chemical master equations. J Stat Phys 17(6):429–468

Gibson MA, Bruck J (2000) Efficient exact stochastic simulation of chemical systems with many species and many channels. J Phys Chem A 104(9):1876–1889

Gikhman II, Skorokhod AV (2004a) The theory of stochastic processes I. Springer, Berlin

Gikhman II, Skorokhod AV (2004b) The theory of stochastic processes II. Springer, Berlin

Gillespie DT (1977) Exact stochastic simulation of coupled chemical reactions. J Phys Chem 81(25):2340–2361

Gillespie DT (2001) Approximate accelerated stochastic simulation of chemically reacting systems. J Chem Phys 115(4):1716–1733

Gillespie CS (2009) Moment-closure approximations for mass-action models. IET Syst Biol 3(1):52–58

Gillespie DT, Petzold LR (2003) Improved leap-size selection for accelerated stochastic simulation. J Chem Phys 119(16):8229–8234

Goss PJE, Peccoud J (1998) Quantitative modeling of stochastic systems in molecular biology using stochastic Petri nets. Proc Natl Acad Sci USA 95:6750–6755

Grima R, Walter NG, Schnell S (2014) Single-molecule enzymology à la Michaelis–Menten. FEBS J 281(2):518–530

Hárs V (1976) A sztochasztikus reakciókinetika néhány kérdéséről (Some problems of stochastic reaction kinetics). Msc, Eötvös Loránd University, Budapest

Hong Z, Davidson DF, Hanson RK (2011) An improved H_2O_2 mechanism based on recent shock tube/laser absorption measurements. Combust Flame 158(4):633–644. https://doi.org/10.1016/j.combustflame.2010.10.002

Iosifescu M, Tăutu P (1973) Stochastic processes and applications in biology and medicine. II. Models. Editura Academiei, New York

Jahnke T, Huisinga W (2007) Solving the chemical master equation for monomolecular reaction systems analytically. J Math Biol 54(1):1–26

Joshi B (2015) A detailed balanced reaction network is sufficient but not necessary for its Markov chain to be detailed balanced. Discret Contin Dyn Syst Ser B 20(4):1077–1105

Juette MF, Terry DS, Wasserman MR, Zhou Z, Altman RB, Zheng Q, Blanchard SC (2014) The bright future of single-molecule fluorescence imaging. Curr Opin Chem Biol 20:103–111

Kelly FP (1979) Reversibility and stochastic networks. Wiley, New York

Kingman JFC (1969) Markov population processes. J Appl Prob 6(1):1–18

Kolmogoroff A (1935) Zur Theorie der Markoffschen Ketten. Math Ann 112:155–160

Krieger IM, Gans PJ (1960) First-order stochastic processes. J Chem Phys 32(1):247–250

Kurtz TG (1970) Solutions of ordinary differential equations as limits of pure jump Markov processes. J Appl Prob 7(1):49–58

Kurtz TG (1972) The relationship between stochastic and deterministic models for chemical reactions. J Chem Phys 57(7):2976–2978

Kurtz TG (1976) Limit theorems and diffusion approximations for density dependent Markov chains. In: Stochastic systems: modeling, identification and optimization, I. Springer, Berlin, pp 67–78

Kurtz TG (1978) Strong approximation theorems for density dependent Markov chains. Stoch Process Appl 6(3):223–240

Lai JYW, Elvati P, Violi A (2014) Stochastic atomistic simulation of polycyclic aromatic hydrocarbon growth in combustion. Phys Chem Chem Phys 16:7969–7979

Lánský P, Rospars JP (1995) Ornstein–Uhlenbeck model neuron revisited. Biol Cybern 72(5):397–406

Lee NK, Koh HR, Han KY, Lee J, Kim SK (2010) Single-molecule, real-time measurement of enzyme kinetics by alternating-laser excitation fluorescence resonance energy transfer. Chem Commun 46:4683–4685

Lente G (2004) Homogeneous chiral autocatalysis: a simple, purely stochastic kinetic model. J Phys Chem A 108:9475–9478

Lente G (2005) Stochastic kinetic models of chiral autocatalysis: a general tool for the quantitative interpretation of total asymmetric synthesis. J Phys Chem A 109(48):11058–11063

Lente G (2010) The role of stochastic models in interpreting the origins of biological chirality. Symmetry 2(2):767–798

Leontovich MA (1935) Fundamental equations of the kinetic theory of gases from the point of view of stochastic processes. Zhur Exper Teoret Fiz 5:211–231

Li G, Rabitz H (2014) Analysis of gene network robustness based on saturated fixed point attractors. EURASIP J Bioinform Syst Biol 2014(1):4

Liggett TM (2010) Continuous time Markov processes: an introduction, vol 113. American Mathematical Society, Providence.

Lipták G, Hangos KM, Pituk M, Szederkényi G (2017) Semistability of complex balanced kinetic systems with arbitrary time delays. arXiv preprint arXiv:170405930

Matis JH, Hartley HO (1971) Stochastic compartmental analysis: model and least squares estimation from time series data. Biometrics, pp 77–102

McAdams HH, Arkin A (1997) Stochastic mechanisms in gene expression. Proc Natl Acad Sci USA 94(3):814–819

Mode CJ (1971) Multitype branching processes: theory and applications, vol 34. American Elsevier, New York.

Mozgunov P, Beccuti M, Horvath A, Jaki T, Sirovich R, Bibbona E (2017) A review of the deterministic and diffusion approximations for stochastic chemical reaction networks. arXiv preprint arXiv:171102567

Munsky B, Khammash M (2006) The finite state projection algorithm for the solution of the chemical master equation. J Chem Phys 124(4):044, 104

Nagypál I, Epstein IR (1986) Fluctuations and stirring rate effects in the chlorite-thiosulphate reaction. J Phys Chem 90:6285–6292

Nagypál I, Epstein IR (1988) Stochastic behaviour and stirring rate effects in the chlorite-iodide reaction. J Chem Phys 89:6925–6928

Norris JR (1998) Markov chains. Cambridge University Press, Cambridge

Øksendal B (2003) Stochastic differential equations, 5th edn. Springer, Berlin

Paulevé L, Craciun G, Koeppl H (2014) Dynamical properties of discrete reaction networks. J Math Biol 69(1):55–72

Pokora O, Lánský P (2008) Statistical approach in search for optimal signal in simple olfactory neuronal models. Math Biosci 214(1–2):100–108

Qian H, Elson EL (2002) Single-molecule enzymology: stochastic Michaelis–Menten kinetics. Biophys Chem 101:565–576

Rathinam M, Petzold LR, Cao Y, Gillespie DT (2003) Stiffness in stochastic chemically reacting systems: the implicit tau-leaping method. J Phys Chem A 119(24):12,784, 11 pp

Rathinam M, Petzold LR, Cao Y, Gillespie DT (2005) Consistency and stability of tau-leaping schemes for chemical reaction systems. Multiscale Model Simul 4(3):867–895

Reddy VTN (1975) On the existence of the steady state in the stochastic Volterra–Lotka model. J Stat Phys 13(1):61–64

Rényi A (1954, in Hungarian) Treating chemical reactions using the theory of stochastic processes. Magyar Tudományos Akadémia Alkalmazott Matematikai Intézetének Közleményei 2:83–101

Robertson HH (1966) In: Walsh JE (ed) The solution of a set of reaction rate equations, Thompson Book, Toronto, pp 178–182

Sakmann B, Neher E (eds) (1995) Single-channel recording, 2nd edn. Plenum Press, New York

Samad HE, Khammash M, Petzold L, Gillespie D (2005) Stochastic modeling of gene regulatory networks. Int J Robust Nonlinear Control 15:691–711

Siegert AJF (1949) On the approach to statistical equilibrium. Phys Rev 76(11):1708

Singer K (1953) Application of the theory of stochastic processes to the study of irreproducible chemical reactions and nucleation processes. J R Stat Soc Ser B 15(1):92–106

Sipos T, Tóth J, Érdi P (1974a) Stochastic simulation of complex chemical reactions by digital computer, I. The model. React Kinet Catal Lett 1(1):113–117

Sipos T, Tóth J, Érdi P (1974b) Stochastic simulation of complex chemical reactions by digital computer, II. Applications. React Kinet Catal Lett 1(2):209–213

Smith G, Golden D, Frenklach M, Moriary N, Eiteneer B, Goldenberg M, Bowman C, Hanson R, Song S, Gardiner W, Lissianski V, Qin Z (2000) Gri-mech 3.0. http://www.me.berkeley.edu/gri_mech

Soai K, Shibata T, Morioka H, Choji K (1995) Asymmetric autocatalysis and amplification of enantiomeric excess of a chiral molecule. Nature 378:767–768

Šolc M (2002) Stochastic model of the n-stage reversible first-order reaction: relation between the time of first passage to the most probable microstate and the mean equilibrium fluctuations lifetime. Z Phys Chem 216(7):869–893

Stoner CD (1993) Quantitative determination of the steady state kinetics of multi-enzyme reactions using the algebraic rate equations for the component single enzyme reactions. Biochem J 291(2):585–593

Tóth J (1981, in Hungarian) A formális reakciókinetika globális determinisztikus és sztochasztikus modelljéről (On the global deterministic and stochastic models of formal reaction kinetics with applications). MTA SZTAKI Tanulmányok 129:1–166

Tóth J (1981) Poissonian stationary distribution in a class of detailed balanced reactions. React Kinet Catal Lett 18(1–2):169–173

Tóth J (1988a) Contribution to the general treatment of random processes used in chemical reaction kinetics. In: Transactions of the Tenth Prague Conference on information theory, statistical decision functions, random processes, held at Prague, from July 7 to 11, 1986, Academia (Publ. House of the Czechosl. Acad. Sci.), Prague, vol 2, pp 373–379

Tóth J (1988b) Structure of the state space in stochastic kinetics. In: Grossmann V, Mogyoródi J, Vincze I, Wertz W (eds) Probability theory and mathematical statistics with applications, Springer, pp 361–369

Tóth J, Érdi P (1992) A sztochasztikus kinetikai modellek nélkülözhetetlensége (The indispensability of stochastic kinetical models). In: Bazsa G (ed) Nemlineáris dinamika és egzotikus kinetikai jelenségek kémiai rendszerekben (Nonlinear dynamics and exotic kinetic phenomena in chemical systems), Jegyzet, Pro Renovanda Cultura Hungariae–KLTE Fizikai Kémiai Tanszék, Debrecen–Budapest–Gödöllő, chap 3, pp 117–143

Tóth J, Rospars JP (2005) Dynamic modelling of biochemical reactions with applications to signal transduction: principles and tools using Mathematica. Biosystems 79:33–52

Tóth J, Török TL (1980) Poissonian stationary distribution: a degenerate case of stochastic kinetics. React Kinet Catal Lett 13(2):167–171

Tóth J, Érdi P, Török TL (1983, in Hungarian) Significance of the Poisson distribution in the stochastic model of complex chemical reactions (A Poisson-eloszlás jelentősége összetett kémiai reakciók sztochasztikus modelljében). Alkalmazott Matematikai Lapok 9(1–2):175–196

Turányi T (1990) Sensitivity analysis of complex kinetic systems. Tools and applications. J Math Chem 5(3):203–248

Turányi T, Tomlin AS (2014) Analysis of kinetic reaction mechanisms. Springer, Berlin

Turner TE, Schnell S, Burrage K (2004) Stochastic approaches for modelling in vivo reactions. Comput Biol Chem 28(3):165–178

Urzay J, Kseib N, Davidson DF, Iaccarino G, Hanson RK (2014) Uncertainty-quantification analysis of the effects of residual impurities on hydrogen–oxygen ignition in shock tubes. Combust Flame 161(1):1–15

Van Kampen NG (2006) Stochastic processes in physics and chemistry, 4th edn. Elsevier, Amsterdam

Vellela M, Qian H (2007) A quasistationary analysis of a stochastic chemical reaction: Keizer's paradox. Bull Math Biol 69(5):1727–1746

Vellela M, Qian H (2009) Stochastic dynamics and non-equilibrium thermodynamics of a bistable chemical system: the Schlögl model revisited. J R Soc Interface 6:925–940

Velonia K, Flomenbom O, Loos D, Masuo S, Cotlet M, Engelborghs Y, Hofkens J, Rowan AE, Klafter J, Nolte RJM, de Schryver FC (2005) Single-enzyme kinetics of CALB catalyzed hydrolysis. Angew Chem Int Ed 44(4):560–564

Wadhwa RR, Zalányi L, Szente J, Négyessy L, Érdi P (2017) Stochastic kinetics of the circular gene hypothesis: feedback effects and protein fluctuations. Math Comput Simul 133:326–336

Weber J, Celardin F (1976) A general computer program for the simulation of reaction kinetics by the Monte Carlo technique. Chimia 30(4):236–237

Weiss S (1999) Fluorescence spectroscopy of single biomolecules. Science 283(5408):1676–1683

Whittle P (1975) Reversibility and acyclicity. In: Perspectives in probability and statistics. Applied probability trust

Whittle P (1986) Systems in stochastic equilibrium. Wiley, Hoboken

Yan CCS, Hsu CP (2013) The fluctuation-dissipation theorem for stochastic kinetics—implications on genetic regulations. J Chem Phys 139(22):224, 109

Zhang J, Hou Z, Xin H (2005) Effects of internal noise for calcium signaling in a coupled cell system. Phys Chem Chem Phys 7(10):2225–2228

Part IV

Selected Addenda

This part is about the use of estimation methods in reaction kinetics, about a short history of software packages written for kinetics and also on the mathematical background helping the reader in reading the book without the need to ferret about definitions and theorems in the internet or in textbooks on different fields of mathematics.

Inverse Problems 11

11.1 Direct and Inverse Problems

Previously (page 195) we mentioned about the difference in scientific and engineering approaches. Here we shall meet another opposition: that of the direct and inverse problems.

A significant part of reaction kinetics and correspondingly the largest part of our book is dedicated to **direct problems**. A direct problem is how the behavior of the deterministic or stochastic model of a reaction or that of a completely specified mechanism can be characterized either quantitatively or qualitatively. A (theoretically important) direct problem is the characterization of the form of the induced kinetic differential equations of reactions. Another one is obtained if given the reaction steps we try to decide if it has an asymptotically stable stationary point, multiple stationary points, a limit cycle, oligo-oscillation, or chaotic behavior. One can ask if the stationary distribution of the stochastic model is unimodal, if the (deterministic or stochastic) model blows up, etc.

It seems to be an exaggeration although it is hardly one that all the work on direct problems is preparatory to **inverse problems**. Generally speaking the solution of an inverse problem gives an answer to the question: What kind(s) of models/mechanisms/reactions can be behind a given set of data? One may be interested in qualitative answers, e.g., when looking for an oscillating reaction, or one may look for quantitative answers, e.g., when one should like to know the exact/approximate/estimated value of a reaction rate coefficient.

The inverse of the abovementioned theoretical direct problem is as follows: given a polynomial differential equation, is there a reaction to induce it? This problem has been solved in Theorem 6.27.

In reaction kinetics the most typical inverse problem in its simplest form is as follows: given the concentration vs. time curves and assuming that the reaction can be described by a mass action type deterministic model, how does an inducing reaction look like, and what are the reaction rate coefficients? A more realistic task is

© Springer Science+Business Media, LLC, part of Springer Nature 2018
J. Tóth et al., *Reaction Kinetics: Exercises, Programs and Theorems*,
https://doi.org/10.1007/978-1-4939-8643-9_11

to have the concentration values at discrete time points only and—what is worse—
with errors added. Before formulating other problems, let us see how the simpler
version is solved by *Mathematica*.

11.2   Parameter Estimation

Suppose we know that the reaction is a simple in- and outflow: $0 \rightleftharpoons X$. Let us
simulate values of concentrations with errors at discrete time points.

```
sol = First[ReplaceAll @@ Concentrations[{0 <=> X},
    {0.33, 0.72}, {0.2}, {0, 20}]]
times=N[Range[0,100]/5]; SeedRandom[100];
data=Transpose[{times,
    (sol /. t -> times) (1 + RandomReal[0.05, 101])}];
lp=ListPlot[data, PlotRange -> All]
```

We added a relatively large (5%) relative error and collected data at 100 time points.
These data will act as a series of measurements. The role of **SeedRandom** is that
the simulated results are the same no matter how many times we repeat them. This
is useful when one does not concentrate on the effect of randomness, but on the
algorithm using the data. The results can be seen in Fig. 11.1.

   Let us define the model to be fitted using the **memo function** construct of
Mathematica.

```
model[a_?NumberQ, b_?NumberQ, c_?] := (model[a,b,c] =
    First[x /. NDSolve[{x'[t] == a-b x[t], x[0]==c},
        x, {t, 20}]]);
```

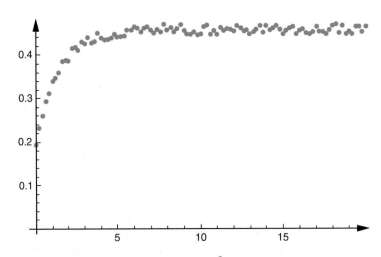

Fig. 11.1 Simulated data taken from the reaction $0 \underset{b}{\overset{a}{\rightleftharpoons}} X$ with $a = 0.33, b = 0.72, x(0) = 0.2$,
and 5% relative error

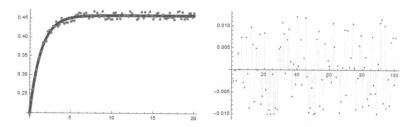

Fig. 11.2 Fitted curve, original data, and residuals concerning reaction $0 \underset{b}{\overset{a}{\rightleftharpoons}} X$

NonlinearModelFit is one of the built-in functions to do the parameter estimation itself. It needs an initial estimate to start.

```
nlm = NonlinearModelFit[data, model[a, b, c][t],
    {{a, 0.1}, {b, 0.1}, {c, 1}}, t]}
```

Let us visualize the goodness of fit by showing the original data, the fitted curve, and also the **residuals** using

```
Row[{Show[
    Plot[nlm[t], {t, 0, 20}, PlotRange -> All,
        PlotStyle -> Directive[Red, Thickness[0.015]]],
        data],
    ListPlot[nlm["FitResiduals"], Filling -> Axis]}]
```

The result can be seen in Fig. 11.2.

The fact that the induced kinetic differential equation of the reaction can explicitly be solved has not been used here. To emphasize this **Concentrations** has solved the induced kinetic differential equation numerically.

Beyond having the estimates of the reaction rate coefficients one gets at the same time quantitative measures of goodness, the list of which is produced by **nlm["Properties"]**. The abundance of properties offered by the program may be perplexing for almost all the users, except possibly those with an extremely good background in statistics. Properties include

```
SingleDeletionVariances, MaxParameterEffectsCurvature,
MeanPredictionConfidenceIntervals, FitCurvatureTable,
StudentizedResiduals, AIC, BIC.
```

How is fitting done? Without knowing the exact details of the implementation of the function **NonlinearModelFit**, the general approach can be found in Chapter VIII of the classical book by Bard (1974) or in the more recent monographs by Seber and Wild (2003) or by Weise (2009). One defines a function of the vector **p** of the parameters, called **objective function**, the sum of squares of differences between measured and model values:

$$Q(\mathbf{p}) := \sum_{t_i \in \text{times}} (\text{measured}(t_i) - \text{model}(t_i, \mathbf{p}))^2. \tag{11.1}$$

Then, the global minimum of this function is looked for starting from an initial estimate of the parameter vector $\mathbf{p}$ and using different strategies (steepest descent, simplex method by Nelder and Mead, simulated annealing, etc.) to modify it.

In an appropriate (although very rare) setting one may be able to prove that a given procedure converges, the convergence rate can be determined and it can be shown that the limit is a statistically good estimate of the parameters. This is the case with the method by Nagy (2011) based on an iterative computation of weighted empirical expectation values and covariance matrices using random samples.

There are a lot of variations and a lot of problems here. As to the variations:

1. Instead of the sum of squares, absolute values or any even power of the differences (between measured values and those obtained from a model) can be used.
2. The terms in the sum may be multiplied by the values of a **weight function** which takes into consideration the importance of the individual measurements. (An often used weight function is the reciprocal of the estimated variance which forces measurement points with the same **relative** error to have the same importance.)
3. The independent variable in our examples is almost always one dimensional (as it is time), but one can meet with multidimensional independent variables as well, e.g., when reaction–diffusion models are treated (when one has time and one or more spatial variables, see Problem 11.4).
4. Instead of concentrations it is only some **features** what are known. These features are global quantitative measures of the reaction, like flame velocity, time up to ignition, period length of oscillation, etc. Then, the objective function is to be redefined in terms of these quantities.

As to the problems:

1. In simple cases one may be given a straight line as the model, and the problem left is only to find its tangent and intersection with the ordinate axis. This is the case of an **explicitly defined model**. In reaction kinetics, however, the typical case is when the model is **implicitly defined**, what we only know about it is that it is a solution of a given differential equation.
2. If the model is linear in the independent variable(s) and the parameters (this is the case of **linear regression**), then the problem is easy to solve from the numerical point of view, and the results admit a clear statistical interpretation. If one only assumes that it is linear in the parameters (e.g., polynomials are fitted), then easy

numerical solvability subsists. If the model is nonlinear in the parameters, then the area of arts opens: practice, endurance, and luck are needed to find physically realistic parameters which also define a well-fitted model. It is the more so if the model is implicitly given. Cf. Problem 11.3.

3. There is a rule of thumb, not a theorem: one should not try to estimate more parameters than the square root of the number of the data. Having fitted a model, the estimated covariance matrix of the parameters shows if the selected parameters can be considered to be independent or not.

4. The **global minimum** of the objective function is needed, but the numerical methods usually only provide **local minima**. (Cf. Singer et al. 2006.) Quite often, the methods get stuck in a local valley of the surface defined by the objective function. A good initial guess of the parameter values is usually a great help. There are methods to find such a guess, see, e.g., Hangos and Tóth (1988) and Kovács and Tóth (2007), based on the fact that the right-hand side of the induced kinetic differential equation of a reaction endowed with mass action type kinetics is linear in the parameters; see Eq. (6.32).

5. Physical (or chemical) restrictions should be taken into consideration on the values of the parameters. This makes the problem of finding the minimum of the objective function more complicated as one arrives at a minimization problem with constraints. On the other side, these restrictions together with possible chemical knowledge on a similar system may give some hint as how to choose the initial estimate.

6. A nonlinear function may even not have a global minimum; see Problem 11.2.

7. The function **NonlinearModelFit** seems to have been designed to accept scalar-valued data. However, this restriction can be bypassed; see Problems 11.4 and 11.5.

8. There are not enough measurements: the time points are not located densely enough, or only some components of the concentration vector are measured.

9. The measurement error (or any other deviation from the values which can be described by the model, e.g., biological variations) is too large.

10. It may happen that it is only possible to determine some combinations (functions) of the parameters and not all of them individually; see Problem 11.7 and some further details on page 333.

11. There are too many measurements. Then, one may calculate a moving average of the data, or take their Fourier transform to get rid of large fluctuations, or even form blocks of data and substitute whole blocks with averages.

11.3 Estimation in the Usual Stochastic Model

Knowing the behavior of the trajectories of the stochastic model of a reaction, one can propose a heuristic algorithm to estimate the reaction rate coefficients, cf. Billingsley (1961). The procedure will be shown on a very simple example.

Example 11.1 Let us consider the reaction $0 \underset{k_{-1}}{\overset{k_1}{\rightleftharpoons}} X$. Suppose we have many measurements with particle number $n_1 \in \mathbb{N}$ showing that the relative frequency of the step $0 \overset{k_1}{\longrightarrow} X$ is $r_1 \in]0, 1[$ and some further measurements with particle number $n_2 \in \mathbb{N}$ showing that the relative frequency of the step $X \overset{k_{-1}}{\longrightarrow} 0$ is $r_2 \in]0, 1[$. Furthermore, suppose that at particle number $n_3 \in \mathbb{N}$, the average waiting time until the next jump (reaction step) is $1/\lambda$ ($\lambda \in \mathbb{R}^+$). Then, it seems to be evident to look for the solutions of the following system of equations:

$$\frac{k_1}{k_1 + k_{-1}n_1} = r_1 \qquad \frac{k_{-1}n_2}{k_1 + k_{-1}n_2} = r_2 \qquad k_1 + k_{-1}n_3 = \lambda. \tag{11.2}$$

Reordering this system one sees that it is a(n overdetermined) linear system for the unknown reaction rate coefficients:

$$\begin{bmatrix} r_1 - 1 & r_1 n_1 \\ r_2 & (r_2 - 1)n_2 \\ 1 & n_3 \end{bmatrix} \begin{bmatrix} k_1 \\ k_{-1} \end{bmatrix} = \begin{bmatrix} 0 \\ 0 \\ \lambda \end{bmatrix}. \tag{11.3}$$

Assuming that $\det \begin{bmatrix} r_1 - 1 & r_1 n_1 \\ r_2 & (r_2 - 1)n_2 \end{bmatrix} = 0$ and $r_2 n_3 \neq (r_2 - 1)n_2$ and $r_2 n_3 \neq (r_2 - 1)n_2$ implies that

$$\mathrm{rank} \begin{bmatrix} r_1 - 1 & r_1 n_1 \\ r_2 & (r_2 - 1)n_2 \\ 1 & n_3 \end{bmatrix} = \mathrm{rank} \begin{bmatrix} r_1 - 1 & r_1 n_1 & 0 \\ r_2 & (r_2 - 1)n_2 & 0 \\ 1 & n_3 & \lambda \end{bmatrix} = 2,$$

therefore (11.3) has a solution; thus one gets the following estimates:

$$\hat{k}_1 = -\lambda \frac{r_1 n_1}{r_1(n_3 - n_1) - n_3} \qquad \hat{k}_{-1} = \lambda \frac{r_1 - 1}{r_1(n_3 - n_1) - n_3}. \tag{11.4}$$

The assumptions imply that the denominator is different from zero, and in this case it can be directly checked that the estimators are positive, but in general the estimators may happen to have —physically meaningless—negative values.

We must admit at the end that the solution of the algebraic (let alone statistical) problems (the solution of some of them can be found in Billingsley 1961) can wait until experimentalists will be able to measure individual particles in more generality than at the moment; see Grima et al. (2014) and the relatively complete literature on the possibility of these kinds of measurements cited by Lente (2013).

11.4 Maximum Likelihood Estimates in a Stochastic Differential Equation Setting

An approach to get statistically good estimates based upon a stochastic kinetic model has been presented in Hangos and Tóth (1988). The starting point of the model is the right-hand side written in the form of (6.32): $\gamma\mathrm{diag}(\mathbf{c}^{\alpha})\mathbf{k}$ shows the linear dependence of the right-hand side of induced kinetic differential equations on the reaction rate coefficients. Then, a stochastic differential equation model may be defined as follows:

$$\mathrm{d}\mathbf{X}(t) = \gamma\,\mathrm{diag}(\mathbf{X}^{\alpha})\mathbf{k}\,\mathrm{d}t + \mathbf{B}(\mathbf{X}(t))\mathrm{d}\mathbf{w}(t) \tag{11.5}$$

where $\mathbf{X}$ is the vector of particle numbers, $\mathbf{k}$ is the vector of reaction rate coefficients, and $\mathbf{B}$ is a positive definite valued matrix function expressing local variance. If $\mathbf{B}(\mathbf{X}) = \mathbf{I}$, i.e., the local variance can be characterized with the identity matrix, then the explicit form of the **maximum likelihood estimate** of the vector of reaction rate coefficients can be shown to be

$$\hat{\mathbf{k}} = \frac{1}{T}\left(\mathbf{\Delta}(\mathbf{X}(T))^{\top}\mathbf{\Delta}(\mathbf{X}(T))\right)^{-1}\int_0^T \mathbf{\Delta}(\mathbf{X}(t))^{\top}\mathrm{d}\mathbf{X}(t) \tag{11.6}$$

with $\mathbf{\Delta}(\mathbf{X}) := \gamma\,\mathrm{diag}(\mathbf{X}^{\alpha})$. The fact that this estimate is a maximum likelihood estimate means that it gives that value $\hat{\mathbf{k}}$ with which the actual measurements are the most probable. The maximum likelihood estimates in general and in this case too have advantageous statistical properties. The estimated value $\hat{\mathbf{k}}$ is normally distributed with mean $\mathbf{k}$ and with the variance $\mathbf{H}$ and the Hessian matrix of the objective function (which is here the likelihood function) with respect to $\mathbf{k}$, also called the **information matrix** when the length of the interval of observation T tends to infinity.

Example 11.2 As the simplest illustration of the method described above, let us consider the elementary reaction $0 \xrightarrow{k} X$. A short calculation shows that in this case one has $\hat{k} = \frac{X(T)-X(0)}{T}$, i.e., the estimate so well-known from many estimation procedures is obtained. It is interesting to note that although the whole trajectory of X is known (measured) on the interval $[0, T]$, it is only the initial value and the final one that is utilized: this pair being a sufficient statistics.

Further examples and discussion can be found in Hangos and Tóth (1988) and see also Kovács and Tóth (2007).

The interested reader will find many useful ideas and algorithms in the following publications on parameter estimation: Bard (1974), Seber and Wild (2003), Singer et al. (2006), and Turányi and Tomlin (2014). A series of sizeable case studies has recently been studied by Villaverde et al. (2015).

11.5   Identifiability and Uniqueness Questions

Estimation should not be the first step in the evaluation process of experimental data. One should know in advance if it is possible to estimate a given reaction rate coefficient from the data one has. This kind of investigation belongs to the field of **identifiability**. Even if we restrict ourselves to the identifiability of reactions, we have to face a vast literature; here we only present the results of a few important papers.

Craciun and Pantea (2008) consider reactions with mass action type kinetics. First, they show the following example for which the reaction rate constants are not uniquely identifiable, even if we are given complete information on the dynamics of concentrations for all chemical species. The right-hand sides of the induced kinetic differential equation of both mechanisms are $\{-9c_{A_0}, 9c_{A_0}, 9c_{A_0}\}$ as it is also shown by

```
RightHandSide[{craciunpantea1}, {3, 3, 3}]
```

and

```
RightHandSide[{craciunpantea1}, {4, 4, 1}]
```

where **craciunpantea1** is the reaction seen in Fig. 11.3. Then, the authors give the obvious definition of identifiability as follows.

Definition 11.3 The reaction rate coefficients in the reaction $\langle \mathcal{M}, \mathcal{R}, \alpha, \beta \rangle$ are **unidentifiable** if there exist $\mathbf{k}_1 \neq \mathbf{k}_2 > \mathbf{0}$ so that for all $\mathbf{c} \in (\mathbb{R}_0^+)^M$

$$\gamma \operatorname{diag}(\mathbf{c}^\alpha)\mathbf{k}_1 = \gamma \operatorname{diag}(\mathbf{c}^\alpha)\mathbf{k}_2 \tag{11.7}$$

holds. It is **identifiable** if it is not unidentifiable.

The authors provided a sufficient condition of identifiability.

Theorem 11.4 *Fix a reactant complex vector. Suppose that all reaction step vectors starting from this reactant complex vector are independent. If this is true for all reactant complex vectors, then the reaction is identifiable.*

In the example of Fig. 11.3, the reaction step vectors $\begin{bmatrix} -1 \\ 2 \\ 0 \end{bmatrix}$, $\begin{bmatrix} -1 \\ 0 \\ 2 \end{bmatrix}$, and $\begin{bmatrix} -1 \\ 1 \\ 1 \end{bmatrix}$ are dependent. However, necessity of the condition is not true (contrary to what the authors believed); see Problem 11.10.

Corollary 11.5 *Generalized compartmental systems (in the wide sense) are identifiable.*

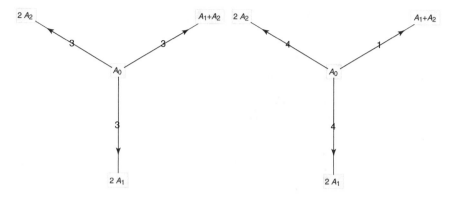

Fig. 11.3 Two mechanisms with identical dynamics

The idea of the paper by Santosa and Weitz (2011) is that among all the reactions having the same right-hand side, they find the realization with the smallest possible number of reaction steps given the concentration vs. time curves in a special case. (The reader of that paper should be careful because of small errors.) Thus, it may be considered as a kind of continuation of the paper by Craciun and Pantea (2008).

It may happen that it is impossible to determine the reaction rate coefficients individually; still it is possible to determine some functions of them. An approach using linear algebra and heuristics and starting from **sensitivity analysis** has been presented by Vajda et al. (1985); see also in Turányi (1990) and Turányi and Tomlin (2014). Analyzing the sensitivity matrix, it may be possible to find, e.g., that one can only determine the ratio or the product of two reaction rate coefficients.

Sedoglavic (2001a,b) had a much more ambitious goal: the right-hand sides are almost arbitrary with him, his model has input and output functions as well, and his problem is which variables or parameters can be determined from the knowledge of the output? To answer this question, he constructed a probabilistic algorithm Sedoglavic (2001b) and wrote a Maple program. A typical result of his program can be seen in the example below which may stimulate some readers to learn his method. The model to be investigated is (Sedoglavic 2001a, p. 106):

Example 11.6

$$\dot{x}_1 = c_2 \frac{\theta_1(x_6 - x_1)u}{\theta_2\theta_3},$$

$$\dot{x}_2 = \frac{(\theta_1 - \theta_6\theta_7)(x_1 - x_2) + \theta_5(x_3 - x_2)}{c_3\theta_3(1 - \theta_2) - \theta_8\theta_7} - \theta_4 x_2,$$

$$\dot{x}_3 = \frac{\theta_5(x_2 - x_3)}{c_4\theta_9},$$

$$\dot{x}_4 = \frac{\theta_6(x_1 - x_4) + \theta_{10}(x_5 - x_4)}{\theta_8},$$

$$\dot{x}_5 = \frac{\theta_{10}(x_4 - x_5)}{\theta_{11}(1 - \theta_8)},$$

$$\dot{x}_6 = \frac{c_1\theta_6\theta_7(x_4 - x_2) + \theta_1(x_2 - x_6)}{\theta_2\theta_3},$$

$$y_1 = \frac{55}{100}\theta_{12}x_1,$$

$$y_2 = \theta_{12}\left(\frac{55}{100}\theta_8x_4 + \theta_{11}(1 - \theta_8)x_5\right).$$

The consequences drawn are the parameters $\theta_1, \theta_3, \theta_5, \theta_7, \theta_9$, and θ_{12} and the variables x_1, x_2, x_3, x_4, x_5, and x_6 cannot be determined. The following transformation leaves the vector field invariant:

$$\begin{aligned}
x_1 &\longrightarrow \lambda x_1, \ x_4 \longrightarrow \lambda x_4, \ \theta_1 \longrightarrow \theta_1/\lambda, \ \theta_7 \longrightarrow \theta_7/\lambda,\\
x_2 &\longrightarrow \lambda x_2, \ x_5 \longrightarrow \lambda x_5, \ \theta_3 \longrightarrow \theta_3/\lambda, \ \theta_9 \longrightarrow \theta_9/\lambda, \qquad (11.8)\\
x_3 &\longrightarrow \lambda x_3, \ x_6 \longrightarrow \lambda x_6, \ \theta_5 \longrightarrow \theta_5/\lambda, \ \theta_{12} \longrightarrow \theta_{12}/\lambda.
\end{aligned}$$

Papers by Meshkat, Di Stefano, and their coworkers also contain relevant information on the topics: Meshkat et al. (2009, 2014) and Meshkat and Sullivant (2014).

Another important definition follows from Craciun and Pantea (2008).

Definition 11.7 Reactions $\langle \mathcal{M}, \mathcal{R}_1, \alpha_1, \beta_1\rangle$ and $\langle \mathcal{M}, \mathcal{R}_2, \alpha_2, \beta_2\rangle$ are **confoundable** if there exists reaction rate coefficients $\mathbf{k}_1$ and $\mathbf{k}_2$, respectively, such that the corresponding right-hand sides are the same for all concentration vectors $\mathbf{c} \in \mathbb{R}^M$. In other words, the reactions are confoundable if there exist two (not necessarily different) vectors of reaction rate coefficients $\mathbf{k}_1, \mathbf{k}_2 > \mathbf{0}$ and stoichiometric matrices $\boldsymbol{\gamma}_1, \boldsymbol{\gamma}_2$ so that for all $\mathbf{c} \in (\mathbb{R}_0^+)^M$

$$\boldsymbol{\gamma}_1\mathrm{diag}(\mathbf{c}^{\alpha_1})\mathbf{k}_1 = \boldsymbol{\gamma}_2\mathrm{diag}(\mathbf{c}^{\alpha_2})\mathbf{k}_2 \qquad (11.9)$$

holds.

Theorem 11.8 *The two reactions $\langle \mathcal{M}, \mathcal{R}_1, \alpha, \beta_1\rangle$ and $\langle \mathcal{M}, \mathcal{R}_2, \alpha, \beta_2\rangle$ are confoundable if and only if the positive cones generated by the reaction step vectors starting from the reactant complex vectors are not disjoint.*

A kind of confoundability can easily be excluded. The mechanism

$$X + Y \xrightarrow{1} 2X \quad X + Y \xrightarrow{1} 2Y$$

can be added to any mechanism without changing its induced kinetic differential equation, because

```
RightHandSide[{X + Y -> 2X, X + Y -> 2Y}, {1, 1}]
```

gives the zero vector. To use the introduced terminology, one can say that any mechanism is confoundable with the one obtained by adding the above mechanism to it.

A weakly reversible example of this type does not exist: (Tóth 1981, p. 48).

Theorem 11.9 *The induced kinetic differential equation of a weakly reversible reaction cannot be the zero polynomial.*

Proof Let the induced kinetic differential equation of the mechanism in question be $\dot{\mathbf{c}} = \mathbf{f} \circ \mathbf{c}$. As $\mathscr{S} = \text{span}(\mathscr{R}_{\mathbf{f}})$ and $\dim(\mathscr{S}) \geq 1$ (Feinberg and Horn 1977, p. 90), it is impossible that $\text{span}(\mathscr{R}_{\mathbf{f}})$ only consists of the $\mathbf{0}$ vector. $\qquad\square$

If two reactions are confoundable, then it is a good idea to use the simpler one or the one with a given property. This way an appropriate realization may reveal properties which otherwise would be hidden. This idea led to a series of papers mainly by Szederkényi, Johnston, Rudan, and his coauthors; see the list of references.

Confoundability is closely related to macroequivalence introduced by Horn and Jackson (1972, p. 111).

Definition 11.10 Mechanisms

$$\langle \mathscr{M}, \mathscr{R}_1, \boldsymbol{\alpha}_1, \boldsymbol{\beta}_1, \mathbf{k}_1 \rangle \text{ and } \langle \mathscr{M}, \mathscr{R}_2, \boldsymbol{\alpha}_2, \boldsymbol{\beta}_2, \mathbf{k}_2 \rangle$$

are **macroequivalent** if the corresponding right-hand sides are the same for all concentration vectors $\mathbf{c} \in (\mathbb{R}_0^+)^M$:

$$\boldsymbol{\gamma}_1 \text{diag}(\mathbf{c}^{\boldsymbol{\alpha}_1})\mathbf{k}_1 = \boldsymbol{\gamma}_2 \text{diag}(\mathbf{c}^{\boldsymbol{\alpha}_2})\mathbf{k}_2 \qquad (11.10)$$

holds.

This concept has slightly been generalized by Csercsik et al. (2012): a mechanism is **dynamically equivalent** with a kinetic polynomial differential equation if its induced kinetic differential equation is the given polynomial equation. With this definition the authors proved the following theorem:

Theorem 11.11 *Suppose the mechanism* $\langle \mathscr{M}, \mathscr{R}, \boldsymbol{\alpha}, \boldsymbol{\beta}, \mathbf{k} \rangle$ *is of deficiency zero and has a single ergodic class. Then no other mechanism with the same complexes is dynamically equivalent with this mechanism.*

Let us note that the existence of a single ergodic class implies that the Feinberg–Horn–Jackson graph consists of a single component (linkage class).

11.6   Inducing Reactions with Given Properties

Theorem 6.27 gives a characterization of kinetic differential equations within the class of polynomial ones. This statement can also be considered as the solution of an inverse problem, and what is more, it raises a series of other inverse problems; see, e.g., Érdi and Tóth (1989, Section 4.7). One can automatically construct the canonical representation (see the proof of the theorem) of a kinetic differential equation; however, it would be desirable to find a realization which is either simple in some sense or obeys some prescribed properties, e.g., the inducing reaction is reversible, is weakly reversible, has a small deficiency, has as few complexes or reaction steps as possible, has an acyclic Volpert graph etc. These problems also raise uniqueness questions.

One of the early approaches to this problem was Tóth (1981, Theorem 9) providing a necessary and sufficient condition that a generalized compartmental system in the narrow sense (see Chap. 3) induces a given differential equation.

Theorem 11.12 *Let $M \in \mathbb{N}; \mathbf{A} \in \mathbb{R}^{M \times M}, \mathbf{Y} \in \mathbb{N}^{M \times M}, and \mathbf{b} \in (\mathbb{R}_0^+)^M$. The differential equation*

$$\dot{\mathbf{x}} = \mathbf{A}\mathbf{x}^{\mathbf{Y}} + \mathbf{b} \tag{11.11}$$

is the induced kinetic differential equation of a

1. *closed*
2. *half open*
3. *open*

generalized compartmental system in the narrow sense if and only if the following relations hold: all the columns of $\mathbf{Y}$ are nonnegative multiples of different elements of the standard basis: $\mathbf{Y} = \begin{bmatrix} y_1 \mathbf{e}_1 & y_2 \mathbf{e}_2 & \dots & y_M \mathbf{e}_M \end{bmatrix}; y_1, y_2, \dots, y_M \in \mathbb{N}_0, y_m \neq y_p$ for $m \neq p$; and

1. $b_i = 0; -a_{mm}, a_{mp}, \beta_m := -\sum_{p=1}^{M} \frac{a_{pm}}{y_p} \in \mathbb{R}_0^+; a_{mm} = \beta_m y_m;$
2. $b_i = 0; -a_{mm}, a_{mp}, -\beta_m \in \mathbb{R}_0^+; a_{mm} \leq \beta_m y_m, \exists m : a_{mm} < \beta_m y_m;$
3. $-a_{mm}, a_{mp}, -\beta_m \in \mathbb{R}_0^+; a_{mm} \leq \beta_m y_m, \exists m : \beta_{mm} > 0;$

and throughout $m, p \in \{1, 2, \dots, M\}, m \neq p$.

Proof The proof of the theorem comes from the comparison of coefficients.   □

The usefulness of the theorem lies in the fact that it assigns a reaction of zero deficiency to a differential equation, because generalized compartmental systems in the narrow sense obviously have zero deficiency; see Theorem 3.17.

Example 11.13 Let us consider an example (Érdi and Tóth 1989, p. 71) leading to an open problem. The equation $\dot{x} = -x + 3y \quad \dot{y} = 3x - y$ is not the induced kinetic differential equation of a generalized compartmental system according to Theorem 11.12 because $\beta_1 = \beta_2 = -2$ and $-1 = a_{mm} \leq \beta_m y_m = -2$ is false for all $m = 1, 2$.

Thus, three complexes (X, Y, and 0) are not enough to induce the given differential equation. However, four suffices, and what is more, it is possible to find an inducing reaction of deficiency zero:

$$Y \xrightarrow{1} 3X \quad X \xrightarrow{1} 3Y,$$

as

```
RightHandSide[{Y -> 3 X, X -> 3 Y}, {1, 1}, {x, y}]
```

gives the differential equation of the example. The example raises the question: suppose the number of monomials on the right-hand side of a kinetic differential equation is N. Does there exist an inducing reaction to this equation with not more than (i.e., exactly) N (or $N + 1$) complexes?

Szederkényi et al. (2011) gave an effective method for the solution of classes of similar problems by rewriting qualitative properties of reactions in terms of a **mixed integer linear programming** (MILP, in short) problem. More specifically, they can compute realizations with given properties.

One of the related theoretical results of central importance by Szederkényi (2010) follows. His aim is to find the $M \times N$ matrix of nonnegative integer components $\mathbf{Y}$ and the $N \times N$ matrix of nonnegative components $\mathbf{K}$ given the product $\mathbf{M} = \mathbf{Y}(\mathbf{K} - \text{diag}(\mathbf{K}^\top)\mathbf{1})$. If this holds, he calls the pair $(\mathbf{Y}, \mathbf{K})$ a **realization** of the matrix $\mathbf{M}$.

Definition 11.14 If the number of zeros in the above setting in $\mathbf{K}$ he has found is minimal, then he has got a **sparse realization,**; if the number of zeros in $\mathbf{K}$ is maximal, then he has got a **dense realization**.

The following properties of dense and sparse realizations of a given chemical reaction network were proved. Suppose $\mathbf{M}$ is given.

Theorem 11.15

1. *The Feinberg–Horn–Jackson graph of the dense realizations is unique.*
2. *The Feinberg–Horn–Jackson graph of any dynamically equivalent realization of a mechanism is the subgraph of the Feinberg–Horn–Jackson graph of the dense realization.*

3. *The Feinberg–Horn–Jackson graph of a mechanism is unique if and only if the Feinberg–Horn–Jackson graphs of the dense and sparse realizations of* **M** *are identical.*

The above results were extended to the case of dynamically equivalent constrained realizations, where a subset of possible reactions is excluded from the network. Here we give some further publications of the group: Rudan et al. (2013, 2014), Johnston et al. (2012a,b,c), and Szederkényi et al. (2012)

Szederkényi et al. (2012) and Johnston et al. (2013) determine weakly reversible realizations, while Szederkényi and Hangos (2011) provide reversible and detailed balanced realizations. A related paper by Schuster and Schuster (1991) finds detailed balanced subnetworks in reactions.

11.7   Exercises and Problems

11.1 Suppose you are given data $(t_1, x_1), \ldots, (t_K, x_K)$, and it is known that they come from measurements fitting to a line passing over the origin. How would you estimate the slope of the line so as to minimize the sum of squares of the differences between measured data and those calculated from the fitted line?

(Solution: page 447)

11.2 Find a function with an infinite number of strict local minima and without a strict global minimum.

(Solution: page 447)

11.3 Show on an example that the estimates of the parameters m and k of the function $t \mapsto m \exp(kt)$ cannot be exactly calculated by taking the logarithm of the function values and that of the data.

(Solution: page 447)

11.4 Generate data taken from the diffusion equation, and add to them a small error. Estimate the diffusion coefficient from the simulated data.

(Solution: page 449)

11.5 `NonlinearModelFit` is originally designed for fitting the parameters of a single (scalar-valued) function, which however may depend on more than one variable, and the number of parameters may also be arbitrary. Try to tame the function to fit the reaction rate coefficients of the Lotka–Volterra reaction given that the concentration vs. time curves are measured with 5% relative error in the time interval $[0, 20]$ and that the original reaction rate coefficients are unity.

Hint: Use the idea of characteristic function. The solution to the previous problem may also help.

(Solution: page 449)

11.6 Apply the method of Sect. 11.4 to the estimation of the reaction $X \xrightarrow{k} 2X$.

(Solution: page 450)

11.7 Suppose the concentration of X is measured for all times without error in the reaction $X \xrightarrow{k_1} Y$ $X \xrightarrow{k_2} Z$. Show that still one can only determine the sum of the coefficients k_1 and k_2.

(Solution: page 450)

11.8 Find a pair of confoundable systems within the class of reversible (stoichiometrically) mass conserving second-order reactions.

(Solution: page 450)

11.9 Show that the induced kinetic differential equation of the two mechanisms in Fig. 11.4 (taken from Csercsik et al. 2012) is the same, i.e., the mechanisms are dynamically equivalent, although the deficiency of the first one is zero, whereas that of the second is two.

(Solution: page 451)

11.10 Show that the induced kinetic differential equation of the mechanisms in Fig. 11.5 (taken from Szederkényi 2009) is the same, i.e., the mechanisms are dynamically equivalent with the same differential equation, or the corresponding reactions are confoundable. Therefore no matter how exactly you measure the concentration vs. time curves, you will be unable to choose between the two mechanisms.

(Solution: page 451)

11.11 Show that the induced kinetic differential equation of the original mechanism and that of its sparse and dense realizations in Fig. 11.6 (taken from Szederkényi 2010) is the same, i.e., the mechanisms are dynamically equivalent with the same differential equation, or the corresponding reactions are confoundable. Therefore no matter how exactly you measure the concentration vs. time curves, you will be unable to choose between the three mechanisms.

(Solution: page 451)

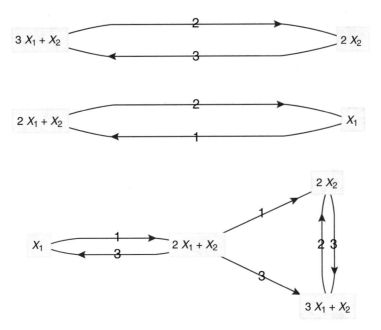

Fig. 11.4 Two mechanisms with identical dynamics: the Feinberg–Horn–Jackson graph of the first one consists of two components

11.8   Open Problems

1. Which are the reactions for which the heuristic estimation procedure given in Sect. 11.3 provides positive estimate for the reaction rate coefficients?
2. Can it be decided from the form of the differential equation that a realization with a minimal number of complexes exists (even if the coefficients are parameters and not numeric constants) as in the case of generalized compartmental systems in the narrow sense? See Problem 6.21.
3. Could you find two reactions with the same induced kinetic differential equation so that the reactions are mass conserving and only contain short complexes?
4. Is it possible to construct an example where the two confoundable reactions are reversible, mass conserving, and second order?
5. What is the necessary and sufficient condition (in terms of the coefficients and exponents) for a kinetic differential equation to be capable be induced by a reaction containing the same number of complexes as the number of different monomials on the right-hand side of the equation?
6. Can you find two mechanisms $\langle \mathcal{M}, \mathcal{R}_1, \alpha_1, \beta_1, \mathbf{k}_1 \rangle$ and $\langle \mathcal{M}, \mathcal{R}_2, \alpha_2, \beta_2, \mathbf{k}_2 \rangle$ so that they are macroequivalent, i.e., their induced kinetic differential equation is the same, but for some other vectors of the reaction rate coefficients, they are not macroequivalent? This example would clearly show that confoundability is a property of mechanisms and not that of reactions.

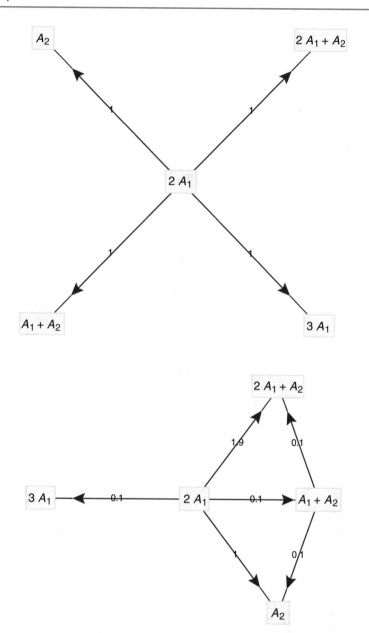

Fig. 11.5 Two mechanisms with identical dynamics

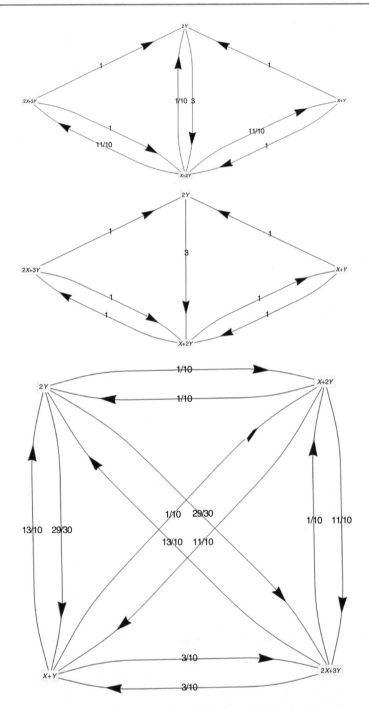

Fig. 11.6 A mechanism and its sparse and dense realizations with identical dynamics

7. Suppose the number of monomials on the right-hand side of a kinetic differential equation is N. Does there exist an inducing reaction to this equation with not more than (i.e., exactly) N (or $N + 1$) complexes?

References

Bard Y (1974) Nonlinear parameter estimation. Academic Press, New York

Billingsley P (1961) Statistical inference for Markov processes. University of Chicago Press, Chicago

Craciun G, Pantea C (2008) Identifiability of chemical reaction networks. J Math Chem 44(1):244–259

Csercsik D, Szederkényi G, Hangos KM (2012) Parametric uniqueness of deficiency zero reaction networks. J Math Chem 50(1):1–8

Érdi P, Tóth J (1989) Mathematical models of chemical reactions: theory and applications of deterministic and stochastic models. Princeton University Press, Princeton

Feinberg M, Horn FJM (1977) Chemical mechanism structure and the coincidence of the stoichiometric and kinetic subspaces. Arch Ration Mech Anal 66(1):83–97

Grima R, Walter NG, Schnell S (2014) Single-molecule enzymology à la Michaelis-Menten. FEBS J 281(2):518–530

Hangos KM, Tóth J (1988) Maximum likelihood estimation of reaction-rate constants. Comput Chem Eng 12(2/3):135–139

Horn F, Jackson R (1972) General mass action kinetics. Arch Ration Mech Anal 47:81–116

Johnston MD, Siegel D, Szederkényi G (2012a) Computing linearly conjugate chemical reaction networks with minimal deficiency. In: SIAM conference on the life sciences—SIAM LS 2012, 7–10 August, San Diego, USA, pp MS1–2–45

Johnston MD, Siegel D, Szederkényi G (2012b) Dynamical equivalence and linear conjugacy of chemical reaction networks: new results and methods. MATCH Commun Math Comput Chem 68:443–468

Johnston MD, Siegel D, Szederkényi G (2012c) A linear programming approach to weak reversibility and linear conjugacy of chemical reaction networks. J Math Chem 50:274–288

Johnston MD, Siegel D, Szederkényi G (2013) Computing weakly reversible linearly conjugate chemical reaction networks with minimal deficiency. Math Biosci 241:88–98

Kovács B, Tóth J (2007) Estimating reaction rate constants with neural networks. Enformatika Int J Appl Math Comput Sci 4(2):515–519

Lente G (2013) A binomial stochastic kinetic approach to the Michaelis-Menten mechanism. Chem Phys Lett 568–569:167–169

Meshkat N, Sullivant S (2014) Identifiable reparametrizations of linear compartment models. J Symb Comput 63:46–67

Meshkat N, Eisenberg M, DiStefano JJ (2009) An algorithm for finding globally identifiable parameter combinations of nonlinear ODE models using Gröbner bases. Math Biosci 222(2):61–72

Meshkat N, Kuo CE, DiStefano J III (2014) On finding and using identifiable parameter combinations in nonlinear dynamic systems biology models and COMBOS: a novel web implementation. PLoS One 9(10):e110, 261

Nagy AL (2011) A new simulated annealing type algorithm for multivariate optimization with rigorous proofs and applications. Master's thesis, Budapest University of Technology and Economics, Institute of Mathematics, Department of Analysis, Budapest

Rudan J, Szederkényi G, Hangos KM (2013) Computing dynamically equivalent realizations of biochemical reaction networks with mass conservation. In: AIP conference proceedings, ICNAAM 2013: 11th international conference of numerical analysis and applied mathematics, 21–27 September, Rhodes, Greece, vol 1558, pp 2356–2359

Rudan J, Szederkényi G, Hangos KM (2014) Efficient computation of alternative structures for large kinetic systems using linear programming. MATCH Commun Math Comput Chem 71(1):71–92

Santosa F, Weitz B (2011) An inverse problem in reaction kinetics. J Math Chem 49:1507–1520

Schuster S, Schuster R (1991) Detecting strictly detailed balanced subnetworks in open chemical reaction networks. J Math Chem 6(1):17–40

Seber GAF, Wild CJ (2003) Nonlinear regression. Wiley, New York

Sedoglavic A (2001a) Méthodes seminumériques en algèbre différentielle; applications à l'étude des propriétés structurelles de systèmes différentiels algébriques en automatique. Docteur en sciences, École polytechnique

Sedoglavic A (2001b) A probabilistic algorithm to test local algebraic observability in polynomial time. In: Proceedings of the 2001 international symposium on symbolic and algebraic computation. ACM, pp 309–317

Singer AB, Taylor JW, Barton PI, Green WH (2006) Global dynamic optimization for parameter estimation in chemical kinetics. J Phys Chem A 110:971–976

Szederkényi G (2009) Comment on "Identifiability of chemical reaction networks" by G. Craciun and C. Pantea. J Math Chem 45:1172–1174

Szederkényi G (2010) Computing sparse and dense realizations of reaction kinetic systems. J Math Chem 47:551–568

Szederkényi G, Hangos KM (2011) Finding complex balanced and detailed balanced realizations of chemical reaction networks. J Math Chem 49:1163–1179

Szederkényi G, Hangos KM, Péni T (2011) Maximal and minimal realizations of reaction kinetic systems: computation and properties. MATCH Commun Math Comput Chem 65:309–332

Szederkényi G, Tuza ZA, Hangos KM (2012) Dynamical equivalence and linear conjugacy of biochemical reaction network models. In: Biological and medical systems. 8th IFAC symposium on biological and medical systems the international federation of automatic control August 29–31, 2012, Budapest, Hungary, vol 1, pp 125–130

Szederkényi G, Hangos KM, Tuza Z (2012) Finding weakly reversible realizations of chemical reaction networks using optimization. MATCH Commun Math Comput Chem 67:193–212

Tóth J (1981, in Hungarian) A formális reakciókinetika globális determinisztikus és sztochasztikus modelljéről (On the global deterministic and stochastic models of formal reaction kinetics with applications). MTA SZTAKI Tanulmányok 129:1–166

Turányi T (1990) Sensitivity analysis of complex kinetic systems. Tools and applications. J Math Chem 5(3):203–248

Turányi T, Tomlin AS (2014) Analysis of kinetic reaction mechanisms. Springer, Berlin

Vajda S, Valko P, Turanyi T (1985) Principal component analysis of kinetic models. Int J Chem Kinet 17(1):55–81

Villaverde AF, Henriques D, Smallbone K, Bongard S, Schmid J, Cicin-Sain D, Crombach A, Saez-Rodriguez J, Mauch K, Balsa-Canto E, Mendes P, Jaeger J, Banga JR (2015) Biopredynbench: a suite of benchmark problems for dynamic modelling in systems biology. BMC Syst Biol 9(1):8

Weise T (2009) Global optimization algorithms-theory and application. Self-published

Past, Present, and Future Programs for Reaction Kinetics

12

12.1 Introduction

Even the smallest reactions have a deterministic model which is too complicated to deal with pen and paper. (However, first, one should not forget about the useful works by Rodiguin and Rodiguina (1964) and by Szabó (1969), and either neglect the results obtained using the qualitative theory of differential equations, e.g., by Kertész (1984), or Boros et al. (2017a,b), or by infinitely many other authors.) Thus, one has to rely on computers. On the other hand, realizations of stochastic models obtained by simulation are useful or even necessary when the model is so simple as to have all the characteristics be calculated by hand. It is very instructive how different of two realizations of such a reaction as $X \rightleftharpoons Y$ might be (see Fig. 12.1). A thorough investigation almost always needs numerical methods of solving induced kinetic differential equations, of calculating stationary points and the Jacobian of the right and sides, of calculating sensitivities, etc. The developments of the theory started about 50 years ago also involve the tools of discrete mathematics, but checking the conditions of a theorem again needs the use of computers.

12.2 Early Endeavors

The mentioned difficulties were realized very early and were attacked even with computers of small (looking back from today) capacities.

Garfinkel et al. (1970) is a very early review of computer applications, especially in biochemical kinetics based on a 192-long reference list. Among the authors we find B. Chance, a pioneer of the topic (and a gold medal winner in sailing at the 1952 Summer Olympics), who used mechanical and analogue computers before the breakthrough of digital ones in the field of biochemistry. Even before the 1970s, it was not rare to treat reactions with a few hundred steps and a few hundred species.

© Springer Science+Business Media, LLC, part of Springer Nature 2018
J. Tóth et al., *Reaction Kinetics: Exercises, Programs and Theorems*,
https://doi.org/10.1007/978-1-4939-8643-9_12

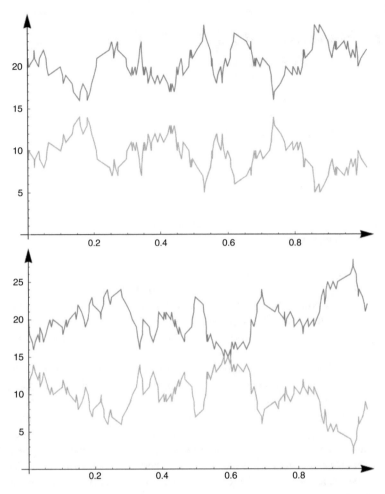

Fig. 12.1 Two realizations of the stochastic model of X $\rightleftharpoons$ Y with the same parameters and initial conditions

We highly recommend the paper especially because of its rich reference list for those interested in the historical development of the field. The paper also reviews classical equilibrium calculations and also curve fitting methods. They also give a detailed description of biochemical systems treated in this way—these pages are surely of no present interest, because of the developments both in biochemistry and computer science. Exotic phenomena such as oscillation are also mentioned. The necessity to automatically transform the reaction steps into a differential equation (called **parsing**) has also been perceived, and the problem of **stiff** equations (see Sect. 9.6.4.3) has also been recognized (the method by Gear (1971) has been published by that time). Importance of computers in teaching has also been realized by the authors.

A few early developments until 1974 are summarized by Hogg (1974) from a relatively narrow perspective: he only used the *Chemical Abstracts*. The fact that inter alia the general program by Érdi et al. (1973), Sipos et al. (1974a) to approximately simulate the stochastic model of reactions with arbitrary complexity has not been included shows that at those times one had not so many tools as today to review the literature.

12.3 Investigation of Structural Properties

The development of programming languages paved the way for symbolic calculations and for treating discrete mathematical objects; therefore it became possible to go beyond numerical calculations. Good reviews are Barnett (2002) and Barnett et al. (2004) mainly dealing with symbolic calculations using the computer, and the authors review work outside kinetics (including biochemistry) as well.

Most of the important results of the Feinberg School are incorporated into a **Toolbox** by Ellison et al. (2011). It is intended to implement in Windows various parts of Chemical Reaction Network Theory (CRNT for short) that have appeared in the literature. The guide that comes with the program does provide some information for newcomers. The theory behind the programs is developed by Martin Feinberg himself and also his students and colleagues, such as Phillipp Ellison, Haixia Ji, Paul Schlosser, Gheorghe Craciun, Guy Shinar, and others.

Version 1.X, written for the Microsoft DOS operating system, contained a **ChemLab** component, which provided numerical solutions (and their graphical display) for the differential equations that derive from mass action systems but was mainly aimed at implementing the theory for networks of deficiencies zero and one.

Version 2.X is written for Windows, and it also extends the power of the old **Network Analyst** component of Version 1.X, which was centered around deficiency-oriented parts of Chemical Reaction Network Theory. It is able to decide whether a network has the injectivity or concordance properties, important in deciding whether a reaction has a single stationary point, or more (see the papers by Banaji and Craciun 2010, 2009; Shinar and Feinberg 2012).

Donnell et al. (2014) describe the related package CoNtRol which is a CRNT tool. It provides a new, fully open-source platform, currently coded in C, Java, Octave, and PHP, to perform computations on reactions. The package has a web-based front-end interfacing with a suite of modular tests, to which users may add new tests in any language. With its array of features, CoNtRol complements existing software tools.

The current functionality of CoNtRol includes a number of necessary and/or sufficient structural tests for multiple equilibria, stable oscillation, convergence to equilibria, and persistence, assuming mass action or more general kinetics. In particular, the following are checked: sufficient conditions for convergence to equilibria based on the theory of monotone dynamical systems (De Leenheer et al. 2006; Angeli 2010; Donnell et al. 2014) conditions of the deficiency zero and deficiency one theorems (Theorems 8.47 and 8.48), structural conditions for persistence (Angeli et al. 2007; Donnell and Banaji 2013) based on examining the siphons (Definition 8.79) of the system, and a large number of sufficient/necessary conditions for injectivity and absence of multistationarity gathered from the literature and developed by Banaji, Pantea, and others. The outputs are cross-referenced to the documentation of CoNtRol, where each conclusion and its implications are described in detail. Some of the multistationarity results of CoNtRol are similar to those of the Chemical Reaction Network Toolbox. The program also draws the Volpert graph of the reaction (see Sect. 3.2).

Tests for multistationarity of CRNs are also implemented in Maple by Feliu and Wiuf (2013). Another program aimed at the same problem is GraTeLPy written by Georg Walther and Matthew Hartley: https://pypi.python.org/pypi/GraTeLPy/.

The ERNEST Reaction Network Equilibria Study Toolbox (Soranzo and Altafini 2009) performs a detailed model analysis of the input reaction by determining the basic system features and by using the Deficiency Zero or Deficiency One Theorems. The toolbox is also capable of running the Deficiency One Algorithm where applicable. However, both of the abovementioned toolboxes assume that the structure of the analyzed network is a priori known; therefore they have no functionality for examining dynamical equivalence. The software, implemented in MATLAB, is available under the GNU GPL free software license from http://users.isy.liu.se/en/rt/claal20/Publications/SoAl09.

It requires the MATLAB Optimization Toolbox.

The most recent and probably most useful page to find programs together with references and theory (also concentrating on Chemical Reaction Network Theory) is probably this: reaction-networks.net/wiki/List_of_references_by_topic.

12.4   Inverse Problems

The program to solve the large set of inverse problems treated by Szederkényi and his coworkers is called **The CRNreals toolbox**. It is available at http://www.iim.csic.es/~gingproc/CRNreals/ together with the associated documentation. The toolbox runs under the popular MATLAB computational environment and uses several free and commercial linear programming and mixed integer linear programming solvers. Szederkényi et al. (2012) describe CRNreals as a toolbox for distinguishability and identifiability analysis of biochemical reaction networks.

12.5   Numerical Treatment of Models

The two major problems fitting this category are the solution of the induced kinetic differential equations and the estimation of reaction rate coefficients.

12.5.1 Solving Induced Kinetic Differential Equations

A general view, neglecting recent developments in formal reaction kinetics, is that practically important reactions contain many elementary steps and species; therefore their models can only be investigated by numerical methods. Therefore it is quite useful that Deuflhard and Bornemann (2002) have collected programs to numerically solve differential equations.

Korobov and Ochkov (2011) show how to use MathCad and Maple to solve induced kinetic differential equations, without constructing automatically the equation itself in the general case. They treat a series of popular reactions like Lotka–Volterra reaction, Brusselator, Belousov–Zhabotinsky reaction and, also some reactions with time-dependent temperature but are not interested in structural analysis of the reactions.

As mentioned above, ChemLab also solves induced kinetic differential equations. CHEMSIMUL by Kirkegaard and Bjergbakke (2000) is another program to solve induced kinetic differential equations.

12.5.2 Parameter Estimation and Sensitivity Analysis

An abundant list of programs and references on programs to numerically solve induced kinetic differential equations and to analyze them from the point of view of sensitivity analysis and to provide methods of parameter estimation can be found in the recent book by Turányi and Tomlin (2014).

PottersWheel (http://www.potterswheel.de/index.html) is aimed at fitting models simultaneously to multiple measurements and to discriminate competing hypotheses. Those interested in fitting reaction rate coefficients to measured data should certainly know the systematic collection of a few benchmark problems at the address //bmcsystbiol.biomedcentral.com/articles/10.1186/s12918-015-0144-4.

Predici (http://www.cit-wulkow.de/images/pdf/Broschueres/Predici11_Overview .pdf) is a commercial product with polymerization as its main topic. Finally, we mention that there exist a few (or, more than enough) other commercial programs that know everything, but one cannot find out what this really means. We neglected these programs.

12.6 Mathematica-Based Programs

The approach by Jemmer (1997) is general. He has constructed a package, which parses an input file containing a set of chemical reactions, rate constants, source and sink terms, and initial conditions. From this information the induced kinetic differential equation is generated. The generality of the proposed protocol is then demonstrated in both symbolic and numerical computations in the analysis of organic reactions and oscillatory combustion reactions and chaotic behavior.

Jemmer (1999) shows a series of examples taken from kinetics with transport processes (advection and diffusion) which he numerically solves via discretization using a previously developed program written in *Mathematica*. The applications come from the fields of chromatography, bacterial population dynamics, polymerization, and plasma chemistry. Using the capabilities of *Mathematica* he obtains symbolic results, as well.

Another general program (also written in *Mathematica* and maintained by Igor Klep, Karl Fredrickson and Bill Helton) is http://www.math.ucsd.edu/~chemcomp/ which is based on the paper by Craciun et al. (2008). The program has two parts, one of them calculates the Jacobian of the right-hand side of the induced kinetic differential equation, and also related quantities, whereas the second one calculates the deficiency of the reaction. The main idea of the programs and the mentioned paper is **homotopy** what in this context means that one calculates the number of stationary points of a simple reaction and draws the consequence that another less simple reaction has the same number of stationary points because the two induced kinetic differential equations are embedded in a family of differential equations the members of which are connected continuously by a parameter. The BIOKMOD by Sánchez (2005) solves induced kinetic differential equations and also fits parameters. We should certainly mention the work by Kyurkchiev et al. (2016) here and also the book by Mulquiney and Kuchel (2003).

12.7   Programs for Teaching

Here and above we certainly do not provide a classification of the programs: now we discuss some of them which have the speciality of having been written in *Mathematica*. Our readers who are also users of *Mathematica* certainly have discovered the short programs called **demonstrations** solving special problems (including precursors of **ReactionKinetics**), e.g.:

- BriggsRauscherMechanismTheChemicalColorClock/
- DynamicBehaviorOfANonisothermalChemicalSystem/
- PharmacokineticModelling/
- HopfBifurcationInTheBrusselator/
- FeinbergHornJacksonGraph
- Volpert Graph
- DescriptiveReactionKinetics

or general problems of reaction kinetics such as

- topic.html?topic=Chemical+Kinetics&limit=200
- VolpertGraphOfChemicalReactions/

(All the above uniform resource locators start with http://demonstrations.wolfram. com/.)

The page http://www.chemie.unibas.ch/~huber/AllChemie/Reaktion1.html uses Maple for the high school problem of stoichiometry: find the stoichiometric coefficients in reaction steps when the atomic structure of the species is known.

Ferreira et al. (1999) used *Mathematica* for teaching purposes. They show some oscillatory and chaotic reactions such as the Lotka reaction, the Lotka–Volterra reaction, and a model of glycolysis. The aim of Francl (2000, 2004) is also teaching; he uses *Mathematica* to numerically solve ordinary differential equations of models like the Lotka–Volterra reaction.

12.8 Miscellanea

There are a lot of programs just to show the possibility of treating chemical kinetic problems using this or that tool. Such a program is, e.g., LARKIN (Deuflhard et al. 1981), capable to automatically generate the right-hand sides of induced kinetic differential equations.

Important early endeavor is seen in the papers by Holmes and Bell (1991, 1992) who construct the induced kinetic differential equations of reactions, check mass conservation, and find the stationary points with their programs written in Maple. A kind of continuation can be found on the web site of **Maplesoft**

http://www.maplesoft.com/applications/view.aspx?SID=4711&view=html&L=G

by D. M. Maede. A few simple example reactions consisting of a single reaction step are shown, their induced kinetic differential equations are solved symbolically and numerically, and finally, concentration vs. time curves are also given. The induced kinetic differential equations are constructed by hand and not automatically.

Huang and Yang (2006) represent a relatively new kind of application of computers: they provide a **computer-assisted rigorous proof** of the existence of a limit cycle and that of chaotic dynamics in a three-variable model of the Belousov–Zhabotinsky reaction.

The group of Olaf Wolkenhauer regularly presents programs for treating deterministic and stochastic models of reactions, although they call their field of interest as Systems Biology. The best way to find useful codes is to search the word "reaction" via http://www.sbi.uni-rostock.de/search/.

The **CHEMKIN** package by Kee et al. (1980) is originally comprised of two major components. The **interpreter** is a **FORTRAN** code which is used to read a symbolic description of an arbitrary, user-specified chemical reaction mechanism. The output of the interpreter is a data file which forms a link to the **gas-phase subroutine library**. This library is a collection of FORTRAN subroutines which may be called to return thermodynamic properties, chemical production rates, derivatives of thermodynamic properties, derivatives of chemical production rates, or sensitivity parameters. Thermodynamic properties are stored in a thermodynamic database. The database is in the same format as the one used in the NASA chemical equilibrium code. Since 1998 CHEMKIN is maintained and developed by Reaction Design (see http://www.reactiondesign.com). KINAL and KINALC developed by T. Turányi and his group is a kind of continuation of CHEMKIN: it helps make a more detailed analysis of mechanisms (see the book by Turányi and Tomlin (2014) and also the page http://respecth.chem.elte.hu/respecth/index.php).

Kintecus (http://www.kintecus.com/) is dealing with combustion models numerically.

Kinpy (https://code.google.com/p/kinpy/) is a source code generator for solving induced kinetic differential equations in Python. Actually, it imports reaction steps and produces automatically the initial value problem to be solved.

Chemical equilibrium calculations usually does not mean finding the stationary point(s) of an induced kinetic differential equation, rather it means calculations based on the definition of the equilibrium constant of a (single) reversible step. Programs to do these kinds of calculations useful for high school students are in abundance everywhere; let us only mention the one written by Akers and Goldberg (2001).

Finally, we mention that Wikipedia seems to exclude reaction kinetics from computational chemistry: http://en.wikipedia.org/wiki/Category:Computational_chemistry_software.

12.9   Stochastic Simulation

Earlier, in Chap. 10, we treated different exact and approximate algorithms to simulate the stochastic model of reactions. Here we only say a few words about the history of simulations. Early papers, like (Schaad 1963; Lindblad and Degn 1967; Hanusse 1973), present algorithms that are based on discrete time, discrete state stochastic models, sometimes based on ad hoc assumptions, and sometimes the model is the discrete skeleton of the usual stochastic model, usually for very simple special cases. The speciality of the papers Érdi et al. (1973) and Sipos et al. (1974a,b) is that they produce the discrete skeleton, and the code has been written in full generality. Rabinovitch (1969) is a paper popularizing the first attempts for educational purposes. The thesis by Hárs (1976) seems to be the first one containing a general algorithm based on the exact model.

12.10   Not a Manual

Here we give some further technical help to the reader who wishes to use our program **ReactionKinetics**. Before that we mention a few packages closest to our one. The *Mathematica* based programs described by Kyurkchiev et al. (2016) seem to have the same functions as our package although not all the details are given in the paper.

Now let us see the technical help to our package. First of all, you copy the file **ReactionKinetics.m** into a directory where *Mathematica* can reach it. It may be something like

Program Files\Wolfram Research\Mathematica\11.1\AddOns\Applications.

A not so nice solution may be to copy it into the directory where your working notebook will be located. This is useful if you only try to use the program temporarily.

Then, to remain on the safe side, you may start your new notebook either with **ClearAll["Global`*"]** or even with **Quit[]**. If you may need the program, then use this: **Get["ReactionKinetics`"]**).

Beyond some administrative data, two palettes will open via

OpenReactionKineticsPalette[ ]

helping enter signs (e.g., that of the stoichiometric matrix y) or many kinds of reaction arrows like $\longleftrightarrow$. **OpenReactionKineticsNames[ ]** helps enter names such as **deficiency** or **atom**.

Now you may be interested in the area covered by our program. Then ask for the **Names["ReactionKinetics`*"]** providing the names of functions, options, etc., or invoke **Information["ReactionKinetics`*"]**. Once you decided that you would like to use a given function, you should like to know how to use it. The usual methods for getting information work (and if not, let us know). For example, the function **Concentrations** returns the concentration-time curves of each species using **DSolve** or **NDSolve** depending on the parametrization of the function. All the necessary information can be obtained by **?Concentrations**. Another important functions are **Simulation** and **SimulationPlot** for which **?Simulation** and **?SimulationPlot** show how to use these functions.

It is also useful to know the list of **Options[ShowFHJGraph]**:

```
{ExternalSpecies -> { }, ComplexColors -> { },
    PlotFunction -> "GraphPlot", Numbered -> False,
    StronglyConnectedComponentsColors -> { }}
```

But, certainly, **?ShowFHJGraph** should also work.

A large (and expanding, especially if you too help us) list of the reactions can also be listed by **Models**, or in a more detailed way by **Reactions**. To use an individual reaction, it is not enough to type its name; one should get it.

```
iva = GetReaction["Ivanova"]
```

Let us mention that mechanisms can be imported using **CHEMKINInput** from CHEMKIN files, and it would be equally possible to use SBML (System Biology Markup Language) format.

A typical function follows.

```
DeterministicModel[iva, {k1, k2, k3}, {x, y, z}]
```

Typical means that many functions take the reaction, the reaction rate coefficients (in case of mass action kinetics), sometimes the initial concentrations, optionally the names of concentrations if one would like to have her/his own notation.

Plotting functions may need different kinds of data.

```
ShowVolpertGraph[{"Glyoxylate Cycle"},
    DirectedEdges -> True, VertexLabeling -> True,
    ImageSize -> 800, GraphLayout->"CircularEmbedding",
    Numbered -> True]
```

One of the most important functions giving descriptions of structural characteristics is **ReactionsData**. It is to be used this way.

```
ReactionsData[iva]["{γ}", "deficiency", "reactionsteporders",
    "complexes"]
```

Do not be afraid of that part of the result which is a sparse matrix; you can also obtain more familiar forms.

```
Through[{Normal, MatrixForm}[ReactionsData[iva]["{γ}"]]]
```

$$\{\{\{-1, 0, 1\}, \{1, -1, 0\}, \{0, 1, -1\}\}, \begin{bmatrix} -1 & 0 & 1 \\ 1 & -1 & 0 \\ 0 & 1 & -1 \end{bmatrix}\}$$

It is not a bad idea to use "X" as a notation for the species X, especially because some of the letters are protected by the package; thus it is safer to avoid using the letters **c, z, P**, etc.

12.11 Exercises and Problems

12.1 Download, try, modify, compare, and extend the programs mentioned in the present chapter. If you find (or you yourself have written) an important one, please let the authors know.

(Solution: page 451)

12.12 Open Problems

Problems
- Nowadays there are programs to numerically solve the induced kinetic differential equations of mechanisms or simulate their stochastic model in full generality. However, these programs are not capable treating **temperature and pressure changes**, mainly because these effects are usually described by ad hoc models. Thus, it would be desirable to have a program which is flexible enough to take into consideration different circumstances even if they change during the reaction.

- Another area where one does not have a general program is the field of **reaction-diffusion processes**. One would need a program to numerically treat mechanisms in general where reaction, convection, and diffusion are considered and the volume is of arbitrary shape in a two- or three-dimensional space. This goal can only be approached step by step as the capacity of computers grows.
- The **parameter estimation** and **sensitivity analysis** programs which also work well for homogeneous reactions (see, e.g., Turányi and Tomlin 2014) should also be extended for these cases. It would be useful to find methods to provide "good" **initial estimates** of the reaction rate coefficients automatically.
- **The theory** of deterministic models including spatial effects (Mincheva and Siegel 2004, 2007) and that of stochastic models (Anderson et al. 2010) is undergoing an explosion in these years. The results should continuously be built in into the programs.

Tools

- First of all, we have to emphasize again a few advantages of *Mathematica* (or, Wolfram Language, if you prefer) from the point of view of the modeler. With a huge inventory of powerful functions in all areas of mathematics and with the possibility of using all the known **programming paradigms** (procedural, functional, term-rewriting, list processing, etc.), it is possible to write a few line codes to solve a problem of high complexity.
- There are a few novelties which are not exclusively characteristic of *Mathematica* but can be well utilized within that environment. One is able to use the **cloud** quite easily both for storage and calculations. The calculating power of **graphics cards** can be utilized without the need of assembly level coding. Many built-in functions of *Mathematica* support **parallel computing** (and some of the calculations are done in parallel without the awareness of the user).
- Recent versions have shown that even realistic problems in the field of solving partial differential equations (on complicated domains in dimensions higher than one) can successfully be attacked by **NDSolve** assuming only first-order reaction steps. Thus, it is desirable to have a code which is able to numerically solve reaction-diffusion equations with as many species as one likes in one, two, and three spatial dimensions endowed with mass action type kinetics, practically with no restrictions.
- And we do not know yet what **quantum computing** will be able to offer to people involved in reaction kinetics.

References

Akers DL, Goldberg RN (2001) Bioeqcalc: a package for performing equilibrium calculations on biochemical reactions. Math J 8(1):86–113

Anderson DF, Craciun G, Kurtz TG (2010) Product-form stationary distributions for deficiency zero chemical reaction networks. Bull Math Biol 72:1947–1970

Angeli A (2010) A modular criterion for persistence of chemical reaction networks. IEEE Trans Autom Control 55(7):1674–1679

Angeli D, De Leenheer P, Sontag ED (2007) A Petri net approach to study the persistence in chemical reaction networks. Math Biosci 210(2):598–618

Banaji M, Craciun G (2009) Graph-theoretic approaches to injectivity and multiple equilibria in systems of interacting elements. Commun Math Sci 7(4):867–900

Banaji M, Craciun G (2010) Graph-theoretic criteria for injectivity and unique equilibria in general chemical reaction systems. Adv Appl Math 44(2):168–184

Barnett MP (2002) Computer algebra in the life sciences. ACM SIGSAM Bull 36(4):5–32

Barnett MP, Capitani JF, von zur Gathen J, Gerhard J (2004) Symbolic calculation in chemistry: selected examples. Int J Quantum Chem 100(2):80–104

Boros B, Hofbauer J, Müller S (2017a) On global stability of the Lotka reactions with generalized mass-action kinetics. Acta Appl Math 151(1):53–80

Boros B, Hofbauer J, Müller S, Regensburger G (2017b) The center problem for the Lotka reactions with generalized mass-action kinetics. Qual Theory Dyn Syst 1–8

Craciun G, Helton JW, Williams RJ (2008) Homotopy methods for counting reaction network equilibria. Math Biosci 216(2):140–149

De Leenheer P, Angeli D, Sontag ED (2006) Monotone chemical reaction networks. J Math Chem 41(3):295–314

Deuflhard P, Bornemann F (2002) In: Marsden JE, Sirovich L, Golubitsky M, Antmann SS (eds) Scientific computing with ordinary differential equations. Texts in Applied Mathematics, vol 42. Springer, New York

Deuflhard P, Bader G, Nowak U (1981) A software package for the numerical simulation of large systems arising in chemical reaction kinetics. In: Ebert KH, Deuflhard P, Jaeger W (eds) Modeling of chemical reaction systems. Springer Series in Chemical Physics, vol 18, Springer, New York, pp 38–55

Donnell P, Banaji M (2013) Local and global stability of equilibria for a class of chemical reaction networks. SIAM J Appl Dyn Syst 12(2):899–920

Donnell P, Banaji M, Marginean A, Pantea C (2014) CoNtRol: an open source framework for the analysis of chemical reaction networks. Bioinformatics 30(11):1633–1634

Ellison P, Feinberg M, Ji H, Knight D (2011) Chemical reaction network toolbox, version 2.3. Available online at http://wwwcrntosuedu/CRNTWin

Érdi P, Sipos T, Tóth J (1973) Stochastic simulation of complex chemical reactions by computer. Magy Kém Foly 79(3):97–108

Feliu E, Wiuf C (2013) A computational method to preclude multistationarity in networks of interacting species. Bioinformatics 29(18):2327–2334

Ferreira MMC, Ferreira WCJ, Lino ACS, Porto MEG (1999) Uncovering oscillations, complexity, and chaos in chemical kinetics using *Mathematica*. J Chem Educ 76(6):861–866

Francl MM (2000) Introduction to the use of numerical methods in chemical kinetics. http://ckw.phys.ncku.edu.tw/public/pub/WebSources/MathSource/library.wolfram.com/infocenter/MathSource/791/index.html

Francl MM (2004) Exploring exotic kinetics: an introduction to the use of numerical methods in chemical kinetics. J Chem Educ 81:1535

Garfinkel D, Garfinkel L, Pring M, Green SB, Chance B (1970) Computer applications to biochemical kinetics. Ann Rev Biochem 39:473–498

Gear CW (1971) The automatic integration of ordinary differential equations. Commun ACM 14(3):176–179

Hanusse P (1973) Simulation des systémes chimiques par une methode de Monte Carlo. C R Acad Sci Ser C 277:93

Hárs V (1976) A sztochasztikus reakciókinetika néhány kérdéséről (Some problems of stochastic reaction kinetics). Msc, Eötvös Loránd University, Budapest

Hogg JL (1974) Computer programs for chemical kinetics: an annotated bibliography. J Chem Educ 51:109–112

Holmes MH, Bell J (1991) The application of symbolic computing to chemical kinetic reaction schemes. J Comput Chem 12(10):1223–1231

Holmes MH, Bell J (1992) An application of Maple to chemical kinetics. Maple Tech Newsl 0(7):50–55

Huang Y, Yang XS (2006) Numerical analysis on the complex dynamics in a chemical system. J Math Chem 39(2):377–387

Jemmer P (1997) Symbolic algebra in mathematical analysis of chemical-kinetic systems. J Comput Chem 18(15):1903–1917

Jemmer P (1999) Symbolic algebra in the analysis of dynamic chemical-kinetic systems. Math Comput Model 30:33–47

Kee RJ, Miller JA, Jefferson TJ (1980) CHEMKIN: a general-purpose, problem-independent, transportable, FORTRAN chemical kinetics code package. Unlimited Release SAND80-8003, Sandia National Laboratories, Livermore

Kertész V (1984) Global mathematical analysis of the Explodator. Nonlinear Anal 8(8):941–961

Kirkegaard P, Bjergbakke E (2000) Chemsimul: a simulator for chemical kinetics. http://www.risoe.dk/rispubl/PBK/pbkpdf/ris-r-1085rev.pdf

Korobov VI, Ochkov VF (2011) Chemical kinetics: introduction with Mathcad/Maple/MCS. Springer, Moscow

Kyurkchiev N, Markov S, Mincheva M (2016) Analysis of biochemical mechanisms using *Mathematica* with applications. Serdica J Comput 10:63–78

Lindblad P, Degn H (1967) A compiler for digital computation in chemical kinetics and an application to oscillatory reaction schemes. Acta Chem Scand 21:791–800

Mincheva M, Siegel D (2004) Stability of mass action reaction-diffusion systems. Nonlinear Anal 56(8):1105–1131

Mincheva M, Siegel D (2007) Nonnegativity and positiveness of solutions to reaction-diffusion systems. J Math Chem 42(4):1135–1145

Mulquiney PM, Kuchel PW (2003) Modelling metabolism with *Mathematica*. CRC Press, Boca Raton, FL, http://www.bioeng.auckland.ac.nz/MCF/invited_users/philip_kuchel.htm, includes CD-ROM

Rabinovitch B (1969) The Monte Carlo method: plotting the course of complex reactions. J Chem Educ 46(5):262

Rodiguin NM, Rodiguina EN (1964) Consecutive chemical reactions: mathematical analysis and development (English edn.). D.Van Nostrand, New York

Sánchez G (2005) Biokmod: a *Mathematica* toolbox for modeling biokinetic systems. Math Educ Res 10(2):1–34. http://diarium.usal.es/guillermo/biokmod/

Schaad J (1963) The Monte Carlo integration of rate equations. J Am Chem Soc 85:3588–3592

Shinar G, Feinberg M (2012) Concordant chemical reaction networks. Math Biosci 240(2):92–113

Sipos T, Tóth J, Érdi P (1974a) Stochastic simulation of complex chemical reactions by digital computer, I. The model. React Kinet Catal Lett 1(1):113–117

Sipos T, Tóth J, Érdi P (1974b) Stochastic simulation of complex chemical reactions by digital computer, II. Applications. React Kinet Catal Lett 1(2):209–213

Soranzo N, Altafini C (2009) Ernest: a toolbox for chemical reaction network theory. Bioinformatics 25(21):2853–2854

Szabó ZG (1969) Kinetic characterization of complex reaction systems. In: Bamford C, Tipper CFH (eds) Comprehensive chemical kinetics, vol 2, Elsevier, Amsterdam, chap 1, pp 1–80

Szederkényi G, Banga JR, Alonso AA (2012) CRNreals: a toolbox for distinguishability and identifiability analysis of biochemical reaction networks. Bioinformatics 28(11):1549–1550

Turányi T, Tomlin AS (2014) Analysis of kinetic reaction mechanisms. Springer, Berlin

Mathematical Background

13

13.1 Introduction

In order to make the book continuously readable but at the same time self-contained, often used but less known fundamental concepts have been collected in this chapter. These include linear algebra complemented with certain operations on vectors and matrices such as the special Hadamard–Schur product, notions and basic properties of directed and undirected graphs, a few theorems from advanced calculus, and also fundamental statements from the theory of ordinary and partial differential equations. Exercises and problems are not missing here either so as to make the material more easy to digest.

The reader is asked to prove some of the statements, but in most cases we refer to the literature.

13.2 Operations on Vectors and Matrices

Here we summarize the definitions and properties of some operations on vectors and matrices which are not so often used but very important for the concise description of reaction kinetics. The main source for this topic is Horn and Jackson (1972, pp. 90). In the following definitions, let $M, N, R \in \mathbb{N}$.

Definition 13.1 The **componentwise product** or **Hadamard product**, also known as the **entrywise product** and the **Schur product** of the matrices $\mathbf{A} = [a_{mr}] \in \mathbb{R}^{M \times R}$ and $\mathbf{B} = [b_{mr}] \in \mathbb{R}^{M \times R}$, is $\mathbf{A} \odot \mathbf{B} := [a_{mr} b_{mr}]$.

Note that now the Hadamard product of vectors has also been defined. Obviously, the set $\mathbb{R}^{M \times R}$ is a **semigroup** with this operation (and is a **ring** together with addition), but one reason why it is not so popular in mathematics might be that there are many **zero divisor**s in this group: the product can often be zero without

© Springer Science+Business Media, LLC, part of Springer Nature 2018
J. Tóth et al., *Reaction Kinetics: Exercises, Programs and Theorems*,
https://doi.org/10.1007/978-1-4939-8643-9_13

any of the factors being zero. However, the concept is very useful to concisely express things both in mathematics and in program codes. Juxtaposition of two lists in *Mathematica* is interpreted as this operation.

Definition 13.2 The **componentwise ratio** of the matrices $\mathbf{A} = [a_{mr}] \in \mathbb{R}^{M \times R}$ and $\mathbf{B} = [b_{mr}] \in (\mathbb{R}^+)^{M \times R}$ (again including the $R = 1$ case) is $\frac{\mathbf{A}}{\mathbf{B}} := \left[\frac{a_{mr}}{b_{mr}} \right]$.

We will also use componentwise exponentiation, logarithm, and powers.

Definition 13.3
1. The **logarithm** of the vector $\mathbf{a} = \begin{bmatrix} a_1 & a_2 & \cdots & a_M \end{bmatrix}^\top \in (\mathbb{R}^+)^M$ is

$$\ln(\mathbf{a}) := \begin{bmatrix} \ln(a_1) & \ln(a_2) & \cdots & \ln(a_M) \end{bmatrix}^\top. \tag{13.1}$$

2. The **exponential function** of the vector $\begin{bmatrix} a_1 & a_2 & \cdots & a_M \end{bmatrix}^\top$ is

$$\exp(\mathbf{a}) := \begin{bmatrix} \exp(a_1) & \exp(a_2) & \cdots & \exp(a_M) \end{bmatrix}^\top. \tag{13.2}$$

3. The **power of the vector** $\mathbf{a} = \begin{bmatrix} a_1 & a_2 & \cdots & a_M \end{bmatrix}^\top \in (\mathbb{R}_0^+)^M$ to the **exponent matrix** (respectively, **exponent vector**) $\mathbf{B} = \begin{bmatrix} b_{11} & b_{12} & \dots & b_{1R} \\ b_{21} & b_{22} & \dots & b_{2R} \\ \dots & & & \\ b_{M1} & b_{M2} & \dots & b_{MR} \end{bmatrix} \in \mathbb{R}^{M \times R}$

is the (column) vector $\mathbf{a}^{\mathbf{B}} := \begin{bmatrix} \prod_{m=1}^{M} a_m^{b_{m1}} \\ \prod_{m=1}^{M} a_m^{b_{m2}} \\ \dots \\ \prod_{m=1}^{M} a_m^{b_{mR}} \end{bmatrix} \in \mathbb{R}^R$.

4. In the definitions above, 0^0 is defined to be 1.

Logarithm and exponential function can also be defined for matrices entrywise. Important properties of these operations, which we often use, are summarized in the following theorems. They are all easily proved using the definitions.

In the following two theorems, let $\mathbf{a}, \mathbf{b}, \mathbf{c} \in \mathbb{R}^M$; $\mathbf{A}, \mathbf{B} \in \mathbb{R}^{M \times R}$; $\mathbf{C} \in \mathbb{R}^{N \times M}$; $\mathbf{D} \in \mathbb{R}^{M \times M}$.

Theorem 13.4
1. *If* $\mathbf{a}, \mathbf{b} \in (\mathbb{R}^+)^M$, *then*
   a. $\left(\frac{\mathbf{b}}{\mathbf{a}} \right)^{\mathbf{A}} = \frac{\mathbf{b}^{\mathbf{A}}}{\mathbf{a}^{\mathbf{A}}} \in (\mathbb{R}^+)^R$,
   b. $\ln\left(\frac{\mathbf{b}}{\mathbf{a}} \right) = \ln(\mathbf{b}) - \ln(\mathbf{a}) \in \mathbb{R}^M$.

2. If $\mathbf{a} \in (\mathbb{R}_0^+)^M$

   a. $\mathbf{a}^{\mathrm{id}_{M \times M}} = \mathbf{a}$,

   b. $\mathbf{a}^{\mathbf{0}_{M \times M}} = \mathbf{1}_M$.

3. If $\mathbf{a} \in (\mathbb{R}_0^+)^M$ and $\mathbf{A} \in \mathbb{R}^{M \times R}$, then

   a. $\ln(\mathbf{a}^{\mathbf{b}}) = \mathbf{b}^{\top} \ln(\mathbf{a}) \in \mathbb{R}$,

   b. $\ln(\mathbf{a}^{\mathbf{A}}) = \mathbf{A}^{\top} \ln(\mathbf{a}) \in \mathbb{R}^R$, and

   c. $\exp(\mathbf{A}^{\top} \ln(\mathbf{a})) = \mathbf{a}^{\mathbf{A}} \in \mathbb{R}^R$.

4. If $\mathbf{a} \in (\mathbb{R}_0^+)^M$, $\mathbf{B} \in \mathbb{R}^{M \times N}$, and $\mathbf{C} \in \mathbb{R}^{N \times R}$, then $(\mathbf{a}^{\mathbf{B}})^{\mathbf{C}} = \mathbf{a}^{\mathbf{BC}} \in \mathbb{R}^R$.

5. If $\mathbf{a}, \mathbf{b} \in (\mathbb{R}_0^+)^M$ and $\mathbf{A} \in \mathbb{R}^{M \times R}$, then

   a. $\mathbf{a}^{\mathbf{B}+\mathbf{C}} = \mathbf{a}^{\mathbf{B}} \odot \mathbf{a}^{\mathbf{C}} \in \mathbb{R}^R$, and more specifically,

   b. $\mathbf{a}^{-\mathbf{A}} \odot \mathbf{a}^{\mathbf{A}} = \mathbf{1}_M$, furthermore,

   c. $(\mathbf{a} \odot \mathbf{b})^{\mathbf{C}} = \mathbf{a}^{\mathbf{C}} \odot \mathbf{b}^{\mathbf{C}} \in \mathbb{R}^R$.

Theorem 13.5 *If* $\mathbf{D} \in \mathbb{R}^{M \times M}$, $\det(\mathbf{D}) \neq 0$ *and* $\mathbf{a}, \mathbf{b} \in (\mathbb{R}^+)^M$, *then* $\mathbf{a}^{\mathbf{D}} = \mathbf{b}$ *implies* $\mathbf{a} = \mathbf{b}^{\mathbf{D}^{-1}}$.

Theorem 13.6 *Let* $J \subset \mathbb{R}$ *be an open interval,* $\mathbf{y} \in \mathscr{C}^1(J, \mathbb{R}^M)$, *and let* $\mathbf{C} \in \mathbb{R}^{M \times M}$. *Then the derivative of* $\mathbf{x} := \mathbf{y}^{\mathbf{C}}$ *is as follows:*

$$\dot{\mathbf{x}} = \mathrm{diag}(\mathbf{y}^{\mathbf{C}}) \cdot \mathbf{C}^{\top} \cdot \frac{\dot{\mathbf{y}}}{\mathbf{y}} = \mathbf{y}^{\mathbf{C}} \odot \left(\mathbf{C}^{\top} \cdot \frac{\dot{\mathbf{y}}}{\mathbf{y}} \right) = \mathbf{x} \odot \left(\mathbf{C}^{\top} \cdot \frac{\dot{\mathbf{y}}}{\mathbf{y}} \right), \tag{13.3}$$

consequently $\frac{\dot{\mathbf{x}}}{\mathbf{x}} = \mathbf{C}^{\top} \cdot \frac{\dot{\mathbf{y}}}{\mathbf{y}}$.

Defining the stochastic model, we used modifications of the power function and also extended it to the case of vectors. Extension of the factorial to vectors is also necessary.

Definition 13.7

1. The **factorial power** of the vector $\mathbf{x} = (x_1, x_2, \ldots, x_M)^{\top} \in \mathbb{N}^M$ with the exponent $\mathbf{a} = (a_1, a_2, \ldots, a_M)^{\top} \in \mathbb{N}^M$ is defined to be

$$[\mathbf{x}]_{\mathbf{a}} := \prod_{m=1}^{M} \frac{x_m!}{(x_m - a_m)!} \mathscr{B}(x_m \geq a_m). \tag{13.4}$$

In *Mathematica* `Times @@ FactorialPower[x, a]` is used to calculate the expression $[\mathbf{x}]_{\mathbf{a}}$.

2. The **factorial** of the vector $\mathbf{x} = (x_1, x_2, \ldots, x_M)^{\top} \in \mathbb{N}^M$ is defined to be

$$\mathbf{x}! := \prod_{m=1}^{M} x_m!, \tag{13.5}$$

or `Times @@ Factorial[x]`.

3. The **binomial coefficient** $\mathbf{x} = (x_1, x_2, \ldots, x_M)^\top \in \mathbb{N}^M$ over $\mathbf{a} = (a_1, a_2, \ldots, a_M)^\top \in \mathbb{N}^M$ is defined to be

$$\binom{\mathbf{x}}{\mathbf{a}} := \frac{\mathbf{x}!}{\mathbf{a}!(\mathbf{x} - \mathbf{a})!} = \frac{[\mathbf{x}]_\mathbf{a}}{\mathbf{a}!}. \tag{13.6}$$

13.3   Linear Algebra

Definition 13.8 Let $\mathscr{V}$ be a linear space. Then, the maximal number of its independent vectors is said to be the **dimension** of $\mathscr{V}$, and it is denoted by $\dim(\mathscr{V})$. The dimension of the linear space spanned by the column vectors of a matrix $\mathbf{A}$ is said to be the **rank** of the matrix, and it is denoted by $\text{rank}(\mathbf{A})$.

Theorem 13.9 (Dimension Theorem) *If $\mathbf{A}$ is a linear map from the linear space $\mathscr{U}$ into the linear space $\mathscr{V}$, then*

$$\boxed{\dim(\text{Ker}(\mathbf{A})) + \dim(\text{Im}(\mathbf{A})) = \dim(\mathscr{U}).} \tag{13.7}$$

Definition 13.10 Let $\mathscr{S} \subset \mathscr{V}$ be a subspace of a linear space $\mathscr{V}$. Then, the **orthogonal complement** denoted by $\mathscr{S}^\perp$ consists of the vectors of $\mathscr{V}$ orthogonal to all the vectors of the subspace $\mathscr{S}$.

Remark 13.11 We also have the following statement:

$$\text{Ker}(\mathbf{A}^\top) = (\text{Im}(\mathbf{A}))^\perp \quad \text{Im}(\mathbf{A}^\top) = (\text{Ker}(\mathbf{A}))^\perp$$

Remark 13.12 An immediate consequence of Theorem 13.9 and of Remark 13.11 is the following fact: if $\mathscr{S} \subset \mathbb{R}^M$ is the linear space spanned by the columns of the matrix $\mathbf{A} \in \mathbb{R}^{M \times R}$, then $\dim(\mathscr{S}^\perp) = M - \text{rank}(\mathbf{A})$.

Definition 13.13 A **row reduction** or **elementary row operation** refers to one of the following three operations:

1. Interchanging two rows of the matrix.
2. Multiplying a row of the matrix by a nonzero scalar.
3. Adding a constant multiple of a row to another row.

The result of a series of appropriately chosen elementary row operations can be a row echelon matrix.

Definition 13.14 A rectangular matrix is in **row echelon form** if it has the following properties:

1. All rows containing only zeros are below all nonzero rows.
2. The first nonzero number (the leading entry) in each row is 1.
3. All entries in a column below a leading entry are zeros.

Lemma 13.15 (Dickson (1913)) *Every set of vectors of natural numbers has finitely many (componentwise) minimal elements.*

Definition 13.16 Let $M, \hat{M} \in \mathbb{N}$; $\hat{M} \leq M$. Then the **generalized inverse** of a matrix $\mathbf{M} \in \mathbb{R}^{\hat{M} \times M}$ is a matrix $\overline{\mathbf{M}} \in \mathbb{R}^{M \times \hat{M}}$ such that $\mathbf{M}\overline{\mathbf{M}}\mathbf{M} = \mathbf{M}$ holds. A **Moore–Penrose inverse** (also called **pseudoinverse**) has three further properties: $\overline{\mathbf{M}}\mathbf{M}\overline{\mathbf{M}} = \overline{\mathbf{M}}$, $(\mathbf{M}\overline{\mathbf{M}})^\top = \mathbf{M}\overline{\mathbf{M}}$, *and* $(\overline{\mathbf{M}}\mathbf{M})^\top = \overline{\mathbf{M}}\mathbf{M}$.

Remark 13.17 Obviously, if $\mathbf{M}$ is of the full rank, then there exists a matrix $\overline{\mathbf{M}}$ such that

$$\mathbf{M}\overline{\mathbf{M}} = \mathrm{id}_{\hat{M}}, \tag{13.8}$$

and this can be shown to be a Moore–Penrose inverse of $\mathbf{M}$.

Lemma 13.18 *Let $M, \hat{M} \in \mathbb{N}$; $M \geq \hat{M}$, and $\mathbf{A} \in \mathbb{R}^{\hat{M} \times \hat{M}}$, $\mathbf{B} \in \mathbb{R}^{M \times M}$, and suppose that the matrix $\mathbf{M} \in \mathbb{R}^{\hat{M} \times M}$ is of the full rank. If*

$$\mathbf{AM} = \mathbf{MB}, \tag{13.9}$$

then all the eigenvalues of the matrix $\mathbf{A}$ are eigenvalues of the matrix $\mathbf{B}$ as well.

Proof Suppose $\mathbf{M}$ is of full rank. Then there exists $\overline{\mathbf{M}} \in \mathbb{R}^{M \times \hat{M}}$ such that $\mathbf{M}\overline{\mathbf{M}} = \mathrm{id}_{\hat{M}}$ holds. We show that no $\lambda \in \mathbb{C}$ that is not an eigenvalue of $\mathbf{B}$ can be an eigenvalue of $\mathbf{A}$. This statement is equivalent to the original one.

If $\lambda \in \mathbb{C}$ is not an eigenvalue of $\mathbf{B}$, then there exists $(\mathbf{B} - \lambda\,\mathrm{id}_M)^{-1}$. We show that in this case there exists $(\mathbf{A} - \lambda\,\mathrm{id}_{\hat{M}})^{-1}$ as well. Furthermore, we explicitly give this inverse as $\mathbf{M}(\mathbf{B} - \lambda\,\mathrm{id}_M)^{-1}\overline{\mathbf{M}}$. Let us check this statement using (13.9)

$$(\mathbf{A} - \lambda\,\mathrm{id}_{\hat{M}})\mathbf{M}(\mathbf{B} - \lambda\,\mathrm{id}_M)^{-1}\overline{\mathbf{M}} = (\mathbf{AM} - \lambda\mathbf{M})((\mathbf{B} - \lambda\,\mathrm{id}_M)^{-1}\overline{\mathbf{M}})$$

$$= (\mathbf{MB} - \lambda\mathbf{M})(\mathbf{B} - \lambda\,\mathrm{id}_M)^{-1}\overline{\mathbf{M}}$$

$$= \mathbf{M}(\mathbf{B} - \lambda\,\mathrm{id}_M)(\mathbf{B} - \lambda\,\mathrm{id}_M)^{-1}\overline{\mathbf{M}} = \mathbf{M}\overline{\mathbf{M}} = \mathrm{id}_{\hat{M}}.$$

$\square$

13.3.1 Norms

Sometimes different **norms** of a vector $\mathbf{m} \in \mathbb{R}^R$ are used. They are defined as follows:

$$\|\mathbf{m}\|_1 := \sum_{r \in \mathcal{R}} |m_r| \tag{13.10}$$

$$\|\mathbf{m}\|_p := \left(\sum_{r \in \mathcal{R}} |m_r|^p \right)^{1/p} \tag{13.11}$$

$$\|\mathbf{m}\|_\infty := \max\{|m_r| \mid r \in \mathcal{R}\} \tag{13.12}$$

13.3.2 Fredholm's Theorem

Finally, we cite (a version of) the Fredholm alternative theorem of linear algebra.

Theorem 13.19 *Let* $\mathbf{A} \in \mathbb{R}^{M \times R}$, $\mathbf{b} \in \mathbb{R}^R$. *The system of linear equations*

$$\mathbf{A}\mathbf{x} = \mathbf{b} \tag{13.13}$$

has a solution if and only if

$$\forall \mathbf{y} : \mathbf{A}^\top \mathbf{y} = \mathbf{0} \longrightarrow \mathbf{b}^\top \mathbf{y} = 0. \tag{13.14}$$

Proof The proof can be seen as the solution to Problem 13.5. $\square$

Remark 13.20 We shall use the theorem in the above form. Let us give a reformulation justifying the name **alternative theorem**. Either (13.13) or

$$\mathbf{A}^\top \mathbf{y} = \mathbf{0} \quad \mathbf{b}^\top \mathbf{y} = 0$$

has a solution.

13.4   Mathematical Programming

In its full generality, linear programming is the problem of finding the supremum (or infimum) of a linear function over a set of vectors defined by linear equalities and inequalities (collectively called **constraints**). Its most common form is

$$\max_{\mathbf{x} \in \mathbb{R}^n} \{\mathbf{c}^\top \mathbf{x} \mid \mathbf{A}\mathbf{x} \leq \mathbf{b}\}, \tag{13.15}$$

where the matrix $\mathbf{A} \in \mathbb{R}^{m \times n}$ and the vectors $\mathbf{b} \in \mathbb{R}^m$ and $\mathbf{c} \in \mathbb{R}^n$ are given. Of course, equality constraints can also be included (utilizing both $\leq$ and $\geq$), but non-equality ($\neq$) constraints and strict inequalities ($>$ or $<$) are not allowed, to avoid the situation when the supremum is finite but is not attained at any $\mathbf{x}$ satisfying the constraints. This is the reason why (4.3) had to be rewritten in the form of (4.8). The use of max in (13.15) is still somewhat misleading, since the supremum may be infinity, which is also not a maximum; nevertheless, this is the customary notation in linear programming. In summary, a linear programming problem may have a "solution" of three different kinds:

- We say that the problem (13.15) is **infeasible** if no $\mathbf{x}$ satisfies the constraints $\mathbf{Ax} \leq \mathbf{b}$; otherwise it is **feasible**.
- If (13.15) is feasible, the function $\mathbf{x} \mapsto \mathbf{c}^\top \mathbf{x}$ may still not be bounded from above on the set $\{\mathbf{x} \mid \mathbf{Ax} \leq \mathbf{b}\}$. In this case we say that the problem is **unbounded**, and we define the maximum to be $+\infty$.
- If the problem is feasible but not unbounded, then there exists a vector $\mathbf{x}$ at which the (finite) maximum is attained.

Remark 13.21 If one leaves the area of linear programming and uses standard concepts of mathematical analysis, then the above few lines can be reformulated as extremum problems. Let us suppose we are given two positive integers, $m, n \in \mathbb{N}$, a matrix $\mathbf{A} \in \mathbb{R}^{m \times n}$, and two vectors $\mathbf{b} \in \mathbb{R}^m, \mathbf{c} \in \mathbb{R}^n$. Let us define the set $\mathscr{D} := \{\mathbf{x} \in \mathbb{R}^n \mid \mathbf{Ax} \leq \mathbf{b}\}$.

1. If $\mathscr{D} = \emptyset$, then no function can be defined on this set, and $\mathbf{c}$ plays no role at all (therefore it is quite an illogical tradition to speak of the infeasibility of problem (13.15).)
2. If $\mathscr{D} \neq \emptyset$, then let $\mathscr{D} \ni \mathbf{x} \mapsto c(\mathbf{x}) := \mathbf{c}^\top \mathbf{x} \in \mathbb{R}$ be the definition of a restriction of the linear function $\mathbb{R}^n \ni \mathbf{x} \mapsto \mathbf{c}^\top \mathbf{x} \in \mathbb{R}$ to the set $\mathscr{D}$. The value of sup(c) may either be $+\infty$, then the problem (13.15) is unbounded, or it may have a finite value. In the latter case, the supremum is a maximum (a fact not immediately following from the continuity of the function c), i.e., there exists (at least) a vector $\mathbf{x}^*$ at which the supremum is taken.

We may if we wish to say in the infeasible case that the supremum of c is $-\infty$.

In *Mathematica*, linear programming problems can be solved using the built-in function **LinearProgramming**. Alternatively, one may use the general-purpose **Maximize** function for symbolic optimization and either **FindMaximum** for local or **NMaximize** for global numerical optimization. Note the difference between their input formats and also between the outputs of these functions.

Theorem 13.22 (Tucker) *Let $\mathscr{S}$ be a subspace of a vector space $\mathbb{R}^M$. Then there exist vectors* $\mathbf{x}, \mathbf{z} \in \mathbb{R}^M$ *such that*

$$\mathbf{x}, \mathbf{z} \in (\mathbb{R}_0^+)^M, \quad \mathbf{x} \in \mathscr{S}, \quad \mathbf{z} \in \mathscr{S}^\perp \quad and \quad \mathbf{x} + \mathbf{z} \in (\mathbb{R}^+)^M.$$

Theorem 13.23 (Haynsworth) *Let* $\mathbf{M} = \begin{bmatrix} A & B^\top \\ B & C \end{bmatrix}$ *be a symmetric matrix. Then*

1. $\mathbf{M} \succ \mathbf{0}$ *if and only if both* $\mathbf{C} \succ \mathbf{0}$ *and* $\mathbf{A} - \mathbf{B}^\top\mathbf{C}^{-1}\mathbf{B} \succ \mathbf{0}$.
2. *If* $\mathbf{C} \succ \mathbf{0}$, *then* $\mathbf{M} \succcurlyeq \mathbf{0}$ *if and only if* $\mathbf{A} - \mathbf{B}^\top\mathbf{C}^{-1}\mathbf{B} \succcurlyeq \mathbf{0}$.

The matrix $\mathbf{A} - \mathbf{B}^\top\mathbf{C}^{-1}\mathbf{B}$ is called the **Schur complement** of $\mathbf{C}$ in $\mathbf{M}$.

13.5   Graphs

Properties of reactions can often be deduced from their *discrete* or *combinatorial* structure (which species take part in which reactions, etc.). This abstract structure is most easily described in terms of graphs. Good introductory texts on graph theory include (Harary 1969; Lovász 2007; Øre 1962).

Definition 13.24 A **directed graph** is an ordered pair $G = (V, E)$ of a finite nonempty set of **vertices** V and a set $E \subset V \times V$ of *ordered* pairs of vertices, called **edge**s. Edges of the form $(v, v) \in E$ are called **loop**s.

Definition 13.25 A(n undirected) **graph** is an ordered pair $G = (V, E)$ of a finite nonempty set of **vertices** V and a set of $E \subset V \times V$ of *unordered* pairs of vertices, called **edge**s. Edges of the form $(v, v) \in E$ are called **loop**s in the case of undirected graphs as well.

Edges of directed graphs are often called **arc**s. Note that the only difference between directed and undirected graphs is that edges have an orientation (a "beginning" and an "end").

Definition 13.26 In both directed and undirected graphs, vertices connected by an edge are said to be **adjacent**. Edges with a common vertex are also called **adjacent**. The **adjacency matrix** of the graph (V, E) is a $|V| \times |V|$ matrix $\mathbf{A}$ with an entry $a_{ij} = 1$, if $(i, j) \in E$; otherwise the entry a_{ij} is zero.

Obviously, in the case of undirected graphs, the adjacency matrix is symmetric and redundant: all the edges are represented twice.

Definition 13.27 Vertices of an edge are said to be **incident** to the edge. The edge (v, w) of an undirected graph connects the two vertices v and w, while the edge

(v, w) of a directed graph begins at the vertex v and ends at the vertex w. The **incidence matrix E** of the directed graph (V, E) with no loops is a $|V| \times |E|$ matrix with an entry

1. $i_{ve} = -1$, if edge e begins at vertex v, that is, if $e = (v, w)$ for some $w \in V$,
2. $i_{ve} = 1$, if edge e ends at vertex v, that is, if $e = (w, v)$ for some $w \in V$,
3. $i_{ve} = 0$ otherwise.

Naturally, every column of the incidence matrix contains precisely one $+1$ and one -1 entry.

Definition 13.28 A **subgraph** of a graph $G = (V, E)$ is a graph $(\overline{V}, \overline{E})$ such that $\overline{V} \subset V$ and $\overline{E} \subset E$ hold. A **spanning subgraph** of G is a subgraph whose vertex set is the same as that of G.

Definition 13.29 A finite sequence of edges $e_1, e_2, \ldots, e_k$ is called a **path** connecting the vertices e_1 and e_k if each pair of consecutive edges have a common vertex. In the case of directed graphs, such a sequence of edges is called a **directed path** if the beginning of each edge is the same as the end of its predecessor. A closed path, i.e., one for which the first and the last edge has a common vertex, is called a **cycle**, respectively, a **directed cycle**. An undirected graph without cycles is a **forest**; it is a **tree**, if it is also connected (see Definition 13.30). A **spanning forest** is a spanning subgraph which is a forest (Fig. 13.1).

Definition 13.30 An undirected graph is said to be **connected** if each pair of its vertices is connected by a(n undirected) path. A directed graph is **weakly connected** if disregarding the directions of its edges a connected undirected graph is obtained. A(n inclusionwise) maximal connected subgraph of an undirected graph is said to be a **connected component** of the graph. A directed graph is **strongly connected** if each pair of its vertices is connected by a directed path. A(n inclusionwise) maximal strongly connected subgraph of a directed graph is said to be a **strong component**

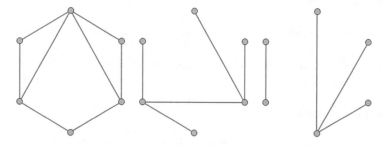

Fig. 13.1 An undirected graph with six vertices (left), along with one of its spanning trees (center) and a spanning forest (right) with two components

of the graph. Strong components of a directed graph from which no edge goes out are called **ergodic components**.

The vertex sets of connected components of an undirected graph partition the vertex set: each vertex belongs to precisely one component. The same holds for the strong components of directed graphs. (See Problem 13.7.)

In some graphs of reactions, the vertices represent either species or reaction steps, and edges only connect species to reactions. Graphs with such special structure are bipartite graphs.

Definition 13.31 A **directed bipartite graph** is a directed graph (V, E) whose vertex set can be partitioned into two sets, V_1 and V_2, such that $E \subset (V_1 \times V_2) \cup (V_2 \times V_1)$. Often, to signify the two classes of vertices, the notation $G = (V_1, V_2, E)$ is used for such graphs.

Since graphs can be considered as special binary relations on finite sets, the standard terminology of relations can also be applied to graphs.

Definition 13.32 A directed or undirected graph (V, E) is **reflexive** if for all $v \in V (v, v) \in E$, i.e., the graphs contain all the possible loops. A directed graph (V, E) is **symmetric** if $(v, w) \in E$ implies $(w, v) \in E$, i.e., if the reverse of every edge is also present in the graph. It is called **transitive**, if for all pairs of edges $(v, w), (w, z) \in E$, we also have $(v, z) \in E$. The **transitive closure** of the directed graph (V, E) is obtained in such a way that whenever there is a directed path beginning in the vertex v and ending in the vertex w, then the edge (v, w) is appended to the set of edges.

Theorem 13.33 *The transitive closure of a directed graph is symmetric if and only if all of its strong components are ergodic.*

Proof See Problem 13.8.   □

13.6   Calculus

Here we review a few theorems which you may not have learned in first year calculus.

Before calculus proper, we cite a trivial, still useful lemma from (Tóth et al. 1997, p. 1533). Let $K, M \in \mathbb{N}$.

Lemma 13.34 *Suppose that*

$$\mathbf{k} \circ \mathbf{x} = \mathbf{l} \circ \mathbf{x}$$

holds for the functions $\mathbf{k}, \mathbf{l} : \mathbb{R}^M \longrightarrow \mathbb{R}^K$ *with some functions* $\mathbf{x} : \mathbb{R} \longrightarrow \mathbb{R}^M$ *for which* $\cup_{\mathbf{x}} \mathscr{R}(\mathbf{x}) = \mathbb{R}^M$ *is true. Then* $\mathbf{k} = \mathbf{l}$ *is also true.*

This lemma is usually used for all the solutions **x** of a differential equation with a right-hand side defined on the whole $\mathbb{R}^M$.

Theorem 13.35 (Brouwer) *Let* $M \in \mathbb{N}$, *and let* $T \subset \mathbb{R}^M$ *be a closed, bounded, and convex set, and* $\mathbf{f} \in \mathscr{C}(T, \mathbb{R}^M)$. *Then, there exists at least one point* $\mathbf{x}_* \in T$ *(called the **fixed point** of* $\mathbf{f}$*) such that* $\mathbf{f}(\mathbf{x}_*) = \mathbf{x}_*$.

It turns out that the concepts of linear dependence and independence can be extended from vectors to nonlinear functions. Let $K, M \in \mathbb{N}; 0 < K < M; \mathbf{a} \in \mathbb{R}^M, \mathbf{u} \in \mathscr{C}^1(B(\mathbf{a}), \mathbb{R}^K)$, where $B(\mathbf{a}) \subset \mathbb{R}^M$ is an open ball centered at $\mathbf{a}$.

Definition 13.36 If rank($\mathbf{u}'(\mathbf{a})$) $= K$ (implying that there exists a ball $B_1(\mathbf{a}) \subset B(\mathbf{a})$ centered at $\mathbf{a}$ so that for all $\mathbf{x} \in B_1(\mathbf{a})$: rank($\mathbf{u}'(\mathbf{x})$) $= K$), then the coordinate functions of $\mathbf{u}$ are said to be **independent** at the point $\mathbf{a}$. If however there exists a ball $B_2(\mathbf{a}) \subset B(\mathbf{a})$ centered at $\mathbf{a}$ so that for all $\mathbf{x} \in B_2(\mathbf{a})$: rank($\mathbf{u}'(\mathbf{x})$) $< K$), then the coordinate functions of $\mathbf{u}$ are said to be **dependent** at the point $\mathbf{a}$.

Lemma 13.37 *Let the coordinate functions of* $\mathbf{u}$ *be independent at the point* $\mathbf{a}$. *Then, there exists a function* $\mathbf{v} \in \mathscr{C}^1(B(\mathbf{a}), \mathbb{R}^{M-K})$ *such that even the coordinate functions of* $\begin{bmatrix} \mathbf{u} \\ \mathbf{v} \end{bmatrix}$ *are independent.*

Proof If $\bar{\mathbf{v}}_{K+1}, \bar{\mathbf{v}}_{K+2}, \ldots, \bar{\mathbf{v}}_M \in \mathbb{R}^M$ are vectors independent from each other and from the row vectors of $\mathbf{u}'(\mathbf{a})$, then with

$$v_{K+1}(\mathbf{x}) := \bar{\mathbf{v}}_{K+1}^\top \mathbf{x}, \ v_{K+2}(\mathbf{x}) := \bar{\mathbf{v}}_{K+2}^\top \mathbf{x}, \ldots, \ v_M(\mathbf{x}) := \bar{\mathbf{v}}_M^\top \mathbf{x},$$

the function $\mathbf{v}(\mathbf{x}) := \begin{bmatrix} v_{K+1}(\mathbf{x}) \\ v_{K+2}(\mathbf{x}) \\ \vdots \\ v_M(\mathbf{x}) \end{bmatrix}$ is an appropriate completion to a **nonlinear basis**.

$\square$

Now a theorem with many applications follows.

Theorem 13.38 *Suppose that the coordinate functions of* $\mathbf{u}$ *are independent at the point* $\mathbf{a}$, *but the coordinate functions of* $\begin{bmatrix} \mathbf{u} \\ w \end{bmatrix}$ *are dependent, where* $w \in \mathscr{C}^1(B(\mathbf{a}), \mathbb{R})$. *Then there exists a function* $W \in \mathscr{C}^1(\mathscr{R}_u, \mathbb{R})$ *such that*

$$w(\mathbf{x}) = W(\mathbf{u}(\mathbf{x})) \quad (\mathbf{x} \in B(\mathbf{a})). \tag{13.16}$$

Proof Let us complete $\mathbf{u}$ to a nonlinear basis $\begin{bmatrix} \mathbf{u} \\ \mathbf{v} \end{bmatrix}$ with $\mathbf{v} \in \mathscr{C}^1(B(\mathbf{a}), \mathbb{R}^{M-K})$, and

let $\begin{bmatrix} \mathbf{p} \\ \mathbf{q} \end{bmatrix} := \begin{bmatrix} \mathbf{u} \\ \mathbf{v} \end{bmatrix}^{-1}$, $\mathbf{p} \in \mathscr{C}^1(\mathbb{R}^M, \mathbb{R}^K), \mathbf{q} \in \mathscr{C}^1(\mathbb{R}^M, R^{M-K})$.

First, we show that $\partial_2 \tilde{W} = 0$, if $\tilde{W} := w \circ \begin{bmatrix} \mathbf{p} \\ \mathbf{q} \end{bmatrix}$. According to the definition of the inverse, one has

$$\begin{bmatrix} \mathbf{u} \\ \mathbf{v} \end{bmatrix} \circ \begin{bmatrix} \mathbf{p} \\ \mathbf{q} \end{bmatrix} = \begin{bmatrix} \mathbf{u} \circ \begin{bmatrix} \mathbf{p} \\ \mathbf{q} \end{bmatrix} \\ \mathbf{v} \circ \begin{bmatrix} \mathbf{p} \\ \mathbf{q} \end{bmatrix} \end{bmatrix} = \begin{bmatrix} \pi \\ \varrho \end{bmatrix}, \tag{13.17}$$

where

$$\mathbb{R}^M \ni \begin{bmatrix} \mathbf{x} \\ \mathbf{y} \end{bmatrix} \mapsto \pi\left(\begin{bmatrix} \mathbf{x} \\ \mathbf{y} \end{bmatrix}\right) := \mathbf{x} \in \mathbb{R}^K, \quad \mathbb{R}^M \ni \begin{bmatrix} \mathbf{x} \\ \mathbf{y} \end{bmatrix} \mapsto \varrho\left(\begin{bmatrix} \mathbf{x} \\ \mathbf{y} \end{bmatrix}\right) := \mathbf{y} \in \mathbb{R}^{M-K}.$$

Upon taking the derivative of (13.17) with respect to the second (vectorial) derivative, we arrive at

$$\left(\partial_1 \mathbf{u} \circ \begin{bmatrix} \mathbf{p} \\ \mathbf{q} \end{bmatrix}\right) \partial_2 \mathbf{p} + \left(\partial_2 \mathbf{u} \circ \begin{bmatrix} \mathbf{p} \\ \mathbf{q} \end{bmatrix}\right) \partial_2 \mathbf{q} = \left(\mathbf{u}' \circ \begin{bmatrix} \mathbf{p} \\ \mathbf{q} \end{bmatrix}\right) \partial_2 \begin{bmatrix} \mathbf{p} \\ \mathbf{q} \end{bmatrix} = \mathbf{0}.$$

As $\mathbf{u}$ and w are dependent, there exists $\mathbf{c} \in \mathbb{R}^K$ so that $w' = \mathbf{c}^\top \mathbf{u}'$; thus

$$\mathbf{0} = \mathbf{c}^\top \left(\mathbf{u}' \circ \begin{bmatrix} \mathbf{p} \\ \mathbf{q} \end{bmatrix}\right) \partial_2 \begin{bmatrix} \mathbf{p} \\ \mathbf{q} \end{bmatrix} = \left(w' \circ \begin{bmatrix} \mathbf{p} \\ \mathbf{q} \end{bmatrix}\right) \partial_2 \begin{bmatrix} \mathbf{p} \\ \mathbf{q} \end{bmatrix} = \partial_2 w \circ \begin{bmatrix} \mathbf{p} \\ \mathbf{q} \end{bmatrix} = \partial_2 \tilde{W},$$

therefore $\tilde{W}$ does not depend on its second (vectorial) variable.

Second, with the well-defined function $W(\mathbf{s}) := \tilde{W}\left(\begin{bmatrix} \mathbf{s} \\ \mathbf{t} \end{bmatrix}\right)$ ($\mathbf{s} \in \mathscr{R}_\mathbf{u}, \mathbf{t} \in \mathbb{R}^{M-K}$), we have $w(\mathbf{x}) = W(\mathbf{u}(\mathbf{x}))$ as required.   □

The following statement is a generalization of a lemma by Higgins (Póta 1981).

Theorem 13.39 (Higgins–Póta) *Let* $k \in \mathbb{N}$; $\lambda_1, \lambda_2, \ldots, \lambda_k \in \mathbb{R}$ *be distinct real numbers; and let* $P_1, P_2, \ldots, P_k$ *not identically zero polynomials of degree* $\mu_1 - 1, \mu_2 - 1, \ldots, \mu_k - 1 \in \mathbb{N}_0$, *respectively, with real coefficients. Then the function*

$$\mathbb{R} \ni t \mapsto g(t) := P_1(t)e^{\lambda_1 t} + P_2(t)e^{\lambda_2 t} + \cdots + P_k(t)e^{\lambda_k t} \tag{13.18}$$

has at most $\mu_1 + \mu_2 + \cdots + \mu_k - 1$ *real zeros.*

Proof See Problem 13.13.   □

Definition 13.40 Let $I \subset \mathbb{N}^M$ be an arbitrary index set. Then the **generating function** g associated to the sequence $\{c_\mathbf{x}\}_{\mathbf{x} \in I}$ of real numbers is defined by the following formula:

$$g(\mathbf{z}) := \sum_{\mathbf{x} \in I} c_\mathbf{x} \mathbf{z}^\mathbf{x} \tag{13.19}$$

where $\mathbf{z} \in \mathbb{C}^M$.

Clearly if $|I| < +\infty$, then g is a multivariate polynomial. In the case when $|I| = +\infty$, it is not so obvious whether the right-hand side of Eq. (13.19) defines a continuous (let alone differentiable) function in some region or not: this depends on the behavior of the considered sequence $\{c_\mathbf{x}\}_{\mathbf{x} \in I}$ and of course on what region of $\mathbb{C}^M$ one would like to define the function g. A useful sufficient condition is given by the Cauchy–Hadamard theorem.

Theorem 13.41 (Cauchy–Hadamard) *Let* $c : \mathbb{N}^M \to \mathbb{R}$ *be a function (with values interpreted as coefficients), and let us consider the formal multidimensional power series*

$$\sum_{\mathbf{0} \leq \mathbf{x} \in \mathbb{N}^M} c_\mathbf{x}(\mathbf{z}-\mathbf{a})^\mathbf{x} := \sum_{x_1 \geq 0,...,x_M \geq 0} c_{x_1,...,x_M}(z_1-a_1)^{x_1} \cdots (z_M-a_M)^{x_M}. \tag{13.20}$$

This series converges with radius of convergence ϱ *if and only if*

$$\lim_{|\mathbf{x}| \to +\infty} \sqrt[|\mathbf{x}|]{|c_\mathbf{x}|\varrho^\mathbf{x}} = 1,$$

i.e., if $\mathbf{z} < \varrho$, *then* (13.20) *converges, and if* $\mathbf{z} \geq \varrho$ *but* $\mathbf{z} \neq \varrho$, *then* (13.20) *diverges.*

13.7   Ordinary and Partial Differential Equations

Differential equations, both ordinary and partial ones, often arise in formal kinetics both as models and as mathematical tools.

13.7.1   Existence, Uniqueness, and Continuous Dependence

Theorem 13.42 (Picard–Lindelöf) *Let* $M \in \mathbb{N}$, *suppose* $T \subset \mathbb{R}^M$ *is an open connected set, and let* $\mathbf{f} \in \mathscr{C}^1(T, \mathbb{R}^M)$, $\mathbf{x}_0 \in \mathbb{R}^M$. *Then, there exists a positive real* *number τ such that the initial value problem*

$$\dot{\mathbf{x}}(t) = \mathbf{f}(\mathbf{x}(t)) \quad \mathbf{x}(0) = \mathbf{x}_0. \tag{13.21}$$

has a unique solution defined (at least) on the interval $] - \tau, \tau[$.

Definition 13.43 The solution of the initial value problem (13.21) is said to be the **maximal solution** if there is no solution which is a proper extension of it.

Theorem 13.44 *Under the notations of Theorem 13.42, the solution of the initial value problem* (13.21) *continuously depends on the initial value* $\mathbf{x}_0$.

Continuous dependence also holds with respect to arbitrary parameters. One can even proceed a step further beyond continuity.

Theorem 13.45 *Suppose that $M, P \in \mathbb{N}$ and that $T \subset \mathbb{R}^M \times \mathbb{R}^P$ is an open connected set; furthermore, let $\mathbf{f} \in \mathscr{C}^1(T, \mathbb{R}^M)$, $\mathbf{x}_0 \in \mathbb{R}^M$, and let the solution of the initial value problem*

$$\dot{\mathbf{x}}(t, \mathbf{p}) = \mathbf{f}(\mathbf{x}(t, \mathbf{p}), \mathbf{p}) \quad \mathbf{x}(0, \mathbf{p}) = \mathbf{x}_0 \tag{13.22}$$

*be denoted by ξ. Then its derivative (the **sensitivity matrix**) defined by $\sigma(t, \mathbf{p}) := \frac{\partial \xi(t, \mathbf{p})}{\partial \mathbf{p}}$ exists and fulfils the following **variational** or **sensitivity equations** around the solution $t \mapsto \xi(t, \mathbf{p})$:*

$$\frac{d\sigma(t, \mathbf{p})}{dt} = \partial_1 \mathbf{f}(\xi(t, \mathbf{p}), \mathbf{p})\sigma(t, \mathbf{p}) + \partial_2 \mathbf{f}(\xi(t, \mathbf{p}), \mathbf{p}). \tag{13.23}$$

Remark 13.46 Note that the sensitivity equations are linear equations in general with time-varying (solution-dependent) coefficients. They form a closed system of differential equations together with the original equations in (13.22). The number of sensitivity equations being $M \times P$ may be quite large. We also remark that part of the literature uses the term "variational equations" and the other part uses "sensitivity equations"; therefore some students may not realize the equality of these seemingly two objects.

Definition 13.47 Consider Eq. (13.21), and suppose a function $\varphi \in \mathscr{C}^1(T, \mathbb{R})$ is given. Then, the **Lie derivative** of φ with respect to the differential equation in (13.21) is the function $\varphi' \mathbf{f} \in \mathscr{C}(T, \mathbb{R})$.

13.7.2 Long Time Behavior

Definition 13.48 Let $M \in \mathbb{N}$, and let $T \subset \mathbb{R}^M$ be a domain, $\mathbf{f} \in \mathscr{C}^1(T, \mathbb{R}^M)$, $\mathbf{x}_0 \in T$; furthermore let the solution of the initial value problem (13.21) be $\mathbb{R}^+ \ni t \mapsto$

$\varphi(t, \mathbf{x}_0) \in \mathbb{R}^M$ (i.e., we suppose it is defined for all $t \in \mathbb{R}^+$). Then, $\mathbf{y} \in \mathbb{R}^M$ is said to be an ω-**limit point** of the point $\mathbf{x}_0$ if there exists a sequence $(t_n)_{n \in \mathbb{N}}$ of real numbers tending to $+\infty$ such that $\lim \varphi(t_n, \mathbf{x}_0) = \mathbf{y}$. The set of all ω-limit points is the ω-**limit set** of the point $\mathbf{x}_0$.

Definition 13.49 For the function $f : \mathbb{R} \longrightarrow \mathbb{R}$, the Lyapunov exponent $\lambda(f)$ is defined to be

$$\lambda(f) := \lim_{t \to +\infty} \frac{1}{t} \ln(|f(t)|) \tag{13.24}$$

if this (finite or infinite) limit exists.

13.7.3 Singular Perturbation

To get approximate description of enzyme reactions (as, e.g., in Chap. 9) or, more generally, reactions taking place on two time scales, we need the following theorem by Tikhonov (1952):

Consider the system of the differential equations

$$\dot{x}(t) = f(x(t), y(t)), \quad \mu \dot{y}(t) = g(x(t), y(t)) \tag{13.25}$$

and the corresponding **degenerate system**

$$\dot{x}(t) = f(x(t), y(t)), \quad y(t) = \varphi(x(t)) \tag{13.26}$$

where φ gives the (unique, isolated) solution to $0 = g(\overline{x}, \overline{y})$, i.e., $0 = g(\overline{x}, \varphi(\overline{x}))$.

Theorem 13.50 *With the above notations, if $\mu \to 0$, then the solution of the system* (13.25) *tends to the solution of* (13.26).

Equation (13.25) is said to be a **singularly perturbed equation**.

13.7.4 Ruling Out Periodicity

There are a few general statements, mainly for two-dimensional systems which give either sufficient or necessary conditions of the existence of periodic solutions of differential equations. It is almost a general rule that sufficient conditions need more work to check. Useful references for the topics are (Farkas 1994; Perko 1996; Tóth and Simon 2005/2009).

The Bendixson–Dulac criterion below gives a sufficient condition for the nonexistence of periodic solutions of two-dimensional differential equations. Recall that a set $E \subset \mathbb{R}^2$ is **simply connected** if every simple closed curve in E can be continuously shrunk in E to a point in E. (Such sets "contain no holes.")

Theorem 13.51 (Bendixson–Dulac) *Let $M \subset \mathbb{R}^2$ be a connected open set, $\mathbf{f} \in \mathscr{C}^1(M, \mathbb{R}^2)$, and let $E \subset M$ be a simply connected open set, $B \in \mathscr{C}^1(E, \mathbb{R}^+)$, such a function with which $D := \mathrm{div}(B\mathbf{f})$ does not change sign and is zero at most at the points of a curve. Then,*

$$\dot{\mathbf{x}} = \mathbf{f} \circ \mathbf{x} \tag{13.27}$$

has no periodic solution fully contained in the set E.

Remark 13.52

- The difficulty of using Theorem 13.51 is that one needs to find an appropriate function B. Problem 8.19 shows an example.
- Bendixson theorem is the special case of the above theorem when $B(x, y) = 1$.
- A multidimensional generalization with reaction kinetic applications can be found in Tóth (1987).

13.7.5 Ensuring Periodicity

Results concerning the existence of periodic solutions for differential equations are collected in this subsection. The first one is the, rather intuitive, assertion that bounded solutions not approaching stationary points must approach a periodic solution. For a formal statement, we need the following definition.

Definition 13.53 Let $K \subset \mathbb{R}^2$ be a connected open set, $\mathbf{f} \in \mathscr{C}^1(K, \mathbb{R}^2)$, and let $\varphi(t, \mathbf{x}_0)$ be the solution of the Eq. (13.27) at time t. The set $L \subset K$ is said to be a **positively invariant set** if for all $\mathbf{x}_0 \in L$ and for all $t \in \mathbb{R}^+$, it is also true that $\varphi(t, \mathbf{x}_0) \in L$.

Theorem 13.54 (Poincaré–Bendixson) *If $L \subset K$ is a bounded and closed positively invariant set of the Eq. (13.27) containing no stationary point, then L does contain a closed trajectory.*

When applying this theorem, the main task is to construct the positively invariant set. Problem 8.22 shows how to do this. The next theorem can be used in higher dimensions as well, but it also requires more calculations. The book by Guckenheimer and Holmes (1983) is a good reference for the topic. Let us start with definitions.

Let $M \in \mathbb{N}$; $\mathbf{f} \in \mathscr{C}^1(\mathbb{R}^M, \mathbb{R}^M)$, and consider the differential equation

$$\dot{\mathbf{x}} = \mathbf{f} \circ \mathbf{x}. \tag{13.28}$$

Definition 13.55 An isolated closed trajectory of Eq. (13.28) is said to be a **limit cycle**.

Isolated here means that none of the trajectories starting from its neighborhood are closed.

Definition 13.56 Equation (13.28) is said to show **conservative oscillation**, if it has constant amplitude periodic solutions depending on the initial condition.

Theorem 13.57 (Andronov–Hopf) *Let $J \subset \mathbb{R}$ be an open interval, $\mathbf{f} = (f, g) \in \mathscr{C}^1(\mathbb{R}^2 \times J, \mathbb{R}^2)$, and consider the differential equation*

$$\dot{\mathbf{x}}(t) = \mathbf{f}(\mathbf{x}(t), p) \quad (p \in J). \tag{13.29}$$

*Suppose that there is a **parameter** $p_0 \in J$ such that for all p close enough to p_0, the equation has a single stationary point $\mathbf{x}_*(p)$, and suppose that the Jacobian $\mathbf{f}'(\mathbf{x}_*(p), p)$ has a pair of complex eigenvalues $\lambda_{1,2}(p) = \alpha(p) \pm i\beta(p)$ so that $\alpha(p_0) = 0$ and $\beta(p_0) > 0$. Furthermore, let us suppose that $\alpha'(p) \neq 0$ in a small neighborhood of p_0 and that*

$$a := \frac{1}{16}(\partial_1^3 f + \partial_1 \partial_2^2 f + \partial_1^2 \partial_2 g + \partial_2^3 g) \tag{13.30}$$

$$+ \frac{1}{16\beta(p_0)}(\partial_1 \partial_2 f(\partial_1^2 f + \partial_2^2 f) - \partial_1 \partial_2 g(\partial_1^2 g + \partial_2^2 g) - \partial_1^2 f \partial_1^2 g + \partial_2^2 f \partial_2^2 g$$

is different from zero at the stationary point. Then, as p passes increasingly through p_0, the stationary point changes its stability: it becomes unstable if $a(\mathbf{x}_*(p_0)) < 0$, and a unique stable limit cycle emerges from it.

13.7.6 First-Order Quasilinear Partial Differential Equations

Here we give a very short summary of the usual solution method of quasilinear partial differential equations (Tóth and Simon 2005/2009, Section 9.2).

Let $M \in \mathbb{N}$ be a natural number, $\Omega \subset \mathbb{R}^{M+1}$ be a **domain** (:=an open and connected set), $\mathbf{f} \in \mathscr{C}(\Omega, \mathbb{R}^M)$, $h \in \mathscr{C}(\Omega, \mathbb{R})$, and let us introduce the following notation:

$$\mathscr{I}_{\mathbf{f},h} := \{u \subset \Omega \mid u \text{ is a function}, \mathscr{D}_u \text{ is a domain}, u \text{ is continuously differentiable}\}. \tag{13.31}$$

The notation $u \subset \Omega$ shows that u is considered to be a (special) relation, i.e., $u \subset \Omega$ is equivalent to saying that $\forall \mathbf{x} \in \mathscr{D}_u : (\mathbf{x}, u(\mathbf{x})) \in \Omega = \mathscr{D}_\mathbf{f} = \mathscr{D}_h$.

It is easy to find such a subset of Ω which is not a function, or which is a function but which is not defined on an open and connected set, or which is not differentiable. The differential equation below will not be defined for such entities.

For an arbitrary element $u \in \mathscr{I}_{\mathbf{f},h}$, one can form the expressions $\mathbf{f}(\mathbf{x}, u(\mathbf{x}))$ and $h(\mathbf{x}, u(\mathbf{x}))$ for all $\mathbf{x} \in \mathscr{D}_u$, because $(\mathbf{x}, u(\mathbf{x})) \in \Omega$. Furthermore, the functions $\Omega \ni \mathbf{x} \mapsto \mathbf{f}(\mathbf{x}, u(\mathbf{x}))$ and $\Omega \ni \mathbf{x} \mapsto h(\mathbf{x}, u(\mathbf{x}))$ (shortly, the functions $\mathbf{f} \circ (\mathrm{id}, u)$ and $h \circ (\mathrm{id}, u)$) are continuous. Therefore, one can raise the question whether the scalar product of the derivative function of u with the function $\mathbf{f} \circ (\mathrm{id}, u)$ is equal to the function $h \circ (\mathrm{id}, u)$. In other words, does it hold for all $\mathbf{x} \in \mathscr{D}_u$ that $u'(\mathbf{x})\mathbf{f}(\mathbf{x}, u(\mathbf{x})) = h(\mathbf{x}, u(\mathbf{x}))$? To formulate it in a more concise way, does the function u obey $(u'\mathbf{f}) \circ (\mathrm{id}, u) = h \circ (\mathrm{id}, u)$? Now we can introduce a formal definition.

Definition 13.58 The map

$$\mathscr{I}_{\mathbf{f},h} \ni u \mapsto (u'\mathbf{f}) \circ (\mathrm{id}, u) = h \circ (\mathrm{id}, u) \in \{\text{True}, \text{False}\} \qquad (13.32)$$

is said to be a **first-order quasilinear partial differential equation**. The elements of the set (of functions) $\mathscr{I}_{\mathbf{f},h}$ for which the map has a **True** value are called **solution**s.

The solution of a quasilinear partial differential equation can be reduced to the solution of a system of ordinary differential equations.

Definition 13.59 The **characteristic differential equation** of the partial differential equation (13.32) is the system

$$\dot{\tilde{\mathbf{x}}} = \mathbf{f} \circ (\tilde{\mathbf{x}}, \tilde{u}) \quad \dot{\tilde{u}} = h \circ (\tilde{\mathbf{x}}, \tilde{u}). \qquad (13.33)$$

The trajectories of the characteristic differential equation are called **characteristic curves**.

Remark 13.60 The solution to (13.33) itself is a parametrization of the characteristic curve. These parametrizations of curves in $\mathbb{R}^{M+1}$ will be treated in a decomposed way, in the form of $(\mathbf{r}, \gamma)$. Thus, the range of the parametrization is a subset of $\mathbb{R}^M \times \mathbb{R}$. As we will see below, the solutions to (13.32) are constructed from characteristic curves.

Theorem 13.61 *The function $\varphi \in \mathscr{I}_{\mathbf{f},h}$ is a solution to (13.32) if and only if for all $\mathbf{x} \in \mathscr{D}_\varphi$, there exists a characteristic curve $(\mathbf{r}, \gamma)$ for which $\mathbf{r}(0) = \mathbf{x}$ and $\gamma = \varphi \circ \mathbf{r}$.*

Let us mention that we have spoken about existence and construction of the solutions but not a single word about their uniqueness.

13.8 Exercises and Problems

13.1 Prove Theorem 13.5.

(Solution: page 451)

13.2 Prove Theorem 13.6.

(Solution: page 452)

13.3 How would you realize the vector operations in *Mathematica*?

(Solution: page 452)

13.4 Prove that the Hadamard product is commutative, associative, and distributive over addition.

(Solution: page 452)

13.5 Prove Theorem 13.19.

(Solution: page 453)

13.6 Let $\mathbf{E}$ be the incidence matrix of the directed graph (V, E), and let $N := |V|$. Prove that $\mathbf{1}_N$ belongs to the left null space of $\mathbf{E}$.

(Solution: page 453)

13.7 Prove that the vertex sets of the connected components of a graph form a partition of the vertex set of the graph. Prove the same for the strong components of directed graphs.

(Solution: page 453)

13.8 Prove the statement of the Theorem 13.33.

(Solution: page 453)

13.9 Suppose a graph is given as G. How do you check using *Mathematica* that it is transitive? If it is not, how do you get its transitive closure?

(Solution: page 453)

13.10 Construct a directed graph for which the number T of ergodic components is larger than the number L of weak components.

(Solution: page 454)

13.11 Show that boundedness, closedness, and convexity are all relevant conditions of the Brouwer's theorem 13.35.

(Solution: page 454)

13.12 Let $K, M \in \mathbb{N}, 0 < K < M$, and let $\mathbf{u}_1, \mathbf{u}_2, \ldots, \mathbf{u}_K \in \mathbb{R}^M$ be independent vectors, and suppose $\mathbf{v} \in \mathbb{R}^M$ is such that $\mathbf{u}_1, \mathbf{u}_2, \ldots, \mathbf{u}_K, \mathbf{v}$ are dependent. Show that $\mathbf{v}$ can be expressed as a linear combination of $\mathbf{u}_1, \mathbf{u}_2, \ldots, \mathbf{u}_K$.

(Solution: page 454)

13.13 Prove Theorem 13.39 (the generalized Higgins lemma).

(Solution: page 455)

13.14 Calculate the sensitivity equation (13.23) around a stationary point $\boldsymbol{\xi}_*(\mathbf{p})$ of (13.22).

(Solution: page 455)

13.15 Solve the partial differential equation of the generating function of the simple inflow (or the Poisson process) $0 \xrightarrow{k_1} X$:

$$\partial_0 G(t, z) = k_1(z - 1)G(t, z) \quad G(0, z) = z^D.$$

(Solution: page 455)

13.16 Solve the partial differential equation of the generating function of the simple linear autocatalysis (or linear birth process) $X \xrightarrow{k_1} 2X$:

$$\partial_0 G(t, z) = k_1(z^2 - z)G(t, z) \quad G(0, z) = z^D.$$

(Solution: page 455)

References

Dickson LE (1913) Finiteness of the odd perfect and primitive abundant numbers with n distinct prime factors. Am J Math 35(4):1913. http://www.jstor.org/stable/2370405

Farkas M (1994) Periodic motions. Springer, New York

Guckenheimer J, Holmes P (1983) Nonlinear oscillations, dynamical systems, and bifurcations of vector fields. Applied mathematical sciences, vol 42. Springer, New York

Harary F (1969) Graph theory. Addison–Wesley, Reading

Horn F, Jackson R (1972) General mass action kinetics. Arch Ratl Mech Anal 47:81–116

Lovász L (2007) Combinatorial problems and exercises. AMS Chelsea Publishing, Providence

Øre O (1962) Theory of graphs, vol 38. AMS Colloquium Publications, Providence

Perko L (1996) Differential equations and dynamical systems. Springer, Berlin

Póta G (1981) On a theorem of overshoot-undershoot kinetics. React Kinet Catal Lett 17(1–2):35–39

Tikhonov AN (1952) Systems of differential equations containing a small parameter at the derivatives. Mat sb 31(73):575–585

Tóth J (1987) Bendixson-type theorems with applications. Z Angew Math Mech 67(1):31–35

Tóth J, Simon LP (2005/2009) Differential equations (Differenciálegyenletek). Typotex, Budapest

Tóth J, Li G, Rabitz H, Tomlin AS (1997) The effect of lumping and expanding on kinetic differential equations. SIAM J Appl Math 57:1531–1556

Solutions

14

14.1 Introduction

1.1 The references we provide may not be the best and the most recent ones, still they may give some help to the interested reader.

As to molecular dynamics, a starting point might be Kresse and Hafner (1993).

Eyring (2004) and Truhlar and Garrett (1980) are classical papers on the transition state theory. Szabó and Ostlund (1996) were published multiple times since the very first version from 1982.

Chapter VIII of the old classics Bard (1974) gives a good introduction into the statistical methods to evaluate kinetic experiments. Weise (2009) is a good modern reference, freely available on the net. A large part of the recent book by Turányi and Tomlin (2014) is also dedicated to this topic.

Epstein and Pojman (1998) and Scott (1991, 1993, 1994) are about exotic phenomena in reaction kinetics.

Interested in more details of Chemical Reaction Network Theory? Then the best starting point is the web site of Feinberg's group: http://www.crnt.osu.edu/home. Then, scholar.google.com might help find papers who cite (possibly develop) the papers by F. J. M. Horn, R. Jackson, and M. Feinberg.

Érdi and Lente (2016) and Van Kampen (2006) contain historical data on stochastic kinetics. Most people dealing with deterministic kinetics seem to be working separately; this situation only starts changing recently; we are not aware of a history of this topics.

© Springer Science+Business Media, LLC, part of Springer Nature 2018
J. Tóth et al., *Reaction Kinetics: Exercises, Programs and Theorems*,
https://doi.org/10.1007/978-1-4939-8643-9_14

14.2   Preparations

2.1 If one extends the concept to reactions having also negative stoichiometric coefficients, then the induced kinetic differential equation of Example 2.13 is

$$\dot{x} = -k_1 xy + k_2 ay - 2k_3 x^2 + k_4 ax - 0.5k_5 xz$$
$$\dot{y} = -k_1 xy - k_2 ay + k_6 mz$$
$$\dot{z} = +2k_4 ax - k_5 xz - 2k_6 mz$$

However, this is the same as the induced kinetic differential equation of the model

$$X + Y \xrightarrow{k_1} 2\,P \qquad Y + A \xrightarrow{k_2} X + P \qquad 2\,X \xrightarrow{k_3} P + A$$
$$X + A \xrightarrow{k_4} 2\,X + 2\,Z \qquad X + Z \xrightarrow{k_5} 0.5\,X + A \qquad Z + M \xrightarrow{2k_6} 0.5\,Y$$

only containing fractional stoichiometric coefficients but no negative ones. Hint: Use **DeterministicModel** of the package.

2.2 One possible way is to ask Google about

$$\texttt{Mathematica program reaction kinetics.}$$

One of the problems with this method is that the word Mathematica appears in some Latin journal names, e.g., *Acta Mathematica, Acta Mathematica Hungarica,* and *Helvetica Mathematica Acta.*

2.3 Please, do not stay there too long; you are allowed to return there from time to time!

Here we give solutions to the toy problems, and at the end, we shall show how the program works on a larger, real-life reaction.

2.4 According to the assumption, the reaction is given in the form

```
robertson = {A -> B, 2B -> B + C -> A + C},
```

the full form of which is

```
List[Rule[A, B],
    Rule[Times[2, B], Rule[Plus[B, C], Plus[A, C]]]].
```

Then, **Union[Level[robertson, {-1}]]** gives you the set of the species and those stoichiometric coefficients which are different from 1. To get rid of the second class, we might proceed this way.

```
species = Cases[Union[Level[robertson, {-1}]],
    Except[_Integer]]
```

The number of species, M is the cardinality of the species set:

```
M = Length[species].
```

Actually, this is beneath the command

```
ReactionsData[robertson]["M", "species"].
```

2.5 According to the assumption, the reaction is given in the form

```
revlv = {X <=> 2X, X + Y <=> 2Y, Y <=> 0},
```

the full form of which is

```
List[
    LeftRightArrow[X, Times[2, X]],
    LeftRightArrow[Plus[X, Y], Times[2, Y]],
    LeftRightArrow[Y,0]
    ].
```

Then,

```
revlv /.
    LeftRightArrow[a_, b_] -> {Rule[a, b], Rule[b, a]}
    // Flatten // Union
```

gives you the set of the reaction **steps** as

```
{0 -> Y, X -> 2X, 2X -> X, Y -> 0, 2Y -> X + Y, X + Y -> 2Y}.
```

The number of reaction steps is the cardinality of this set: **R=Length[steps]**.
Actually, this is beneath the command

```
ReactionsData[revlv]["R", "reactionsteps"].
```

Finally, let us mention that **ToReversible["Lotka-Volterra"]** gives the reversible version of the irreversible Lotka–Volterra reaction.

2.6 Let **bimol = {A + B -> C}**. Then,

```
{α} = Transpose[List[Coefficient[bimol[[1, 1]], #]&
    /@ {A, B, C}]]
{β} = Transpose[List[Coefficient[bimol[[1, 2]], #]&
    /@ {A, B, C}]].
```

Actually, this is beneath the command **ReactionsData[bimol][α, β]**.

2.7 Having seen the solution of previous problems, the present one is easy to solve.

```
ReactionsData[{E + S1 <=> ES1 <=> E*S1 -> E* + P1,
    E* + S2 <=> E*S2 <=> ES2 -> E + P2] [{γ}]
```

Then, the result is obtained by **MatrixRank[%]**.

14.3   Graphs of Reactions

3.1 The example in Fig. 14.1 shows that the requirement is not equivalent to weak reversibility.

3.2 The immediate solution with the program is to use

```
WeaklyReversibleQ[reaction],
```

and in this case, it is enough to pass **reaction** to the function; the user does not have the task of creating the Feinberg–Horn–Jackson graph. If one wants to use the built-in function **TransitiveClosureGraph**, then she/he has to write another function, say, **SymmetricGraphQ**, which may be defined as follows:

```
SymmetricGraphQ[g_] :=
    (a = AdjacencyGraph[g]; a == Transpose[a])
```

3.3 One possible solution is that one constructs the Feinberg–Horn–Jackson graph:
rg = ReactionsData[{"Robertson"}] ["fhjgraphedges"]. Next,

```
TableForm[IncidenceMatrix[rg],
    TableHeadings -> {VertexList[rg], EdgeList[rg]},
    TableAlignments -> Right]
```

does the job.

3.4

1. In the case of the Michaelis–Menten reaction, one has $N = 3, L = 1, S = 2$; thus $\delta = 3 - 1 - 2 = 0$.

Fig. 14.1 A Feinberg–Horn–Jackson graph with all the vertices on at least one directed cycle which, however, is not weakly reversible

2. In the case of the Lotka–Volterra reaction, one has $N = 6, L = 3, S = 2$; thus $\delta = 6 - 3 - 2 = 1$. Note that the same result is obtained if the reverse reaction step is added to some or all the reactions steps.
3. In the case of the Robertson reaction, one has $N = 5, L = 2, S = 2$; thus $\delta = 5 - 2 - 2 = 1$.

This is how one gets the above results using the line:

```
Map[ReactionsData[#]["deficiency"]&,
    {{"Michaelis-Menten"}, {"Robertson"}}]
```

In the case of the Lotka–Volterra reaction, one should not forget to specify the external species as follows:

```
ReactionsData[{"Lotka-Volterra"}, {"A", "B"}]
    ["deficiency"]
```

3.5 There are two edges between X and X $\longrightarrow$ 2 X via the complexes X and 2 X and also between Y and X + Y $\longrightarrow$ 2 Y via the complexes X + Y and 2 Y.

3.6

1. If the number of reaction steps R is 1, then one has $N = 2, L = 1, S = 1$; thus $\delta = 2 - 1 - 1 = 0$.
2. Let the reaction be a compartmental system, and consider the lth linkage class. Let the number of complex vectors in this linkage class be N_l, as they are independent, except possibly the zero vector, $S_l = N_l - 1$; therefore $\delta_l = N_l - 1 - S_l$. For the whole compartmental system, one has $N = \sum N_l, S = \sum S_l$; therefore $\sum N_l - L - \sum S_l = \sum (N_l - 1 - S_l) = 0$.
3. The argument above holds for reactions where the complex vectors (except possibly the zero vector) are independent, as in the case of generalized compartmental systems in the narrow sense or in the case when the complexes have no common species (Siegel and Chen 1995, Lemma 5). Example 3.14 shows that the statement does not hold for compartmental systems in the wide sense.

3.7

1. A one species reaction with positive deficiency is 0 $\longrightarrow$ X $\longrightarrow$ 2 X.
2. A first-order reaction with positive deficiency is X $\longrightarrow$ Z + X   Y $\longrightarrow$ Z + Y ($N = 4, L = 2, S = 1, \delta = 1$).
3. A first-order reaction which is not a compartmental system still having zero deficiency is X $\longrightarrow$ 2 X.

When other authors speak of first-order reactions, they actually mean compartmental systems, i.e., systems where all the complexes are not longer than one, not only the reactant ones. A relevant difference is shown by the first example.

In all the problems below, suppose a reaction is given in the form of a list of irreversible steps as **reaction**, and try to find an algorithm and a *Mathematica* code to solve the problems.

3.8

1. The set of complexes are

```
complexes[reaction_] := Union[Flatten[
    Map[ReplaceAll[h_[x___] -> List[x]], reaction]]],
```

and their number is **N = Length[%]**.

2. A nice form (although different from the one the program provides) of the Feinberg–Horn–Jackson graph of the Lotka–Volterra reaction is obtained by

```
GraphPlot[
    {"A" + "X" -> 2"X", "X" + "Y" -> 2"Y",
        "Y" -> "B"}, DirectedEdges -> True,
    VertexLabeling -> True].
```

3. We may assume that the list of species and that of reaction steps are known. A graph has to be constructed with (weighted) directed edges using the condition if a species is needed to or is produced in a given reaction step. This is a way how to draw edges corresponding to the stoichiometric coefficients in the reactant complexes.

```
Graph[DeleteCases[Flatten[Table[If[{α}[[m, r]] != 0,
    Labeled[DirectedEdge[X[m], r], {α}[[m, r]]]],
    {m, M}, {r, R}]], Null], VertexLabels -> "Name"]
```

Complete the code by adding the edges corresponding to the stoichiometric coefficients in the product complexes.

4. The number of connected components L of the Feinberg–Horn–Jackson graph may be calculated by

```
Length @ WeaklyConnectedComponents @ fhj,
```

if **fhj** is the Feinberg–Horn–Jackson graph of the reaction, e.g.,

```
Length[WeaklyConnectedComponents[Graph[
    {"A" + "X" -> 2"X", "X" + "Y"-> 2"Y",
    "B" -> "Y"}]]].
```

Alternatively, one can use our functions.

```
ReactionsData[
    {"A" + "X" -> 2"X", "X" + "Y" -> 2"Y",
    "B" -> "Y"}, {"A", "B"}]
    ["fhjweaklyconnectedcomponents"]
```

The results are different, because our function takes into consideration the external species.

5. To calculate the deficiency, we also need the rank of the matrix γ of the reaction step vectors, which is **MatrixRank[γ]**.

3.9 One possibility might be

```
glyoxalateshort =
    {oxaloacetate + "acetyl-CoA" <=> citrate
        <=> isocitrate <=>  succinate + glyoxylate,
    glyoxylate + "acetyl-CoA" <=> malate
        <=> oxaloacetate}
```

If one uses appropriate options such as

```
ShowVolpertGraph[glyoxalateshort,
    Numbered -> True, ImageSize -> 800
    GraphLayout -> "CircularEmbedding"],
```

then Fig. 14.2 is obtained, which is similar to textbook figures.

3.10 In this case, $\mathcal{M} = \{Cl_2, Cl^*, CH_4, {}^*CH_3, HCl, CH_3Cl\}$. Then, in the first step, species belonging to the initial set $\mathcal{M}_0 := \{Cl_2, CH_4\}$ will be assigned the index 0, and also reaction steps which can take place only using reactants from the set of the initial species, thus, $\mathcal{R}_0 := \{Cl_2 \longrightarrow 2\,Cl^*\}$ receives also index 0. Next, species being produced from reaction steps with index 0, i.e., species $\mathcal{M}_1 := \{Cl^*\}$ will receive index 1, and also reaction steps which can take place only using reactants

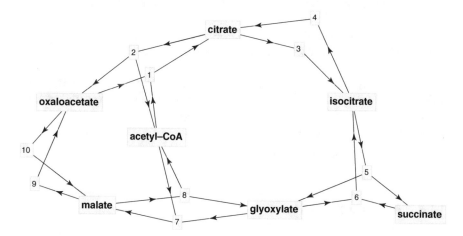

Fig. 14.2 A simplified version of the glyoxylate cycle as shown in textbooks on biochemistry

with indices 0 or 1 will receive index 1. This means that the step

$$\mathscr{R}_1 := \{CH_4 + Cl^* \longrightarrow {}^*CH_3 + HCl\}$$

receives index 1 as well. Next, species $\mathscr{M}_2 := \{{}^*CH_3, HCl\}$ receive index 2 as well. Finally, reaction step $\mathscr{R}_2 := \{{}^*CH_3 + Cl_2 \longrightarrow CH_3Cl + Cl^*\}$ receives index 2 as well, and species $\mathscr{M}_3 := \{CH_3Cl\}$ receives index 3. There is no species or reaction step with index $+\infty$. Should you use the program, you would perhaps produce a nice table by asking this:

```
VolpertIndexing[
    {Cl₂ -> 2Cl*, CH₄ + Cl* -> *CH₃ + HCl,
    *CH₃ + Cl₂ -> CH₃Cl + Cl*},
    Cl₂, CH₄}, Verbose -> True]
```

3.11 As above, again $\mathscr{M} = \{Cl_2, Cl^*, CH_4, {}^*CH_3, HCl, CH_3Cl\}$. Then, in the first step, species belonging to the '$\mathscr{M}_0 := \{Cl^*, CH_4\}$ will be assigned the zero index, and also reaction steps which can take place only using reactants from the set of the initial species, thus, $\mathscr{R}_0 := \{CH_4 + Cl^* \longrightarrow {}^*CH_3 + HCl\}$ receives also zero index. Next, species being produced from reaction steps with index zero, i.e., species $\mathscr{M}_1 := \{{}^*CH_3, HCl\}$, will receive index 1, and also reaction steps which can take place only using reactants with indices 0 or 1 would receive index 1. But there is no such reaction step; the process of indexing stops; all the remaining species and reaction steps receive the index $+\infty$: $\mathscr{M}_{+\infty} := \{Cl_2, CH_3Cl\}$ $\mathscr{R}_{+\infty} := \{Cl_2 \longrightarrow 2Cl^*, {}^*CH_3 + Cl_2 \longrightarrow CH_3Cl + Cl^*\}$.

3.12 A possible cycle is CO_2-L_3-$NaHCO_3$-L_2-CO_2, showing that it is possible that the deficiency of a reaction is zero; still there is a cycle in its S–C–L graph.

3.13 The S–C–L graph of the reaction $X \longrightarrow Y, 2Y \longrightarrow 2X$ consists of a single component, whereas the number of its linkage classes is 2.

3.14 As cyclicity of the Feinberg–Horn–Jackson graph implies the cyclicity of the Volpert graph one should look for a reaction with acyclic Feinberg–Horn–Jackson graph. It is easy to verify that the simple example $X \longrightarrow Y$ $X + Y \longrightarrow Z$ will do. If you need a support, use

```
AcyclicGraphQ[Graph[{X -> Y, X + Y -> Z}]].
```

14.4   Mass Conservation

4.1 The reaction $A \rightleftharpoons B$ is mass producing, but not strongly so. At the same time, it is mass consuming, but not strongly so.

4.2 A reaction is strongly mass producing and strongly mass consuming if there are two different vectors ρ_1 and ρ_2 satisfying (4.5) and (4.7), respectively. The reaction $A \longrightarrow B$ serves as a simple example that this is possible. This reaction is strongly mass producing with the vector $\rho_1 = (1, 2)^\top$ and strongly mass consuming with $\rho_2 = (2, 1)^\top$. More generally, Theorem 4.16 guarantees that if a reaction with an acyclic Volpert graph does not have the empty complex among its reactant or product complexes, then it is strongly mass producing and strongly mass consuming at the same time.

4.3 A simple implementation may be the following:

```
MQ[gamma_,m_,r_] := Quiet[First[Maximize[{lambda,
    Transpose[gamma].Array[rho[#]&, m]}]
        == ConstantArray[0, r],
    Thread[Array[rho[#]&, m]
        >= lambda ConstantArray[1, m]],
    Array[rho[#]&, m].ConstantArray[1, m]
        == 1},
    Join[Array[rho[#]&, m], {lambda}]]] > 0];
```

Both reactions are mass conserving.

The number of species is 9 (respectively, 5), $\text{rank}(\gamma)$ is 6 (respectively, 3), and the number of independent mass conservation relations is $9 - 6 = 3$ (respectively, $5 - 3 = 2$) as stated. They can be obtained using **MassConservationRelations**.

An alternative solution follows.

```
FindInstance[Join[Thread[Array[ρ[#]&, 9].ReactionsData[
    {A -> 2M + N2, A + M -> CH4 + B,
    2M -> C2H6, M + B -> MED,
    M + A -> C, M + C -> TMH}
    ][γ] == 0],
    Thread[Array[ρ[#]&, 9] > 0]],
    Array[ρ[#]&, 9], Integers],
```

and similarly

```
FindInstance[Join[Thread[Array[ρ[#]&, 5].ReactionsData[
    {A + B <=> C -> 2D <=> B + E}][γ] == 0],
    Thread[Array[ρ[#]&, 5] > 0]],
    Array[ρ[#]&, 5], Integers].
```

In both cases, one obtains a positive vector of masses.

4.4

1. Elementary row operations do not change the range space of a matrix; hence pivoting on $\gamma^\top$ does not change the consistency of the system $\gamma^\top \rho = 0$.

2. First, we can simplify the calculations by making the steps irreversible, using Theorem 4.18. Pivoting on the $\boldsymbol{\gamma}^\top$ of A $\longrightarrow$ B $\longrightarrow$ C $\longrightarrow$ A + 2 D, we get

$$\boldsymbol{\gamma}^\top = \begin{bmatrix} -1 & 1 & 0 & 0 \\ 0 & -1 & 1 & 0 \\ 1 & 0 & -1 & 2 \end{bmatrix} \sim \begin{bmatrix} 1 & -1 & 0 & 0 \\ 0 & -1 & 1 & 0 \\ 0 & 1 & -1 & 2 \end{bmatrix} \sim \begin{bmatrix} 1 & -1 & 0 & 0 \\ 0 & 1 & -1 & 0 \\ 0 & 0 & 0 & 2 \end{bmatrix},$$

and before pivoting on the third row, we find that it violates the condition.

3. Pivoting on the $\boldsymbol{\gamma}^\top$ of A $\longrightarrow$ B + D, C $\longrightarrow$ A + D, C $\longrightarrow$ D, we get

$$\boldsymbol{\gamma}^\top = \begin{bmatrix} -1 & 1 & 0 & 1 \\ 1 & 0 & -1 & 1 \\ 0 & 0 & -1 & 1 \end{bmatrix} \sim \begin{bmatrix} 1 & -1 & 0 & -1 \\ 0 & 1 & -1 & 2 \\ 0 & 0 & -1 & 1 \end{bmatrix} \sim \begin{bmatrix} 1 & -1 & 0 & -1 \\ 0 & 1 & -1 & 2 \\ 0 & 0 & 1 & -1 \end{bmatrix},$$

without encountering any violating rows. Note however that the *reduced* row echelon form of this matrix is

$$\boldsymbol{\gamma}^\top \sim \begin{bmatrix} 1 & 0 & 0 & 0 \\ 0 & 1 & 0 & 1 \\ 0 & 0 & 1 & -1 \end{bmatrix},$$

with two violating rows, which proves that the reaction is not mass conserving. Nevertheless, even this stronger necessary condition is insufficient, as the example

$$\boldsymbol{\gamma}^\top = \begin{bmatrix} 1 & 0 & 0 & -1 & 1 & 1 \\ 0 & 1 & 0 & 1 & -1 & 1 \\ 0 & 0 & 1 & 1 & 1 & -1 \end{bmatrix},$$

shows. The corresponding reaction passes the test because $\boldsymbol{\gamma}^\top$ is already in reduced row echelon form, but it is not mass conserving, since the rows of $\boldsymbol{\gamma}^\top$ sum to $\begin{bmatrix} 1 & 1 & 1 & 1 & 1 & 1 \end{bmatrix}^\top$.

4.5 The scalar inequality

$$\boldsymbol{\rho}^\top \boldsymbol{\gamma}(\cdot, r) = 0 \tag{14.1}$$

can have at most $M - 1$ linearly independent solution for all $r \in \mathscr{R}$; therefore for all $r \in \mathscr{R}$, there exists $\boldsymbol{\rho}_r \in \mathbb{R}^M$ such that $\boldsymbol{\rho}_r^\top \boldsymbol{\gamma}(\cdot, r) < 0$ holds. Let us define with arbitrary positive constants c_r the vector $\boldsymbol{\rho} := \sum_{r \in \mathscr{R}} c_r \boldsymbol{\rho}_r \in \mathbb{R}^M$. This vector does fulfill (4.18).

14.5   Decomposition of Reactions

5.1 Since in this case again $\mathbf{d} = \begin{bmatrix} 3 & 1 \end{bmatrix}^{\top}$, the product complexes are the same as those given in the Example 5.3. The only step without direct catalysis is $H + H_2O \longrightarrow OH + H_2$.

5.2 The columns of a matrix form a minimal linearly dependent set if and only if its null space is one dimensional, spanned by a vector with all nonzero components. A one-line implementation that avoids the explicit use of local variables may be the following (MLD stands for minimal linearly dependent):

```
MLD[M_] := Length[#]==1 && Min[Abs[#[[1]]]]>0 &
    [NullSpace[Transpose[M]]]
```

5.3 A simple implementation might be the following:

```
DependenceType[A_, set_] :=
    Switch[NullSpace[Transpose[A[[set]]]],
    { }, "I",
    _?(Length[#] >= 2 || Min[Abs[#]] == 0 &), "D",
    _, "S"]

MDSIter[{{set_, type : Alternatives["U", "S", "I"]},
    A_, m_, sol_}/; set === Range[First[set], m]]:=
    {{set, type}, A, m, sol}
MDSIter[{{set_, "D"}, A_, m_, sol_}
    /; set === Range[First[set], m]]:=
    {{Delete[set, -2], "U"}, A, m, sol}
MDSIter[{{{T___, t_, m_}, type_}, A_, m_, sol_}]:=
    {{{T, t+1}, "U"}, A, m, sol}
MDSIter[{{{T___, t_}, "I"}, A_, m_, sol_} /; t < m]:=
    {{{T, t, t+1}, "U"}, A, m, sol}
MDSIter[{{{T___, t_}, "D"}, A_, m_, sol_} /; t < m]:=
    {{{T, t+1}, "U"}, A, m, sol}
MDSIter[{{{T___, t_}, "S"}, A_, m_, sol_} /; t < m]:=
    {{{T, t+1}, "U"}, A, m, sol}
MDSIter[{{set_, "U"}, A_, m_, sol_}]:=
    (t = DependenceType[A, set]; {{set, t}, A, m,
    If[t === "S", Append[sol, set], sol]})

MinimalDependentSubsets[A_] := Last[FixedPoint[MDSIter,
    {{{1}, "U"}, A, Length[A] + 1, {}}]]
```

The algorithm maintains a data structure of the form

```
{{set, type}, A, m, sol},
```

where **A** is the matrix whose rows are the vectors of interest (their number is **m-1**), **sol** is a list containing all minimal dependent subsets found so far, and **set** is the index set of the next subset to be tested. The value of **type** is

- **"S"** (for "simplex") if **set** is known to be a minimal linearly dependent set,
- **"I"** if it is known to be independent,
- **"D"** if it is known to be linearly dependent but not minimal,
- is **"U"** for "unknown," if it has not been tested yet.

A linear algebraic test to decide a set's type is implemented in **DependenceType**. The function **MinimalDependentSubsets** starts out with the singleton **"1"** as the first set to be tested. Its type is **"I"** and naturally **sol={}**. The function iteratively changes **set** by adding or removing a vector whenever it is found to be independent or non-minimally dependent, until the last set is reached.

5.4 There are 213 minimally dependent subsets among the given species, as the result of

```
Length[MinimalDependentSubsets[Most /@ Transpose[
    ToAtomMatrix[szalkai][[2]]]]]
```

is 213, where **szalkai** is the list of the reaction steps given.

5.5 The Volpert indexing algorithm terminates in three iterations. The indices of the species are $0, 0, 1, 1, 0, 2$, the indices of the reaction steps are $0, 1, 1, 0, 1, 2$. Each species and reaction step has a finite index; therefore none of them can be eliminated. Note that the atomic structure of the species plays no role here. Here is how you get the result with the program.

```
VolpertIndexing[
    {"H" + "O2" -> "O" + "OH", "O" + "H2" -> "H" + "OH",
    "OH" + "H2" -> "H" + "H2O", 2 "H" -> "H2",
    "H" + "OH" -> "H2O", "H2O" -> "H" + "OH"},
    {"H", "H2", "O2"}]
```

5.6 A possible solution follows.

```
MyOmittable1[gamma_, b_] :=
Block[{m, r, zs, optval, optz},
    {m, r} = Dimensions[gamma];
    zs = Array[z, r];
    1 - FixedPoint[Function[χ,
    {optval, optz} = Minimize[{χ.zs,
    And @@ Thread[gamma.zs == b] &&
    And @@ Thread[zs >= 0] && χ.zs >= 1}, zs];
    χ (1 - Boole[Thread[(zs /. optz)==0]])],
    Array[1 &, r]]]
```

5.7 A covering decomposition set corresponding to the entire reaction, that is, $\mathscr{R}' = \mathscr{R}$, is a set of decompositions that contains every reaction step that may take part in a decomposition. Hence, the omittable reaction steps are those reaction steps that are **not** contained in any of the decompositions returned by the function `CoveringDecompositionSet`:

```
MyOmittable2[gamma_, b_] :=
    Complement[Range[Length[First[gamma]]],
    Union[Last /@ Position[CoveringDecompositionSet
    [b, gamma], _?Positive]]]
```

5.8 Using the notations of Sect. 5.2.2, if $\mathbf{z}$ is the vector corresponding to a decomposition, then $\sum_{i=1}^{R} z_i$ is the number of reaction steps. Furthermore, $\mathbf{z}$ satisfies (5.2). Hence, the optimal solution of the linear program $\min_{\mathbf{z} \in \mathbb{R}^R} \{\sum_{i=1}^{R} z_i \mid \boldsymbol{\gamma}\mathbf{z} = \mathbf{b}, \mathbf{z} \geq 0\}$ is a lower bound on the number of reaction steps in the decompositions of an overall reaction. This lower bound is exact if the optimal solution to this problem is an integer vector.

14.6   The Induced Kinetic Differential Equation

6.1 Let us start from Eq. (6.8). The factor $\mathbf{k} \odot \mathbf{c}^{\alpha}$ is the componentwise product of the vectors $\mathbf{k}$ and $\mathbf{c}^{\alpha}$. The power $\mathbf{c}^{\alpha}$ is a vector of the rth component which is $\prod_{p=1}^{M} c_p(t)^{\alpha(p,r)}$; thus the rth component of the Hadamard–Schur product of these two factors is $k_r \prod_{p=1}^{M} c_p(t)^{\alpha(p,r)}$. If the obtained Hadamard–Schur product is multiplied by the matrix $\boldsymbol{\gamma}$, one obviously gets the same expression as in Eq. (6.7).

6.2 The additivity assumption means that the following equations should hold

$$w(x_1 + x_2, y, z) = w(x_1, y, z) + w(x_2, y, z) \tag{14.2}$$

$$w(x, y_1 + y_2, z) = w(x, y_1, z) + w(x, y_2, z) \tag{14.3}$$

$$w(x, y, z_1 + z_2) = w(x, y, z_1) + w(x, y, z_2). \tag{14.4}$$

These equations imply that

$$w(x, y, z) = k(y, z)x = l(z, x)y = m(x, y)z.$$

Using (14.3), one can show that $k(y, z) = \kappa(z)y$. Using (14.4), one can show that $\kappa(z) = \lambda z$; thus $w(x, y, z) = \lambda xyz$ with $\lambda > 0$, if one should like to avoid the three-variable zero function.

6.3 In the case of mass action kinetics, the reaction rate of the reaction step r in Eq. (2.1) at the concentration vector $\mathbf{c}$ of the species being $k_r \mathbf{c}^{\alpha(\cdot,r)}$ is positive if all the components c_m of the vector $\mathbf{c}$ corresponding to the indices for which $\alpha(m, r) > 0$ is positive. If, however, any of the components c_m of the vector $\mathbf{c}$ corresponding to the indices for which $\alpha(m, r) > 0$, is zero, then the product is zero.

6.4 We give a general solution but illustrate it on the example of the Lotka–Volterra reaction. One needs the complex vectors and the reaction steps.

```
{Y, r} = ReactionsData[{"Lotka-Volterra"},
    {"A", "B"}]["complexes", "fhjgraphedges"]
```

Next, compare the columns of the complex matrix with the reactant and complex vectors, and form a nice table.

```
TableForm[Boole[Outer[SameQ, Y, Last /@ r]]
    -Boole[Outer[SameQ, Y, First /@ r]],
    TableHeadings -> {Y, r},
    TableAlignments -> {Right, Top}]
```

Using finally **TeXForm[%]** can be used to give a table, now without table headings.

$$
\begin{array}{ccc}
-1 & 0 & 0 \\
1 & 0 & 0 \\
0 & -1 & 0 \\
0 & 1 & 0 \\
0 & 0 & -1 \\
0 & 0 & 0
\end{array}
$$

6.5 First of all:

$$\mathbf{Y} \in \mathbb{N}^{M \times N}, \quad \mathbf{E} \in \{-1, 0, 1\}^{N \times R}, \quad \boldsymbol{\gamma} \in \mathbb{Z}^{M \times R}.$$

Also, $(\mathbf{YE})_{mr} = \sum_{n=1}^{N} Y_{mn} E_{nr}$, and this sum is just the stoichiometric coefficient $\gamma(m, r)$.

6.6 One has $N = 3$ complexes in this reaction: $\mathscr{C} = \{E + S, C, E + P\}$ which can be obtained in the following way.

```
Y = ReactionsData[{"Michaelis-Menten"}]["complexes"]
```

The reaction steps are as follows.

```
r = ReactionsData[{"Michaelis-Menten"}]["fhjgraphedges"]
```

Now let us ask if a complex is the reactant or product complex of a reaction step.

```
gg[a_, b_] := Which[a===b[[1]], -1, a===b[[2]], 1, True, 0]
```

```
TableForm[Outer[gg, Y, r],
    TableHeadings -> {Y, r},
    TableAlignments -> {Right, Top}]
```

And finally, the pure table (without headings) is $\begin{bmatrix} -1 & 1 & 0 \\ 1 & -1 & -1 \\ 0 & 0 & 1 \end{bmatrix}$.

6.7 The reaction steps of the Mole reaction are

```
mole = GetReaction["Mole"]
```

Let us collect the complex vectors, and calculate the "pure monomials"

```
Times @@@ (Power[{x, y}, #]& /@
    Union @ Flatten[Transpose /@ ReactionsData[mole]
    [{α}, {β}], 1])
```

The result will be $\{1, y, x, xy, x^2y^2\}$, as expected.

6.8 Starting from the definition of the complex formation vector, one gets

$$
\begin{aligned}
\mathbf{Yg(c)} &= \mathbf{Y}(\mathbf{R(c)} - \mathbf{R}^\top(\mathbf{c}))\mathbf{1}_N \\
&= \sum_{n \in \mathcal{N}} \mathbf{y}_n \sum_{q \in \mathcal{N}} (r_{nq}(\mathbf{c}) - r_{qn}(\mathbf{c})) \\
&= \sum_{n \in \mathcal{N}} \sum_{q \in \mathcal{N}} r_{nq}(\mathbf{c})\mathbf{y}_n - \sum_{n \in \mathcal{N}} \sum_{q \in \mathcal{N}} \mathbf{y}_n r_{qn}(\mathbf{c}) \\
&= \sum_{n \in \mathcal{N}} \sum_{q \in \mathcal{N}} r_{nq}(\mathbf{c})\mathbf{y}_n - \sum_{n \in \mathcal{N}} \sum_{q \in \mathcal{N}} \mathbf{y}_q r_{nq}(\mathbf{c}) \\
&= \sum_{n \in \mathcal{N}} \sum_{q \in \mathcal{N}} r_{nq}(\mathbf{c})(\mathbf{y}_n - \mathbf{y}_q)
\end{aligned}
$$

as stated. The equivalence of Eqs. (6.22) and (6.24) can be seen similarly.

6.9 In the general case, we can use the final formula (6.32):

```
{γ}.DiagonalMatrix[Times @@ c^{α}]
```

where α and γ can be calculated using **ReactionsData**. Multiply this matrix with the vector of reaction rate coefficients to get the right-hand side of the induced kinetic differential equation of any reaction.

6.10 As the induced kinetic differential equation of the reaction in Fig. 6.5 is

$$\dot{x} = x(-a - bx + cy) \quad \dot{y} = y(-e - cx - fy)$$

one can easily find the quantities in the definition of the generalized Lotka–Volterra system.

6.11 An example which leads to a Kolmogorov system which is not of the generalized Lotka–Volterra form is given by the induced kinetic differential equation of the reaction $2\,X + Y \xrightarrow{1} X + 2\,Y$ which is $\dot{x} = -x^2 y = x(-xy) \quad \dot{y} = x^2 y = y(xy)$, a Kolmogorov system, and not a generalized Lotka–Volterra system. Such a second-order reaction does not exists, why?

6.12 The solution to the induced kinetic differential equation $\dot{a} = -a, \dot{b} = a - b, \dot{c} = b$ of the reaction $A \xrightarrow{1} B \xrightarrow{1} C$ with the initial condition $a(0) = a_0, b(0) = b_0, c(0) = c_0$ is

$$a(t) = a_0 e^{-t},$$
$$b(t) = e^{-t}(a_0 t + b_0),$$
$$c(t) = a_0 + b_0 + c_0 - e^{-t}(a_0 t + a_0 + b_0) \quad (t \in \mathbb{R}^+).$$

This can be obtained using

```
ReplaceAll @@ Concentrations [{"A" -> "B" -> "C"}, {1, 1},
    {a0, b0, c0}, {a, b, c}, t].
```

If $a_1 > a_0, b_1 > b_0, c_1 > c_0$, then one has

$$0 < (a_1 - a_0)e^{-t},$$
$$0 < e^{-t}((a_1 - a_0)t + b_1 - b_0),$$
$$0 < (a_1 - a_0)(1 - e^{-t}(1 + t)) + (b_1 - b_0)(1 - e^{-t}) + c_1 - c_0 \quad (t \in \mathbb{R}^+),$$

which shows that the given induced kinetic differential equation is a monotone system.

6.13 The trajectories of the Lotka–Volterra reaction go along a closed curve determined by the initial condition. Taking another initial condition which is larger componentwise, one gets trajectories going along another closed curve which contains the previous one; therefore the differences of the coordinate functions of the solutions will not be of the constant sign: Actually they are periodic functions taking on both negative and positive values; thus the Lotka–Volterra reaction is not a monotone system; see Figs. 14.3 and 14.4. Is the reaction monotone with respect to any other cone?

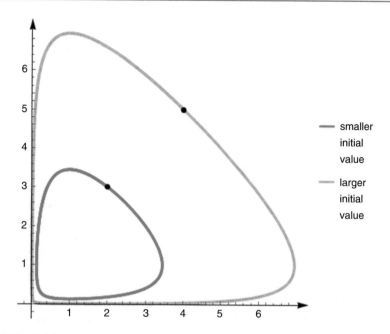

Fig. 14.3 Trajectories of the irreversible Lotka–Volterra model with $k_1 = 1, k_2 = 1, k_3 = 1$ and with the initial conditions $x(0) = 2, y(0) = 3$ and $x(0) = 4, y(0) = 5$, respectively

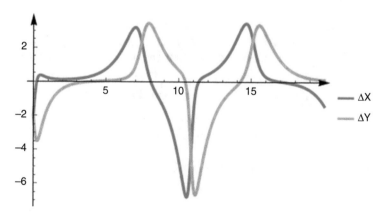

Fig. 14.4 Concentration differences in the irreversible Lotka–Volterra model with $k_1 = 1, k_2 = 1, k_3 = 1$ when started from the initial conditions $x(0) = 2, y(0) = 3$ and $x(0) = 4, y(0) = 5$, respectively

6.14 It follows from the hypothesis that $f(x, y) = \sum_{n=0}^{\infty} a_n(y)x^n$, where, for each y, $a_n(y) = 0$ for all but finitely many n. Since $\mathbb{R}$ is not a countable union of finite sets, there exists an integer N such that the set $\mathbb{F} := \{y \mid a_n(y) = 0 \quad \text{for all} \quad n > N\}$

is infinite. Denoting by φ_1 the restriction of f to $\mathbb{R} \times \mathbb{F}$, we have

$$\varphi_1(x, y) = \sum_{n=0}^{N} a_n(y)x^n, \qquad (x, y) \in \mathbb{R} \times \mathbb{F}. \tag{14.5}$$

Choosing $N + 1$ distinct values $x_0, x_1, \dots, x_N$, substituting them for x in (14.5), and solving the resulting system of equations, we obtain

$$a_n(y) = \sum_{j=0}^{N} c_{jn}\varphi_1(x_j, y) \qquad (y \in \mathbb{F}, n = 0, \dots, N), \tag{14.6}$$

where the c_{jn} are real constants. Thus, the function g, defined on $\mathbb{R} \times \mathbb{R}$ by

$$g(x, y) := \sum_{n=0}^{N} \sum_{j=0}^{N} c_{jn}\varphi_1(x_j, y)x^n, \tag{14.7}$$

is a polynomial. Moreover, (14.5) and (14.6) show that for each $x' \in \mathbb{R}$, the polynomial $f(x', y) - g(x', y)$ has a zero at each point of $\mathbb{F}$ and hence is equal to zero for all y. For essentially the same proof of the general case, see Carroll (1961).

6.15 The induced kinetic differential equation of the autocatalytic reaction $2\,\mathrm{X} \xrightarrow{1} 3\,\mathrm{X}$ is $\dot{x} = x^2$. Let us look for its solution in the form of a Taylor series: $x(t) = \sum_{q=0}^{+\infty} c_q \frac{t^q}{q!}$. Upon substitution this form into the induced kinetic differential equation one gets for the coefficients $c_q = q c_0^{q+1}$ $(q = 1, 2, \dots)$. Taking into the consideration the initial condition $x(0) = x_0$, one obtains

$$\begin{aligned} x(t) &= c_0 + c_1 t + c_2 \frac{t^2}{2!} + c_3 \frac{t^3}{3!} + \dots \\ &= c_0(1 + c_0 t + (c_0 t)^2 + (c_0 t)^3 + \dots) \\ &= \frac{c_0}{1 - c_0 t} = \frac{x_0}{1 - x_0 t}. \end{aligned}$$

The series is convergent for $t < x_0$. This is the same result what we would obtain by the usual method of integration or by using **Concentrations**. We emphasize that although the method works in full generality, the domain of convergence should be determined individually.

   We show a solution by the program (*Mathematica*, not **ReactionKinetics**) because the idea can be used in more complicated (nonpolynomial) cases too. First, let $\mathtt{x[t\_]}$ $\mathtt{:=}$ $\mathtt{Sum[c_k\ t^k/k!,\ \{k,\ 0,\ 5\}]}$ + $\mathtt{O[t]}^6$. Then the coefficients can be obtained this way.

```
First @ Solve @ LogicalExpand[x'[t]  ==  x[t]^2].
```

6.16 Let z be defined by $\dot{z} = 0$ $z(0) = 1$, and let us introduce the following variables:

$$\omega_{0,0} := z \quad \omega_{0,1} := y \quad \omega_{1,0} := x \quad \omega_{1,1} := xy \quad \omega_{2,0} := x^2.$$

Then we have the following homogeneous quadratic equations in the new variables:

$$\dot{\omega}_{0,0} = 0$$
$$\dot{\omega}_{0,1} = B\omega_{1,0}\omega_{0,0} - \omega_{2,0}\omega_{0,1}$$
$$\dot{\omega}_{1,0} = A\omega_{0,0}^2 + \omega_{1,0}\omega_{1,1} - (B+1)\omega_{0,0}\omega_{1,0}$$
$$\dot{\omega}_{1,1} = A\omega_{0,0}\omega_{0,1} + \omega_{1,1}^2 - (B+1)\omega_{0,0}\omega_{1,1}$$
$$\dot{\omega}_{2,0} = 2A\omega_{0,0}\omega_{1,0} + 2\omega_{1,1}\omega_{2,0} - 2(B+1)\omega_{0,0}\omega_{2,0}.$$

6.17 Let $M \in \mathbb{N}$, $P \in \mathbb{N}_0$ and suppose the polynomial differential equation

$$\dot{x}_m = \sum_{|\alpha| \leq P} a_{m,\alpha} \mathbf{x}^\alpha \quad (m \in \mathcal{M})$$

(where $\alpha \in \mathbb{N}_0^M$, $|\alpha| := \sum_{m \in \mathcal{M}} \alpha_m$) contains no negative cross effect, which means that

$$\alpha = \begin{bmatrix} \alpha_1 & \alpha_2 & \cdots & \alpha_{m-1} & 0 & \alpha_{m+1} & \cdots & \alpha_M \end{bmatrix}^\top \text{ implies } a_{m,\alpha} \geq 0. \tag{14.8}$$

Let us introduce the following new variables

$$\omega_\beta := \mathbf{x}^\beta \quad (\beta \in \mathbb{N}_0^M, |\beta| \leq P - 1).$$

Then,

$$\dot{\omega}_\beta = \sum_{m \in \mathcal{M}} \beta_m \mathbf{x}^{\beta - \mathbf{e}_m} \sum_{|\alpha| \leq P} a_{m,\alpha} \mathbf{x}^\alpha,$$

where $\mathbf{e}_m$ is the mth element of the standard basis.

If $\alpha_m = 0$, then $a_{m,\alpha} \geq 0$ because of (14.8), and then the mth term can be replaced by the quadratic terms $\beta_m \sum_{|\alpha| \leq P} a_{m,\alpha} \omega_\alpha \omega_{\beta - \mathbf{e}_m}$. As all the coefficients here are nonnegative, the presence of negative cross effect is excluded. (Terms where $\beta_m = 0$ are missing, therefore the problem that $\omega_{\beta - \mathbf{e}_m}$ is undefined causes no problem.)

If $\alpha_m > 0$, then one can have the quadratic terms $\beta_m \sum_{|\alpha| \leq P} a_{m,\alpha} \omega_{\alpha - \mathbf{e}_m} \omega_\beta$, and as these terms form part of the right-hand side of $\dot{\omega}_\beta$, and ω_β is present in each term, the presence of negative cross effect is excluded again.

Fig. 14.5 The flux of OH in
the reaction (4.13)

6.18 You may not know the name of the function to be used, find it this way:
`?*Atom*`. Then the solution is as follows.

`AtomConservingQ[{"HCOOH"+"CH3OH" -> "HCOOCH3"+"H2O"}],`

giving the answer `True`.

6.19 The reaction (4.13) is simple enough to set up the (generalized) atomic matrix
of the species as follows:

$$
\mathbf{Z} = \begin{matrix}
 & \text{HCOOH} & \text{CH}_3\text{OH} & \text{HCOOCH}_3 & \text{H}_2\text{O} \\
\text{H} & 1 & 0 & 1 & 1 \\
\text{O} & 0 & 0 & 1 & 0 \\
\text{CO} & 1 & 0 & 1 & 0 \\
\text{OH} & 1 & 1 & 0 & 1 \\
\text{CH}_3 & 0 & 1 & 1 & 0
\end{matrix}. \tag{14.9}
$$

The flux of the hydroxyl can be seen on the very simple Fig. 14.5.

6.20 As all the steps are reversible, we have two choices: We can either construct
the flux graph with all the fluxes, or we can only draw a graph with net fluxes.
Start with the first one. A straightforward calculation gives the flux graphs for the
hydrogen and the oxygen atoms, respectively. Now instead of the two-way arrow
connecting H and OH, we draw a single arrow pointing from H to OH if $W :=
(w_1(\mathbf{c}(t)) - w_{-1}(\mathbf{c}(t)))/2 + (w_{-3}(\mathbf{c}(t)) - w_3(\mathbf{c}(t)))/6$ is positive with the weight
W, and an arrow in the opposite direction if the difference is negative, then the
weight will be $-W$. Thus, it will be a dynamic figure, changing with time, best
realized using `Manipulate` (Fig. 14.6).

6.21 The three complexes (assuming mass action kinetics) can only be $2\,\text{X}$, $\text{X}+\text{Y}$
and $2\,\text{Y}$. Construct the induced kinetic differential equation of the reversible triangle
reaction with these complexes as vertices, and compare the coefficients with those
of

$$
x' = ay^2 - bxy, \quad y' = bx^2 - axy \tag{14.10}
$$

an immediate contradiction arises with the fact that $a, b, c > 0$ should hold.

6.22 The usual model to describe this process is

$$
\dot{c}(t) = k_0 e^{-\frac{A}{RT(t)}} c(t) \quad \dot{T}(t) = -k_0 e^{-\frac{A}{RT}} c(t) Q
$$

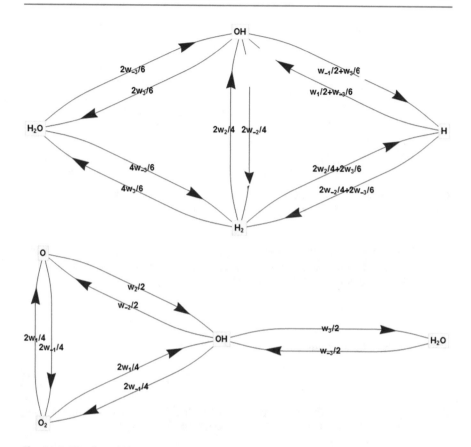

Fig. 14.6 The flux of H and O atoms in the reaction (6.60)

with $Q < 0$. This system of equations can be solved using **ParametricNDSolve**:

```
ndstemp = ParametricNDSolve[
    {c'[t] == E^{-1000/T[t]}c[t],
    T'[t] == -Q E^{-1000/T[t]}c[t],
    c[0] == 1, T[0] == 300}, {c, T}, {t, 0, 1000},
        {Q}]
```

This is how, e.g., the concentration vs. time function can be plotted.

```
temperatureautoc = Plot[Evaluate[c[-100][t]/.ndstemp],
    {t, 0, 50}, PlotStyle -> Directive[{Thick, Red}],
    PlotRange -> All, AxesLabel ->
    {Style["t", Italic, Bold, 16],
    Style["c(t)", Italic, Bold, 16]},
    PlotLegends -> Placed[{"Q = -100}, Above]]
```

And the result can be seen in Fig. 14.7.

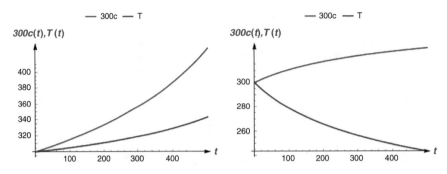

Fig. 14.7 The change of concentration and temperature in a reaction described by (6.57)–(6.58) with $k_0 = R = 1$, $A = 1000$, $a = 0$, $T(0) = 300$, $c(0) = 1$, and $Q = -100$ and $Q = 600$

6.23 The equation for the temperature shows that its derivative is never zero; therefore the function T is strictly monotonous; thus it is invertible. Let us define the function C by the formula $C := c \circ T^{-1}$; then one has $C' \circ T \dot{T} = \dot{c}$, or

$$C'(T) = -b.$$

This gives the simple relation

$$C(T) = -b(T - T_0) + c_0, \tag{14.11}$$

if the initial conditions are $c(0) = c_0$ and $T(0) = T_0$. To get an explicit form for the function T, let us substitute (14.11) into the equation for T to get

$$\dot{T} = e^{-\frac{a}{T}} \left(-b(T - T_0) + c_0 \right)^d.$$

This equation has the solution for $d = 1$

$$T(t) = \varphi^{-1} \left(-t - \frac{\mathrm{Ei}(\frac{a}{T_0}) - e^{\frac{ab}{bT_0+c_0}} \mathrm{Ei}\left(a\left(\frac{1}{T_0} - \frac{b}{bT_0+c_0}\right)\right)}{b} \right)$$

with

$$\varphi(\vartheta) := \frac{-\mathrm{Ei}(\frac{a}{\vartheta}) + e^{\frac{ab}{bT_0+c_0}} \mathrm{Ei}\left(a\left(\frac{1}{\vartheta} - \frac{b}{bT_0+c_0}\right)\right)}{b},$$

where $\mathrm{Ei}(z) := \int_{-z}^{+\infty} \frac{e^{-t}}{t}\, dt$. Let us remark that the function $T \mapsto C(T)$ can be calculated even in the generalized Arrhenius case for decompositions of any order; however, we do not see how the individual functions c and T can symbolically be determined in these more complicated cases. There is no hindrance to calculate anything numerically.

14.7 Stationary Points

7.1 The result of

```
gb = GroebnerBasis[
    {x(1-2x+3y+z), y(1+3x-2y+z), {z(2+y-z)}, {x, y, z}]
```

being

$$\{2z^4 - 15z^3 + 27z^2 - 10z, \ yz - z^2 + 2z,$$

$$2y^3 + y^2 - y - 4z^2 + 8z, \ 3xz^2 - 6xz - z^3 + 7z^2 - 10z,$$

$$3xy - 2y^2 + y + z^2 - 2z, \ 2x^2 - xz - x - 2y^2 + y + z^2 - 2z\},$$

one can solve $2z^4 - 15z^3 + 27z^2 - 10z = 0$ for z to get $z_1 = 0, z_2 = 1/2, z_3 = 2, z_4 = 5$; then substitute the solutions into $yz - z^2 + 2z = 0$ to find the values of $y_1 = -2, y_2 = -3/2, y_3 = 0, y_4 = 0$; and finally one gets $x_1 = -3/2, x_2 = -1, x_3 = 0, x_4 = 1/2, x_5 = 3/2$; thus the solutions are

x_*	y_*	z_*
$-\frac{3}{2}$	$-\frac{3}{2}$	$\frac{1}{2}$
-1	-1	0
0	0	2
0	$\frac{1}{2}$	0
0	3	5
$\frac{1}{2}$	0	0
$\frac{3}{2}$	0	2
0	0	0

7.2 . First, a mass-consuming reaction

$$\sum_{m \in \mathcal{M}} \alpha(m, r) X(m) \rightarrow \sum_{m \in \mathcal{M}} \beta(m, r) X(m) \quad (r \in \mathcal{R})$$

can be transformed into a stoichiometrically mass conserving one by adding a single dummy species, say Y, in the following way. Find a vector ρ showing that the reaction is mass consuming. This means that for some reaction steps, one has $\sum_{m \in \mathcal{M}} \alpha(m, r) \rho(m) > \sum_{m \in \mathcal{M}} \beta(m, r) \rho(m)$. Let us add as many molecules of Y to the right sides of these reaction steps as missing from the equality, i.e., let the stoichiometric coefficient of the dummy species on the right-hand side of the reaction step be $\sum_{m \in \mathcal{M}} \alpha(m, r)\rho(m) - \sum_{m \in \mathcal{M}} \beta(m, r)\rho(m)$. Then, adding 1 as the last component of the mass vector ρ will show that the amended reaction will be stoichiometrically mass conserving. Find its nonnegative stationary point and discard the component of this stationary point corresponding to the species Y, and you obtain a nonnegative stationary point of the original system. (The newly defined stoichiometric coefficient may be assumed to be an integer, why?)

7.3

- **The smart solution**
  As the induced kinetic differential equation of the Horn–Jackson reaction is

$$\dot{x} = -\dot{y} = (y - x)(2\varepsilon x^2 + (2\varepsilon - 1)xy + 2\varepsilon y^2),$$

  the stationary points are for which either $x_* = y_*$ or $(2\varepsilon(x_*)^2 + (2\varepsilon - 1)x_* y_* + 2\varepsilon(y_*)^2) = 0$ holds. The investigation of this quadratic polynomial shows that it has two additional positive real roots for $0 < \varepsilon < \frac{1}{6}$; thus there are three stationary points in each positive reaction simplex; see Fig. 7.3.
- **The routine solution**
  The less smart solution can be based upon the fact that the cubic equation $y^3 + py + q = 0$ has three real root if and only if $\frac{p^3}{27} + \frac{q^2}{4} < 0$. Use the fact that $\dot{x} + \dot{y} = 0$ to reduce the equation for the stationary solution vector to a single cubic equation, transform it into one without quadratic term, and then apply the mentioned criteria for the existence of three real roots.

A final remark: Check it by linear stability analysis that the unique stationary state (if we have only a single one) is relatively stable and if we have three stationary states than the middle one is asymptotically stable and the other two are unstable.

7.4 This command will give those stationary points under different conditions which fulfill linear first integrals:

```
Rest @ StationaryPoints[{X + Y -> U, Y + Z -> V},
    {k1, k2}, {x0, y0, 0, z0, 0}, {x, y, u, z, v}]
```

7.5 Equality of the creation and annihilation rates of the complexes at the positive stationary concentration vector $\begin{bmatrix} x_* & y_* \end{bmatrix}^\top$ is equivalent to writing $y_* = \frac{k_1}{k_{-1}} x_* = \sqrt{\frac{k_2}{k_{-2}}} x_*$; thus the necessary condition of complex balancing is

$$k_1^2 k_{-2} = k_{-1}^2 k_2. \tag{14.12}$$

We show that it is sufficient as well. Suppose that (14.12) holds, then the equation for the stationary point is as follows:

$$0 = -k_1 x_* + k_{-1} y_* - 2k_2 x_*^2 + 2\frac{k_2 k_{-1}^2}{k_1^2} y_*^2$$

$$= (-k_1 x_* + k_{-1} y_*) + 2\frac{k_2}{k_1^2}(-k_1^2 x_*^2 + k_{-1}^2 y_*^2)$$

$$= (-k_1 x_* + k_{-1} y_*)\left(1 + 2\frac{k_2}{k_1^2}(k_1 x_* + k_{-1} y_*)\right),$$

and this can only hold if the first factor is zero, proving the first equality of complex balancing. But this, together with stationarity, implies that the second equality also holds.

7.6 The reaction is of the form

$$\sum_{m \in \mathcal{M}} \alpha(m, r) X_m \underset{k_{-r}}{\overset{k_r}{\rightleftharpoons}} \sum_{m \in \mathcal{M}} \beta(m, r) X_m \quad (r \in \mathcal{M})$$

with the induced kinetic differential equation: $\dot{c} = \gamma(k \odot c^\alpha - k_- \odot c^\beta)$. The stationary point(s) should fulfill $\gamma(k \odot c_*^\alpha - k_- \odot c_*^\beta) = 0 \in \mathbb{R}^M$ or, equivalently, $k \odot c_*^\alpha = k_- \odot c_*^\beta$. This last equation can be reformulated as $\gamma^\top \cdot \ln c_* = \ln \kappa$, having a single solution which can be transformed back to show the existence of a single positive stationary point (Erle 2000). Along this proof some nonnegative solutions may have been lost. Note that in the case of $M > P$ supposing that γ is of the full rank, one obtains in a similar way the existence of positive stationary states which however form an $M - P$ parameter variety.

7.7 First of all, the given reaction rates fulfill Conditions 1 on page 6. Next, the induced kinetic differential equation of the reaction is

$$\dot{x} = -x^2 + y - 2x + 2y^2, \quad \dot{y} = x^2 - y + 2x - 2y^2,$$

and the complex formation vector is $g(x, y) = \left[-x^2 + y \; x^2 - y \; -x + y^2 \; x - y^2\right]^\top$. That is, $\left[1\; 1\right]^\top \in E \cap C$ and, e.g., $\left[2\,(\sqrt{65} - 1)/4\right]^\top \in E \setminus C$. (More generally, all the points of the form $\left[x_* \; \frac{1}{4}\left(\sqrt{8x_*^2 + 16x_* + 1} - 1\right)\right]$ with $0 < x_* \neq 1$ belong to $E \setminus C$.)

7.8 Let us calculate the Lie derivative of the function $\left[x \; y \; z \; u \; v\right] \mapsto \frac{x^{k_2}}{z^{k_1}}$ with respect to the induced kinetic differential equation

$$\dot{x} = -k_1 xy \quad \dot{y} = -k_1 xy - k_2 yz \quad \dot{z} = -k_2 yz \quad \dot{u} = k_1 xy \quad \dot{v} = k_2 yz$$

(i.e., the scalar product of the function and the right-hand side) to get zero in the open first orthant. Independence follows from the fact that the matrix

$$\begin{bmatrix} 1 & 0 & 0 & 1 & 0 \\ 0 & 1 & 0 & 1 & 1 \\ 0 & 0 & 1 & 0 & 1 \\ \frac{k_2 x^{k_2-1}}{z^{k_1}} & 0 & \frac{k_1 x^{k_2}}{z^{k_1+1}} & 0 & 0 \end{bmatrix}$$

is of the full rank.

7.9 Although the positive reaction simplexes are unbounded, the simple example $X \rightleftharpoons 0$ shows that the presence of the zero complex does not prohibit the existence of a (what is more, unique) positive stationary point. The example also shows that even in the case when for all $\rho > 0$ one has $\rho^\top y \not\leq 0$, it is still possible to have a positive stationary point.

7.10 If the vector $\begin{bmatrix} a_* & b_* & c_* \end{bmatrix}$ is a positive stationary point, then $-k_1 a_* b_* + k_{-1} c_* = 0$ meaning that "all" the reaction pairs (the single one) proceed with the same rate in both reactions; thus the reaction is detailed balanced. Furthermore, the deterministic model of the reaction (7.23) is

$$\dot{a} = -k_1 ab + k_{-1}c \quad \dot{b} = -k_1 ab + k_{-1}c \quad \dot{c} = k_1 ab - k_{-1}c$$
$$a(0) = a_0 \qquad\qquad b(0) = b_0 \qquad\qquad c(0) = c_0$$

which simplifies to

$$\begin{aligned} \dot{a}(t) &= -k_1 a(t)(a(t) - a_0 + b_0) + k_{-1}(-a(t) + a_0 + c_0) \\ &= -k_1 a(t)^2 + (k_1 a_0 - k_1 b_0 - k_{-1})a(t) + k_{-1}(a_0 + c_0) \\ &= -k_{-1}\left(Ka(t)^2 - (K(a_0 - b_0) - 1)a(t) - a_0 - c_0 \right) \end{aligned} \qquad (14.13)$$

with $K := \frac{k_1}{k_{-1}}$.

If the reaction starts from nonnegative initial concentrations a_0, b_0, c_0 for which either $a_0, b_0 > 0$, or $c_0 > 0$, the unique positive (relatively **globally** asymptotically stable) stationary concentration

$$a_* = \frac{1}{2K}(-1 + K(a_0 - b_0) + r)$$

$$b_* = \frac{1}{2K}(-1 + K(b_0 - a_0) + r))$$

$$c_* = \frac{1}{2K}(1 + K(a_0 + b_0 + 2c_0) - r),$$

where

$$K := \frac{k_1}{k_{-1}}, r := \sqrt{1 + 2K(a_0 + b_0 + 2c_0) + K^2(a_0 - b_0)^2}$$

will be attained, as, e.g., the explicit form of the solutions shows.

7.11 The induced kinetic differential equation of the reversible triangle reaction being

$$\dot{a} = -k_1 a + k_2 b - k_6 a + k_5 c$$
$$\dot{b} = k_1 a - k_2 b - k_3 b + k_4 c$$
$$\dot{c} = k_3 b - k_4 c + k_6 a - k_5 c$$

together with the mass conservation relation

$$a(t) + b(t) + c(t) = a_0 + b_0 + b_0 =: m$$

implies that the unique, relatively asymptotically stable vector of positive stationary concentrations—if at least one of the initial concentrations a_0, b_0, c_0 is positive—are as follows:

$$a_* = m\frac{A}{A + B + C} \quad b_* = m\frac{B}{A + B + C} \quad c_* = m\frac{C}{A + B + C} \tag{14.14}$$

with

$$A := (k_2 + k_3)(k_4 + k_5) - k_3 k_4$$
$$B := (k_4 + k_5)(k_6 + k_1) - k_5 k_6$$
$$C := (k_6 + k_1)(k_2 + k_3) - k_1 k_2.$$

Relative global asymptotic stability can either be seen from the explicit form of the solution as given by

`Concentrations[ToReversible["Triangle"], Array[k[#], 6],{a0, b0, c0}],`

(not a royal way!) or by linear stability analysis of the two equations obtained by reduction using mass conservation.

From now on the explicit form (14.14) of the stationary point will not be used in the arguments. The fact that the reaction is detailed balanced at a positive stationary point $\begin{bmatrix} a_* & b_* & c_* \end{bmatrix}^\top$ can be expressed as

$$k_1 a_* = k_2 b_* \quad k_3 b_* = k_4 c_* \quad k_5 c_* = k_6 a_*. \tag{14.15}$$

Taking the product of these equalities and simplifying with the (positive) product, $a_* b_* c_*$ implies that

$$k_1 k_3 k_5 = k_2 k_4 k_6 \tag{14.16}$$

holds.

Conversely, suppose that (14.16) holds, and calculate the difference of the reaction rates $k_1 a_* - k_2 b_*$ with any stationary point. (The other equalities can be investigated similarly.) As the stationary points fulfill the system of equations

$$0 = -k_1 a_* + k_2 b_* - k_6 a_* + k_5 c_* \quad 0 = k_1 a_* - k_2 b_* - k_3 b_* + k_4 c_*,$$

they can be expressed as $a_* = \frac{c_*(k_2 k_4 + k_2 k_5 + k_3 k_5)}{k_1 k_3 + k_2 k_6 + k_3 k_6}$, $\quad b_* = \frac{c_*(k_1 k_4 + k_1 k_5 + k_4 k_6)}{k_1 k_3 + k_2 k_6 + k_3 k_6}$. Now the difference, using the condition (14.16), is

$$k_1 \frac{c_*(k_2 k_4 + k_2 k_5 + k_3 k_5)}{k_1 k_3 + k_2 \frac{k_1 k_3 k_5}{k_2 k_4} + k_3 \frac{k_1 k_3 k_5}{k_2 k_4}} - k_2 \frac{c_*(k_1 k_4 + k_1 k_5 + k_4 \frac{k_1 k_3 k_5}{k_2 k_4})}{k_1 k_3 + k_2 \frac{k_1 k_3 k_5}{k_2 k_4} + k_3 \frac{k_1 k_3 k_5}{k_2 k_4}} = 0$$

Q.E.D.

7.12 The induced kinetic differential equation of the Wegscheider reaction being

$$\dot{a} = -k_1 a + k_2 b - k_3 a^2 + k_4 a b \quad \dot{b} = k_1 a - k_2 b + k_3 a^2 - k_4 a b$$

—which simplifies to

$$\dot{a} = -k_1 a + k_2 (a_0 + b_0 - a) - k_3 a^2 + k_4 a (a_0 + b_0 - a)$$
$$= -(k_3 + k_4) a^2 - (k_1 + k_2 - k_4 (a_0 + b_0)) a + k_2 (a_0 + b_0).$$

—together with the mass conservation relation $a(t) + b(t) = a_0 + b_0 =: m$ imply that, unless all the initial concentrations are zero, the unique positive (relatively asymptotically stable) stationary concentration vector is as follows:

$$a_* = \frac{k_1 + k_2 - k_4 m - r}{-2(k_4 + k_3)} \quad b_* = \frac{k_1 + k_2 + k_4 m + 2 k_3 m - r}{2(k_4 + k_3)} \tag{14.17}$$

with $r := \sqrt{(k_1 + k_2 - k_4 m)^2 + 4 k_2 m (k_3 + k_4)}$, assuming that $k_1 + k_2 < k_4 m$. (Which is the sign to be changed in the opposite case?)

From now on the explicit form (14.17) of the stationary point will not be used in the arguments. The fact that the reaction is detailed balanced at a positive stationary point $\begin{bmatrix} a_* & b_* \end{bmatrix}^\top$ can be expressed as $k_1 a_* = k_2 b_* \quad k_4 b_*^2 = k_3 a_* b_*$. Taking the product of these equalities and simplifying with the (positive) product, $a_* b_*^2$ implies that

$$k_1 k_4 = k_2 k_3 \qquad (14.18)$$

holds.

Conversely, suppose that (14.18) holds, and calculate the difference of the reaction rates $k_1 a_* - k_2 b_*$ with any stationary point. (The other equalities can be investigated similarly.) As the stationary points fulfill the equation

$$0 = -k_1 a_* + k_2 b_* - k_3 a_*^2 - k_4 a_* b_*,$$

b_* can be expressed as $b_* = a_* \frac{k_1 - k_3 a_*}{k_2 - k_4 a_*}$. Now the difference, using the condition (14.18)

$$k_1 a_* - k_2 a_* \frac{k_1 - k_3 a_*}{k_2 - k_4 a_*} = k_1 a_* - k_2 a_* \frac{k_1 - k_3 a_*}{k_2 - \frac{k_2 k_3}{k_1} a_*} = 0$$

Q.E.D.

In the second part, we have used a similar method to the one applied in the case of the triangle reaction. However, there exists a simpler alternative here. As

$$-k_1 a_* + k_2 b_* - k_3 a_*^2 - \frac{k_2 k_3}{k_1} a_* b_* = \frac{k_1 + k_3 a_*}{k_1} (k_2 b_* - k_1 a_*),$$

this expression can only be zero, if the second factor is zero.

7.13 Simple calculations (or the use of the function **RightHandSide**) show that both has $\dot{x} = -3x^3 + 1.5y^3$ $\dot{y} = 3x^3 - 1.5y^3$ as its induced kinetic differential equation. The second one has been obtained by Szederkényi and Hangos (2011) who formulated a linear programming problem, the solution of which gives a complex balanced or detailed balanced realization of a given kinetic differential equation.

7.14 As the induced kinetic differential equation of the Lotka–Volterra reaction is $\dot{x} = k_1 x - k_2 xy$ $\dot{y} = k_2 xy - k_3 y$, it has two nonnegative stationary points: $\begin{bmatrix} 0 & 0 \end{bmatrix}^{\top}$ and $\begin{bmatrix} k_3/k_2 & k_1/k_2 \end{bmatrix}^{\top}$. A simple verification shows that a nonlinear first integral is $\begin{bmatrix} x & y \end{bmatrix}^{\top} \mapsto k_3 \ln(x) + k_1 \ln(y) - k_2 x - k_2 y$. Elementary calculations show that its minimum is just the positive stationary point.

7.15 As the induced kinetic differential equation of the Ivanova reaction is

$$\dot{x} = x(-k_1 y + k_3 z) \quad \dot{y} = y(k_1 x - k_2 z) \quad \dot{z} = z(-k_3 x + k_2 y)$$

it has two first integrals; the linear one is $\begin{bmatrix} x & y & z \end{bmatrix} \mapsto x + y + z$. This implies that the unique positive stationary point is $\begin{bmatrix} x_* & y_* & z_* \end{bmatrix} = \frac{x_0 + y_0 + z_0}{k_1 + k_2 + k_3} \begin{bmatrix} k_2 & k_3 & k_1 \end{bmatrix}$. It is on the level set of the nonlinear first integral $\begin{bmatrix} x & y & z \end{bmatrix} \mapsto x^{k_2} y^{k_3} z^{k_1}$ if

$$k_1^{k_1} k_2^{k_2} k_3^{k_3} \left(\frac{x_0 + y_0 + z_0}{k_1 + k_2 + k_3} \right)^{k_1 + k_2 + k_3} = x_0^{k_2} y_0^{k_3} z_0^{k_1}$$

also holds.

7.16 The deficiency of the Lotka–Volterra reaction being 1, Theorem 7.34 can be applied. The nonterminal complexes are as follows: X, X + Y, Y. The first two only differs in Y; the second one only differs in X; thus both stationary concentrations are independent from the initial ones.

An easy calculation also shows that the coordinates of the only positive stationary point of the reaction being $\begin{bmatrix} \frac{k_3}{k_2} & \frac{k_1}{k_2} \end{bmatrix}$ are the same; no matter what the initial conditions are, they only depend on the reaction rate coefficients.

7.17 The deficiency is $\delta = N - L - S = 8 - 3 - 3 = 2$, thus one cannot apply Theorem 7.34. However, the stationary points obey the equations

$$0 = k_3 x_* - 2k_4 x_*^2 + k_1 y_* - k_2 x_* y_*, \quad 0 = -k_1 y_* - k_2 x_* y_* + 2k_5 z_*, \quad 0 = k_3 x_* - k_5 z_*.$$

First of all, $z_* = \frac{k_3 x_*}{k_5}$, $y_* = \frac{2k_3 x_*}{k_1 + k_2 x_*}$; thus we can either have the trivial solution or we can proceed to get positive solutions. As for x_* we have a quadratic equation containing none of the initial concentrations having a single positive solution; thus all the components of all the nonnegative stationary points are independent of the initial concentrations.

7.18 The deficiency calculated either by hand or by the function **ReactionsData** turns out to be one. The nonterminal complexes $EI_p + I$ and EI_p only differ in I; therefore the mechanism shows absolute concentration robustness for the species I. Could you prove that the mechanism does have a positive stationary point?

7.19 Upon solving $0 = -x_* y_* + f(1 - x_*)$ for y_* and substituting the result into $0 = x_* y_* + y_* ((1 - f_2/f) y_0 - y_*) + (f - f_2) y_0 - f y_*$, the cubic equation

$$x_*^3 - x_*^2 + (f + (1 - f_2/f) y_0) x_* - f = 0 \tag{14.19}$$

is obtained for y_*. The number of its real positive roots depends on the sign of the coefficient $(f + (1 - f_2/f) y_0)$ of the first degree term: it can be one, two, or three depending on the value of f if the value of other parameters is fixed. For studying the parameter dependence, one may use the definitions as follows. Let

```
lilipota[f2_, y0_, f_] :=
    f x^3 - f x^2 + x(f^2 + (f - f2)y0) - f^2;
```

```
Manipulate[ParametricPlot[(XX[f2_, y0_, {φ}_] :=
    x /. NSolve[lilipota[f2, y0, 10^{φ}] == 0, {x}];
    Distribute[{{φ}, XX[f2, y0, {φ}]}, List]), {{φ},-3,-1},
    PlotRange -> {{-3, -1}, {0, 1.2}}],
    {{f2, 1/135}, 0.0009, 0.0075, 0.0001},
    {{y0, 10/27}, 0.24, 0.38, 0.01}].
```

Cf. Li and Li (1989) and also Póta (2006, pp. 4, 83, 84). When manipulating, figures of the form of a **pitchfork**, a **mushroom**, and also an **isola** will appear, respectively. Now, you may start reproducing and manipulating the figures by Ganapathisubramanian and Showalter (1984) as well.

14.8 Transient Behavior

8.1 Let the sensitivity matrix be $\mathbf{S} := \left[\frac{\partial c_m}{\partial k_r}\right]_{M \times R}$. Then starting from $\dot{\mathbf{c}} = \boldsymbol{\gamma} \cdot \text{diag}(\mathbf{c}^{\alpha}) \cdot \mathbf{k}$, one gets (check it using coordinates if you wish!)

$$\dot{\mathbf{S}} = \boldsymbol{\gamma} \cdot \text{diag}(\mathbf{c}^{\alpha}) + \boldsymbol{\gamma} \cdot \left(\mathbf{c}^{\alpha} \odot (\boldsymbol{\alpha}^{\top} \cdot \mathbf{S} \cdot \text{diag}(\mathbf{c})^{-1}) \odot \mathbf{k}\right). \qquad (14.20)$$

Note that this is an inhomogeneous linear differential equation for the sensitivities. However, as the coefficients are time dependent, linearity does not imply that even given the concentration vs. time curves, one can solve Eq. (14.20) symbolically. An important exception is when the solution in question is a stationary solution: In that case the coefficients are time independent. How does (14.20) simplify in the case of compartmental systems?

8.2 Nil, namely, the sensitivity equations simplify to

$$\dot{s}_{11} = k_1 s_{11}, \quad \dot{s}_{21} = -k_3 s_{21}$$
$$\dot{s}_{12} = k_3 s_{12}, \quad \dot{s}_{22} = -k_3 s_{22}$$
$$\dot{s}_{13} = k_1 s_{13}, \quad \dot{s}_{23} = -k_3 s_{23}$$

having constant zero solutions.

8.3 The right-hand side of the differential equation

$$\dot{x} = y^2 \boxed{-2yz} + z^2, \quad \dot{y} = x - y, \quad \dot{z} = x + y - z. \qquad (14.21)$$

contains a negative cross effect where shown; still the vector on the right-hand side always points into the interior of the first orthant.

8.4 The induced kinetic differential equation of a single species mechanism is of the form

$$\dot{c} = a_0 + a_1 c + \cdots + a_{R-1} c^{R-1} \tag{14.22}$$

where $R \in \mathbb{N} \setminus \{0\}$; $a_0 \in \mathbb{R}_0^+$; $a_1, \ldots, a_{R-1} \in \mathbb{R}$. If $R = 1, 2$, then Eq. (14.22) is linear; therefore its solutions cannot blow up. If $R > 2$ and the right-hand side has no positive root, then the solution can be obtained from

$$\frac{\dot{c}}{a_0 + a_1 c + \cdots + a_{R-1} c^{R-1}} = 1$$

by integration:

$$\int_0^t \frac{\dot{c}(s)}{a_0 + a_1 c(s) + \cdots + a_{R-1} c(s)^{R-1}} \, ds = \int_{c(0)}^{c(t)} \frac{1}{a_0 + a_1 c + \cdots + a_{R-1} c^{R-1}} \, dc = t.$$

Convergence of the integral when its upper limit tends to $+\infty$ implies the statement.

8.5 Being mass consuming means that there is a positive vector $\rho \in \mathbb{R}^M$ such that $\rho^\top \gamma \lneq 0^\top$ holds. Nonnegativity of the concentration vs. time curves and the integrated form of the induced kinetic differential equation together with the above condition imply $0 \le \rho^\top c(t) = \rho^\top c(0) \lneq 0$, showing that all the concentrations are bounded.

8.6 If one applies Theorem 8.19 with $\omega_1 = 1, \omega_2 = 1, \omega_3 = 0,$ one has $A(\omega) = \begin{bmatrix} 3/4 & 0 & 0 \\ 0 & 1 & 0 \\ 0 & 0 & 1/2 \end{bmatrix}$, $b(\omega) = 0, c(\omega) = 1, \lambda(\omega) = 1/2, \Delta(\omega) = -1,$ which means that the solutions of the Eq. (8.65) blow up for any choice of the initial conditions. One can also apply Theorem 8.21, but (8.14) does not hold with the given components of ω. However, with $\omega_1 = 2, \omega_2 = 1, \omega_3 = 0,$ the inequality (8.14) does hold.

8.7 If one applies Theorem 8.19 with $\omega_1 = 1, \omega_2 = 1, \omega_3 = 0,$ one has $A(\omega) = \begin{bmatrix} 1/2 & 0 & 0 \\ 0 & 1 & 0 \\ 0 & 0 & 1 \end{bmatrix}$, $b(\omega) = 0, c(\omega) = 0, \lambda(\omega) = 1/2, \Delta(\omega) = 0,$ which means that the solutions of the Eq. (8.66) blow up for those choices of the initial concentrations x_0, y_0, z_0 for which $x_0 + y_0 > 0$ holds. Note that here any real value is allowed for z_0, and one of the two other initial values can also be negative as far as $x_0 + y_0 > 0$ holds.

For arbitrary nonnegative initial values, we note first that the solutions are nonnegative. Using the differential equation and the inequality between the quadratic and arithmetic means, we have

$$(x + y)^{\cdot} \geq (x + y)^{\cdot} + \dot{z} = \frac{x^2}{2} + y^2 \geq \frac{x^2}{2} + \frac{y^2}{2} \geq \left(\frac{x + y}{2}\right)^2,$$

and the beginning and the end show that $x + y$ blows up under the given conditions.

8.8 If one applies Theorem 8.19 with $\omega_1 = 1$, $\omega_2 = 1$, $\omega_3 = 0$, one has $\mathbf{A}(\omega) = \begin{bmatrix} 1/2 & 0 & 0 \\ 0 & 1 & 0 \\ 0 & 0 & 1 \end{bmatrix}$, $\mathbf{b}(\omega) = \begin{bmatrix} -1 & 0 & 0 \end{bmatrix}^T$, $c(\omega) = 0$, $\lambda(\omega) = 1/2$, $\Delta(\omega) = 1/2$, which means that the solutions of the Eq. (8.67) blow up for those choices of the initial concentrations x_0, y_0, z_0 for which $x_0 + y_0 > 1 + \sqrt{2}$ holds.

After this series of problems, the reader might use **ParametricNDSolve** to find initial values leading to blowup which are not covered by the mentioned theorem.

8.9 The solution of the induced kinetic differential equation

$$\dot{x} = -x + 4y \quad \dot{y} = x - y$$

with the initial concentration $\begin{bmatrix} x_0 & y_0 \end{bmatrix}^T$ being

$$\begin{bmatrix} 1/2e^{-3t}((1 + e^{4t})x_0 + 2(-1 + e^{4t})y_0) \\ 1/4e^{-3t}((-1 + e^{4t})x_0 + 2(1 + e^{4t})y_0) \end{bmatrix}$$

tends to $\begin{bmatrix} +\infty & +\infty \end{bmatrix}^T$ as $t \to +\infty$. Note that this is **not** in contradiction with the zero-deficiency theorem.

8.10 The simplest example is A $\longrightarrow$ B. However, Feinberg and Horn (1977) on page 85 give an example with $T = L = 2$.

8.11 A simple example is $2\,A \longleftarrow A + B \longrightarrow 2\,B$, with unequal reaction rate coefficients. Here we have $\mathscr{S} = \mathscr{S}^*, T = 2, L = 1, \delta = 1$; thus $T > L$, but $T > L > \delta$ does not hold.

8.12 The stoichiometric matrix of the reaction

$$X \xrightarrow{k} Y \quad X \xrightarrow{1} Z \quad Y + Z \xrightarrow{1} 2\,X$$

is $\boldsymbol{\gamma} := \begin{bmatrix} -1 & -1 & 2 \\ 1 & 0 & -1 \\ 0 & 1 & -1 \end{bmatrix}$; therefore the only solution to $\boldsymbol{\rho}^\top \boldsymbol{\gamma} = \mathbf{0}$ is $\begin{bmatrix} 1 & 1 & 1 \end{bmatrix}^\top$. The induced kinetic differential equation of the reaction is

$$\dot{x} = -(k+1)x + 2yz \quad \dot{y} = kx - yz \quad \dot{z} = x - yz.$$

If we are looking for numbers a, b, c such that $a\dot{x} + b\dot{y} + c\dot{z} = 0$, then for the case $k \neq 1$, we shall only find the components of $\boldsymbol{\rho}$ above, but in the case of $k = 1$, any triple for which $2a = b + c$ holds will do. Let us choose, e.g., $a = 2, b = 1, c = 3$.

8.13 Consider the reversible Lotka–Volterra reaction; it is weakly reversible, and its deficiency is 1 as shown by

`ReactionsData[ToReversible[lv]]["deficiency"]`

The conditions that the reaction is complex balanced at the stationary point $\begin{bmatrix} x_* & y_* \end{bmatrix}^\top$ means that the stationary point can only be $x_* = \frac{k_1}{k_{-1}}$ $y_* = \frac{k_{-3}}{k_3}$ and $k_2 \frac{k_1 k_{-3}}{k_{-1} k_3} = k_{-2} \frac{k_{-3}^2}{k_3^2}$ should also hold. (Actually, we arrived at the condition of detailed balance.) Could you find a reaction which is not weakly reversible and is still complex balanced for some reaction rate coefficients?

8.14 The induced kinetic differential equation of the reaction (8.68) is

$$\dot{a}(t) = k_1 p - k_2 a(t) b(t)^2 \quad \dot{b}(t) = k_2 a(t) b(t)^2 - k_3 b(t) \qquad (14.23)$$

with $k_1, k_2, k_3, p \in \mathbb{R}^+$. We try to simplify Eq. (14.23) using a **diagonal transformation** of the variables (or **change of units**) introducing new variables by

$$\tau := \vartheta t, x(\tau) := \xi a(t), y(\tau) := \eta b(t),$$

where the parameters $\vartheta, \xi, \eta \in \mathbb{R}^+$ will be determined later, in an appropriate way. As

$$\xi \dot{a}(t) = x'(\tau)\vartheta \quad \eta \dot{b}(t) = y'(\tau)\vartheta, \qquad (14.24)$$

in terms of the new variables, one has the following system:

$$x'(\tau) = \frac{\xi}{\vartheta}(k_1 p - k_2 a(t) b(t)^2) = \frac{\xi}{\vartheta}\left(k_1 p - \frac{k_2}{\xi \eta^2} x(\tau) y(\tau)^2\right) \qquad (14.25)$$

$$y'(\tau) = \frac{\eta}{\vartheta}(k_2 a(t) b(t)^2 - k_3 b(t)) = \frac{\eta}{\vartheta}\left(\frac{k_2}{\xi \eta^2} x(\tau) y(\tau)^2 - \frac{k_3}{\eta} y(\tau)\right). \qquad (14.26)$$

Let us introduce $\mu := \frac{\xi k_1 p}{\vartheta}$, and let us choose the transformation parameters in the following way:

$$\vartheta := k_3, \xi := \eta := \sqrt{k_2/k_3},$$

then we have

$$x' = \mu - xy^2 \quad y' = xy^2 - y. \tag{14.27}$$

Equation (14.27) is said to be the **dimensionless from** of Eq. (14.23). It clearly shows that the qualitative behavior of the trajectories only depend on a single (combined) parameter $\mu = \frac{\sqrt{k_2/k_3}k_1 p}{k_3}$. The general theory and use of these kinds of transformations are treated in textbooks on dimensional analysis, as, e.g., in Szirtes (2007).

Let us turn to the investigation of stationary points. There is a unique positive stationary point of (14.27): $[x_* \ y_*] = [1/\mu \ \mu]$. The eigenvalues of the Jacobian of the right-hand side at the stationary point can be calculated from

$$\begin{vmatrix} -\mu^2 - \lambda(\mu) & -2\frac{1}{\mu}\mu \\ \mu^2 & 1 - \lambda(\mu) \end{vmatrix} = \lambda(\mu)^2 - \lambda(\mu)(1 - \mu^2) + \mu^2 = 0,$$

and are

$$\lambda_\pm(\mu) = \frac{1 - \mu^2 \pm \sqrt{1 - 6\mu^2 + \mu^4}}{2}.$$

The expression under the root sign is negative, if $\sqrt{2} - 1 < \mu < \sqrt{2} + 1$; therefore in this interval, the eigenvalues are complex. Their real part is zero, if $\mu = 1$. The derivative of λ being $-\mu \pm \frac{-3\mu + \mu^3}{\sqrt{1 - 6\mu^2 + \mu^4}}$ is different from zero (it is $-1 \pm i$). Also, using the notations of the Theorem 13.57, one has $a = -\frac{3}{8}$; therefore the conditions of the mentioned theorem hold; a stable limit cycle emerges as the parameter μ crosses the value 1. Use **Manipulate** to follow this phenomenon.

Note that $\mu = 1$ means that one has the following relationship between the reaction rate coefficients and the input flow rate of the external species P: $k_1^2 k_2 p^2 = k_3^2$. Could you have similar consequences with $x^a y^b$ and y^c instead of xy^2 and y with some restrictions on the exponents?

8.15 The induced kinetic differential equation of the reaction (8.46) (obtained either by **DeterministicModel** or by hand) is

$$\dot{x} = k_1(\beta_1 - 1)x - k_2 xy$$
$$\dot{y} = -k_2 xy + \beta_2 k_3 z - k_4 y$$
$$\dot{z} = k_2 xy - k_3 z.$$

Thus the coordinates of the unique positive stationary point are as follows (again, either by hand—recommended—or using **StationaryPoints**):

$$x_* = \frac{k_4}{k_2(\beta_2 - 1)} \qquad y_* = \frac{k_1(\beta_1 - 1)}{k_2} \qquad z_* = \frac{k_1 k_4}{k_2 k_3} \frac{\beta_1 - 1}{\beta_2 - 1}.$$

The origin is a stationary point as well. The Jacobian of the right-hand side is

$$\begin{bmatrix} k_1(\beta_1 - 1) - k_2 y & -k_2 x & 0 \\ -k_2 y & -k_2 x - k_4 & k_2 \beta_3 \\ k_2 y & k_2 x & -k_3 \end{bmatrix}.$$

This implies that two of the eigenvalues are positive, the third one is negative.

8.16 Using **StationaryPoints** gives you the stationary points of (8.50), of which one can pick the single positive one. This calculation can also be carried out by hand even with varying reaction rate coefficients to get

$$\begin{bmatrix} a_* & b_* & c_* & d_* & e_* \end{bmatrix} = \begin{bmatrix} a_* & \frac{k_3}{k_2} a_* & \frac{k_1}{k_2} a_* & \frac{k_5}{k_6} a_* & \frac{k_4}{k_5} a_* \end{bmatrix}$$

where $a_* = \frac{a_0 + b_0 + c_0 + d_0 + e_0}{1 + \frac{k_3}{k_2} + \frac{k_1}{k_2} + \frac{k_5}{k_6} + \frac{k_4}{k_5}}$. And the characteristic polynomial of the Jacobian is of the form $-\lambda(A + B\lambda^2 + C\lambda^4)$ so that $0 < A < B < C$ and $B^2/4AC < 0$; thus it has a zero eigenvalue and four pure imaginary eigenvalues which means that linear stability analysis is not enough to decide its stability.

Let the nth maximum on one of the components, say, x be ξ_n. It is instructive to plotting ξ_{n+1} as a function of x_n and see that it is not random.

8.17 See Pintér and Hatvani (1977–1980).

8.18 Observe that $\mathbf{1}^\top \mathbf{A} = \mathbf{0}^\top$. This shows that $\mathbf{A}$ is rank-deficient, so it has a zero eigenvalue. The example $\begin{bmatrix} -1 & 1 & 0 & 0 \\ 1 & -1 & 0 & 0 \\ 0 & 0 & -2 & 2 \\ 0 & 0 & 2 & -2 \end{bmatrix}$ shows that multiplicity of zero can also be larger than one. Moreover, as a closed compartmental system is mass conserving, all the positive reaction simplexes are bounded, and the trajectories stay in these bounded reaction simplexes. Had we have a single eigenvalue with positive real parts, we could specify an initial concentration defining a solution which is unbounded.

8.19 Let $B(x, y) := \frac{1}{xy}$ and then $D(x, y) := \text{div}(B\mathbf{f})(x, y) = \frac{a}{y} + \frac{B}{x}$ should be zero; thus $a = B = 0$. If either b or A is zero, then x or y is monotonous. Suppose none of them is zero, then the positive stationary point which can only be $x_* = -\frac{C}{A}$, $y_* = -\frac{c}{b}$ should hold which also implies that C and A and c and b, respectively, are of the opposite signs. The eigenvalues of the Jacobian at the stationary point are $\pm\sqrt{cC}$. If $cC > 0$, then the stationary point is a saddle, which cannot be surrounded by a closed orbit. If $cC < 0$, then the stationary point is a center, and we actually arrive at the Lotka–Volterra model.

8.20 The positive stationary point is $x^* = \frac{\delta}{\beta}$, $y^* = \frac{d}{b}$. The Jacobian of the right-hand side at the stationary point being $\begin{bmatrix} 0 & -b\frac{\delta}{\beta} \\ -\beta\frac{d}{b} & 0 \end{bmatrix}$ has the real eigenvalues $\pm\sqrt{d\delta}$ of the opposite sign, thus the stationary point is a saddle. It would have been enough to require that d and δ is of the same sign, and $b, \beta \neq 0$.

8.21 As the induced kinetic differential equation of the mechanism is

$$\dot{x} = a + x^2 y - (b+1)x - xy^2$$
$$\dot{y} = bx - x^2 y,$$

the divergence is $2xy - (b+1) - y^2 - x^2 = -(b+1) - (x+y)^2 < 0$; therefore according to Theorem 13.51, the mechanism has no periodic solutions.

8.22 We follow the arguments of Ault and Holmgreen (2003). As the induced kinetic differential equation of the Brusselator $0 \underset{1}{\overset{1}{\rightleftharpoons}} X \overset{b}{\longrightarrow} Y \quad 2X + Y \overset{a}{\longrightarrow} 3X$ is (6.47)

$$\dot{x} = 1 - (b+1)x + ax^2 y \quad \dot{y} = bx - ax^2 y,$$

the unique stationary point is $(x^*, y^*) = (1, \frac{b}{a})$. The Jacobian at the stationary point is $\begin{bmatrix} b-1 & a \\ -b & -a \end{bmatrix}$. Both of the eigenvalues are positive if $b > (1 + \sqrt{a})^2$; therefore in this case—the only one we are considering here—the stationary point is an unstable focus. Let us draw a sufficiently small circle around the unstable stationary point, and then this circle will form the inner boundary of the bounded closed positively invariant set: the trapping region. The outer boundary of the the trapping region will be a pentagon; let us define all of its sides. First, let us introduce the notations:

$$\mathbb{R}^2 \ni (p, q) \mapsto f(p, q) := 1 - (b+1)p + ap^2 q$$
$$\mathbb{R}^2 \ni (p, q) \mapsto g(p, q) := bp - ap^2 q.$$

1. The trajectories cross the line $\{(p, q) \in (\mathbb{R}_0^+)^2 | p = \frac{1}{b+1}\}$ inward, because in the points of this line $f(p, q) > 0$.

2. The subset $\{(p, q) \in \mathbb{R}^{+2} | b - apq = 0\} \subset \{(p, q) \in \mathbb{R}^{+2} | g(p, q) = 0\}$ of the y-nullcline and the vertical line defined above intersect each other at the point $\left(\frac{1}{b+1}, \frac{b(b+1)}{a}\right)$; therefore the trajectories cross the line

$$\left\{\left(p, \frac{b(b+1)}{a}\right) \in (\mathbb{R}_0^+)^2 \,\middle|\, p \in \left[\frac{1}{b+1}, +\infty\right[\right\}$$

inward.

3. If $p > 1$, then

$$(1, 1) \cdot (f(p, q), g(p, q))^\top = 1 - (b+1)p + ap^2q + bp - ap^2q = 1 - p < 0,$$

thus a line section with the normal vector $(1, 1)$ with abscissas larger than 1 will also be appropriate as a part of the boundary, if we take the section from the previously defined horizontal line section until the y-nullcline

$$\{f(p, q) \in \mathbb{R}^{+2} | b - apq = 0\}.$$

4. The next line section is vertical, starting from the abovementioned intersection point until the abscissa.

5. The final section is part of the abscissa which closes the boundary.

The details of the calculation can be followed in Fig. 14.8.

8.23 Let us fix the value of the parameter a, and consider b as a changing parameter. The critical parameter value will obviously be $b = a + 1$; here the eigenvalues of the Jacobian are purely imaginary and different from zero: $\lambda_{1,2}(a + 1) = \pm i \sqrt{a}$; see the solution of the previous problem. At the critical parameter value, the derivative of the real part of the eigenvalue is different from zero: $\Re(\lambda)'(a + 1) = \frac{1}{2} \neq 0$.

The *Mathematica* demonstration Várdai and Tóth (2008) shows how the limit cycle emerges.

8.24 Let us calculate the divergence of the right-hand side of (8.49) to get $-(k_1 + k_2 + k_3 + k_4 + k_5 + k_6) < 0$; thus the application of Bendixson–Dulac Theorem 13.51 with the Dulac function $B(a, b) := 1$ implies that (8.49) cannot have periodic solutions.

8.25

1. The eigenvalues of the coefficient matrix in case I are $0, \frac{-3-i\sqrt{3}}{2}, \frac{-3+i\sqrt{3}}{2}$; therefore after a short calculation (e.g., using the function **Concentrations**),

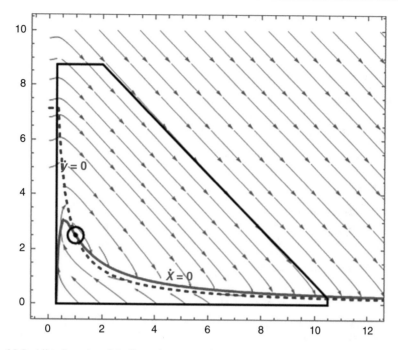

Fig. 14.8 All trajectories of the Brusselator go to the trap. Here $a = 1, b = 2.5$

we have

$$a(t) = \frac{1}{3}\left(1 + 2e^{-3t/2}\cos\left(\frac{\sqrt{3}t}{2}\right)\right)$$

and (using **TrigFactor**)

$$\dot{a}(t) = \frac{2e^{-3t/2}\cos(\pi/6 + \sqrt{3}t/2)}{\sqrt{3}}.$$

The last function has an infinite number of zeros. The explicit form of the solutions also shows that the second derivative of a at times τ when $\dot{a}(\tau) = 0$—being equal to $-\dot{c}(\tau)$—cannot be zero, which implies that the infinitely many arguments are times where a shows a local extrema.

2. The lack of oscillations is a consequence of the Póta–Jost theorem (Theorem 8.64), because the eigenvalues being $-3, -3, 0$ are real. Let us remark that the reaction is detailed balanced, having the deficiency zero, and fulfilling the circuit conditions (Feinberg 1989).

3. The reaction is not detailed balanced with this set of reaction rate coefficients. However, as the eigenvalues of the coefficient matrix are the real numbers $-4, -3, 0$, again one can apply the Póta–Jost theorem to get the result that none

of the concentration vs. time curves can have more than $1 = 3 - 2$ strict local extrema.

8.26 The result of the line

```
SymmetricMatrixQ[D[RightHandSide[
    {2X -> 3X + 3Y, 2Y -> 3X + 3Y, X + Y -> 3X + 3Y},
    {k2, k2, 3 k2}, {x, y}], {{x, y}, 1}]]
```

being `True` shows the statement.

8.27 The result of the line

```
Div[RightHandSide[
    {2X -> X + Y -> 2X + Y, X + Y -> 2Y -> 2X + Y},
    {1, 3, 2, 1/2}, {x, y}], {x, y}]
```

being zero shows the statement.

8.28 Upon introducing the definition

```
rhs = RightHandSide[{2X -> X + Y -> 0, 2Y -> X + Y},
    {k, 2k, k}, {x, y}],
```

the result of the expressions

```
Div[Reverse[rhs], {x, y}]
Div[{1, -1}rhs, {x, y}]
```

both being zero shows the statement. As

$$-kx^2 - 2kxy + ky^2 + i(kx^2 - 2kxy - ky^2) = k(i-1)z^2 \tag{14.28}$$

with $z := x + iy$, the induced kinetic differential equation of the reaction can be transformed into the equation for the complex valued function z as $\dot{z} = k(i-1)z^2$, and the real and implicit part of its solution gives the concentration of x and y, respectively.

$$x(t) = \frac{x_0 + kt(x_0^2 + y_0^2)}{1 + 2kt(x_0 + y_0 + ktx_0^2 + kty_0^2)} \tag{14.29}$$

$$y(t) = \frac{y_0 + kt(x_0^2 + y_0^2)}{1 + 2kt(x_0 + y_0 + ktx_0^2 + kty_0^2)} \tag{14.30}$$

8.29 Comparing the coefficients of the corresponding polynomials shows that in all the three cases, it is only the zero polynomial which can occur as the right-hand side of the corresponding induced kinetic differential equations.

8.30 According to the assumption, the model to describe the reaction is as follows:

$$\dot{b} = -k_1 b \quad \dot{p} = k_1 b - k_2 p \quad \dot{l} = k_2 p$$
$$b(0) = 1 \qquad p(0) = 0 \qquad l(0) = 0, \tag{14.31}$$

where b, p, l denote the quantity of bismuth, polonium, and lead, respectively. The reaction rate constants k can be calculated from the half-lives T from the equation

$$\frac{1}{2} = \frac{c(t+T)}{c(t)} = \frac{c(0)e^{-k(t+T)}}{c(0)e^{-kt}} = e^{-kT}$$

giving $k = \frac{\ln(2)}{T}$. (Note that the half-life does depend neither on the initial concentration nor on the time of measurement.) Therefore

$$k_1 = \frac{\ln(2)}{5}(\approx 0.139) \quad k_2 = \frac{\ln(2)}{138}(\approx 0.005).$$

Solving (14.31) gives

$$b(t) = e^{-k_1 t} = 2^{-\frac{t}{5}} \tag{14.32}$$

$$p(t) = \frac{k_1}{k_2 - k_1}(e^{-k_1 t} - e^{-k_2 t}) = -\frac{138}{133}(2^{-\frac{t}{138}} - 2^{-\frac{t}{5}}) \tag{14.33}$$

$$l(t) = = \frac{k_1(1 - e^{-k_2 t}) - k_2(1 - e^{-k_1 t})}{k_1 - k_2} = 1 - \frac{138}{133}2^{-\frac{t}{138}} + \frac{5}{133}2^{-\frac{t}{5}}. \tag{14.34}$$

The derivative of p is zero at $t_* = \frac{\ln(k_1/k_2)}{k_1 - k_2}(\approx 24.833)$, and it is easy to show that this is a maximum, because the second derivative of p being

$$-k_1 k_2 \left(\frac{k_1}{k_2}\right)^{\frac{k_2}{k_2 - k_1}}$$

is negative. (Instead of calculating the derivative, one can also solve the equation $0 = \dot{p} = k_1 b - k_2 p$.) As the final value of l, i.e., $\lim_{t \to +\infty} l(t)$, is 1, (in the long run, everything becomes lead), thus one has to solve the equation $\frac{k_1(1 - e^{-k_2 t}) - k_2(1 - e^{-k_1 t})}{k_1 - k_2} = \frac{1}{2}$ what one can only do numerically to arrive at $t_{**} \approx 145.347$. (Using the fact that $k_1 \gg k_2$, one can obtain an approximate symbolic solution by using Taylor-series expansion. However, dozens of terms are needed to get a good approximation with this method.)

8.31 The induced kinetic differential equation for the vector of concentrations now has the form

$$\dot{\mathbf{c}}(t) = \sum_{r \in \mathcal{R}} \boldsymbol{\gamma}(\cdot, r) w_r(\mathbf{c}(t), T(t)), \tag{14.35}$$

Table 14.1 Constants in the temperature-dependent description of reactions of Eq. (14.36)

Notation	Meaning	Unit
ρ	Mass density	$\mathrm{kg\,m^{-3}}$
C_p	Specific heat capacity at constant pressure	$\mathrm{J\,K^{-1}\,kg^{-1}}$
ΔH_r	Standard specific enthalpy of reaction	$\mathrm{J\,kg^{-1}}$
t_{res}	$:= \frac{pV}{\upsilon RT_a}$ residence time	s
p	Pressure	Pa
R	Universal gas constant	$\mathrm{J\,mol^{-1}K^{-1}}$
υ	Molar flow rate	$\mathrm{mol^{-1}}$
χ	Heat transfer coefficient	$\mathrm{W\,m^{-2}\,K^{-1}}$
S	Surface area	$\mathrm{m^2}$
T_a	Ambient temperature	K
V	Volume	$\mathrm{m^3}$

and the time evolution equation for the temperature is

$$\rho C_p \dot{T}(t) = \left(\sum_{r \in \mathcal{R}} w(\mathbf{c}(t), T(t))(-\Delta H_r) \right) - \left(\frac{\rho C_p}{t_{res}} + \chi S \right) \frac{T(t) - T_0}{V},$$

$$(14.36)$$

where the meaning of the parameters together with their units can be found in Table 14.1.

8.32

1. Since $f(c) = c^2$, we have $\mathcal{R}_\mathbf{f} = \mathbb{R}_0^+$ (which is not even a linear space), and $\mathcal{S}^* = \mathbb{R}$.
2. Here $\mathcal{R}_f$ is a one-dimensional linear space, whereas $S = 2$.
3. In this case $\mathbf{f}(\mathbf{c}) = \begin{bmatrix} k_1 c_1 - k_2 c_1 c_2 \\ k_2 c_1 c_2 - k_3 c_2 \end{bmatrix}$. Let us characterize those real numbers a and b for which $\mathbf{f}(\mathbf{c}) = \begin{bmatrix} a \\ b \end{bmatrix}$ can be solved. A short calculation shows that such numbers should fulfill the inequality $(k_2(a + b) - k_1 k_3)^2 \geq 4bk_1 k_2 k_3$, which is the outer part of a parabola; see Fig. 14.9. Finally, $\mathcal{S}^* = \mathbb{R}^2$.
4. Here $\mathbf{f}(\mathbf{c}) = \begin{bmatrix} -k_1 c_1 + k_2 c_2^2 + 3k_3 c_2^2 \\ k_1 c_1 - 2k_2 c_2^2 - 2k_3 c_2^2 \end{bmatrix}$; thus we wonder if

$$\begin{bmatrix} -k_1 & k_2 + 3k_3 \\ k_1 & -2(k_2 + k_3) \end{bmatrix} \begin{bmatrix} c_1 \\ c_2 \end{bmatrix} = \begin{bmatrix} a \\ b \end{bmatrix} \qquad (14.37)$$

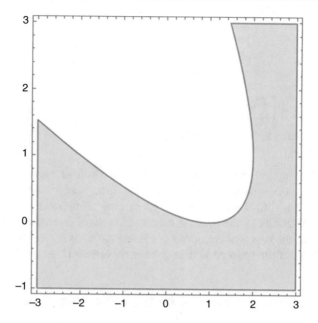

Fig. 14.9 The image of the right-hand side of the irreversible Lotka–Volterra model with $k_1 = 1, k_2 = 1, k_3 = 1$

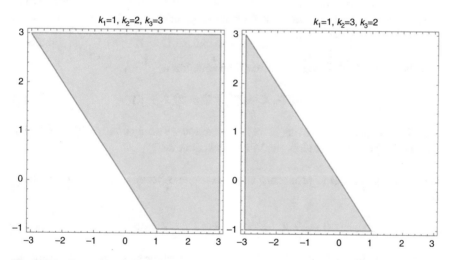

Fig. 14.10 The image of the right-hand side of the Johnston–Siegel model with $k_1 = 1, k_2 = 2, k_3 = 3$ and $k_1 = 1, k_2 = 3, k_3 = 2$

has a solution with a nonnegative second component: $c_2^2 = \frac{a+b}{k_3-k_2}$. (Note that the determinant of the linear system (14.37) is not zero.) This is the case for half planes; see Fig. 14.10. Finally, $\mathscr{S}^* = \mathbb{R}^2$.

5. Here

$$\mathscr{R}_f = \mathscr{S}^* = \left\{ a \begin{bmatrix} -1 \\ 1 \end{bmatrix} \middle| a \in \mathbb{R} \right\},$$

because $\mathbf{f(c)} = \begin{bmatrix} -k_1 c_1 + 4k_2 c_2^2 \\ k_1 c_1 - 4k_2 c_2^2 \end{bmatrix}$, and $-k_1 c_1 + 4k_2 c_2^2$ can take any real values, as it can be seen by substituting $c_2 = 0$.

8.33 The assumption that the two concentration vs. time curves in the Lotka–Volterra reaction take their maxima at the same time would imply that these two functions have a zero derivative at the given time. This however contradicts to the unique solvability of the induced kinetic differential equation of the reaction. The periods of the two concentrations can never differ, as it can be seen from the following form of the induced kinetic differential equation: $\frac{\dot{x}}{x} = k_1 - k_2 y$ $\frac{\dot{y}}{y} = k_2 - k_3 x$.

8.34 In the case of the equation of the harmonic oscillator $\dot{x} = y$ $\dot{y} = -x$, the exponent matrix $\mathbf{B} = \begin{bmatrix} -1 & 1 \\ 1 & -1 \end{bmatrix}$ is not invertible. In the case of the Lorenz equation (6.40), one has $M = 3, R = 4$; therefore the exponent matrix is not even quadratic. In the case of the given polynomial equation, $\lambda = 0$ $\mathbf{A} = \begin{bmatrix} 1 & 0 \\ 0 & -1 \end{bmatrix}$ $\mathbf{B} = \begin{bmatrix} -1 & 2 \\ 1 & -1 \end{bmatrix}$. Thus, one can choose $\mathbf{C} = \mathbf{B}^{-1} = \begin{bmatrix} 1 & 2 \\ 1 & 1 \end{bmatrix}$, and with this choice, the Lotka–Volterra form of the equation is

$$\dot{X} = X(-X - 2Y) \quad \dot{Y} = Y(X + Y). \tag{14.38}$$

Additional problem: Prove that there is no reaction with only three complexes which has (14.38) as its induced kinetic differential equation.

8.35 We use the same procedure as in Example 3.32. All the complexes of the reaction

$$0 \rightleftharpoons X \quad X + Y \rightleftharpoons 2Y \quad Y \rightleftharpoons 0$$

are short, and they—0, X, X + Y, 2 Y, Y—can be represented one after another as shown in Fig. 14.11.

Fig. 14.11 Complex graphs of the individual complexes of the Lotka–Volterra reaction

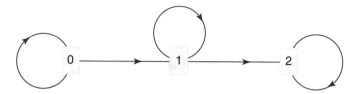

Fig. 14.12 Complex graph of the Lotka–Volterra reaction

Having the complex graphs of the complexes, one is able to put together the complex graph of the whole reaction; see in Fig. 14.12. This figure contains two odd dumbbells, in accordance with the possibility of existence of periodic solutions as stated in the corollary.

8.36 Follow Póta (1985), who argues in a very similar way to Póta (1983). The Dulac function is $B(x, y, z) := \frac{1}{xyz}$. The lengthy analysis reveals that the only equation in three variables fulfilling the requirements and leading to oscillation can only be

$$\dot{x} = -\alpha_1 xy + \alpha_2 zx \quad \dot{y} = \beta_1 xy - \beta_2 yz \quad \dot{z} = -\gamma_1 zx + \gamma_2 yz,$$

which is only slightly more general than the induced kinetic differential equation of the Ivanova reaction $X + Y \longrightarrow 2Y \quad Y + Z \longrightarrow 2Z \quad Z + X \longrightarrow 2X$. Póta (1985) also shows that if terms describing a continuously stirred tank reactor are added to each of the starting equations, then periodic reactions are excluded. Finally, let us mention that the paper relies on a theorem by Demidowitch which has later been corrected, extended, and applied to some kinetic differential equations; see Tóth (1987).

8.37 The induced kinetic differential equation of three species second-order reactions is of the form

$$\dot{x} = a_0 + a_1 x + a_2 y + a_3 z + a_4 xy + a_5 yz + a_6 xz + a_7 x^2 + a_8 y^2 + a_9 z^2$$
$$\dot{y} = A_0 + A_1 x + A_2 y + A_3 z + A_4 xy + A_5 yz + A_6 xz + A_7 x^2 + A_8 y^2 + A_9 z^2$$
$$\dot{z} = \alpha_0 + \alpha_1 x + \alpha_2 y + \alpha_3 z + \alpha_4 xy + \alpha_5 yz + \alpha_6 xz + \alpha_7 x^2 + \alpha_8 y^2 + \alpha_9 z^2.$$

Assuming the fact that $\psi(x, y, z) := xyz$ is a first integral means that

$$0 = yz(a_0 + a_1 x + a_2 y + a_3 z + a_4 xy + a_5 yz + a_6 xz + a_7 x^2 + a_8 y^2 + a_9 z^2)$$
$$+ xz(A_0 + A_1 x + A_2 y + A_3 z + A_4 xy + A_5 yz + A_6 xz + A_7 x^2 + A_8 y^2 + A_9 z^2)$$
$$+ xy(\alpha_0 + \alpha_1 x + \alpha_2 y + \alpha_3 z + \alpha_4 xy + \alpha_5 yz + \alpha_6 xz + \alpha_7 x^2 + \alpha_8 y^2 + \alpha_9 z^2).$$

Comparing the coefficients leads to the form

$$\dot{x} = x(a_1 + a_4 y + a_6 z + a_7 x)$$

$$\dot{y} = y(A_2 + A_4 x + A_5 z + A_8 y)$$

$$\dot{z} = -z(a_1 + A_2 + y(a_4 + A_8) + x(a_4 + A_7) + z(a_6 + A_5))$$

showing that this equation is indeed a generalized Lotka–Volterra system.

With an appropriate choice of the coefficients, the Ivanova reaction can shown to be a special case. Note that the absence of negative cross effect played no role here.

8.38 Consider the system (6.35). Then, simple substitution shows that (8.37) is fulfilled with $H_m(\mathbf{x}) := x_m$ with the corresponding cofactor $k_m(\mathbf{x}) := f_m(\mathbf{x})$.

8.39 The induced kinetic differential equation of the Wegscheider reaction is $\dot{x} = -k_1 x + k_{-1} y - k_2 x^2 + k_{-2} xy$ $\dot{y} = k_1 x - k_{-1} y + k_2 x^2 - k_{-2} xy$. Given the initial concentrations so that $x(0) + y(0) = m$, one has to solve the quation $0 = -k_1 x + k_{-1}(m - x) - k_2 x^2 + k_{-2} x(m - x)$. Analyzing this equation it turns out that it always has a single positive solution for all possible positive values of the reaction rate coefficients and m.

8.40 Instead of direct calculations, one can say that the reversible triangle reaction has a deficiency zero, and it is weakly reversible; thus the zero-deficiency theorem ensures the statement.

8.41 Carrying out the steps described in Sect. 6.6 for the H atoms, one has

Pair of species	Flux rate of H molecules
$CH_3 \longrightarrow C_3H_7$	$3/10 \times 7 \times 0 = 0$
$CH_3 \longrightarrow C_4H_8$	$3/10 \times 8 \times w = 2.4w$
$CH_3 \longrightarrow H_2$	$3/10 \times 2 \times w = 0.6w$
$C_3H_7 \longrightarrow CH_3$	$7/10 \times 3 \times 0 = 0$
$C_3H_7 \longrightarrow C_4H_8$	$7/10 \times 8 \times w = 5.6w$
$C_3H_7 \longrightarrow H_2$	$7/10 \times 2 \times w = 1.4w$

Similarly, for the C atoms, one obtains

Pair of species	Flux rate of C molecules
$CH_3 \longrightarrow C_3H_7$	$1/4 \times 7 \times 0 = 0$
$CH_3 \longrightarrow C_4H_8$	$1/4 \times 4 \times w = w$
$CH_3 \longrightarrow H_2$	$1/4 \times 0 \times w = 0$
$C_3H_7 \longrightarrow CH_3$	$3/4 \times 1 \times 0 = 0$
$C_3H_7 \longrightarrow C_4H_8$	$3/4 \times 4 \times w = 3w$
$C_3H_7 \longrightarrow H_2$	$3/4 \times 0 \times w = 0$

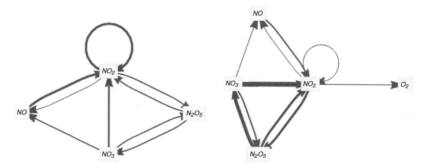

Fig. 14.13 The graph of N and O fluxes in the Ogg reaction (2.11). The thickness of the edges is proportional to the fluxes in column 3 of Tables 14.2 and 14.3

Table 14.2 Flux rate of N molecules

Pair of species	Flux rate of N molecules
$N_2O_5 \longrightarrow NO_2$	$2/4 \times 1 \times w_1 = w_1/2$
$N_2O_5 \longrightarrow NO_3$	$2/4 \times 1 \times w_1 = w_1/2$
$NO_2 \longrightarrow N_2O_5$	$1/4 \times 2 \times w_{-1} = w_{-1}/2$
$NO_2 \longrightarrow NO$	$1/4 \times 1 \times w_2 = w_2/4$
$NO_2 \longrightarrow NO_2$	$1/4 \times 1 \times w_2 = w_2$
$NO_3 \longrightarrow N_2O_5$	$1/4 \times 2 \times w_{-1} = w_{-1}/2$
$NO_3 \longrightarrow NO_2$	$1/4 \times 1 \times w_2 = 1/4 \times 2 \times w_3 = w_2/4 + w_3/2$
$NO_3 \longrightarrow NO$	$1/4 \times 1 \times w_2 = w_2/4$
$NO \longrightarrow NO_2$	$1/4 \times 2 \times w_3 = w_3/2$

8.42 The reaction in question is as follows:

$$N_2O_5 \underset{k_{-1}}{\overset{k_1}{\rightleftharpoons}} NO_2 + NO_3 \overset{k_2}{\longrightarrow} NO_2 + NO + O_2 \quad NO_3 + NO \overset{k_3}{\longrightarrow} 2\,NO_2 \quad (14.39)$$

Now one has to take into consideration all of the steps. Let us consider the flux of N atoms first.

As to the O atoms, one gets Table 14.3.

Figure 14.13 shows the graphs of the element fluxes assuming that all the reaction rate coefficients and concentrations are unity.

8.43 The deterministic model of the Lotka–Volterra reaction (as, e.g., given by the function **DeterministicModel**) is

$$x' = k_1 x - k_2 xy \quad y' = k_2 xy - k_3 y. \quad (14.40)$$

Table 14.3 Flux rate of O molecules

Pair of species	Flux rate of O molecules
$N_2O_5 \longrightarrow NO_2$	$5/10 \times 2 \times w_1 = w_1$
$N_2O_5 \longrightarrow NO_3$	$5/10 \times 3 \times w_1 = 3w_1/2$
$NO_2 \longrightarrow N_2O_5$	$2/10 \times 5 \times w_{-1} = w_{-1}$
$NO_2 \longrightarrow NO_2$	$2/10 \times 2 \times w_2 = 2w_2/5$
$NO_2 \longrightarrow NO$	$2/10 \times 1 \times w_2 = w_2/5$
$NO_2 \longrightarrow O_2$	$2/10 \times 2 \times w_2 = 2w_2/5$
$NO_3 \longrightarrow N_2O_5$	$3/10 \times 5 \times w_{-1} = 3w_{-1}/5$
$NO_3 \longrightarrow NO_2$	$3/10 \times 2 \times w_1 + 3/8 \times 4 \times w_2 = 3w_1/5 + 3w_2/2$
$NO_3 \longrightarrow NO$	$3/10 \times 1 \times w_1 = 3w_2/10$
$NO_3 \longrightarrow O_2$	$3/10 \times 2 \times w_1 = 3w_2/5$

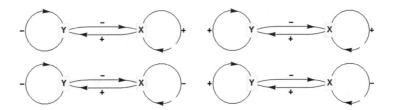

Fig. 14.14 Influence diagrams of the Lotka–Volterra reaction in different subdomains of the first quadrant

As

$$\mathbf{f}'(x, y) = \begin{bmatrix} k_1 - k_2y & -k_2y \\ k_2x & k_2x - k_3 \end{bmatrix}, \tag{14.41}$$

the influence diagrams in the interior points of four domains determined by the nullclines of the individual variables can be constructed as shown in Fig. 14.14.

8.44 In all three cases, we assume that the residence time in the reactor is ϑ, and the concentrations in the feed are $a^f, b^f \in \mathbb{R}^+$.

• The induced kinetic differential equation of the reaction

$$A + B \underset{k_{-1}}{\overset{k_1}{\rightleftharpoons}} 2A \quad A \underset{a^f/\vartheta}{\overset{1/\vartheta}{\rightleftharpoons}} 0 \underset{1/\vartheta}{\overset{b^f/\vartheta}{\rightleftharpoons}} B$$

being

$$\dot{a} = k_1ab - k_1a^2 + (a^f - a)/\vartheta \quad \dot{b} = -k_1ab + k_1a^2 + (b^f - b)/\vartheta,$$

the stationary states a_* and b_* should obey $b_* = a^f + b^f/a_*$ and

$$0 = -k_1 a_*(a^f + b^f - a_*) + k_{-1}a_*^2 + (a_* - a^f)/\vartheta =: P_2(a_*).$$

The quadratic polynomial P_2 is negative at 0 and has a positive leading coefficient; therefore it cannot have two positive roots.

- $A + 2B \underset{k_{-1}}{\overset{k_1}{\rightleftharpoons}} 3A \quad A \underset{a^f/\vartheta}{\overset{1/\vartheta}{\rightleftharpoons}} 0 \underset{1/\vartheta}{\overset{b^f/\vartheta}{\rightleftharpoons}} B$

  Now we have

$$\dot{a} = 2k_1 ab^2 - 2k_1 a^3 + (a^f - a)/\vartheta \quad \dot{b} = -2k_1 ab^2 + 2k_1 a^3 + (b^f - b)/\vartheta;$$

thus for a_* one has

$$P_3(a_*) := 2a_*^3(-k_1 + k_{-1})/\vartheta^2 + 4a_*^2 k_1(a^f + b^f)/\vartheta^2$$
$$+ a_*(-2k_1(a^f + b^f)^2 + 1/\vartheta)/\vartheta^2 - k_2 = 0.$$

The assumption that $P_3(a_*) = p_3(a_* - \lambda_1)(a_* - \lambda_2)(a_* - \lambda_3)$ holds with two or three positive λs can easily be shown to contradict to the sign pattern of the coefficients of P_3.

- $2A + B \underset{k_{-1}}{\overset{k_1}{\rightleftharpoons}} 3A \quad A \underset{a^f/\vartheta}{\overset{1/\vartheta}{\rightleftharpoons}} 0 \underset{1/\vartheta}{\overset{b^f/\vartheta}{\rightleftharpoons}} B$

  A similar calculation as above leads for the first component of the stationary state to a polynomial equation $P_3(a_*) = 0$, such that the sign of the leading term and that of the first-order term is positive, and the signs of the two other terms are negative as follows:

$$P_3(a_*) = p_3 a_*^3 - p_2 a_*^2 + p_1 a_* - p_0; \quad (p_0, p_1, p_2, p_3 \in \mathbb{R}^+).$$

It can be easily shown that such a polynomial can have three positive roots, e.g., in the case if $p_0 = 6$, $p_1 = 11$, $p_2 = 6$, and $p_3 = 1$. The simple problem remains to show that the reaction rate coefficients, the feed concentrations, and the residence time can be chosen so as to have the prescribed coefficients.

If one should like to map the part of the space of parameters $k_1, \ldots, \vartheta$ where this can happen, one can apply the necessary and sufficient condition for a cubic polynomial to have three positive roots which can be constructed from the negativity of the discriminant and the Routh–Hurwitz criterion.

What happens if some of the feed concentrations are zero? Is it possible to have the answer in some cases using the deficiency one theorem (Theorem 8.48)? One may use **DeterministicModel** and **StationaryPoints**.

8.45 Verify that

$$\psi_1(\mathbf{c}) := \sum_{m=1}^{5} c_m \quad \psi_2(\mathbf{c}) := \prod_{m=1}^{5} c_m^{a_m}$$

(with $a_1 := k_2 k_5, a_2 := k_3 k_5, a_3 := k_1 k_5, a_4 := k_2 k_6, a_5 := k_2 k_4$) are first integrals of the induced kinetic differential equation of the reaction (8.50); thus all the trajectories stay in a bounded set as $t \to +\infty$.

14.9   Approximations of the Models

9.1 In the case of the reversible bimolecular reaction, one can either start from A, B, or C.

```
VolpertIndexing[{"A" + "B" <=> "C"}, {"A", "B"}]
```

shows that $c(t) = t d(t)$ near the origin with a continuous function d, and

```
VolpertIndexing[{"A" + "B" <=> "C"}, {"C"}]
```

shows that $a(t) = t e(t)$ and $b(t) = t f(t)$ near the origin with continuous functions e and d.

9.2 If $\boldsymbol{\rho}^\top \boldsymbol{\gamma} \leq \mathbf{0}$ holds with $\boldsymbol{\rho} \in (\mathbb{R}_0^+)^M$, then using the induced kinetic differential equation of the reaction, one has

$$\boldsymbol{\rho}^\top \dot{\mathbf{c}} = \boldsymbol{\rho}^\top \boldsymbol{\gamma} (\mathbf{w} \circ \mathbf{c}) = \sum_{r \in \mathcal{R}} w_r \circ \mathbf{c} \sum_{m \in \mathcal{M}} \rho_m \gamma(m, r) \leq 0. \tag{14.42}$$

Taking the integral of both sides, one gets

$$\sum_r \int_0^t w_r(\mathbf{c}(s)) \, ds \sum_m \rho_m \gamma(m, r) = \sum_m \rho_m c_m(t) - \sum_m \rho_m c_m(0) \leq 0 \tag{14.43}$$

showing the summability of all the reactions for which strong inequality holds. The sufficient condition is obviously fulfilled if the reaction is strongly mass consuming; see Eq. (4.7).

9.3 One has to find nonnegative solutions to $\boldsymbol{\rho}^\top \boldsymbol{\gamma} \leq \mathbf{0}$ so that some of the inequalities hold in the strict sense.

Lotka–Volterra reaction: $\begin{bmatrix} 1 & 2 & 0 \end{bmatrix}^\top$ is a solution with which the first two reaction steps can be shown to be summable.

Reversible bimolecular reaction: The condition cannot be applied.

Michaelis–Menten reaction: $\begin{bmatrix} 2 & 2 & 4 & 1 \end{bmatrix}^\top$ shows that the third step is summable.

Robertson reaction: $\begin{bmatrix} 1 & 1 & 2 \end{bmatrix}^\top$ is a solution with which the second reaction step can be shown to be summable.

9.4 The Jacobian of the right-hand side of (9.6) is $\begin{bmatrix} -k_1e_0 + k_1c & k_1s + k_{-1} \\ k_1e_0 - k_1c & -k_1s - (k_{-1} + k_2) \end{bmatrix}$.

Evaluated at the stationary point—at the origin—it is $\begin{bmatrix} -k_1e_0 & k_{-1} \\ k_1e_0 & -(k_{-1} + k_2) \end{bmatrix}$.

Finally, its characteristic equation $\lambda^2 + (k_1e_0 + k_{-1} + k_2)\lambda + k_1k_2e_0 = 0$ obviously has two negative real solutions, implying that the origin is an asymptotically stable node.

9.5 The induced kinetic differential equation of the reaction S $\rightleftharpoons$ C $\longrightarrow$ P is $\dot{s} = -k_1s + k_{-1}c, \dot{c} = k_1s - (k_{-1} + k_2)c, \dot{p} = k_2c$ with the initial condition $s(0) = s_0, c(0) = p(0) = 0$. Assuming $c \ll s$, the mass conservation relation $s+c+p = s_0$ can approximately be written as $p \approx s_0 - s$; therefore $|\dot{s}| = |\dot{p}|$, which means that $k_{-1}c - k_1s \approx k_2c$, implying that $\dot{c} \approx 0$. This last approximate equality gives us $c \approx \frac{k_1s}{k_{-1}+k_2}$, or $\dot{p} \approx \frac{k_1k_2s}{k_{-1}+k_2}$. This expression gives a simple approximation of the product formation rate provided by the solution of the full induced kinetic differential equation of the reaction.

9.6 The result of the substitution is

$$S' = \frac{(1-\alpha)(\beta - 1)S}{1 - \alpha + \alpha S}.$$

To find the symbolic solution, the best is to use **DSolve** which gives

`S[t] -> (1−α)/αProductLog[α/(1 − α)Exp[t(β − 1)+C[1]/(1 − α)]].`

The constant can also be determined, and finally one has (Schnell and Mendoza 1997)

$$S(t) = \frac{1 - \alpha}{\alpha} W \left(\frac{\alpha}{1 - \alpha} s_0 e^{\frac{\alpha s_0}{1-\alpha}} e^{t(\beta-1)} \right),$$

where W is the same as the **ProductLog** function of *Mathematica*. The domain of the obtained solution is to be investigated as soon as the values of the parameters are given.

9.7 The simplest way to find the unique stationary point (that is also positive) is to use **NSolve**. **StationaryPoints** which gives the same answer when you have constructed an inducing reaction.

It is hopeless to compute a Gröbner basis of this system without a computer. The expression

```
GroebnerBasis[{2 + x - x^2 + y^2, 3 + x y - y^3 + x z,
    2 + x^3 - y z - z^2}, {x, y, z}]
```

yields three polynomials of degrees 11 and 12, with coefficients having up to 25 digits. The univariate polynomial of the triangularized system is a polynomial in z, and it has two real roots, but only one of them is nonnegative. (*Mathematica* cannot give a closed form formula for the roots, but it can determine symbolically whether they are real or nonnegative.) Plugging this number into the second equation, we can determine the values of y: This time there is only one real root, and it is nonnegative. (Again, there is no closed formula for this number, but it can be expressed as the root of a degree-12 polynomial.) Finally, the correct values of x can be determined by substituting the possible values of x and y into the third polynomial of the basis. Again, there is only one real root x, and it is nonnegative.

The situation is the same with any permutation of the variables.

9.8 Let us suppose that $\tilde{\mathbf{h}}$ is also a generalized inverse of $\mathbf{h}$, i.e., $\mathbf{h} \circ \tilde{\mathbf{h}} = \mathrm{id}_{\mathbb{R}^{\hat{M}}}$ also holds. Then,

$$(\mathbf{h'f}) \circ \overline{\mathbf{h}} = (\mathbf{h'f}) \circ \overline{\mathbf{h}} \circ \mathrm{id}_{\mathbb{R}^{\hat{M}}} = (\mathbf{h'f}) \circ \overline{\mathbf{h}} \circ (\mathbf{h} \circ \tilde{\mathbf{h}}) = (\mathbf{h'f}) \circ (\overline{\mathbf{h}} \circ \mathbf{h}) \circ \tilde{\mathbf{h}}$$

$$= (\hat{\mathbf{f}} \circ \mathbf{h}) \circ (\overline{\mathbf{h}} \circ \mathbf{h}) \circ \tilde{\mathbf{h}} = (\hat{\mathbf{f}} \circ \mathbf{h}) \circ \tilde{\mathbf{h}} = (\mathbf{h'f}) \circ \tilde{\mathbf{h}}$$

proves independence.

9.9 Instead of applying Theorem 13.38, we use a slightly different solution. Let us introduce $\bar{u} := x + y$ and then $(a + c)x + (b + d)y = (a + c)x + (b + d)(\bar{u} - x)$. The derivative of this expression with respect to x being $(a + c) - (b + d)$ should be zero.

9.10 An immediate calculation (possibly using **Eigensystem**) gives the eigenvalues 2 and $x_1 + x_2 + 2$ corresponding to the eigenvectors $\begin{bmatrix} -x_2 & x_1 \end{bmatrix}^\top$ and $\begin{bmatrix} 1 & 1 \end{bmatrix}^\top$, respectively.

9.11

Triangle reaction   The relevant quantities here are as follows:

$$\mathbf{K} = \begin{bmatrix} -(k_1 + k_{-3}) & k_{-1} & k_3 \\ k_1 & -(k_{-1} + k_2) & k_{-2} \\ k_{-3} & k_2 & -(k_3 + k_{-2}) \end{bmatrix} \quad \mathbf{M} = \begin{bmatrix} 1 & 0 & 0 \\ 0 & 1 & 1 \end{bmatrix},$$

and the condition (9.23) for the unknown matrix $\hat{\mathbf{K}} = \begin{bmatrix} -a & b \\ a & -b \end{bmatrix}$ is now

$$\begin{bmatrix} -(k_1 + k_{-3}) & k_{-1} & k_3 \\ k_1 + k_{-3} & -k_{-1} & -k_3 \end{bmatrix} = \mathbf{MK} = \hat{\mathbf{K}}\mathbf{M} = \begin{bmatrix} -a & b \\ a & -b \end{bmatrix}\begin{bmatrix} 1 & 0 & 0 \\ 0 & 1 & 1 \end{bmatrix} = \begin{bmatrix} -a & b & b \\ a & -b & -b \end{bmatrix}$$

implying $a = (k_1 + k_{-3})$ $b = k_3 = k_{-1}$. (The last equality is necessary for exact lumpability in this case.) The condition for detailed balancing in this zero-deficiency case reduces to the single circuit condition : $k_1 k_2 k_3 = k_{-1} k_{-2} k_{-3}$. This implies for the lumped system that $k_1 k_2 = k_{-2} k_{-3}$, although the reaction rate coefficients k_2 and k_{-2} have no role in the lumped reaction.

Square reaction   The relevant quantities here are as follows:

$$\mathbf{K} = \begin{bmatrix} -(k_1 + k_{-4}) & k_{-1} & 0 & k_4 \\ k_1 & -(k_{-1} + k_2) & k_{-2} & 0 \\ 0 & k_2 & -(k_{-2} + k_3) & k_{-3} \\ k_{-4} & 0 & k_3 & -(k_4 + k_{-3}) \end{bmatrix} \quad \mathbf{M} = \begin{bmatrix} 1 & 0 & 0 & 0 \\ 0 & 1 & 0 & 0 \\ 0 & 0 & 1 & 1 \end{bmatrix},$$

and the condition (9.23) for the unknown matrix $\hat{\mathbf{K}} = \begin{bmatrix} -d - g & b & c \\ d & -b - h & f \\ g & h & -c - f \end{bmatrix}$

is now

$$\begin{bmatrix} -(k_1 + k_{-4}) & k_{-1} & 0 & k_4 \\ k_1 & -(k_{-1} + k_2) & k_{-2} & 0 \\ k_{-4} & 0 & k_2 & -k_{-2} & -k_4 \end{bmatrix} = \mathbf{MK} = \hat{\mathbf{K}}\mathbf{M}$$

$$= \begin{bmatrix} -d - g & b & c & c \\ d & -b - h & f & f \\ g & h & -c - f & -c - f \end{bmatrix}$$

implying $b = k_{-1}$, $c = 0 = k_4$, $d = k_1$, $f = k_{-2} = 0$, $g = k_{-4}$, $h = k_2$, . The condition for detailed balancing in this zero-deficiency case would reduce to the single circuit condition, but the original reaction is not even weakly reversible and so is the lumped reaction.

9.12 As the coefficient matrix of the induced kinetic differential equation of the reaction in Fig. 9.8 is $\mathbf{K} = \begin{bmatrix} -13 & 2 & 4 \\ 3 & -12 & 6 \\ 10 & 10 & -10 \end{bmatrix}$, our task is to solve the equation (linear and overdetermined) in the new coefficients

$$\mathbf{MK} = \begin{bmatrix} 1 & 1 & 0 \\ 0 & 0 & 1 \end{bmatrix} \begin{bmatrix} -13 & 2 & 4 \\ 3 & -12 & 6 \\ 10 & 10 & -10 \end{bmatrix} = \begin{bmatrix} -10 & -10 & 10 \\ 10 & 10 & -10 \end{bmatrix}$$

$$= \begin{bmatrix} -\hat{k}_{21} & \hat{k}_{12} \\ \hat{k}_{21} & -\hat{k}_{12} \end{bmatrix} \begin{bmatrix} 1 & 1 & 0 \\ 0 & 0 & 1 \end{bmatrix} = \begin{bmatrix} -\hat{k}_{21} & -\hat{k}_{21} & \hat{k}_{12} \\ \hat{k}_{21} & \hat{k}_{21} & -\hat{k}_{2} \end{bmatrix} = \hat{\mathbf{K}}\mathbf{M}.$$

One reaction having the linear induced kinetic differential equation with the coefficient matrix $\hat{\mathbf{K}}$ is $\hat{A}(1) \underset{10}{\overset{10}{\rightleftharpoons}} \hat{A}(2)$. We had to use this rather cautious wording as we know that the reaction is not uniquely determined by its induced kinetic differential equation; see Sect. 11.6. Check that more generally, the induced kinetic differential equation of the reversible triangle reaction in Fig. 9.8 can be exactly lumped with the above lumping matrix $\mathbf{M}$ if $k_{CA} = k_{CB}$ holds and all the other reaction rate coefficients are arbitrary.

9.13 The induced kinetic differential equation of the reaction (9.42) is

$$\dot{x}_1 = -(k_{21} + k_{31} + k_{41})x_1 + k_{12}x_2 + k_{13}x_3 + k_{14}x_4$$

$$\dot{x}_2 = k_{21}x_1 - (k_{12} + k_{32} + k_{42})x_2 + k_{23}x_3 + k_{24}x_4$$

$$\dot{x}_3 = k_{31}x_1 + k_{32}x_2 - (k_{13} + k_{23} + k_{43})x_3 + k_{34}x_4$$

$$\dot{x}_4 = k_{41}x_1 + k_{42}x_2 + k_{43}x_3 - (k_{14} + k_{24} + k_{34})x_4.$$

Let us introduce the notations $\kappa := k_{41} = k_{42} = k_{43}$, $\lambda := k_{14} + k_{24} + k_{34}$, and let

$$\begin{bmatrix} y_1 \\ y_2 \end{bmatrix} := \mathbf{M} \begin{bmatrix} x_1 \\ x_2 \\ x_3 \\ x_4 \end{bmatrix},$$

or, $y_1 := x_1 + x_2 + x_3$ and $y_2 := x_4$; then one has

$$\dot{y}_1 = -\kappa y_1 + \lambda y_2 \quad \dot{y}_2 = \kappa y_1 - \lambda y_2; \qquad (14.44)$$

as stated. Note that one has for the initial conditions: $y_1(0) = x_1(0) + x_2(0) + x_3(0)$ and $y_2(0) = x_4(0)$.

9.14 Let the induced kinetic differential equation be $\dot{c} = 1$, and introduce $\hat{c} = -c$; then the lumped differential equation is $\dot{\hat{c}} = -1$ which is not kinetic.

9.15 More precisely, we will show that the lumped differential equation can be considered as the induced kinetic differential equation of a reaction of the mentioned types. Let us start with the induced kinetic differential equation of five species

$$\dot{x} = Ax + ay + bz + cu + dv + \xi$$
$$\dot{y} = ex + By + fz + gu + hv + \eta$$
$$\dot{z} = ix + jy + Cz + ku + lv + \zeta$$
$$\dot{u} = mx + ny + oz + Du + pv + \upsilon$$
$$\dot{v} = qx + ry + sz + tu + Ev + v$$

with all the coefficients—except possibly A, B, C, D, E—being nonnegative, and, without restricting generality, let the proper lumping matrix be

$$\mathbf{M} := \begin{bmatrix} 1 & 1 & 0 & 0 & 0 \\ 0 & 0 & 1 & 1 & 1 \end{bmatrix}. \tag{14.45}$$

With the notation

$$\mathbf{K} := \begin{bmatrix} A & a & b & c & d \\ e & B & f & g & h \\ i & j & C & k & l \\ m & n & o & D & p \\ q & r & s & t & E \end{bmatrix}, \quad \varphi = \begin{bmatrix} \xi \\ \eta \\ \zeta \\ \upsilon \\ v \end{bmatrix}, \tag{14.46}$$

we have that the off-diagonal elements of $\mathbf{K}$ are nonnegative, and $\varphi \geq \mathbf{0}$. Furthermore, in the case of a closed compartmental system, the diagonal elements of $\mathbf{K}$ are equal to the negative of the column sums; in the case of half-open systems, they are not more than the negative of the column sums. In both cases $\varphi = \mathbf{0}$. In case of open compartmental systems, φ can also have positive components. Let us calculate $\hat{\mathbf{K}}$ using the representation (9.26):

$$\hat{\mathbf{K}} = \begin{bmatrix} \frac{1}{2}(A + B + a + e) & \frac{1}{3}(b + c + d + f + g + h) \\ \frac{1}{2}(i + j + m + n + q + r) & \frac{1}{3}(C + D + E + k + l + o + p + s + t) \end{bmatrix}.$$

Furthermore, the transform of the constant term is $\hat{\varphi} = \begin{bmatrix} \xi + \eta & \zeta + \upsilon + v \end{bmatrix}^{\mathsf{T}}$. These formulas show the statement on the form of the lumped system.

9.16 Lump the reaction $X \overset{1}{\longrightarrow} 0 \ 3\,Y \overset{\frac{1}{3}}{\longrightarrow} 0$ with the function $h(x, y) := -2y^2$ to obtain the deterministic model of the reaction $2\,U \longrightarrow 3\,U$.

9.17 In the case of the original system, one has

$$\text{rank} \begin{bmatrix} 1 & 1 & 0 & -k & 0 & k^2 \\ 1 & 0 & 0 & 0 & 0 & 0 \\ 1 & -1 & 0 & k & 0 & -k^2 \end{bmatrix} = 2 < 3.$$

In the lumped system,

$$\text{rank} \begin{bmatrix} 3 & 2 & 0 & -2k \\ 3 & -2 & 0 & 2k \end{bmatrix} = 2.$$

The lumping matrix is $\begin{bmatrix} 2 & 1 & 0 \\ 0 & 1 & 2 \end{bmatrix}$.

9.18 The simple bimolecular reaction $A + B \rightleftharpoons C$ has the linear first integrals defined by $\omega_1 := \begin{bmatrix} 1 & 0 & 1 \end{bmatrix}^T$, $\omega_2 := \begin{bmatrix} 1 & -1 & 0 \end{bmatrix}^T$ of which the second one is not even a nonnegative vector. However, the positive vector $\omega_1 := \begin{bmatrix} 1 & 1 & 2 \end{bmatrix}^T$ also defines a first integral for the induced kinetic differential equation of the reaction.

9.19 See Frank (2008, Chapter 8).

9.20 Let us consider again the reaction in Fig. 14.15. Let all the reaction rate constants be unity. Now we are going to show that the vector $\rho = \begin{bmatrix} 2 & 1 & 4 & 4 & 5 & 2 & 2 & 1 & 1 \end{bmatrix}^T$ is orthogonal to the right-hand side of the induced kinetic differential equation

$$\begin{aligned}
a' &= -2ab + c & b' &= -2ab + c & c' &= ab - 2c \\
d' &= c - de + f & e' &= -f + de & f' &= ab - g + j^2 \\
g' &= g - h + j^2 & h' &= 2h - 4j^2 & j' &= c + f - de
\end{aligned}$$

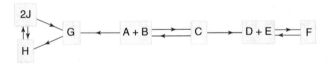

Fig. 14.15 The Feinberg–Horn–Jackson graph of an example by (Feinberg and Horn 1977, page 89)

of the reaction, but it is not orthogonal from the left to the matrix

$$
\gamma =
\begin{array}{c|cccccccccc}
 & 1 & 2 & 3 & 4 & 5 & 6 & 7 & 8 & 9 & 10 \\
\hline
A & 0 & 0 & 0 & 1 & -1 & -1 & 0 & 0 & 0 & 0 \\
B & 0 & 0 & 0 & 1 & -1 & -1 & 0 & 0 & 0 & 0 \\
C & 0 & 0 & -1 & -1 & 1 & 0 & 0 & 0 & 0 & 0 \\
D & 1 & -1 & 1 & 0 & 0 & 0 & 0 & 0 & 0 & 0 \\
E & -1 & 1 & 0 & 0 & 0 & 0 & 0 & 0 & 0 & 0 \\
F & 0 & 0 & 0 & 0 & 0 & 1 & -1 & 0 & 0 & 1 \\
G & 0 & 0 & 0 & 0 & 0 & 0 & 1 & -1 & 1 & 0 \\
H & 0 & 0 & 0 & 0 & 0 & 0 & 0 & 2 & -2 & -2 \\
J & 1 & -1 & 1 & 0 & 0 & 0 & 0 & 0 & 0 & 0
\end{array}
\tag{14.47}
$$

of the elementary reaction vectors. Indeed, $\rho^\top \left[a'\, b'\, c'\, d'\, e'\, f'\, g'\, h'\, j' \right]^\top = 0$, and $\rho^\top \gamma = \left[0\, 0\, 1\, -1\, 1\, -1\, 0\, 0\, 0\, 0 \right]^\top \neq \mathbf{0}^\top$. Use `ReactionsData` to check the calculations. Let us mention that Problem 13.10 gives another, much simpler example.

9.21 If the vector $\omega \in \mathbb{R}^M$ defines a linear first integral for the (9.37), then one has $\omega^\top \mathbf{f} = 0$; therefore by induction

1. $\omega^\top \mathbf{x}^{n+1} = \omega^\top \mathbf{x}^n = \omega^\top \mathbf{x}^0$,
2. $\omega^\top \mathbf{x}^{n+1} = \omega^\top (\mathbf{x}^n - n \cdot 10^6 \cdot \mathbf{f}(\mathbf{x}^n)) = \omega^\top \mathbf{x}^n = \omega^\top \mathbf{x}^0$,

respectively. Obviously, the sequences defined above can only be close to the solution accidentally.

9.22 We restrict the problem for the autonomous scalar case. Furthermore, it is enough to show that a single step of the method has the given property. Let $f \in \mathscr{C}^1(\mathbb{R}, \mathbb{R})$, and let us suppose that the positive half line is an invariant set of the differential equation $y'(x) = f(y(x))$. It means that $f(\eta) \geq 0$, if $\eta > 0$. The solution starting from the point (ξ, η) with $\eta > 0$ is approximated on the interval $[\xi, \xi + h]$ $(h > 0)$ by the Euler method as $[\xi, \xi + h] \ni x \mapsto \eta + hf(\eta)(x - \xi) > 0$.

9.23 Let $\mathbf{f} \in \mathscr{C}^1(\mathbb{R}^M, \mathbb{R}^M)$, and let us suppose that $\mathbf{y}^*$ is a stationary point of the differential equation $\mathbf{y}'(x) = \mathbf{f}(\mathbf{y}(x))$, i.e., $\mathbf{0} = \mathbf{f}(\mathbf{y}^*)$. The Euler method with a step $h \in \mathbb{R}^+$ gives $\mathbf{y}^* + h\mathbf{f}(\mathbf{y}^*) = \mathbf{y}^*$.

Conversely, if $\mathbf{y}^* + h\mathbf{f}(\mathbf{y}^*) = \mathbf{y}^*$ with a positive h, then $\mathbf{f}(\mathbf{y}^*) = \mathbf{0}$, or $\mathbf{y}^*$ is a stationary point of the differential equation.

9.24 The Euler method in this case is $x^0 := x_0$, $y^0 := y_0$; and $x^{n+1} := x^n + hy^n$, $y^{n+1} = y^n - hx^n$. Then,

$$(x^{n+1})^2 + (y^{n+1})^2 = (x^n + hy^n)^2 + (y^n - hx^n)^2 = (1 + h^2)((x^n)^2 + (y^n)^2)$$

meaning that the numerical approximation will take samples from an expanding spiral.

9.25 The induced kinetic differential equation will be the same, but the reaction rate of the first reaction becomes $10^6 \frac{\mu mol}{dm^3 sec}$ in the new units; therefore the induced kinetic differential equation ceases to be stiff any more.

9.26 The mass action type induced kinetic differential equation of the reaction (9.49) is

$$[\dot{Br_2}] = -k_1[Br_2] - k_3[H][Br_2] + k_5[Br]^2 \tag{14.48}$$

$$[\dot{Br}] = 2k_1[Br_2] - k_2[H_2][Br] + k_3[H][Br_2] + k_4[H][HBr] - 2k_5[Br]^2 \tag{14.49}$$

$$[\dot{H_2}] = -k_2[H_2][Br] + k_4[H][HBr] \tag{14.50}$$

$$[\dot{H}] = k_2[H_2][Br] - k_3[H][Br_2] - k_4[H][HBr] \tag{14.51}$$

$$[\dot{HBr}] = k_2[H_2][Br] + k_3[H][Br_2] - k_4[H][HBr]. \tag{14.52}$$

Put 0 instead of the derivative in Eqs. (14.49) and (14.51), and solve the emerging algebraic system to get

$$[Br]^2 = \frac{k_1}{k_5}[Br_2] \quad [H] = \frac{k_2[H_2]}{k_3[B_2] + k_4[HBr]} \sqrt{\frac{k_1[Br_2]}{k_5}}.$$

Putting the second expression into $[\dot{HBr}] = 2k_3[H][Br_2]$, one gets finally

$$[\dot{HBr}] \approx \frac{2k_2[H_2]}{[B_2] + \frac{k_4}{k_3}[HBr]} \sqrt{\frac{k_1}{k_5}} [Br_2]^{3/2}$$

as the approximate rate of the overall reaction $H_2 + Br_2 \longrightarrow 2\,HBr$.

14.10 The Stochastic Model

10.1 The master equation is given by

```
MasterEquation[gene,
    {λp1, λm1, λ2, λ3, γp, γm},
    {t, x, y, u, v}, p].
```

The equation for the probability generating function is obtained similarly

```
ProbabilityGeneratingFunctionEquation[gene,
    {λp1, λm1, λ2, λ3, γp, γm},
    {t, z1, z2, z3, z4}].
```

The following command gives us the solution.

```
SolveProbabilityGeneratingFunctionEquation[gene,
    {λp1, λm1, λ2, λ3, γp, γm},
    {i0, a0, m0, p0}, Method -> "MatrixExponential"]
```

10.2 Let $X(0) = x_0 > 0$ be fixed. The master equation of the outflow then reads as $\dot{p}_x(t) = kp_{x+1}(t) - kp_x(t)$, where $k > 0$, $x \in \mathbb{N}_0$ and $p_x(0) = \mathfrak{B}(x = x_0)$. For the probability generating function G, one has $\dot{G}(t, z) = k(1 - z)\frac{\partial G}{\partial z}(t, z)$, where $G(0, z) = z^{x_0}$ and $G(t, 1) = 1$. This is a linear partial differential equation; the method of characteristics (Theorem 13.61) yields $G(t, z) = \left((z - 1)\exp(-kt) + 1\right)^{x_0}$. Taylor-series expansion then results in $G(t, z) = \sum_{x=0}^{x_0} p_x(t)z^x = \sum_{x=0}^{x_0} \binom{x_0}{x}\exp(-ktx)(1 - \exp(-kt))^{x_0-x}z^x$; hence the transient probability distribution is binomial. From here the solutions to the moment equations can be obtained from the appropriate moments of the binomial distribution with parameters x_0 and $\exp(-kt)$. For example, the first moment as a function of time is $\mathbf{E}(X(t)) = x_0 \exp(-kt)$. Since the reaction is of the first order, it is consistent in mean; hence the solution to its first moment equation coincides with that of the induced kinetic differential equation. The second moment is $\mathbf{E}(X(t)^2) = x_0(1 + (x_0 - 1)\exp(-kt))\exp(-kt)$, and one can proceed similarly. These formulas can be verified by putting them into the moment equations which is left to the reader.

10.3 The master equation of the given reaction is

$$\dot{p}(t, x, y) = k_1(x - 1)(y + 1)\mathfrak{B}(x \geq 2, y \geq 0)p(t, x - 1, y + 1)$$
$$+ k_2(x + 1)(y - 1)\mathfrak{B}(x \geq 0, y \geq 2)p(t, x + 1, y - 1)$$
$$- (k_1xy + k_2xy)\mathfrak{B}(x \geq 1, y \geq 1)p(t, x, y).$$

In this form it is rather a formidable task to find its solution. However, the partial differential equation for the probability generating function reads as

$$\dot{G}(t, z_1, z_2) = (k_1 z_1 - k_2 z_2)(z_1 - z_2)\frac{\partial^2 G}{\partial z_1 \partial z_2}(t, z_1, z_2).$$

Though this latter equation has a more compact form, it is still sufficiently hard to say anything about its solution. Note that the reaction is mass conserving, i.e., the total mass of the two species remains constant. It implies that for the coordinate processes, it also holds that $X(t) + Y(t) = x_0 + y_0$, that is, the change in the number of molecules of the species X and Y takes place along a line segment in the (closed) first quadrant, analogously to the deterministic setting. The internal states of each line segment form a transient communicating class. The two end points of the line segments are absorbing. In the chemical language, this means that starting the process at any of the internal states, one of the species becomes extinct. Hence the two stationary distributions are concentrated on one of the end points of the line segment.

The fact that X and also Y are (sub or super)martingales comes from Theorem 10.15.

Using the sole mass conservation relation, the induced kinetic differential equation of this reaction translates to the equation

$$\dot{x}(t) = (k_1 - k_2)x(t)(x_0 + y_0 - x(t)), \tag{14.53}$$

which is the well-known **logistic equation** (here x denotes the deterministic concentration of the species X). If $k_1 = k_2$ then this has constant solutions, while in the stochastic case, still a nontrivial martingale process arises. On the other hand, if $k_1 > k_2$, the only asymptotically stable and unstable stationary points of the reduced equation (14.53) are $x_* = x_0 + y_0$ and $x_{**} = 0$, respectively. (If $k_1 < k_2$, 0 is asymptotically stable and $x_0 + y_0$ is unstable.) This phenomenon appears in a more delicate way in the stochastic model.

Notice that this reaction can be considered as a slightly modified version of the Ehrenfest model (Ehrenfest and Ehrenfest 1907; Vincze 1964).

10.4 We invoke the formula (10.14) which describes the change of units between the reaction rate coefficients. The *Mathematica* function

```
avogadronumber = UnitConvert[Quantity[1, "AvogadroConstant"]];
ksto[reactionstep_, kdet_, V_] :=
    kdet (avogadronumber V)^
    (1 - ReactionsData[{reactionstep}]
        ["reactionsteporders"])
```

employs a transformation also on the values of the corresponding reaction rate coefficients **ksto** and **kdet**. We only show the second application:

```
Inner[ksto[#1, #2, 10^{-12} "L"]&,
ReactionsData[{"Lotka-Volterra"},{"A","B"}][
    "reactionsteps"],
    {5("s")^{-1},
    40"L"Quantity[1, ("Moles")^{-1}]("s"){-1},
    0.5("s")^{-1}}, List]
```

10.5 The states of a communicating class can reach one another with positive probability. The reaction $0 \longrightarrow X$ has communicating classes $U_k = \{k\}$ for $k \in \mathbb{N}_0$. All the classes are transient. On the other hand, making the reaction reversible, that is, in the case of $0 \rightleftharpoons X$, we only have a single communicating class, $U_0 = \mathbb{N}_0$, and this class is recurrent. Slightly more complicated version of the previous reaction is $X \rightleftharpoons 2X$. In this case when no molecules are present, the reaction cannot proceed. Hence $U_0 = \{0\}$, and $U_1 = \mathbb{N}\setminus\{0\}$ are the communicating classes. Both involve recurrent states. Finally, the reaction $X + Y \rightleftharpoons 2X$ $X + Y \rightleftharpoons 2Y$ involves infinitely many communicating classes: $U_{k+1} = \{(i, j)|i + j = k\}$ (for $k \geq 2$) in addition to $U_0 = \{(0, 0)\}$, $U_1 = \{(0, 1)\}$ and $U_2 = \{(1, 0)\}$. Also in these cases, all the classes are recurrent.

10.6 The *Mathematica* function is

```
ProbabilityGeneratingFunctionEquation[
    alpha_, beta_, rates_, vars_] :=
    D[g @@ vars, First[vars]] ==
    Total[rates MapThread[(Times @@ (Rest[vars]^#1) -
    Times @@ (Rest[vars]^#2))*
    Derivative[0, Sequence @@ #2][g] @@ vars&,
    Transpose /@ {beta, alpha}]]
```

where **vars** is the user-given list of variables, e.g., **vars={t,z1,z2}** can be given in the case of a two-species reaction. The first entry of this list is used as the time variable. One can make attempts to solve the obtained equation in very special cases using the built-in *Mathematica* function **DSolve**.

10.7 Consider the reaction $X + Y \longrightarrow 2X$ $X + Y \longrightarrow 2Y$ with the assumption that the two reaction rate coefficients are the same for the deterministic and the stochastic model as well. Then the induced kinetic differential equation of the reaction is $\dot{x}(t) = \dot{y}(t) = 0$. It is also not hard to see that for the induced kinetic Markov process (X, Y), it holds that $\frac{d}{dt}\mathbf{E}X(t) = \frac{d}{dt}\mathbf{E}Y(t) = 0$. It follows that the process is consistent in mean with this choice of reaction rate coefficients. However, as we have already noticed in Problem 10.3, (X, Y) is a nontrivial process.

10.8 Consider the reaction $X + Y \xrightarrow{k_1} 3X + 3Y$ which is not stoichiometrically mass conserving. In this case $\gamma(\cdot, 1) = \begin{bmatrix} 2 & 2 \end{bmatrix}^\top$; hence $\mathbf{D}_2((x, y)) = k_1 xy \begin{bmatrix} 4 & 4 \\ 4 & 4 \end{bmatrix}$ which is a singular matrix for all $x, y \in \mathbb{N}$.

10.9 Equation (10.40) has in the first case the following form

$$k_1 \pi(x - 1) + k_2[x - a]_a \, \pi(x - a) + k_{-2}[x - a + 1]_{a-1} \pi(x - a + 1)$$
$$= \pi(x)\big(k_1 + k_2[x]_a + k_{-2}[x]_{a-1}\big),$$

where $x \in \mathbb{N}$. The stationary distribution is Poissonian if and only if $a = 1$. In this case the mean is $(k_1 + k_{-2})/k_2$.

The second example leads to two stationary distributions concentrated on the two closed communicating classes of the process. The first one is the product of two Poisson distributions, which can be written as

$$\pi(x, y) = \mathcal{B}(x \geq 1) \frac{1}{\exp(k_1/k_{-1}) - 1} \frac{1}{x!} \left(\frac{k_1}{k_{-1}} \right)^x$$
$$\times \mathcal{B}(y \geq 0) \exp(-k_2/k_{-2}) \frac{1}{y!} \left(\frac{k_2}{k_{-2}} \right)^y \quad (x \geq 0, y \geq 0),$$

while the second one is concentrated on $0 \times \mathbb{N}_0$, that is,

$$\pi(x, y) = \mathcal{B}(x = 0) \mathcal{B}(y \geq 0) \exp(-k_2/k_{-2}) \frac{1}{y!} \left(\frac{k_2}{k_{-2}} \right)^y \quad (x \geq 0, y \geq 0).$$

is also stationary.

10.10 First of all we need a function which calculates the propensity functions (λ_r).

```
Propensity[X_, alpha_, rates_] := rates *
  (Times @@ FactorialPower[X, #]& /@ Transpose[alpha]).
```

Now, the direct reaction method can be implemented by the following *Mathematica* function.

```
DirectMethod[X0_, alpha_, beta_, rates_, maxiter_, maxtime_]
    := NestWhileList[(# + {RandomReal[
    ExponentialDistribution[Lambda]], First[
    RandomChoice[lambdas -> Transpose[beta - alpha],1]]
    })&,
    {0, X0}, (First[#] <= maxtime && (Lambda =
    Total[lambdas = Propensity[Last[#], alpha, rates]])
    > 0)&, 1, maxiter]
```

10.11 One can use the function **Concentrations** and the solution of the previous problem to compare the results.

10.12 Recall the definition of the generating function from (10.25), and further define the function:

$$\widetilde{G}_a(t, z_2) := \frac{\partial^a (z_1^{a-1} G)}{\partial z_1^a}(t, 1, z_2).$$

Using this function one can transform the original equation to a simpler one, which becomes first order, namely,

$$\dot{\widetilde{G}}_a(t, z_2) = (k_2 z_2 - k_2 + a k_1) z_2 \frac{\partial \widetilde{G}_a}{\partial z_2}(t, z_2),$$

with initial condition

$$\widetilde{G}_a(0, z_2) = x_0(x_0 + 1) \cdots (x_0 + a - 1) z_2^{y_0}, \ (X(0), Y(0)) = (x_0, y_0).$$

Using the method of characteristics, we can obtain the solution

$$\widetilde{G}_a(t, z_2) = x_0(x_0+1) \cdots (x_0+a-1) \left(\frac{k_2 - a k_1}{k_2 + (k_2 - a k_1 - k_2 z_2) \exp((k_2 - a k_1)t)/z_2} \right)^{y_0}$$

The required combinatorial moments are then obtained by substituting $z_2 = 1$ into the previous display. A careful examination of the denominator of the solution then gives the bound (10.51).

The induced kinetic differential equation of the reaction is $\dot{x} = k_1 x y$ and $\dot{y} = k_2 y$, which implies that $y(t) = y_0 \exp(k_2 t)$ and $x(t) = x_0 \exp\left(y_0 k_1(\exp(k_2 t) - 1)/k_2\right)$, where $t \geq 0$. In this case the absence of blow-up comes from the fact that the reaction only contains a first-order autocatalysis. Note furthermore that $E(Y(t)) = y(t)$.

These and similar transformations on the generating function can be useful to treat similar problems; see Becker (1973b) and Table 14.4.

10.13 Using the shorthand notation $m_1(t) := EX(t)$ and $m_2(t) := EX(t)^2$ ($t \geq 0$), the first two moment equations of the induced kinetic Markov process with stochastic mass action type kinetics are

$$\dot{m}_1(t) = (k_1 + k_2) m_1(t) - k_2 m_2(t),$$

$$\dot{m}_2(t) = (k_1 - k_2) m_1(t) + (2k_1 + 3k_2) m_2(t) - 2k_2 \boxed{EX(t)^3}.$$

Table 14.4 Reactions treated by Dietz and Downton (1968), Becker (1970, 1973a,b)

Reaction steps	
$X \longleftarrow 0 \rightleftharpoons Y \longleftarrow X + Y$	
$2X_2 \longleftarrow X_2 \rightleftharpoons 0 \rightleftharpoons X_1 \longrightarrow 2X_1$	$X_1 + X_2 \longrightarrow X_2$
$2X_2 \longleftarrow X_2 \rightleftharpoons 0 \rightleftharpoons X_1 \longrightarrow 2X_1$	$2X_1 + X_2 \longleftarrow X_1 + X_2 \longrightarrow X_2$
$X + Y \longrightarrow Y \longrightarrow 2Y$	$Y \longrightarrow 0$
$X \longrightarrow 2X$	$X + Y \longrightarrow Y \longleftarrow 0$
$X \longleftarrow 0 \rightleftharpoons Y \longleftarrow X + Y$	
$X + Y_1 \longrightarrow Y_1 \longrightarrow 0 \longleftarrow Y_2 \longleftarrow X + Y_2$	
$X_i + Y \longrightarrow Y \rightleftharpoons 0$	
$X_i + Y_i \longrightarrow Y_i \rightleftharpoons 0 \rightleftharpoons Y_i$	

This shows that the moment equations are not closed, since the third moment appears on the right-hand side of the second equation. Using the program these can be obtained by issuing the following commands:

```
MomentEquations[{X <=> 2X}, {1}, {k1, k2}]
MomentEquations[{X <=> 2X}, {2}, {k1, k2}]
```

Now, assuming that

$$\mathbf{E}X(t)^3 = \mathbf{E}X(t) + 3(\mathbf{E}X(t))^2 + (\mathbf{E}X(t))^3 \tag{14.54}$$

and substituting it into the previous display, we arrive at

$$\dot{\widetilde{m}}_1(t) = (k_1 + k_2)\widetilde{m}_1(t) - k_2\widetilde{m}_2(t),$$

$$\dot{\widetilde{m}}_2(t) = (k_1 - 3k_2)\widetilde{m}_1(t) - 6k_2\widetilde{m}_1(t)^2 - 2k_2\widetilde{m}_1(t)^3 + (2k_1 + 3k_2)\widetilde{m}_2(t), \tag{14.55}$$

where $(\widetilde{m}_1, \widetilde{m}_2)$ denotes the functions which are hoped to approximate the solutions of the original moment equations.

Now, this latter, Eq. (14.55), is a closed one and can be solved by **NDSolve** numerically. Note that the only stationary distribution of the induced kinetic Markov process with stochastic mass action type kinetics of the reaction $X \longleftrightarrow 2X$ is the Poisson distribution with mean and variance k_1/k_2. One can check that the stationary points of the Eq. (14.55) are $\begin{bmatrix} 0 & 0 \end{bmatrix}^\top$ and $\begin{bmatrix} k_1/k_2 & k_1/k_2 + (k_1/k_2)^2 \end{bmatrix}^\top$ in agreement with the assumption (14.54) which the Poisson distribution makes exact.

10.14 For simplicity, assume that $k_{ii} = 0$ for all $1 \leq i \leq M$. The partial differential equation for the probability generating function G is

$$\dot{G}(t, z) = \sum_{1 \leq i, j \leq M} k_{ij}(z_j - z_i)\frac{\partial G}{\partial z_i}(t, z). \tag{14.56}$$

Using the method of characteristics, we are to determine those curves along which G is constant. For this we should solve the following system of ordinary differential equations:

$$\dot{\xi}_i(s) = -\xi_i(s) \sum_{j=1}^{M} k_{ij} + \sum_{j=1}^{M} k_{ij}\xi_j(s),$$

where $\xi_i(0) = z_i$. Now, define the matrix $\mathbf{A} = [a_{ij}]_{i,j=1}^{M}$, where $a_{ij} = k_{ij}$ if $i \neq j$ and $a_{ii} = -\sum_{j=1}^{M} k_{ij}$, i.e., $\mathbf{A} = \mathbf{K} - \text{diag}(\mathbf{K}^\top \mathbf{1})$. Notice that $\mathbf{A}$ is actually the infinitesimal generator matrix evaluated at $\mathbf{x} = \mathbf{1}$. With this notation the previous equation becomes more transparent, namely, $\dot{\boldsymbol{\xi}} = \mathbf{A}\boldsymbol{\xi}$, where $\boldsymbol{\xi} = \begin{bmatrix} \xi_1 & \xi_2 & \cdots & \xi_M \end{bmatrix}^\top$. The general solution to this first-order linear and homogeneous ordinary differential equation is simply $\boldsymbol{\xi}(s) = \exp(\mathbf{A}s)\mathbf{z}$ for all $s \geq 0$. Hence we obtained that $G(t, \boldsymbol{\xi}(s)) = \mathbf{z}^{\mathbf{x}_0}$ holds for all $s, t \geq 0$. It then implies that

$$G(t, \mathbf{z}) = \left(\exp(-\mathbf{A}t)\mathbf{z} \right)^{\mathbf{x}_0}$$

is the (only) solution to (14.56) which satisfies the initial condition $G(0, \mathbf{z}) = \mathbf{z}^{\mathbf{x}_0}$ and the boundary condition $G(t, \mathbf{1}) \equiv 1$. Note that in this case, G is a multivariate polynomial.

10.15 The reaction in question is a closed compartmental system. Hence the results of the previous problem can be applied. In the case of $k_{12} = k_{21} = 1$, it follows that

$$\exp(-\mathbf{A}t) = \exp(-t) \begin{bmatrix} \cosh(t) & \sinh(t) \\ \sinh(t) & \cosh(t) \end{bmatrix}.$$

That is

$$G(t, \mathbf{z}) = \exp(-2t)\left(\frac{1}{2}\left(z_1^2 + z_2^2\right)\sinh(2t) + z_1 z_2 \cosh(2t) \right), \tag{14.57}$$

where we have taken into consideration that $\mathbf{x}_0 = \begin{bmatrix} 1 & 1 \end{bmatrix}^\top$.

10.16 The first conditional moment velocity is

$$\mathbf{D}_1(\mathbf{x}) = \begin{bmatrix} -k_1 x \\ k_1 x - k_2 y \\ k_2 y \end{bmatrix},$$

where $\mathbf{x} = \begin{bmatrix} x & y \end{bmatrix}^\top$, while the second one is

$$\mathbf{D}_2(\mathbf{x}) = \begin{bmatrix} k_1 x & -k_1 x & 0 \\ -k_1 x & k_1 x + k_2 y & -k_2 y \\ 0 & -k_2 y & k_2 y \end{bmatrix}.$$

10.17 First, notice that a spanning tree of the Feinberg–Horn–Jackson graph of a single linkage class closed compartmental system with M species has $M - 1$ edges, hence the rank of the stoichiometric matrix γ is $M - 1$. It follows that $\rho = \begin{bmatrix} 1 & 1 & \cdots & 1 \end{bmatrix}$ is the only solution to $\rho^\top \gamma = 0$, which means that the only mass conservation relation is the total number of species. On the other hand, note that Theorem 3.17 implies that a compartmental system has deficiency zero. Then using the weak reversibility condition together with Theorem 7.15, one can apply Theorem 10.21. Hence, the only stationary distribution for the jth closed communicating class $U_j = \{\mathbf{x} = \begin{bmatrix} x_1 & x_2 & \cdots & x_M \end{bmatrix}^\top \in \mathbb{N}_0^M \mid x_1 + x_2 + \cdots + x_M = j\}$ ($j \in \mathbb{N}_0$) is the **multinomial distribution**

$$\pi_{U_j}(\mathbf{x}) = \frac{j!}{x_1! x_2! \cdots x_M!} (c_1)_*^{x_1} (c_2)_*^{x_2} \cdots (c_M)_*^{x_M} \quad (\mathbf{x} \in U_j),$$

where $\mathbf{c}_* = \begin{bmatrix} (c_1)_* & \cdots & (c_M)_* \end{bmatrix}^\top$ is the complex balanced stationary point of the induced kinetic differential equation with mass action type kinetics.

10.18 Since each column vector of α and β is one of the standard base vectors, it follows that the characteristic equation is a homogeneous linear system of differential equations; hence we can use matrix exponential to obtain its solution. In terms of *Mathematica* functions, one possible solution follows:

```
SolveCCSProbabilityGeneratingFunctionEquation
     [alpha_,beta_,rratecoeffs_,init_]  :=
Module[{zb, bb, Ab},
      zb = Array[Subscript[z,#]&, Length[alpha]];
      {bb, Ab} = CoefficientArrays[
      alpha.(rratecoeffs*
      MapThread[(Times@@(zb^#1)-Times@@(zb^#2))&,
      Transpose /@ {beta, alpha}]), zb];
      Times@@((MatrixExp[Ab t].zb)^init)
]
```

14.11 Inverse Problems

11.1 Now the objective function is $Q(m) := \sum_{k=1}^{K}(mt_k - x_k)^2$. The necessary condition of the minimum is that $Q'(m) = \sum_{k=1}^{K}(mt_k - x_k)t_k = m\sum_{k=1}^{K}t_k^2 - \sum_{k=1}^{K}x_k t_k = 0$ what can easily be solved for the estimate $\hat{m} = \frac{\sum_{k=1}^{K}x_k t_k}{\sum_{k=1}^{K}t_k^2}$ which is always defined if one has at least one measurement at a time different from zero. In order to check the usual sufficient condition, let us calculate the second derivative, $Q''(m) = \sum_{k=1}^{K}t_k^2 > 0$ which shows (again using the abovementioned slight restriction about the measurements) that the function Q has a minimum at $\hat{m}$.

11.2 The function $p \mapsto e^{-0.1p}(2 + \sin(p))$ has an infinite number of minima and maxima and has no global minimum. (When proving it, *Mathematica* might be useful.) All its minima are positive, and they tend to zero as the location of the minima tend to $+\infty$. Some of the minima can numerically be calculated this way:

```
FindMinimum[Exp[-0.1p](2 + Sin[p]), {p, #}]& /@
    Range[200].
```

The location of extrema can also be calculated symbolically, if not otherwise, using *Mathematica* to get $2\left(k\pi + \arctan\left(\frac{1}{12}(-1 - \sqrt{97})\right)\right) \quad (k \in \mathbb{Z})$.
    Without "strict" the sin function would also solve the problem.

11.3 The simulated measurements may be obtained this way.

```
kk = 0.01; mm = 2.5; times = N[Range[0, 100]];
    SeedRandom[37];
data = {#, mm Exp[-kk #]
    (1+0.05 RandomVariate[NormalDistribution[]])}&
    /@ times;
nlm = NonlinearModelFit[data, μ Exp[-κ t], {μ, κ}, t]
```

And `nlm["BestFitParameters"]` provides

$\{\mu \to 2.47389, \kappa \to 0.00986956\}.$

Now let us use for fitting the logarithm of our "measurements":

```
f[{t_, x_}] := {t, Log[x]}; logdata = Map[f, data];
lm = LinearModelFit[logdata, t, t]
```

And now the result is $\{0.908942, -0.00997263\}$ which should be transformed back to get `Exp[0.908942]=2.4817]`$\neq$`2.47389`.
    The difference, although small in this case in both the parameters, is existent. With today's computers one is not forced (not even allowed) to get a **final** estimate of parameters this way. However, linearizations of this and similar kinds are very useful to provide an **initial** estimate to start nonlinear estimation.

11.4 One may do this. Let the initial mass distribution in the interval $[0, 1]$ be given by the function $[0, 1] \ni x \mapsto f(x) := \sin(2\pi x)^2$. In order to simulate measurements, one has to solve the diffusion equation:

```
sol = ParametricNDSolveValue[{D[u[t, x], t] ==
    d D[u[t, x], {x, 2}],
    u[0, x] == f[x], u[t, 0] == f[0], u[t, 1] == f[1]},
    u, {t, 0, 10}, {x, 0, 1}, {d}];
```

One gets simulated data by adding a small normally distributed noise to the discrete sample taken from the solution of the diffusion equation. The exact value of the diffusion coefficient is taken to be 0.0271.

```
data = Table[{t, x, sol[0.0271][t, x] +
    RandomVariate[NormalDistribution[0, 0.07]]},
    {t, 0, 10, 0.5}, {x, 0.1, 0.9, 0.1}];
```

The fitting starts from the initial estimate 0.01 of the diffusion coefficient.

```
nlm = NonlinearModelFit[Flatten[data, 1], sol[d][t, x],
    {{d, 0.01}}, {t, x}]
```

`nlm["BestFitParameters"]` gives d -> 0.0261931, a value quite close to the exact value. If one wants to see the data together with the measured values, then one may do this.

```
Show[Plot3D[nlm[t, x], {t, 0, 10}, {x, 0, 1},
    Mesh -> 40,
    AxesLabel -> {"Time", "Space", "MassDensity"}],
    Graphics3D[{Red, PointSize[Medium],
        Point[Flatten[data, 1]]}]]
```

And the final result may look like Fig. 14.16.

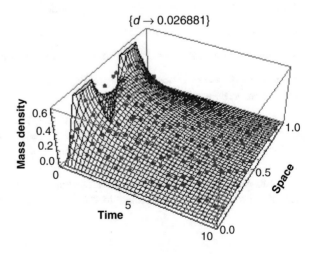

Fig. 14.16 Mass density measurements and fitted surface

Before and after fitting, one may wish to plot temporal or spatial sections of the mass density *u* or even **Animate** or **Manipulate** them.

11.5 Simulation can again be done in the same way as in the simple case of X $\longrightarrow$ 0 above:

```
predatorPrey = ParametricNDSolveValue[{
    x'[t] ==  x[t] (α − β y[t]),
    y'[t] == -y[t] (γ − δ x[t]),
    x[0] == x0, y[0] == y0}, {x, y}, {t, -50, 100},
    {α, β, γ, δ, x0, y0}];
data = Table[Prepend[
    Through[predatorPrey[1, 1, 1, 1, 2, 3][t]]
    (1 + 0.05 RandomReal[{-1, 1}]), t],
    {t, 0, 20, 0.2}];
predata = Most /@ data;
preydata = {#[[1]], #[[3]]}& /@ data;
```

Now the main trick is the introduction of a two-variable function *z* which compresses the coordinate functions in itself: $z(1, t) := x(t)$ $z(2, t) := y(t)$, and with this we rewrite the data in the following form:

```
indexedData =
    Flatten[Function[{i},
    Map[ {i, #[[1]], #[[i+1]]}&, data]] /@ Range[2], 1];
```

The model to be used for fitting is

```
model[{α}_?NumericQ, {β}_?NumericQ, {γ}_?NumericQ,
    {δ}_?NumericQ, x0_?NumericQ, y0_?NumericQ]
    [i_?NumericQ, t_?NumericQ] :=
    Part[Through[predatorPrey[
        {α}, {β}, {γ}, {δ}, x0, y0][t]], i]
```

When fitting, we provide some relatively good initial estimates of the parameters to **NonlinearModelFit**.

```
nlm = NonlinearModelFit[indexedData,
    model[{α}, {β}, {γ}, {δ}, x0, y0][i, t],
    {{{α}, 1.5}, {{β}, 1.5}, {{γ}, 1.5},
    {{δ}, 1.5}, {x0, 2.5}, {y0, 3.5}}, {i, t},
    PrecisionGoal -> 4, AccuracyGoal -> 4]
```

And the results can be obtained as below and be seen in Fig. 14.17.

```
fit[i_, t_] = nlm["BestFit"];
nlm["ParameterTable"]
```

Alternatively, one can use **Concentrations** to get simulated data.

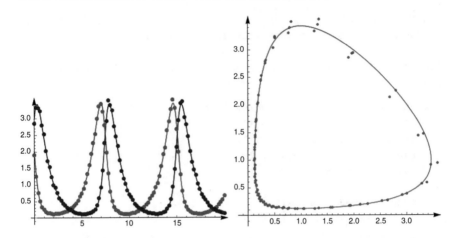

Fig. 14.17 Fitted solution and trajectory and simulated data

Let us mention that the solution of the two problems above are due to our colleague R. Csikja.

11.6 In this case the calculations lead to $\hat{k} = \frac{X(T)^2 - X(0)^2}{2TX(T)^2}$. To heuristically judge the goodness of this estimate, substitute the deterministic concentration of X into the formula to get $\hat{k} \approx k\frac{e^{2kT}-1}{2kT}$. This expression tends to k if kT tends to zero.

11.7 Solving the induced kinetic differential equation of the reaction by hand or by the program

```
Concentrations[{X -> Y, X -> Z}, {k1, k2},
    {x0, y0, z0}, {x, y, z}, t][[2, 1]]
```

gives $t \mapsto x(0)e^{-(k_1+k_2)t}$, and this function only depends on the sum of the two reaction rate coefficients. The situation is more advantageous if one also measures either $t \mapsto y(t)$ or $t \mapsto z(t)$ as well: In this case both reaction rate coefficients can be calculated. This topic is treated in detail in the works Vajda and Rabitz (1994) and Vajda and Várkonyi (1982). Denis-Vidal et al (2014) treats the problem for a special combustion model.

11.8 A possible example is the pair (Tóth 1981, p. 49)

$$2\,X \underset{3}{\overset{3}{\rightleftharpoons}} X+Y \underset{2}{\overset{2}{\rightleftharpoons}} 2\,Y \quad \text{and} \quad X+Y \underset{1}{\overset{1}{\rightleftharpoons}} 2\,X \underset{1}{\overset{1}{\rightleftharpoons}} 2\,Y$$

because their common induced kinetic differential equation is

$$\dot{x} = -\dot{y} = -3x^2 + xy + 2y^2.$$

11.9 Simple calculation (or `DeterministicModel`) shows that the induced kinetic differential equation of both mechanisms in Fig. 11.4 are

$$\dot{x}_1 = -6x_1^3x_2 + 9x_2^2 - 2x_1^2x_2 + x_1 \quad \dot{x}_2 = +2x_1^3x_2 - 3x_2^2 - 2x_1^2x_2 + x_1$$

This is not in contradiction to Theorem 11.11 because the number of ergodic classes in the first reaction is larger than 1.

11.10 One can use the program to verify that the right-hand sides are the same, both

```
RightHandSide[szeder1, Array[1&, 4], {x, y}]
RightHandSide[szeder2, {1, 1/10, 1/10, 19/10, 1/10, 1/10}, {x, y}]
```

lead to the same result: $\{-2x^2, 3x^2\}$. The reaction

$$X + Y \xrightarrow{1} 2X \quad X + Y \xrightarrow{1} 2Y$$

is mass conserving and second order. It can be added to any mass conserving second-order reaction without effecting the induced kinetic differential equation.

11.11 Either by hand or by using `DeterministicModel`, one gets as the induced kinetic differential equation of all the three mechanism as

$$\dot{x} = -3x^2y^3 + xy^2 - xy + 3y^2 \quad \dot{y} = 2xy - 2x^2y^3.$$

14.12 Reaction Kinetic Programs

12.1 This is what we shall have been doing by the preparation of the next edition of the book.

14.13 Mathematical Background

13.1 Since $a^D = b$ implies $D^\top \log(a) = \log(b)$, therefore $\log(a) = (D^\top)^{-1}\log(b) = (D^{-1})^\top \log(b)$ from which the statement follows.

13.2 If $\mathbf{y} := \mathbf{x}^{\mathbf{C}^{-1}}$, then $\mathbf{x} = \mathbf{y}^{\mathbf{C}} = \left[y^{\mathbf{c}_{\cdot 1}} \ y^{\mathbf{c}_{\cdot 2}} \ \cdots \ y^{\mathbf{c}_{\cdot M}} \right]^{\top}$, and one has

$$
\dot{\mathbf{x}} =
\begin{bmatrix}
c_{11}\dot{y}_1 \frac{y^{\mathbf{c}_{\cdot 1}}}{y_1} + c_{21}\dot{y}_2 \frac{y^{\mathbf{c}_{\cdot 1}}}{y_2} + \cdots + c_{M1}\dot{y}_M \frac{y^{\mathbf{c}_{\cdot 1}}}{y_M} \\
c_{12}\dot{y}_1 \frac{y^{\mathbf{c}_{\cdot 2}}}{y_1} + c_{22}\dot{y}_2 \frac{y^{\mathbf{c}_{\cdot 2}}}{y_2} + \cdots + c_{M2}\dot{y}_M \frac{y^{\mathbf{c}_{\cdot 2}}}{y_M} \\
\cdots \\
c_{1M}\dot{y}_1 \frac{y^{\mathbf{c}_{\cdot M}}}{y_1} + c_{2M}\dot{y}_2 \frac{y^{\mathbf{c}_{\cdot M}}}{y_2} + \cdots + c_{MM}\dot{y}_M \frac{y^{\mathbf{c}_{\cdot M}}}{y_M}
\end{bmatrix}
\tag{14.58}
$$

$$
=
\begin{bmatrix}
y^{\mathbf{c}_{\cdot 1}} \left(\frac{c_{11}}{y_1}\dot{y}_1 + \frac{c_{21}}{y_2}\dot{y}_2 + \cdots + \frac{c_{M1}}{y_M}\dot{y}_M \right) \\
y^{\mathbf{c}_{\cdot 2}} \left(\frac{c_{12}}{y_1}\dot{y}_1 + \frac{c_{22}}{y_2}\dot{y}_2 + \cdots + \frac{c_{M2}}{y_M}\dot{y}_M \right) \\
\cdots \\
y^{\mathbf{c}_{\cdot M}} \left(\frac{c_{1M}}{y_1}\dot{y}_1 + \frac{c_{2M}}{y_2}\dot{y}_2 + \cdots + \frac{c_{MM}}{y_M}\dot{y}_M \right)
\end{bmatrix}
\tag{14.59}
$$

$$
=
\begin{bmatrix}
y^{\mathbf{c}_{\cdot 1}} \left(\frac{\mathbf{c}_{\cdot 1}}{\mathbf{y}} \right)^{\top} \cdot \dot{\mathbf{y}} \\
y^{\mathbf{c}_{\cdot 2}} \left(\frac{\mathbf{c}_{\cdot 2}}{\mathbf{y}} \right)^{\top} \cdot \dot{\mathbf{y}} \\
\cdots \\
y^{\mathbf{c}_{\cdot M}} \left(\frac{\mathbf{c}_{\cdot M}}{\mathbf{y}} \right)^{\top} \cdot \dot{\mathbf{y}}
\end{bmatrix}
=
\begin{bmatrix}
y^{\mathbf{c}_{\cdot 1}} \left(\frac{\mathbf{c}_{\cdot 1}}{\mathbf{y}} \right)^{\top} \\
y^{\mathbf{c}_{\cdot 2}} \left(\frac{\mathbf{c}_{\cdot 2}}{\mathbf{y}} \right)^{\top} \\
\cdots \\
y^{\mathbf{c}_{\cdot M}} \left(\frac{\mathbf{c}_{\cdot M}}{\mathbf{y}} \right)^{\top}
\end{bmatrix}
\cdot \dot{\mathbf{y}},
\tag{14.60}
$$

thus

$$
\dot{\mathbf{y}} = \frac{\mathbf{y}}{\mathbf{y}^{\mathbf{C}}} \odot (\mathbf{C})^{-1} \cdot \dot{\mathbf{x}}.
\tag{14.61}
$$

13.3 Possible solutions follow.

```
power[a_?VectorQ, b_] := Times @@ (a^b)
a_?VectorQ⊙b_?VectorQ} :=a b
div[a_?VectorQ, b_?VectorQ]} := a/b
log[a_?VectorQ] := Log[a]
exp[a_?VectorQ] := Exp[a]
```

Complete the programs with type and positivity checks. Then, try them to see that they work when one thinks they should and they don't otherwise.

13.4 Immediately follows from the definition.

13.5 Let us suppose first that $\mathbf{Ax} = \mathbf{b}$ has a solution, and let $\mathbf{y}$ such that $\mathbf{A}^{\top}\mathbf{y} = 0$, and calculate $\mathbf{y}^{\top}\mathbf{b}$:

$$
\mathbf{y}^{\top}\mathbf{b} = \mathbf{y}^{\top}\mathbf{Ax} = \mathbf{x}^{\top}\mathbf{A}^{\top}\mathbf{y} = 0.
$$

Second, let us suppose that $\forall \mathbf{y} : \mathbf{A}^\top \mathbf{y} = 0$ implies $\mathbf{b}^\top \mathbf{y} = 0$. This, however, is equivalent to saying that $\mathbf{b}$ lies within the space spanned by the rows of $\mathbf{A}^\top$, or by the columns of $\mathbf{A}$, a necessary and sufficient condition for the solvability of $\mathbf{Ax} = \mathbf{b}$.

13.6 One has to prove that $\mathbf{1}_N^\top \mathbf{E} = \mathbf{0}_R$, an immediate consequence of the fact that all the columns of $\mathbf{E}$ contains exactly one entry 1 and exactly one entry -1.

13.7 Rephrasing what we need to prove, we want to show that the relation "being in the same component" is an equivalence relation on the vertices for undirected graphs and that the same holds for strong components of directed graphs. But this is immediate from the definitions: If in an undirected graph u and v are connected by a path, and v and w are also connected by a path, then the concatenation of these paths is a path connecting u and w, so u and w are in the same component. The argument is the same for directed paths and strong components.

13.8 Every ergodic component is a strong component; therefore all we need to show is that every strong component is ergodic if and only if the transitive closure of the graph is symmetric.

First, let us suppose that the transitive closure of a graph G is symmetric, i.e., that there is a path from a vertex v to u if and only if there is a path from u to v. This means that for every edge (u, w) in G, there is a path connecting w and u; therefore, w is in the same strong component as u. Hence, the strong component containing u is ergodic. As the same holds for every vertex u, we have shown that every strong component is ergodic. (Note that we have tacitly used the result of Problem 13.7, namely, that every vertex u belongs to precisely one strong component.)

Conversely, suppose that every strong component is ergodic. Now if there is a path from u to v, then u and v are in the same strong component (by assumption), therefore there is a path connecting v to u. Hence, the transitive closure is symmetric.

13.9 `TransitiveClosureGraph[G]` answers the second question. As to the first one, one has two options. Either use `SymmetricQ` from the package **Combinatorica** (be careful with this package as there is an overlap of its functions and the built-in ones) or write a function of your own, e.g., like this:

```
weaklyreversibleQ[reaction_] := Module[{a},
    a = AdjacencyMatrix[TransitiveClosureGraph[
    ReactionsData[{reaction}]["fhjgraphedges"]]];
    a == Transpose[a]]
```

Compare this with the function `WeaklyReversibleQ` of the package.

13.10 Consider the graph in Fig. 14.18. Here there are three strong components, $\{1, 2, 3\}$, $\{4, 5\}$, and $\{6\}$, there is one weak component, this is the whole graph, and the ergodic components are $\{1, 2, 3\}$ and $\{6\}$; therefore the number L of the weak components is smaller then the number $T = 2$ of ergodic components.

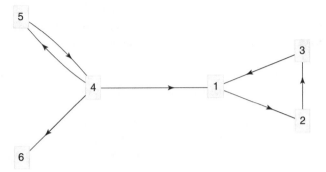

Fig. 14.18 A directed graph with one weak component and two ergodic components

An even simpler example can also be given. Consider the Feinberg–Horn–Jackson graph of the parallel reaction C ⟵ A ⟶ B. This graph has three strong components (each vertex is a strong component by itself), only two of which are ergodic because there are edges leaving {A}.) The corresponding undirected graph has a single component, because the graph (in the undirected sense) is connected.

13.11

1. The function $T := \mathbb{R} \ni x \mapsto x + 1$ is continuous, T is closed and convex, but unbounded; T has no fixed points.
2. The function $T :=\,]-1, 1[\ni x \mapsto (x + 1)/2$ is continuous, T is bounded and convex, but not closed; T has no fixed points.
3. The annulus with the boundaries of the circles of radius 1 and 2 with the origin as their centers in the plane is bounded and closed, but not convex. No nontrivial rotation (although they are continuous) around the origin has any fixed points.

13.12 As $\mathbf{u}_1, \mathbf{u}_2, \ldots, \mathbf{u}_K, \mathbf{v}$ are dependent, there are real numbers $c_1, c_2, \ldots, c_K$, c_{K+1} not all of which are zero such that $c_1 \mathbf{u}_1 + c_2 \mathbf{u}_2 + \cdots + c_K \mathbf{u}_K + c_{K+1} \mathbf{v} = \mathbf{0}$. Since $\mathbf{u}_1, \mathbf{u}_2, \ldots, \mathbf{u}_K$ are independent, c_{K+1} cannot be zero; thus $\mathbf{v} = -\frac{1}{c_{K+1}}(c_1 \mathbf{u}_1 + c_2 \mathbf{u}_2 + \cdots + c_K \mathbf{u}_K)$.

13.13 We apply induction on k. The assertion holds if $k = 1$: This is equivalent to the fundamental theorem of algebra. Now suppose that the claim is proven for a fixed $\hat{k} \geq 1$ and consider the case of $k = \hat{k} + 1$. We denote the number of zeros of a function h by $Z(h)$.

We invoke the following corollary of Rolle's theorem: If a μ times differentiable function h has $Z(h)$ zeros, then its μth derivative $h^{(\mu)}$ has at least $Z(h) - \mu$ zeros. Applying this to the function $h = t \mapsto g(t)e^{-\lambda_k t}$, which has the same number of zeros as g, we obtain that $Z(g) - \mu_k = Z(h) - \mu_k \leq Z(h^{(\mu_k)})$, and at the same time, we have $Z(h^{(\mu_k)}) \leq \mu_1 + \cdots + \mu_{k-1} - 1$ by the inductive hypothesis, since

$h^{(\mu_k)}$ is also of the form (13.18) with $k-1$ terms. Rearranging the inequalities gives $Z(g) \leq \sum_{i=1}^{k} \mu_i - 1$, completing the proof.

13.14 Simple substitution into Eq. (13.23) shows that now the sensitivity equations form the inhomogeneous linear, constant coefficient equation (cf. Tóth et al 1997)

$$\frac{d\sigma(t, \mathbf{p})}{dt} = \partial_1 \mathbf{f}(\xi_*(\mathbf{p}), \mathbf{p})\sigma(t, \mathbf{p}) + \partial_2 \mathbf{f}(\xi_*(\mathbf{p}), \mathbf{p})$$

which can even be explicitly solved if needed.

13.15 Actually, this is a separable (what is more: linear) ordinary differential equation, the general solution to which is $Me^{k_1 t(z-1)}$. From the initial condition: $M = z^D$.

One can also calculate the generating function under the general (stochastic) initial condition, i.e., under the condition $G(0, z) = F(z)$. Then we have $G(t, z) = F(z)e^{k_1 t(z-1)}$, i.e., the number of particles is the sum of a variable with the initial distribution and a variable with Poisson distribution with the parameter $k_1 t$ independent from the first one.

13.16 This is a (first order) partial differential equation, the solution to which is
$$\left(\frac{z}{z - e^{k_1 t(z-1)}} \right)^D.$$
One can also calculate the generating function under the general (stochastic) initial condition, i.e., under the condition $G(0, z) = F(z)$. Then we have $G(t, z) = F\left(\frac{z}{z - e^{k_1 t(z-1)}} \right)$.

References

Ault S, Holmgreen E (2003) Dynamics of the Brusselator. http://www.bibliotecapleyades.net/archivos_pdf/brusselator.pdf

Bard Y (1974) Nonlinear parameter estimation. Academic Press, New York

Becker NG (1970) A stochastic model for two interacting populations. J Appl Prob 7(3):544–564

Becker N (1973a) Carrier-borne epidemics in a community consisting of different groups. J Appl Prob 10(3):491–501

Becker NG (1973b) Interactions between species: some comparisons between deterministic and stochastic models. Rocky Mt J Math 3(1):53–68

Carroll FW (1961) A polynomial in each variable separately is a polynomial. Am Math Mon 68(1):42

Denis-Vidal L, Cherfi Z, Talon V, Brahmi EH (2014) Parameter identifiability and parameter estimation of a diesel engine combustion model. J Appl Math Phys 2(5):131–137

Dietz K, Downton F (1968) Carrier-borne epidemics with immigration. I: immigration of both susceptibles and carriers. J Appl Prob 5(1):31–42

Ehrenfest P, Ehrenfest T (1907) Über zwei bekannte Einwande gegen das Boltzmannsche H-Theorem. Phys Z 8:311–314

Epstein I, Pojman J (1998) An introduction to nonlinear chemical dynamics: oscillations, waves, patterns, and chaos. Topics in physical chemistry series, Oxford University Press, New York, http://books.google.com/books?id=ci4MNrwSlo4C

Érdi P, Lente G (2016) Stochastic chemical kinetics. Theory and (mostly) systems biological applications. Springer Series in Synergetics. Springer, New York

Erle D (2000) Nonoscillation in closed reversible chemical systems. J Math Chem 27(4):293–302

Eyring H (2004) The activated complex in chemical reactions. J Chem Phys 3(2):107–115

Feinberg M (1989) Necessary and sufficient conditions for detailed balancing in mass action systems of arbitrary complexity. Chem Eng Sci 44(9):1819–1827

Feinberg M, Horn FJM (1977) Chemical mechanism structure and the coincidence of the stoichiometric and kinetic subspaces. Arch Ratl Mech Anal 66(1):83–97

Frank J (2008) Numerical modelling of dynamical systems. Lecture notes, Author, http://homepages.cwi.nl/~jason/Classes/numwisk/index.html

Ganapathisubramanian N, Showalter K (1984) Bistability, mushrooms, and isolas. J Chem Phys 80(9):4177–4184

Kresse G, Hafner J (1993) Ab initio molecular dynamics for liquid metals. Phys Rev B 47(1):558

Li R, Li H (1989) Isolas, mushrooms and other forms of multistability in isothermal bimolecular reacting systems. Chem Eng Sci 44(12):2995–3000

Pintér L, Hatvani L (1977–1980) Solution of problem 10 of the 1979 Miklós Schweitzer competition. Mat Lapok 28(4):349–350

Póta G (1983) Two-component bimolecular systems cannot have limit cycles: a complete proof. J Chem Phys 78:1621–1622

Póta G (1985) Irregular behaviour of kinetic equations in closed chemical systems: oscillatory effects. J Chem Soc Faraday Trans 2 81:115–121

Póta G (2006) Mathematical problems for chemistry students. Elsevier, Amsterdam

Schnell S, Mendoza C (1997) Closed form solution for time-dependent enzyme kinetics. J Theor Biol 187:207–212

Scott SK (1991, 1993, 1994) Chemical chaos. International series of monographs on chemistry, vol 24. Oxford University Press, Oxford

Siegel D, Chen YF (1995) The S-C-L graph in chemical kinetics. Rocky Mt J Math 25(1):479–489

Szabó A, Ostlund NS (1996) Modern quantum chemistry: introduction to advanced electronic structure theory. Courier Dover Publications, Mineola

Szederkényi G, Hangos KM (2011) Finding complex balanced and detailed balanced realizations of chemical reaction networks. J Math Chem 49:1163–1179

Szirtes T (2007) Applied dimensional analysis and modeling. Butterworth-Heinemann, Burlington

Tóth J (1981) A formális reakciókinetika globális determinisztikus és sztochasztikus modelljéről (On the global deterministic and stochastic models of formal reaction kinetics with applications). MTA SZTAKI Tanulmányok 129:1–166, in Hungarian

Tóth J (1987) Bendixson-type theorems with applications. Z Angew Math Mech 67(1):31–35

Tóth J, Li G, Rabitz H, Tomlin AS (1997) The effect of lumping and expanding on kinetic differential equations. SIAM J Appl Math 57:1531–1556

Truhlar DG, Garrett BC (1980) Variational transition-state theory. Acc Chem Res 13(12):440–448

Turányi T, Tomlin AS (2014) Analysis of kinetic reaction mechanisms. Springer, Berlin

Vajda S, Rabitz H (1994) Identifiability and distinguishability of general reaction systems. J Phys Chem 98(20):5265–5271

Vajda S, Várkonyi P (1982) A computer program for the analysis of structural identifiability and equivalence of linear compartmental models. Comput Programs Biomed 15(1):27–44

Van Kampen NG (2006) Stochastic processes in physics and chemistry, 4th edn. Elsevier, Amsterdam

Várdai J, Tóth J (2008) Hopf bifurcation in the brusselator. http://demonstrations.wolfram.com/HopfBifurcationInTheBrusselator/, from The Wolfram Demonstrations Project

Vincze I (1964) Über das Ehrenfestsche Modell der Wärmeübertragung. Arch Math 15(1):394–400

Weise Th (2009) Global optimization algorithms-theory and application. Self-published 2

Glossary

Here we give a short, informal description of terms which are precisely defined in the text, but our use of which might be a slightly different from the general one.

circumstances Given a mechanism, physical quantities such as volume pressure and temperature—these form together the circumstances—are usually fixed.

detailed balance This term is mainly used to express the corresponding property in the deterministic model of reactions.

formal reaction kinetics The mathematical theory of deterministic and stochastic kinetics as opposed to Chemical Reaction Network Theory which was up to recent times understood as the theory of deterministic models.

kinetically mass conserving The reaction has this property if the induced kinetic differential equation of it has a linear first integral with a positive coefficient vector.

mechanism The reaction together with the kinetics and (the usually fixed) circumstances. In this book almost always mass action type kinetics is used.

microscopic reversibility This term is only used to express the corresponding property in the stochastic model of reactions.

model We use this expression as loosely as usual; it may mean a reaction, a mechanism, or an induced kinetic differential equation.

reaction Reaction (sometimes complex chemical reaction) is the set of reaction steps kinetics, reaction rate coefficients, and circumstances excluded.

reaction step A single physical process, reversible reactions are represented as pairs of (irreversible) reaction steps.

stoichiometrically mass conserving The reaction has this property if the orthogonal complement of the stoichiometric space contains a positive vector.

© Springer Science+Business Media, LLC, part of Springer Nature 2018
J. Tóth et al., *Reaction Kinetics: Exercises, Programs and Theorems*,
https://doi.org/10.1007/978-1-4939-8643-9

Index

© Springer Science+Business Media, LLC, part of Springer Nature 2018

J. Tóth et al., *Reaction Kinetics: Exercises, Programs and Theorems*,
https://doi.org/10.1007/978-1-4939-8643-9

Printed in the United States
By Bookmasters